Theoretical Models of Chemical Bonding
Part 4

Theoretical Treatment of Large Molecules and Their Interactions

Editor: Z. B. Maksić

With contributions by
J. G. Angyán, A. van der Avoird, E. J. Baerends,
R. Bonaccorsi, Ch. L. Brooks, III, R. Cammi,
R. E. Christoffersen, P. Ch. Hiberty,
G. M. Maggiora, V. Magnasco, R. McWeeny,
G. Náray-Szabó, J. D. Petke, R. A. van Santen,
S. Scheiner, S. Shaik, A. J. Stone, O. Tapia,
J. Tomasi

With 104 Figures and 52 Tables

Springer-Verlag Berlin
Heidelberg GmbH

Professor Dr. Zvonimir B. Maksić

Theoretical Chemistry Group
The "Rudjer Bošković" Institute
41001 Zagreb, Bijenička 54, Croatia/Yugoslavia
and
Faculty of Natural Sciences and Mathematics
University of Zagreb
41000 Zagreb, Marulićev trg 19, Croatia/Yugoslavia

ISBN 978-3-642-63492-5

Library of Congress Cataloging-in-Publication Data
Theoretical treatment of large molecules and their interactions /
  editor, Z. B. Maksić ; with contributions by J. G. Angyán ... [et
  al.].
    p.   cm. — (Theoretical models of chemical bonding ; pt. 4)
  ISBN 978-3-642-63492-5   ISBN 978-3-642-58177-9 (eBook)
  DOI 10.1007/978-3-642-58177-9
    1. Macromolecules. 2. Molecular association. I. Maksić, Z. B.
(Zvonimir B.) II. Ángyán, János, 1956–   . III. Series.
QD381.T43 1991
541.2'2—dc20   91-14223 CIP

© Springer-Verlag Berlin Heidelberg 1991
Originally published by Springer-Verlag Berlin Heidelberg New York in 1991
Softcover reprint of the hardcover 1st edition 1991

VA51/3020-543210 — Printed on acid-free paper

To the memory of my parents
Olivera and Branko Maksić

Preface

The French chemist Marcelin Berthelot put forward a classical and by now an often cited sentence revealing the quintessence of the chemical science: "La Chimie cree son objet". This is certainly true because the largest number of molecular compounds were and are continuously synthesized by chemists themselves. However, modern computational quantum chemistry has reached a state of maturity that one can safely say: "La Chimie Theorique cree son objet" as well. Indeed, modern theoretical chemistry is able today to provide reliable results on elusive systems such as short living species, reactive intermediates and molecules which will perhaps never be synthesized because of one or another type of instability. It is capable of yielding precious information on the nature of the transition states, reaction paths etc. Additionally, computational chemistry gives some details of the electronic and geometric structure of molecules which remain hidden in experimental examinations. Hence, it follows that powerful numerical techniques have substantially enlarged the domain of classical chemistry. On the other hand, interpretive quantum chemistry has provided a conceptual framework which enabled rationalization and understanding of the precise data offered either by experiment or theory. It is modelling which gives a penetrating insight into the chemical phenomena and provides order in raw experimental results which would otherwise represent just a large catalogue of unrelated facts. One should point out here that by models we imply simple physical models, obtained in a controlled way, which reflect the essence of systems under study. It is also noteworthy that models can be developed in a completely ab initio fashion without any resort to experiment thus supporting once more a previous statement that theoretical chemistry creates its own object of research. However, modern investigations of the structure and properties of matter at the molecular level require joint efforts of computational and inter-pretive quantum chemistry combined with empirical chemical research. Concomitantly, the main impetus for the Series „Theoretical Models of Chemical Bonding" was a need to increase understanding and the cooperation of researchers of these conventionally and traditionally separated disciplines of chemical investigation.

The fourth volume deals with giant molecular systems of biological interest and their interactions. Since they are out of grasp of the rigorous ab initio methods, penetrating wisdom and suitable models are the only means in exploring their properties. This is not necessarily a drawback because an approximate picture offerred by an easy to understand model may be more useful than a very accurate but obscure solution provided by a plain quantum theory of numbers. Consequently, the emphasis in the book is laid on the qualitative aspects of the theory which does not imply that necessary mathematics is avoided. The chapters are intended to give "the state of the art" survey of the most important scientific findings in the respective fields. Their overlapping is not zero, which is a desirable feature because it increases the readability of the text. In particular, the interested reader will find that electrostatic interactions are an essential ingredient of several chapters. This is not surprising since electrostatics is, figuratively speaking, Ariadne's thread which enables a pathway to be traced through a labyrinth of a large number of interactions and degrees of freedom in extended (almost infinite) systems like biopolymers, molecular crystals, solutions etc.

The present volume begins with an extensive description of the important fragmentation approach in studying gigantic chemical systems. The underlying idea is a reduction of the insoluble problems involving perhaps thousands of atoms to a problem of interactions of small units or building-blocks. The choice of the (transferable) basic units is not unique, meaning that their definition is a delicate decision. Attention is focused here mainly on the functional groups since other possibilities were discussed in detail in vol. 2 of the Series.

A brief chapter is devoted to a combined quantum and classical (semiclassical) approach in investigating structure, motion and reactivity of proteins and nucleic acids which in turn provide a key to understanding biological systems. This is followed by the molecular fragment FSGO basis set methodology applied to examining the electronic excited states of individual biomolecules (chlorophyll, porphyrin etc.) and their aggregates.

The long range molecular interactions, their classical electrostatic components and a general partitioning of weak interactions into meaningful physical contributions are then considered in great detail. The formalism is illustrated by applications to some simple VdW and hydrogen-bonded dimers. Then a review of accurate ab initio procedures suitable for description of hydrogen-bonded systems is given.

The following block of chapters is dedicated to chemical reactivity. It commences with the concept of extramolecular electrostatic potential which yields important clues in locating reactive sites in molecules. Then the model of curve-crossing/avoided crossing diagrams based on the VB approach is scrutinized illustrating a power of a proper mixture of qualitative ideas and particular quantitative procedures in considering chemical reactivity. This block is concluded by a survey of theoretical treatments of chemisorption and in particular by a discussion of bonding between metal particles and metal adsorbents. A deep analysis reveals a key role of the symmetry principles and rehybridization which are of utmost importance for understanding chemical reactivity on boundary surfaces.

The last part of the book addresses the questions of bulk and spectral properties of molecular crystals and the chemically extremely important problem of solvent effects involving their influence on molecular properties and reactivity.

To reiterate, if this book and the whole Series contribute to a better understanding of the modern theoretical chemistry and to an increased collaboration between molecular scientists working in different fields, then our efforts will be greatly rewarded.

Finally, it is my pleasant duty to thank all authors for their scholarly written articles. I am also grateful to the Alexander von Humboldt-Stiftung for financial support since a part of editing has been performed during my stay at the Organisch-chemisches Institut der Universität Heidelberg. Thanks go to Professor R. Gleiter too for his hospitality and stimulating discussions.

Z. B. Maksić

Table of Contents

Chemical Fragmentation Approach to the Quantum Chemical Description of Extended Systems

János G. Ángyán and Gábor Náray-Szabó

CHINOIN Pharmaceutical and Chemical Works, H-1325 Budapest, P.O. Box 110, Hungary

1 Introduction

The proper description of the electronic structure of extended systems by state-of-the-art methods remains a challenging task [1] of quantum chemistry. In the past few years the notion of "large molecular systems" has undergone a considerable evolution, and species, which seemed almost impossible to treat by ab initio quantum chemistry are now in the domain of routine calculations. This development is mainly due to the revolution of computer technology (vector and/or parallel supercomputers) and new computational techniques, which are better adapted to the new generation of computers, like the direct SCF method [2]. The application of advanced computational techniques made it possible to undertake such spectacular calculations like the ab initio study of the C_{60} Buckminsterfullerene [3] or the largest system ever studied by ab initio SCF calculations, the $C_{150}H_{30}$ molecule [4]. It seems that in the very near future several theoretical chemistry laboratories will be in the position to perform routinely calculations on molecules or molecular aggregates, containing 500/600 electrons [5, 6].

Although further, presently unforeseeable development is quite possible in the near future, we must be avare of the fact that simple brute force calculations on very large systems cannot replace approximate models. These can be specifically designed to give a better understanding of the chemistry and physics of the phenomena taking place in extended systems, such as imperfect solids, surfaces or in bio-macromolecular environments. Since these large systems are always built up from relatively simple building blocks, this kind of model can be extremely efficient when we are able to exploit our experience about the constituting fragments of the system.

The existence of such *building blocks* or *fragments* is a fundamental principle in experimental structural and physical chemistry [7]. For example, individual molecules carry well-defined spectroscopical and structural properties. These properties are essentially preserved although they are more-or-less influenced by the various chemical environments: in crystalline phase, in a liquid, or in different solutions. This feature that a considerable number of molecular properties remain practically the same in quite different circumstances, is called *transferability* [8]. The transferability of molecular properties is associated with the weakness of the intermolecular forces as compared to the intramolecular interactions, which govern these molecular properties. Molecules are very suitable building blocks of the matter provided their size does not exceed a certain limit. Small molecules may be considered as building blocks in molecular crystals, liquids, solutions and in general in various condensed phase systems.

Unfortunately the size of the molecules, i.e. the group of atoms connected together by a continuous series of covalent chemical bonds, may be exceedingly large. Macromolecules consist of several thousand atoms, all linked together by chemical bonds, and in the extreme case of covalent crystals or amorphous materials the size of the "molecule" may even tend to infinity. Obviously, for such very large molecules one should find some *intramolecular* constituents which could play the role of building units [9]. The choice of the most appropriate building blocks is a delicate task, since there is no unique criterion which could guide us in designating these fragments.

The concept of transferable intramolecular fragments is used in many fields of experimental chemistry [7]. For example the concept of *bond increments*, which is a practical way to interpret molecular dipole moments [10] or polarizabilities [11] is intimately related to the description of the electronic structure in terms of two-electron groups or localized bonds [12]. This latter topic is the main subject of a review paper by Surján [13], therefore we focus our attention to other treatments, which consider larger fragments, like functional groups, as relevant subunits. Nevertheless the concept of localized bonds remains an essential ingredient of the theory and it will prove to be useful for the definition of larger fragments, too. These can be defined as the monomeric units of polymers or biopolymers, like amino acid residues in proteins or nucleotides in nucleic acids. Although such building blocks seem to be quite natural, in quantum chemical practice, one is faced with the problem of proper treatment of the junction region between the monomeric units.

There is another aspect of the quantum chemical utilization of subunits in extended systems. The interesting or relevant chemical or physical events quite often take place in a well-defined local region of a larger system. For example, an enzymatic reaction is essentially localized on the active site of the enzyme, and the role of the remaining part of the macromolecule, although not negligible, remains of secondary importance [14].

There are other cases, where the idea of transferable submolecular building blocks is exploited. The concept of the *chromophore* in spectroscopy reflects the empirical observation that the appearance of certain characteristic absorption bands always indicate the presence of well-defined fragments, like carbonyl groups or aromatic rings, in the molecule [15]. Spectroscopic measurements in larger molecules or in extended systems, e.g. molecular crystals [16] or biopolymers can be conveniently interpreted by the concept of chromophores, while the effect of the chemical environment involves a relatively small perturbation. The optical absorption [15] and circular dichroism [17] spectra of even small and medium sized molecules can also be systematized with respect to the characteristic properties of the chromophores.

The notion of *functional groups* is used mainly in the discussion of organic reactivity [18] and in vibrational spectroscopy [19]. The underlying idea, i.e. the presence of more-or-less transferable properties associated with a given group of atoms, is always the same.

The "classical" examples for the situation of an interacting large environment (solvent) and a small portion of interest (solute) composite system are just the liquid solutions [20]. However the methodological arsenal of the theory of solvent effects can be used in many analogous situations. Crystal defects [21], adsorption sites [22], heterogeneous catalysis [23] and many other interesting chemical phenomena may be treated by very similar techniques.

The concept of chemical fragmentation can be exploited in quantum chemical calculations in two alternative ways. An a posteriori analysis of the results obtained for the total system in terms of fragments may deepen our understanding about the way the subunits are linked together. A typical example of this kind of approach is the quantitative PMO analysis for the interpretation of conformational preferences in organic molecules [24, 25]. Nevertheless, such an approach has no practical advantages for large systems, since a complete calculation of the supersystem is required. On the other hand, an a priori exploitation of our knowledge about the nature of the constituting subunits may facilitate greatly the treatment of the whole supersystem and, at the same time, offers some conceptual advantages too.

Although it is inevitable to rely on certain results of a posteriori analyses, we will be concerned mainly with the second approach and try to overview the various possibilities for the systematic exploitation of the concept of chemical fragmentation for the construction of wave functions of large, extended systems. Of course, these wave functions are always of an approximate, sometimes very approximate, character.

Actually the requirements with respect to the wave function quality may strongly depend on the nature of the phenomenon under study. A high-level description inluding electron correlation is necessary, for example, to describe a bond fission process. A much poorer wave function quality is surely enough to estimate the electrostatic and/or inductive interaction of some distant parts of the system. This suggests that in many cases it is advantageous to partition the extended system to different regions. One needs the highest accuracy in the description of the subsystem of interest, and lesser precision might be sufficient for parts of the system which play a secondary role. Such a computational strategy seems to be extremely fruitful for the treatment of various cases of catalytic reactions, like enzymatic processes, adsorption, heterogeneous catalysis, solid state reactivity. In all these cases two main underlying ideas will be used [26]: (a) the total wave function of the extended system is constructed from nearly transferable fragments; (b) the accuracy and the level of description decreases with the distance from the chemically interesting relevant fragment of the extended system. This philosophy allows one to build a model, which is physically well-founded, may be improved at will, and the approximation involved can be controlled relatively easily.

2 Heuristic Basis: The Concept of Fragments

As it has been emphasized in the Introduction, the concept of fragments or building blocks is widely used not only in the molecular description of extended systems, but also in relatively small or medium sized polyatomic molecules. The basic idea is always to exploit a certain kind of transferability of the properties of these fragments, which can be atoms, bonds, substituents, functional groups, and so on. Before turning to our main topic, i.e. how to use the concept of fragments for the construction of the wave function for extended systems, let us see what can we learn from the various kinds of a posteriori analyses (mostly available for medium size molecules). Then hopefully we shall be in the position to give a more precise idea of the fragment concept.

2.1 A Posteriori Fragment Analysis

2.1.1 The Topological Partitioning Method

Dalton's atomic hypothesis, that a molecule is an ensemble of atoms linked together by a network of chemical bonds, is a fundamental principle of chemistry and physics. This atomic hypothesis, which may serve as a basis for the definition of a posteriori chemical fragmentation models, has been rigorously expounded by Bader and his coworkers [27, 28]. The Bader's theory is based on the topological analysis [29, 30] of the properties of the electronic charge density distribution function of the molecule containing n electrons

$$\varrho(r) = \langle \Psi | \sum_{i}^{n} \delta(r - r_i) | \Psi \rangle. \tag{2.1}$$

Bader's atoms are defined through the partitioning of $\varrho(r)$ in three-dimensional real space. The atomic regions are bounded by atomic surfaces, which satisfy the following zero flux condition:

$$\nabla \varrho(r)\, n(r) = 0, \tag{2.2}$$

where $n(r)$ is the vector normal to the surface at point r and $\nabla \varrho(r)$ is the gradient vector field of the molecular charge density. Such surfaces usually encompass one nucleus, where the electron density attains its maximal value and all the points where the gradient vector $\nabla \varrho(r)$ directed towards this nucleus, which is an attractor in the topological sense. According to recent calculations of Gatti et al. [31] in Li clusters one obtains domains enclosed by zero-flux surfaces *without* any nucleus. This may be an interpreted as an indication of the special properties of metallic bonds [32].

The beauty of the above topological definition of the atom in a molecule lies in the fact that it coincides with the rigorous quantum mechanical definition of an open subsystem[1] [27, 33, 34]. In particular, the atomic action integral, which is defined through the atomic one-particle Lagrangian density, is zero within the atomic volume:

$$W_{12}[\Psi, \Omega] = -\frac{\hbar}{4m} \int_{t_1}^{t_2} dt \int_{\Omega} dr\, \nabla^2 \varrho(r, t) = 0. \tag{2.3}$$

It has also been shown that in the atomic regions the virial theorem holds [27, 33]. That is why Bader's partitioning of the electron density is often cited as the "virial partitioning" method [35, 36]. Bader's theory leads to an unequivocal formulation of the transferability idea: the ensemble of atomic properties, in particular the contribution of an atom to the total energy, may be the same in different molecules, provided the charge density of the atomic region is identical [27]. The chemical bonds can be identified with the paths of maximal electron density, which interconnect the various nuclei. This point and several other chemical applications of Bader's theory have been recently reviewed by Wiberg [37].

2.1.2 Localized Bonds and Loge Theory

The various localization methods, which exploit the fact that the determinantal wave function is invariant to a unitary transformation, have been discussed in details by Surján [13]. These methods lead to two-electron bonds for localizable systems, and

---

[1] We speak about an open subsystem in the sense that it freely exchanges charge and momentum with its surroundings through its boundaries.

this particular topic is covered by the above review. However it should be mentioned that certain systems cannot be localized effectively. Typical examples are the aromatic π-systems, where the index of "localizability" differs considerably from the ideal value of 1 [8].

The failure of efforts to localize certain systems to two-electron fragments calls our attention to the necessity of using larger bulding blocks than just a simple two-electron bond. The difficulties in the proper treatment of aromatic systems by the PCILO method, which uses exclusively such two-electron bonds (e.g. Kekulé structures for aromatic systems) are well known [38], and lead to the development of the "Extended PCILO Method", where this constraint has been released [39].

Daudel proposed an interesting concept for the analysis of the many-electron wave functions, which is known as the "loge theory" [35]. According to Daudel, it is possible to define domains (loges) of the three-dimensional space where the fluctuation of the number of electrons is minimized. In particular the 2-electron loges coincide practically with the conventional localized orbitals ("quasi-loges"). As we shall see this criterium is of crucial importance for an efficient way of constructing the many-electron wave function from building blocks. This property allows us to neglect those configurations, where electrons are transferred from one group to the other, thus reducing considerably the computational effort needed.

2.1.3 Fragment Orbital Analysis

A straightforward way to analyse intramolecular interactions is to evaluate the interaction energies (and their components) between the orbitals belonging to various fragments of the molecule. These fragments represent usually different functional groups. A successful implementation of such an analysis in ab initio quantum chemistry is the quantitative PMO method, originally proposed by Whangbo et al. [24].

The fragment orbitals are determined through the partition of the Fock matrix of the full problem, by assigning the basis orbitals to different fragments. The diagonalization of the fragment blocks of the Fock matrix leads to a set of "canonical" fragment orbitals, and the molecular orbitals of the total molecule can be expressed as a linear combination of these fragment orbitals. The interactions between the occupied orbital of fragment A with the virtual one of fragment B (stabilizing) and between the occupied orbitals of both fragments (destabilizing) allow the rationalization and interpretation of many conformational and stereochemical problems [40, 41].

Whereas this scheme works pretty well if the dominant interactions arise between the π-type fragment orbitals, it fails in the case of σ-interactions. The reason for this failure is easy to understand. The connection of neighbouring fragments is usually made of a σ-bond, which is artificially broken by the fragmentation. This results in an occupation number significantly less than 2 for almost all of the σ-type orbitals in both fragments. Bernardi and Bottoni proposed extending the qualitative PMO analysis to σ-interactions by using localized fragment orbitals [25, 42]. A subsequent Boys localization of the canonical fragment orbitals allows one to concentrate the effect of bond breaking to only one σ-orbital, which appears with an occupancy of 1, while the remaining σ-orbitals have an occupancy near to 2.

The lesson to draw from these experiences is that the use of localized orbitals may be an adequate tool to separate the regions of strong (covalent) interactions of the fragments from the set of orbitals involved only in weak (non-covalent) interactions, governing the chemical event under study.

2.2 A Priori Use of Fragments

2.2.1 Atoms or Modified Atoms in Molecules

It is a common chemical experience that for many properties the atoms preserve their identity in a certain sense even after molecule formation. It is therefore not surprising that this concept is present in various fields of theoretical chemistry either as a prescription for the calculation of the molecular wave function, or as a conceptual framework to systematize experimental data.

An example, where this idea is directly exploited for the construction of the wave function is the atoms-in-molecules (AIM) method of Moffit [43, 44]. In this approach the ground and excited-state wave functions of the isolated atoms are used for the expansion of the total wave function. As noted by Schipper [45] this procedure is not flexible enough for a satisfactory description of the strong covalent bonds. The formation of the molecule cannot be considered as a "small" perturbation to the free-atom wave function, therefore highly excited and ionized atomic wave functions would also be needed for a satisfactory description of the molecule.

Another, different way of utilizing the atomic hypothesis was realized in the Chemical Hamiltonian Approach [46] which exploits the LCAO model of quantum chemistry. It is important to stress that the LCAO (linear combination of atomic orbitals) expansion of the molecular wave functions is itself, in a certain sense, based on the concept of "building blocks". In effect, the idea behind the LCAO method is the chemical experience that the basic building blocks of the molecules are the atoms. Thus the atomic orbitals obtained from atomic wave functions could be suitable for the expansion of molecular orbitals.

The Chemical Hamiltonian Approach uses the atomic orbital basis set for the definition of an atomic partition of the Hamiltonian operator. An atomic subsystem of a molecule consists of a nucleus and the set of basis functions centered on it. Accordingly, the partition of the finite basis set corresponds to the physical partition of a molecule into atomic subsystems. This raises the problem of non-orthogonality of the basis functions belonging to the different subsystems (atoms, in the present case) and also the problem of basis set superposition error (BSSE), which is a consequence of the finiteness of the basis set [46].

The atomic hypothesis was investigated in an exhaustive manner by Maksić. He tested the the Independent Atoms in Molecule (IAM) and the Modified Atoms in Molecule (MAM) models for some simple physical observables as diamagnetic shielding, diamagnetic susceptibility and ESCA shifts. It seems that these parameters can be considered as pretty sensitive probes of the wave function quality. In this context it might be surprising that for certain properties even the IAM model is able to give reliable estimates. Nevertheless, in most cases one should take into account that the atoms are considerably modified by their molecular environment [47–50].

2.2.2 Definition of Larger Fragments

Without going into the mathematical details we shall qualitatively overview and compare the possibilities for defining polyatomic fragments in quantum chemical calculations. Since the present-day techniques always use finite basis expansions for the development of the molecular wave function, the crucial question is how to partition this basis set in the most effective way.

From a purely formal point of view the molecules are assemblies of atoms and bonds, consequently the definition of fragments may be done in terms of these constituents. We can imagine fragments as ensembles of (a) atoms, (b) bonds. Evidently in the first case the bonds linking the fragments should be broken, while in the second case the atomic centres at the borderline of two fragments should be artificially partitioned.

It may seem that the more physical choice is to define a fragment as the set of atoms with the associated atomic basis functions. This works quite well, if there are no covalent bonds to be broken, i.e. the fragments are separate molecules. However such a treatment is not very practical if the fragments are linked by strong covalent interaction (cf. Fig. 1).

It is clear that if we are seeking a good zeroth order approximation to the molecular wave function as an appropriate superposition of fragment wave functions, the interaction of the fragments should be as small as possible or, at least, this interaction should be limited to a small region of the extended system, without affecting larger portions of the molecule. As already mentioned above, the best way to achieve this goal is to use the bond concept. One can define strictly localized molecular orbitals (SLMOs) representing one-centre lone pairs, two-centre σ-bonds or many-centre π-systems [51–57]. The SLMOs serve as a basis set and, as it will be shown later, their use offers a chemically meaningful partition of one-electron properties and of the total electronic energy into bond or (larger) fragment contributions.

One cannot mention the a priori use of localized orbitals, without emphasising the importance of the hybrid orbitals. It is well-known that hybridization is an extremely fruitful concept in chemistry and in the rationalization of molecular structure and bonding [7]. As it was shown by Maksić, hybridization often takes into account the most essential modifications of the atomic properties, and it can be a very good starting point for a MAM (modified atoms in molecules) model [50]. The SLMOs

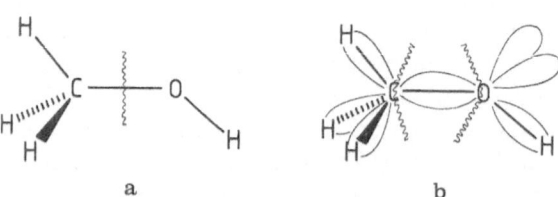

                        a   b

Fig. 1 a, b. The two possible ways of defining fragments in a covalently bonded system. **a)** The fragments are defined by a set of bonded atoms. Splitting one chemical bond is unavoidable. **b)** The fragments are defined by a set of localized bonds. Bonds belonging to the same atom may be classified as belonging to different fragments. It may be necessary to somehow partition the nuclear charges of these atoms

can be built most easily from appropriate hybrid orbitals. Nevertheless, this tool has its own drawbacks too. While the definition of the hybrids is relatively straightforward, in extended basis sets its construction is quite controversial.

In the context of using localized orbitals as the building blocks of the fragments, we are faced with another problem: what to do with the nuclei. It is a common procedure to consider a certain portion of the nuclear charge to belong to the bond, thus forming electrically neutral subunits [46]. It is useless to say that such a prescription is quite arbitrary, therefore the problem needs further study.

To conclude we can say that the practically useful procedures of fragmentation should be based upon some kind of basis set partitioning. If the fragments are defined primarily as sets of atoms one obtains strongly interacting open-shell fragments that form covalent bonds. If the fragments are defined as assemblies of bonds, one should either artificially partition certain nuclei or, which is perhaps the most physical solution, consider a certain number of two-electron bonds as the junctions between the fragments and give up any transferability in these regions.

2.2.3 The Prototype Molecule Approach

The prototype molecule (or cluster[1]) approach to the quantum chemical treatment of extended systems has proved to be a valuable tool, especially if the chemical phenomenon of interest is mainly of local character. The prototype molecule models were used e.g. in the quantum chemistry of silicates [58], zeolites [59] and enzymes [14]. In the solid state, quantum chemistry calculations on prototype molecules [60] represent an important alternative to crystal orbital type methods. Although these latter calculations may be very important as starting points even for the description of local phenomena which violate the exact translational symmetry, the cluster approaches have the advantage of providing a direct space representation of the wave function.

Nevertheless, the prototype molecule models raise again the problem of the appropriate definition of a "relevant part" of the extended system. Moreover, if long-range electrostatic forces play an important role, their explicit consideration may also be unavoidable.

As concerns the definition of the prototype molecule, at least two main cases can be distinguished. If the extended system consists of subsystems which are held together by nonbonded interactions (multipolar electrostatic forces, induction and dispersion interactions), it is possible to designate closed shell chemical subsystems without breaking covalent chemical bonds, and the prototype molecules can be the constituent molecules themselves. The situation becomes much more complicated, when the subunits are held together by covalent interactions, chemical bonds. In this case there are no natural frontiers for the representative subsystems, and it is unavoidable that a certain number of chemical bonds are broken. The prototype molecules are formed by saturating the broken (dangling) bonds by some appropriately selected atoms.

---

[1] In the literature the "cluster model" terms is perhaps more frequently used. However, the term "cluster" should be reserved for designating the experimentally existing atomic aggregates. Prototype molecules are those molecular systems, which may or may not exist experimentally, but they are designed to represent a characteristic portion of an extended system, such as a biological macromolecule, protein, crystal etc.

In the context of semiempirical procedures several attempts have been made to use "pseudoatoms" which mimic the hybridization and electronegativity of the usually polyvalent atom at the frontier of the cluster [61–63].

Another possibility is to saturate dangling bonds of the representative cluster with hydrogen atoms. The prototype molecule thus obtained is relatively easy to handle with standard quantum chemical methods, and one can hope that the artefact effects are small or negligible, if the model system is sufficiently large. In a few cases, comparison of the electronic charge density maps in the finite models saturated by hydrogen atoms and the similar maps in accurate three-dimensional models confirm the use of prototype molecules [64].

3 Theoretical Framework: Group Function Theory

In the following we shall be concerned with the problem of the electronic structure of molecular systems containing N electrons and M nuclei. The validity of the usual Born-Oppenheimer, or clamped nuclei, approximation will be assumed, that is we shall investigate the distribution of electrons in the field of the fixed nuclei. In principle the approximate solution of the Schrödinger equation of all the electrons provides us with the different electronic states of the molecule, once the position of the nuclei and the number of electrons is given. Essentially this is the procedure followed in everyday routine calculations of ab initio quantum chemistry, where we do not take into account the a priori knowledge about the properties of the different fragments of the total composite system.

The group function (GF) [65–67], or generalized product, method is based on the philosophy that the wave function of any composite system can be developed in terms of isolated group states. In this respect its philosophy is basically different from that of the molecular orbital (MO) theory. As it has been stressed recently by Schipper [45], the MO model is biased towards the united atom limit, instead of the isolated groups or isolated atoms limit. This fact may explain the well-known failure of the RHF-MO theory in the correct description of dissociation processes. A pictorial illustration of the relationship between the molecular orbitals and the united atom MOs has been shown for example in Ref. [68]. The united atom description [69] could perhaps be mathematically acceptable, since a basis may be complete even if it is expanded around a single centre, nevertheless, for obvious reasons it is quite unacceptable for the representation of large molecules, or extended systems. In contrast to the MO theory the GF method is biased towards the dissociated group limit, therefore it is much better adapted to describing extended systems.

Although in the following we shall mainly use the group function model for the treatment of extended systems with spatially separated fragments of functional groups, it should be mentioned that the first applications of the idea of separable electron groups were in the theoretical justification of the π-electron theories of quantum chemistry [70]. The most important results were summarized by the so-called σ-π

separability conditions. These involve that the electrons can be divided to two separate groups, the n_σ σ-electrons and the n_π π-electrons, fulfilling that $n = n_\sigma + n_\pi$, where n is the total number of electrons in the molecule, i.e. there is no electron interchange between the two groups. The total wave function is supposed to be written as the antisymmetrized product of the wave functions of the σ and π-electron groups, which are separately normalized and expanded on the basis of orthonormal Slater determinants, built up from an orthonormal set of one-electron basis orbitals. It is quite important to note that the one-electron orbitals for the σ and π-electrons form mutually exclusive sets, which are orthogonal to each other. This construction guarantees the *strong orthogonality* of the σ and π wave functions [71].

Strong orthogonality plays a crucial role in the theory of separable electron groups. Note that the σ-π separability in planar conjugated molecules is a fortunate case, since the orthogonality is ensured by symmetry: the orbitals for these electron groups belong to different irreducible representations of the C_s point group.

Another example, where rigorous formulation of the separability conditions were attempted is the core pseudopotential method in quantum chemistry [72]. This treatment is based on the separability of the core and valence electron groups of atoms and molecules. However, here the fulfillment of the orthogonality constraint is much more problematic: the basis functions belonging to the core and valence regions are not orthogonal by symmetry, their overlap is non-negligible. Nevertheless for a proper description one ought to ensure the orthogonality of the valence electrons to the core region, otherwise these former would "collapse" into the low-energy regions normally occupied by the core electrons. This problem can be avoided either by working with an explicitly orthogonalized basis (valence orbitals Schmidt orthogonalized to the core [73]), or by adding some extra conditions to the variational problem, which take into account the orthogonality requirements. The latter approach has been advocated by Huzinaga and Cantu [74], who introduced the coupling (or screening) operator method for pseudopotentials. This technique is of great importance in the non-orthogonal formulations of the group function theory (see below).

The chemical notion of bonds and functional groups suggests a further possibility for the partitioning of the valence electrons to reflect experimentally observed transferability of bond or group properties in terms of localized bonds, or assemblies of bonds corresponding to functional groups. The direct determination of the localized fragment orbitals in solids or in molecules may be based on the Adams-Gilbert equations [75–78], which are closely related to the non-orthogonal group function method [83].

3.1 Group Function Theory of McWeeny

Perhaps the most systematic theoretical formulation of the problem of separable electron groups is due to McWeeny [65–67]. As mentioned above, the GF theory has served as the basis for various quantum chemical models and also for efficient treatments of the electron correlation problem in the framework of geminal methods [13, 79–81]. In the present context we emphasise rather the possible applications for the treatment of large molecules or aggregates.

In the GF theory [65–67] the total N-electron wave function Ψ is expressed in the form of a linear combination

$$\Psi = \sum_{\varkappa} C_{\varkappa} \Psi_{\varkappa} \tag{3.1}$$

of the $\Psi_{\varkappa}$ configuration functions. $\Psi_{\varkappa}$ is the antisymmetrized product of the $\Phi_{Aa}(1 \dots N_A)$ group functions

$$\Psi_{\varkappa} = \mathscr{A} \, |\Phi_{Aa}(1 \dots N_A) \, \Phi_{Bb}(N_A + 1 \dots N_A + N_D) \dots| \, . \tag{3.2}$$

The group functions $\Phi_{Aa}(1 \dots N_A)$ are N_A-electron determinants and describe a many-electron system in its a-th electronic state. (The index $\varkappa$ stands for a collection of a, b, ... etc. indices.) The operator $\mathscr{A}$ is the normalized intergroup antisymmetrizer. In the closed-shell case Φ_{Aa} is built up from doubly filled molecular orbitals ψ_{Ai}

$$\Phi_{Aa} = \det |\psi_{Ai}(1) \, \bar{\psi}_{Ai}(2) \dots \psi_{Aj}(N_A)| \tag{3.3}$$

with

$$\psi_{Ai} = \sum_{\mu=1}^{m_a} c_{i\mu}^A \chi_{\mu}^A \, . \tag{3.4}$$

Of course, more general forms of the group functions are possible including fully correlated N_A-electron wave functions which can always be expanded as linear combination of determinants.

It is important to stress that only an approximate Ψ wave function can be obtained if the partition of the electrons among different groups is kept fixed in all the configuration functions $\Psi_{\varkappa}$. This approximation can be called the simplified group function (SGF) method [45], in contrast to the "full" method where all the possible partitions of the N electrons are allowed. Nevertheless the SGF approximation is quite good in all the cases where there are no important covalent interactions. This is the case for most of the van der Waals complexes and also for interacting two-electron bonds [81], or for fragments defined as an ensemble of localized bonds.

In general one assumes that the group functions are strongly orthogonal, i.e.

$$\int d\tau_1 \, \Phi_{Aa}^*(1, \dots i, j, \dots) \, \Phi_{Bb}(1, \dots, k, l, \dots) = 0 \tag{3.5}$$

for $A \neq B$. This condition is a mathematical formulation of the physical individuality of the different electron groups. (A discussion of the "strong" and "weak" ortho-gonality in the case of geminal-type wave functions can be found in Ref. [12]. It is to be noted that in the case of more than two electrons the weak orthogonality is not equivocally defined.) The strong orthogonality is automatically satisfied if the molecular orbitals fulfill the following condition

$$\int d\tau_1 \, \psi_{Ai}^*(1) \, \psi_{Bj}(1) = \delta_{ij} \delta_{AB} \, . \tag{3.6}$$

This can be guaranteed by construction if the ψ_{Ai} orbitals of the different groups are built up from mutually exclusive sets of orthonormal atomic basis functions χ_{μ}^A. The atomic basis functions of the semiempirical ZDO theories automatically meet this requirement. In ab initio calculations a preliminary Löwdin-orthogonalization of the basis set may be necessary [82].

The energy of a non-degenerate ground state can usually be approximated by using a single $\Psi_{\varkappa}$ configuration function

$$E = \langle \Psi_{\varkappa} | \mathcal{H} | \Psi_{\varkappa} \rangle = \sum_A \langle \Phi_{Aa} | \mathcal{H}_{\text{eff}}^A | \Phi_{Aa} \rangle$$

$$- \tfrac{1}{2} \sum_{A,B} \{ J^{AB}(aa, bb) - K^{AB}(aa, bb) \} + E_{\text{nuc}}, \tag{3.7}$$

where $\mathcal{H}_{\text{eff}}^A$ is the effective Hamiltonian of the A-the group including the J^B average Coulomb and K^B exchange potential of the other groups

$$\mathcal{H}_{\text{eff}}^A = \sum_{i=1}^{N_A} h(i) + \tfrac{1}{2} \sum_{i,j}^{N'_A} r_{ij}^{-1} + \sum_{i=1}^{N_A} \sum_{B(\neq A)} \{ J^B(i) - K^B(i) \} . \tag{3.8}$$

The matrix elements $J^B(aa, bb)$ and $K^B(aa, bb)$ of the J^B and K^B operators are defined as

$$J^{AB}(aa, bb) = \int\int d\tau_1 \, d\tau_2 \, r_{12}^{-1} P^A(aa \mid 1; 1) \, P^B(bb \mid 2; 2) ,$$

$$K^{AB}(aa, bb) = \int\int d\tau_1 \, d\tau_2 \, r_{12}^{-1} P^A(aa \mid 2; 1) \, P^B(bb \mid 1; 2) , \tag{3.9}$$

where $P^A(aa \mid 1; 2)$ is the first-order density matrix element of the group A. Provided that we use single determinantal group functions of closed-shell fragments, the total energy can be written in terms of the spatial orbitals ψ_{Ai}

$$E = \sum_A \left\{ \sum_i 2(\psi_{Ai} | \mathcal{H}_{\text{eff}}^A | \psi_{Ai}) + \sum_{ij} 2(\psi_{Ai}\psi_{Ai} | \psi_{Aj}\psi_{Aj}) - (\psi_{Ai}\psi_{Aj} | \psi_{Aj}\psi_{Ai}) \right\}$$

$$+ \sum_A \sum_B \sum_i \sum_j \{ 2(\psi_{Ai}\psi_{Ai} | \psi_{Bj}\psi_{Bj}) - (\psi_{Ai}\psi_{Bi} | \psi_{Bj}\psi_{Ai}) \} . \tag{3.10}$$

The group functions can be optimized by imposing the condition that the first-order polarization energy of the system vanishes

$$E_{\text{pol}} = - \sum_A \sum_{a'(\neq a)} \frac{|\langle \Phi_{Au} | \mathcal{H}_{\text{eff}}^A | \Phi_{Aa'} \rangle|^2}{E(Aa \to Aa')} = 0 . \tag{3.11}$$

This condition leads to a generalized form of the local Brillouin theorem of the usual SCF method

$$\langle \Phi_{Aa} | \mathcal{H}_{\text{eff}}^A | \Phi_{Aa'} \rangle = 0 \tag{3.12}$$

and it is equivalent with the requirement that the group energies, $E_A = \langle \Phi_{Aa} | \mathcal{H}_{\text{eff}}^A | \Phi_{Aa} \rangle$, be stationary

$$\delta E_A = 0 , \quad \forall A . \tag{3.13}$$

The above conditions lead to a set of equations for the group functions, which are coupled by the intergroup Coulomb and exchange potentials. The group functions, obtained by an iterative self-consistent solution of these equations, yield also the stationary value of the total energy of the system.

In practical applications some groups can be held frozen in the iterative procedure. This is the case in core pseudopotential calculations where the core electron wave function represented by a suitable analytical form, remains fixed and the valence electron wave function is optimized in the effective field of the core.

It is important to note that the group functions obtained by the above procedure are not exact eigenvectors of the effective group Hamiltonian, $\mathscr{H}_{\text{eff}}^A$, since the full variational space is not available for a particular group function Φ_{Aa}. The variational procedure should be restricted to a particular subspace so that Φ_{Aa} remains orthogonal to Φ_{Bb}, $B \neq A$, B representing a subspace spanned by another group. This condition may be automatically guaranteed by the construction of the group molecular orbitals. If this is not the case and the orbitals are possibly distorted into the "forbidden" space, appropriate techniques, using projection operators, should be applied in the solution of the orbital equations [74–78, 83].

3.2 Non-Orthogonal Group Function Theory

This point is at the origin of one of the difficulties one meets when the strong orthogonality constraint is relaxed. As mentioned above, the non-vanishing overlap of the one-electron functions belonging to different groups

$$\int d\tau_1 \, \psi_{Ai}^*(1) \, \psi_{Bj}(1) = S_{Ai, Bj} \qquad (A \neq B) \tag{3.14}$$

means that the individuality of the groups is lost and the subspaces belonging to them become ill-defined.

Nevertheless, by the introduction of the reciprocal (or biorthogonal) orbitals

$$\hat{\psi}_{Ai} = \sum_{Bj} S_{Bj,Ai}^{-1} \psi_{Bj} \tag{3.15}$$

the formal simplicity of the energy expression (3.10) can be recovered

$$E = \sum_A \left\{ \sum_i 2(\psi_{Ai}|\mathscr{H}_{\text{eff}}^A|\tilde{\psi}_{Ai}) + \sum_{ij} 2(\psi_{Ai}\tilde{\psi}_{Ai}|\psi_{Aj}\tilde{\psi}_{Aj}) - (\psi_{Ai}\tilde{\psi}_{Aj}|\psi_{Aj}\tilde{\psi}_{Ai}) \right\}$$

$$\times \sum_A \sum_B \sum_i \sum_j \left\{ 2(\psi_{Ai}\tilde{\psi}_{Ai}|\psi_{Bj}\tilde{\psi}_{Bj}) - (\psi_{Ai}\tilde{\psi}_{Bj}|\psi_{Bj}\tilde{\psi}_{Ai}) \right\}. \tag{3.16}$$

but, in fact, the $\tilde{\psi}_{Ai}$ orbitals mix up global effects with intragroup interactions.

Whereas in the strong orthogonal case the total energy could be made stationary by optimizing the E_A group energies this is no longer true in the non-orthogonal case. Nevertheless, according to Mehler's proposition [83–87], the $\delta E_A = 0$ condition can be used to obtain the wave function of the individual interacting molecules in aggregates. As previously mentioned, the variation of the non-orthogonal group functions should be restricted to the appropriate subspace. This has been done by adding a further variational constraint requiring that the $\langle \psi_{Ai} | \psi_{Bj} \rangle = S_{Ai,Bj}$ intergroup overlap integrals remain constant throughout the variation of orbitals of A. The resulting effective Hartree-Fock equations are the following

$$\left\{ 2\mathscr{F}^A - \sum_{Bj} \left(|\psi_{Bj}\rangle \langle \tilde{\psi}_{Bj}| \, \mathscr{F}^A + \mathscr{F}^A \, |\tilde{\psi}_{Bj}\rangle \langle \psi_{Bj}| \right) \right.$$

$$\left. + \sum_{k} \left(|\psi_{Ak}\rangle \langle \tilde{\psi}_{Ak}| \, \mathscr{F}^A \, |\tilde{\psi}_{Ak}\rangle \langle \psi_{Ak}| \right) \right\} |\psi_{Ai}\rangle = \varepsilon_{Ai} \, |\psi_{Ai}\rangle . \qquad (3.17)$$

In addition to the usual Group Function Fock operator, $\mathscr{F}^A$, two correction terms appear. The first one is the screening (or coupling) operator which represents a repulsive potential. It role is to prevent electrons of subsystem A to collapse into regions occupied by other groups. The second correction term is just a level shift which makes possible to interpret the ε_{Ai} eigenvalues as orbital energies. This Non-Orthogonal Group Function (NOGF) model is very closely related to the Adams-Gilbert equations [75–77], and the Huzinaga-Cantu screening operator technique [74].

3.3 Local Space Approximation

An interesting approach to the problem of calculating the local wave function of a fragment in interaction with its surrounding has been proposed by Kirtman and de Melo [88]. Their approach is based on the density matrix formulation of the HF problem. The zeroth order approximation to the density matrix of the total system is a simple direct sum of the density matrices of the fragments, provided that the AO basis is orthogonal by construction or it is properly orthogonalized.

The first-order corrections due to the interaction are calculated so that the idempotency of the density matrix be preserved. According to McWeeny [94] the most general form of this first-order correction to R_0 is

$$\Delta R^{(1)} = V + V^\dagger \qquad (3.18a)$$

with

$$V = (1 - R_0) X R_0 , \qquad (3.18b)$$

where X is an arbitrary symmetric matrix. The so-called local space approximation [88–93] means that the matrix X is of purely local space character

$$X \Rightarrow X_Q = QXQ , \qquad (3.19)$$

where Q is the matrix of the projection operator Q to the local space

$$Q = \sum_{\mu\nu} |\chi_\mu\rangle \, S_{\mu\nu}^{-1} \, \langle \chi_\nu| . \qquad (3.20)$$

Using the variation condition one can derive a linear equation for X_Q which, in turn, can be used to evaluate the first-order correction to the density matrix. This leads to an iterative scheme involving matrix manipulations in the local space. It is important to note that Kirtman's treatment describes not only inductive but delocalization effects, as well. A practical limitation is that one should work with orthogonal AO basis sets therefore the calculations have been done with semiempirical ZDO model Hamiltonians [91], or with explicitly Löwdin-orthogonalized basis sets in ab initio calculations.

4 Calculation of Electrostatic Potentials from Strictly Localized Fragments

The transferability of the localized two-electron bonds or eventually of the orbitals associated with larger fragments may be a good starting point to construct zeroth-order approximations to the molecular wave function. In most of the cases this type of wave function represents only an intermediary step towards a more refined description, which would take into account the mutual interaction of the fragments. Before discussing the methods, which consider explicitly the fragment interactions, let us examine briefly the possible applications of the simple approximation, which uses directly the superposition of fragment wave functions.

There are quite a wide range of procedures, which use the strict transferability of fragment, bond, or atomic properties for the calculation of the aggregate values of various characteristics. These are mostly empirical or semiempirical approaches, although their relationship with the quantum chemical transferability concept is quite straightforward. Perhaps the best-known example is the transferability of the bond dipole moment [8], which has been extensively studied by quantum chemical tools as well [95]. The transferability of bond polarizabilities is a more controversial subject which has been thoroughly discussed, for example, by Claverie [96].

Applications of the bond polarity and bond polarizability concept for predicting the properties of really large molecular systems are relatively scarce. Nevertheless, the most widespread and successful exploitation of the bond transferability idea in connection with the molecular charge distributions has been done for the evaluation of electrostatic potentials and fields around biological macromolecules.

4.1 Atomic Monopole Models

The assumption of complete transferability of the charge distribution of the fragments in extended systems, like biopolymers, is often accompanied by simplified representations of the charge distributions. Atomic point charge models seem to be a reasonable compromise between accuracy and computational ease. There are a large variety of methods ranging from empirical fitting to relatively accurate ab initio Hartree-Fock calculations for the determination of the point charge sets characterizing the fragments, e.g. the amino acid residues, in large systems.

First of all, one has to stress that the definition of atomic charges, corresponding to a given molecular charge density distribution, is not unique. Numerous population analysis schemes have been proposed, usually in order to "improve" the classical Mulliken population analysis [97]. In spite of its obvious failure in some cases (e.g. if using very large or unbalanced basis sets) the Mulliken charges remain quite popular. Recently it has been shown that the Mulliken population analysis has a privileged role in the LCAO model: this charge distribution has an associated operator [98].

A class of improved population analysis schemes has been designed to reproduce the total dipole moment of the molecule when calculated by point charges. Such a "dipole moment conserving" procedure was proposed e.g. by Jug [99] and by Thole and van Duijnen [100]. A more general multipole fitted scheme has also been derived [101]. A slightly different approach is to determine "potential derived" atomic charges which are fitted to reproduce the values of the electrostatic potential outside the van der Waals envelope of the molecule [102, 103].

It follows from the strict transferability of the fragment charge distribution that no charge flow is allowed from one building block to another. The electroneutrality of the full system, e.g. of the complete biopolymer, can be guaranteed most easily if the fragments themselves are separately electroneutral. It follows from this condition that the boundaries of the fragments should be selected along the least polar bonds.

At this point one is faced with the problem of how to treat the dangling bonds, along which the fragments have been separated. The most simple method is to add hydrogen atoms to saturate them. The net charge obtained for the fragments in such model calculations is usually not an integer value, therefore compensatory charges should be distributed over the fragment atoms.

In order to appreciate the differences in the atomic charges obtained for the same amide fragment ($NHCOC^{\alpha}H^{\alpha}$) of peptides and proteins a number of point charge sets were compared by Rullman [104] (Table 1). The charges used in various empirical force fields come from different quantum chemical calculations. For example, the ECEPP [105] and CHARMM [106] charges were determined on the basis of CNDO/2 studies, while the AMBER [107] and OPLS [108] charges came from approximately scaled ab initio SCF calculations. Rullman and van Duijnen [109] used the dipole moment conserving population analysis [100] for the HF SCF wave function with the Mehler-Paul basis set [110].

It is important to note that an uncertainty of 0.02 atomic units in the atomic charge may correspond to electrostatic energy errors in the order of 4 kJ/mol which may be chemically significant [104].

Table 1. Partial charges of the $NHCOC^{\alpha}_{n}H^{\alpha}_{n}$ group in empirical force fields (in millielectrons)

	N	H	C	O	C^{α}	H^{α}	total
GROMOS	−280	280	380	−380	0	0	0
ECEPP	−356	176	450	−384	64	20	−30
CHARMM	−360	260	480	−480	100	0	0
AMBER	−463	252	616	−504	35	48	−16
OPLS	−570	370	500	−500	200	0	0

4.2 Multipole Expansions

Alberte Pullman and her coworkers did the first large-scale studies on biological macromolecules, namely on nucleic acids, in the seventies [111]. According to their method, one divides the macromolecule into appropriate subunits and determines the charge distribution of these fragments by fairly good quality ab initio calculations.

Biopolymers, like nucleic acids or proteins, are especially well-suited for this approach, since these macromolecules are built up from a restricted number of monomeric units, like nucleotide bases, phosphates, sugars (nucleic acids) or from the 20 naturally occurring amino acids (proteins). This feature makes it possible to set up a "library" of charge densities, wave functions, atomic charges or other relevant properties.

Since the full ab initio evaluation of all electrostatic integrals needed for the determination of the potential would be extremely long, further approximations are necessary to speed up calculations. Pullman and coworkers used the overlap multipole (OMTP) approximation, which replaces the contribution of each atomic orbital product by a multipole expansion of the potential up to quadrupoles [112]. However the original procedure had a weak point: the treatment of junctions of the monomeric units. Technically the simplest way is to cut the bonds linking different monomeric units and saturate the dangling bonds with hydrogen atoms. The model monomers thus obtained were used for the construction of the library, and the aggregate properties could be calculated by simple superposition of the individual contributions coming from the subunits in an appropriate relative geometry. Nevertheless one may suspect that the presence of extra hydrogens in the region of intermonomer bonds may lead to artefact effects, which are quite difficult to control. Anyway, one may assume that such effects are not dominant at a reasonable distance from the critical regions [112]. This method has several drawbacks. First, even the most judicious choice of the cutting out procedure leaves some doubts about the importance of the artefact effects due to the presence of additional hydrogen atoms. Second, the macromolecule has to be built up of fragments with fixed geometry, which gives rise to difficulties if the subunit geometries are more-or-less flexible. This is extremely important in proteins, where the amino acid residues show considerable conformational flexibility. These problem can be circumvented by the use of localized molecular orbitals as basic subunits.

Tomasi and his coworkers considered the localized two-electron bonds as directly transferable from one molecule to another [113, 114]. Model calculations on simple molecules with subsequent localization lead to a library of the charge distributions, $\varrho^0(r)$, of the various fragments corresponding to either specific bonds or small functional groups. These charge distributions consist of, in addition to the contribution of one or more electron pairs, the core electrons (represented by point charges) and the appropriate amount of nuclear charges to make the whole fragment electrically neutral [113]. In order to facilitate the potential calculations, $\varrho^0(r)$ has been replaced by a collection of point charges:

$$\varrho^0(r) \approx \sum_i q_i \delta(r - r_i) \tag{4.1}$$

which reproduce the potential with sufficient accuracy. Lavery and his coworkers followed essentially the same idea in their LMTP (multipole expansion for localized orbitals) procedure [115, 116]. The appropriate localized orbitals are calculated according to the Boys' localization criterion and they are supposed to be insensitive with respect to the conformational changes. The contribution of the core electrons and the nuclei is represented by point charges, while the potential of the bonds is developed in multipole series around the centroids of the localized orbitals, truncated after the quadrupolar [115] or the octupolar [116] terms. The latter representation yielded quite accurate electrostatic potentials: at distances larger than 2.5 Å the error was less than 1–2 kcal/mol.

Exact multipolar representation of the electronic charge distribution, expressed on Gaussian basis, is possible by using the well-known product theorem of Gaussian functions. This approach was first proposed by Hall [117] for spherical floating Gaussians, and it was later generalized by Stone [118]. General procedures to reduce such many-centre multipole distributions to multipoles situated on a limited number of chemically meaningful centres, like atoms and bonds, was proposed by Stone and Alderton [119], and Vigné-Maeder and Claverie [120]. These atomic multipoles are very useful for designing good-quality, anisotropic atom-atom pair potentials [121]. A recent review with a full account of the relevant literature can be found in Ref. [122]. The atomic multipoles seem to be reasonably transferable, and they might be used in the future for the simulation of electrostatic properties in biological macromolecules as well. We can mention that recently a set of atomic multipoles was determined by the method of Stone for the amino-acid residues [123].

The above cited methods use point multipolar representation of the electron density associated with atoms or localized orbitals. Since the radial part of the electronic charge distribution is neglected, the electrostatic potentials become very poor in the close proximity of the molecule, e.g. in the lone pair regions, including contacts with hydrogen bonded substrates or solvents.

4.3 The Bond Increment Method

The bond increment (BI) method, developed in our laboratory [124–126], offers the possibility of taking into account the spreading out of the charge cloud by using a more elaborate expression of the electrostatic potential, then the simple point charge or point multipole representation. The bond increment method starts from the following approximate wave function:

$$\Psi = \det |\varphi_1(1) \, \bar{\varphi}_1(2) \ldots \bar{\varphi}_i(2i)| \,, \tag{4.2}$$

which is constructed as the antisymmetrized product of the φ_i strictly localized molecular orbitals (SLMOs).

The SLMOs, according to the usual chemical picture, can be σ, lone pair or π orbitals, and they assume the following simple mathematical structure, respectively:

$$\varphi_i^\sigma = c_{Ai} h_{Ai} + c_{Bi} h_{Bi}, \qquad \varphi_i^{lp} = h_{Ai}, \qquad \varphi_i^\pi = \sum_{A=1}^{M_{\pi_i}} c_{Ai} u_{Ai}^{np_z}. \tag{4.3}$$

Here h_{Ai} stands for a normalized hybrid orbital centered at atom A:

$$h_{Ai} = b_i^{ns} u_A^{ns} + b^{np_x} u_A^{np_x} + b^{np_y} u_A^{np_y} + b^{np_z} u_A^{np_z}, \tag{4.4}$$

where $u_A^{np_x}$, etc. are normalized Slater orbitals. When $A = H$ (hydrogen) $h_{Ai} = u_A^{1s}$. M_{π_i} denotes the number of π centers in the i-th π-type molecular orbital. Note that these SLMOs are non-orthogonal, the interbond overlap integrals

$$S_{ij} = \langle \varphi_i | \varphi_j \rangle \tag{4.5}$$

are nonzero for $i \neq j$. Nevertheless, the electrostatic potential is calculated by neglecting the effect of this non-orthogonality. In the spirit of this approximation all three-centre integrals, involving the overlap charge distribution of two orbitals on different atomic centres, are neglected as well. The resulting expression for the electrostatic potential (and, mutatis mutandis all one-electron properties) involves a sum for the *bond increments*:

$$V(r) = -2 \sum_i^N \sum_A^{M_i} c_{Ai}^2 \langle h_{Ai} | r^{-1} | h_{Ai} \rangle + \sum_A^M Z_A^{eff} |r - R_A|^{-1}$$

$$= \sum_i^N V_i^{BOND}(r) + \sum_A^M V_A^{NUC}(r). \tag{4.6}$$

The bond increment method has been applied in this form to numerous chemical and biochemical problems, like the interpretation of organic reaction mechanisms [127], protonation energies [128] and catalytic effects in enzymatic reactions [129].

One can rearrange the (4.6) expression and collect the terms according to atomic centres, instead of bonds. Such a transcription provides us a convenient link to the atomic multipole models, discussed in Sects. 4.1 and 4.2.

Let n_A denote the number of bonds (hybrid orbitals) belonging to atom A. Then the terms of the potential expression can be grouped according to individual atomic contributions as:

$$V(r) = \sum_A^M \left\{ Z_A^{eff} |r - R_A|^{-1} - 2 \sum_{k=1}^{n_A} c_{Ak}^2 \langle h_{Ai} | r^{-1} | h_{Ai} \rangle \right\}. \tag{4.7}$$

The advantage of this form is that the potential can be expressed with the atomic multipole moments, $Q_\lambda^\mu(A)$:

$$Q_\lambda^\mu(A) = -\sqrt{\frac{4\pi}{2\lambda + 1}}\, 2 \sum_{k=1}^{n_A} c_{Ak}^2 \int dr\, r^{\lambda+2} \int d\Omega\, \mathscr{Y}_\lambda^\mu(\Omega) |h_{Ak}(r, \Omega)|^2. \tag{4.8}$$

The zeroth-order moment, the atomic charge is defined as

$$Q_0^0(A) = Z_A^{eff} - 2 \sum_{k=1}^{n_A} c_{Ak}^2 \tag{4.9}$$

and assuming a standard *sp* basis set the highest nonvanishing multipole moment is quadrupole.

The potential at the point r is the sum of atomic multipole contributions. A very convenient feature of the resulting expression is that the asymptotical value of the potential is expressed with point multipoles, while at shorter distances an exponentially vanishing correction factor, $f(2\zeta r_A, \lambda)$, appears:

$$V(r) = \sum_A \sum_{\lambda=0}^{2} \sum_{\mu=-\lambda}^{\lambda} Q_\lambda^\mu(A) \frac{\mathscr{Y}_\lambda^\mu(\Omega)}{r_A^{\lambda-1}} \sqrt{\frac{4\pi}{2\lambda+1}} \{1 - e^{-2\zeta r_A} f(2\zeta r_A, \lambda)\}.$$

(4.10)

The actual values of the atomic multipole moments depend on the coefficients of the hybrid orbitals in the individual bond orbitals (bond polarities), on the orientation of the hybrids and on the degree of hybridization. Since these latter two parameters depend significantly on the geometrical arrangement of the atoms [47], the bond increment method may be an adequate tool for constructing the zeroth order wave function, which nevertheless describes the main trends in the conformational and geometrical dependence of the atomic charge distributions and consequently of the electrostatic potential.

A promising approach may be to fit the polarities of the bond orbitals, in order to reproduce ab initio atomic multipole moments, obtained in various geometries. Notice that such a method would be a generalization of the quite popular potential-fitted point charge models [102]. One can ask, why do we construct the zeroth order wave function with the intermediary of SLMOs, instead of taking directly the atomic multipoles from accurate ab initio calculations? In effect, this would be a good possibility for molecules with conformationally rigid subsystems. Unfortunately, while the point charge models are quite insensitive to geometrical (conformational) changes, the higher multipoles (the anisotropy of the atomic charge distributions) should follow the local geometrical changes.

Since the conformational dependence of higher atomic multipoles is not known analytically, we have to make recourse to the appropriately oriented SLMOs (and hybrid orbitals), which give rise to the atomic multipoles.

The atomic multipole expansion of the BI electrostatic potential is extremely useful, when the long-range, purely point-multipolar part of the potential yields an important contribution. This is the case in crystals, where the multipolar sums (up to the quadrupolar potential) are conditionally convergent lattice sums. Special techniques, like Ewald summation method [130, 131] and its generalizations [132] are needed to handle properly these infinite sums. Recently we applied the multipolar BI method, coupled with Ewald summation for the evaluation of electrostatic potentials and fields in zeolite cavities [133] and for the prediction of the IR frequency sequence of the different acidic sites in H-faujasite [134].

The bond increment model includes approximations, which are justified mainly by economy considerations. For example, it is supposed that the hybrid orbitals have standard *s*-characters and they are oriented towards the neighbouring atoms. In many test cases these simplifications did not lead to serious errors. However, an especially delicate case is that of oxygens with two lone pairs. It turned out that at distorted

X-O-Y angles the s-characters should be readjusted such that the orthogonality of the hybrids on the same centre be maintained.

Considering the approximate character of the model, an empirically motivated rescaling of the results has also been attempted. This was motivated mainly by the failure of the BI model to reproduce correctly the negative potential regions above π-systems. The following very simple formula has been proposed [133]

$$V(r) = \sum_{i}^{N} (1 - \beta_i) V_i^{\text{BOND}}(r) + \sum_{A}^{M} V_A^{\text{NUC}}(r),\qquad\qquad (4.11)$$

where the β_i parameters are to be fitted to ab initio (e.g. STO-3G) potential values and they should satisfy the condition:

$$\sum_{i} \beta_i = 0.\qquad\qquad (4.12)$$

In fact, a considerable improvement could be achieved in a whole series of molecules [133]. Whereas for the molecules containing only first-row elements, the β_i parameters were fairly transferable for a given type of bond, this transferability was not valid for seond-row elements.

The common weakness of the transferable models is that they do not take into account the modification of the subsystem charge distributions under the influence of the electrostatic field of the remainder of the system. In the following we shall overview the models and methods which are designed to describe this phenomenon.

5 Inductive Interaction of Strictly Localized Fragments

5.1 Iterative Methods

The average electrostatic field of the neighbouring groups may lead to the distortion of the charge distributions of the individual fragments. This effect is usually referred to as the inductive interaction and probably this is the most important correction to the zeroth order ("superposition of transferable fragments") description of polar systems. The general mathematical framework of the treatment of this inductive interactions has been discussed in Sect. 3. Nevertheless, in practical calculations further approximations have been introduced in order to make the group function equations tractable even for larger systems.

In the case of intermolecular interactions between polar molecules Otto and Ladik proposed the so-called mutually consistent field (MCF) method [135–136]. They discussed various aspects of the MCF approach in a series of papers [137–139], and compared it with the conventional SCF supermolecule and perturbational calculations.

The main advantage of the MCF method as compared to the supermolecule calculations is that the dimensionality of the variational calculation can be considerably reduced. Nevertheless a large portion of the intermolecular integrals are still needed, unless one does not introduce additional approximations. Otto and Ladik achieve this goal in their first paper [135] by simply neglecting the intermolecular exchange potential and by approximating the intermolecular Coulomb integrals as electrostatic interaction with the Mulliken point charges of the partner molecules.

Although this latter approximation was found to be unsatisfactory from a numerical aspect and it was later abandoned by the authors, it is interesting to discuss it briefly. It can be shown, that one can arrive to this Mulliken point charge approximation by introducing the following "asymmetrical Mulliken integral approximation" [140]:

$$(\mu_A \nu_A \mid \lambda_B \sigma_B) \approx \tfrac{1}{2} S_{\mu\nu}\{(\mu_A \nu_A \mid \lambda_B \lambda_B) + (\mu_A \nu_A \mid \sigma_B \sigma_B)\} \tag{5.1}$$

where μ_A and λ_B stand for the atomic basis functions χ_μ^A and χ_ν^B, belonging to the fragments A and B, respectively. If the two-center integrals are replaced by the Coulomb point charge interactions:

$$(\mu_A \nu_A \mid \lambda_B \lambda_B) \approx (\mu_A \mid \frac{1}{|R_B - r|} \mid \nu_A) \tag{5.2}$$

one really obtains the scheme adapted in Ref. [135]. This asymmetry has some consequences concerning the variational character of the resulting wave function. Due to the above mentioned asymmetry the interaction energy of molecule A in the field of molecule B is different if the roles are reversed, i.e. we consider molecule B in the field of A. It means that the energy expression is not unique and the variational principle cannot be applied unequivocally for the total system [140].

The MCF model has been refined to take into account the exchange potential by a simplified local X_α representation:

$$V_{X_\alpha}^B(r) = -6\alpha \left\{ \frac{3}{4\pi} \varrho^B(r) \right\}^{1/3}, \tag{5.3}$$

where $\varrho_B(r)$ is the electronic density of molecule B.

The MCF method was successfully applied, for example, to the study of hydration of the glycine zwitterion with 6 and 12 water molecules [138]. It has been concluded that mutual polarization effects were quite important in the first hydration shell, but presumably the water molecules in more distant hydration shells are not polarized appreciably by the solute therefore their effect can be represented by the electrostatic potential of the corresponding unperturbed charge distributions.

A practically identical model has been proposed independently by Weinstein and his coworkers: the "interaction field modified Hamiltonian" (IFMH) model [141]. These authors discussed in details the case of the HF ... HCOF complex and compared the IFMH results with perturbational calculations. It is interesting to note that in the course of the IFMH calculation the first iteration and the fully converged results yield practically identical interaction energy at least for the actual case. On the other hand, individual components of the interaction energy differ considerably in the two calculations, which shows the importance of charge reorganizations.

The models discussed above are applicable to intermolecular interactions. There are several methods which have been elaborated along similar lines for the treatment of intramolecular interactions of fragments. The bond orbital type approaches, where the fragments are two-electron bonds, have been reviewed by Surján [13]. Some procedures consider larger fragments. One of them, the Extended PCILO method [39], was motivated by the difficulties in the original PCILO approximation. The "fragments in molecule" (FIM) model of Klessinger [142, 143] is designed to describe functional groups in larger molecules. This is a variant of the "molecules in molecules" (MIM) approach, originally proposed by Longuet-Higgins and Murrell [144], elaborated mainly for the calculation of the electronic spectrum of multichromophoric systems [145–148]. In a series of papers von Niessen attempted to develop an ab initio MIM method, that is an accurate treatment of intramolecular fragments of large molecules [149]. In this approach the fragments, which are transferable in the first approximation, are constructed from localized orbitals. The interaction region of the fragments is also defined by localized orbitals and it is treated by appropriate non-orthogonal functions.

Although they go beyond the purely inductive approaches, one should mention here the Linear Combination of Fragment Orbitals procedures. The first breakthrough of ab initio calculations on large systems was brought by the molecular fragment approach of Christoffersen [150]. His method uses a floating spherical Gaussian orbital (FSGO) basis set following the proposition of Frost and coworkers [151]. In the first step one should select appropriate molecular fragments that appear in the larger system. These fragments do not correspond necessarily to actual stable molecular species, since they are merely useful constructs that describe a moiety of anticipated molecular environments. The geometry of the fragments correspond to that observed in the large molecule and may be different from that in the isolated species. The fragments may include closed-shell (CH_4, NH_3, H_2O, HF, H_3O^+, etc.) and open-shell ($\cdot CH_3$, $\cdot NH_2$, $\cdot NH_3^+$, $\cdot OH$, etc.) moieties whose wave functions can be determined by the optimization of both linear and nonlinear variational parameters of the FSGO basis functions. Once a fragment library is created this can be used to construct the large molecule. The basic assumption is that the molecular orbitals of the large system can be written as a linear combination of fragment orbitals χ_k^A

$$\varphi_i = \sum_A^P \sum_k^{N_A} c_{ki}^A \chi_k^A , \qquad (5.4)$$

where the summation runs over the P fragments and all the orbitals of each fragment. The SCF equations in this approach are straightforward generalizations of the usual SCF procedure with the only difference that the wave functions are fixed linear combinations of FSGOs extending over several centres. One should mention that recently a pseudopotential FSGO method was proposed [152, 153]. The most restrictive feature of this model lies in the subminimal nature of the FSGO basis set. This may explain the modest popularity of the method.

5.2 Description of Molecular and Covalent Crystals

A possible application of the idea of mutually consistent group functions is the description of polar molecular crystals. By the virtue of the translational symmetry the quantum chemical calculations can be restricted to a unit cell (or to an asymmetric unit) of the crystal and electrostatic (and induction) interactions can be taken into account in a self-consistent manner.

We should be aware of the fact that such a local description does not supply for the correct wave function of the crystal, which ought to respect the translational symmetry. Nevertheless, orbitals localized to the unit cell can be excellent starting points for the construction of a good quality tight-binding wave function of the crystal, either in the framework of a perturbational treatment [154, 155] or using them as basis function to build up Bloch orbitals [156].

5.2.1 Self-Consistent Madelung Potential (SCMP) Method

Many scientists are interested in the genuine properties of a molecule embedded in a crystal. It is expected that in the course of the crystal formation the electrostatic potential of the crystal (the so-called *Madelung potential*) has a considerable effect on the charge distribution, polarizability and even on the geometry and force constants of the constituent molecules. There are several heuristically derived procedures following this idea [157–161] which can be classified as *self-consistent Madelung potential* (SCMP) methods [158]. In this section we shall derive an effective Schrödinger equation for a molecular subsystem in a molecular crystal, corresponding to the SCMP model [140].

The whole crystal can be constructed by elementary translations (and eventually by appropriate space-group operations) of a basic unit (unit cell or asymmetric unit). We shall call this basic unit the "motif". It is choosen as a closed shell molecule or molecular complex or even a neutral assembly of ions, containing n electrons and M nuclei. Let N be the number of unit cells within the Born-Kármán boundaries and denote by Q the elementary translation vectors pointing from the origin to the L-th cell. (For the sake of notational simplicity we shall assume that each unit cell contains only one motif. Generalization to several motifs per unit cell is straightforward.)

The total Hamiltonian of the crystal is constituted of the Hamiltonians of free motifs (H_L^0) and interaction terms:

$$H = \sum_{L=1}^{N} H_L^0 + \tfrac{1}{2} \sum_{L=1}^{N} \sum_{L,'(\neq L)} V_{LL'} \,. \tag{5.5}$$

The interaction Hamiltonian $V_{LL'}$ can be written in the Longuet-Higgins notation as:

$$V_{LL'} = \int\int dr\, dr'\, \varrho_L(r)\, |r - r'|^{-1}\, \varrho_{L'}(r') \,. \tag{5.6}$$

Here we introduced the operator of the total (nuclear and electronic) charge density of the L-th subsystem:

$$\varrho_L(r) = \sum_{A=1}^{M} Z_A \delta(r - R_A(L)) - \sum_{i=1}^{n} \delta(r - r_i(L)) \tag{5.7}$$

and the notation $R_A(L)$ and $r_i(L)$ indicates that the A-th nucleus and the i-th electron belongs to the L-th unit cell.

The electronic wave function of the crystal (Ψ) is assumed to be built up from the antisymmetrized and normalized Φ_L wave functions of the motifs. These functions are supposed to be strictly localized on their respective sites. This means that Φ_L can be expanded on a set of basis functions, which are centered in the region of the L-th unit cell and no delocalization occurs to neighbouring cells. Moreover, the wave functions of the motifs are considered to be mutually strongly orthogonal.

The approximate total wave function Ψ of the crystal is constructed as a simple Hartree product of the constituent wave functions, i.e. Ψ is not antisymmetric with respect to the electron exchange between different motifs. Notice that due to the lack of exchange forces and overlap repulsion this wave function is not applicable at short intermolecular distances. Nevertheless, if we impose a reasonable separation of the motifs (e.g. based on experimental crystal structures), these effects can be neglected as far as we are interested in the Madelung-effect on the charge distributions. There are some recent attempts to take into account the exchange and overlap effects by appropriate pseudopotentials [162–164].

Using this Hartree-product wave function, the total energy of the crystal is:

$$\langle \Psi | H | \Psi \rangle = N \langle \Phi_1 | H_1^0 | \Phi_1 \rangle$$
$$+ \frac{N}{2} \sum_{L=1}^{N} \int \int dr\, dr' \langle \Phi_1 | \varrho_1(r) | \Phi_1 \rangle |r - r'|^{-1} \langle \Phi_L | \varrho_L(r') | \Phi_L \rangle. \quad (5.8)$$

By the virtue of the translational symmetry the charge densities (as well as the wave functions) of the first and L-th motifs are related to each other by the trivial transformation:

$$\varrho_1(r) = \varrho_L(r - Q_L). \quad (5.9)$$

Therefore the total energy per motif (unit cell) can be expressed using exclusively the wave function of the first motif (in the following the label of the motif will be simply omitted):

$$\mathcal{E} = \langle \Phi | H^0 | \Phi \rangle + \frac{1}{2} \sum_{L'}^{N} \int \int dr\, dr' \langle \Phi | \varrho(r) | \Phi \rangle |r - r' + Q_L|^{-1} \langle \Phi | \varrho(r') | \Phi \rangle$$
$$= \langle \Phi | H^0 | \Phi \rangle + \frac{1}{2} \sum_{L'}^{N} \int \int dr\, dr' \langle \Phi | \varrho(r) | \Phi \rangle G(r, r') \langle \Phi | \varrho(r') | \Phi \rangle. \quad (5.10)$$

In the last equation an effective interaction kernel has been defined by the following lattice sum [130–132]:

$$G(r, r') = \sum_L |r - r' + Q_L|^{-1}. \quad (5.11)$$

The effective Schrödinger equation of the SCMP model can be derived by applying the variational theorem to the energy functional, $\mathscr{E}(\Phi)$. Then the condition for the energy minimum is:

$$\delta\mathscr{E}(\Phi) = \langle\Phi\mid\Phi\rangle^{-1}\left\{\langle\delta\Phi\mid H^0 + \tfrac{1}{2}\sum_{L'}^{N}\int\int dr\,dr'\,\langle\Phi\mid\varrho(r)\mid\Phi\rangle\,G(r,r')\,\varrho(r')\mid\Phi\rangle\right.$$

$$+ \left.\langle\Phi\mid\tfrac{1}{2}\sum_{L'}^{N}\int\int dr\,dr'\,\langle\delta\Phi\mid\varrho(r)\mid\Phi\rangle\,G(r,r')\,\varrho(r')\mid\Phi\rangle\right\}\times$$

$$\times\left\{\langle\Phi\mid H^0 + \tfrac{1}{2}\sum_{L'}^{N}\int\int dr\,dr'\,\langle\Phi\mid\varrho(r)\mid\Phi\rangle\,G(r,r')\,\langle\Phi\mid\varrho(r')\mid\Phi\rangle\right\}\langle\delta\Phi\mid\Phi\rangle$$

$$+ \text{complex conjugate} = 0\,. \tag{5.12}$$

This leads to the following nonlinear Schrödinger equation provided that $G(r,r') = G(r',r)$ and the charge density operators appearing from the left and right of the interaction kernel in Eq. (5.12) are the same:

$$\{H^0 + \int\int dr\,dr'\,\langle\Phi\mid\varrho(r)\mid\Phi\rangle\,G(r,r')\,\varrho(r')\}\mid\Phi\rangle = E\mid\Phi\rangle\,. \tag{5.13}$$

The special properties of nonlinear Schrödinger equations, like Eq. (5.13) has been discussed by Sanhueza et al. [165], and a special perturbation theory has been developed for their solution [166].

An important property of Eq. (5.13) is that the expectation value of the operator

$$H_{\text{eff}} = H^0 + \int\int dr\,dr'\,\langle\Phi\mid\varrho(r)\mid\Phi\rangle\,G(r,r')\,\varrho(r') \tag{5.14}$$

is not equal to the value of the energy $\mathscr{E}$. The following relationship holds if the effective interaction kernel is symmetric with respect to the interchange of variables and the two charge density operators are the same:

$$\mathscr{E} = \langle\Phi\mid H_{\text{eff}}\mid\Phi\rangle - \tfrac{1}{2}\int\int dr\,dr'\,\langle\Phi\mid\varrho(r)\mid\Phi\rangle\,G(r,r')\,\langle\Phi\mid\varrho(r')\mid\Phi\rangle\,. \tag{5.15}$$

Unfortunately the symmetry requirements for the perturbation operator has not been respected in several previous applications of the SCMP model [157–161] which may have lead to inconsistencies in the calculations. These come from an uncritical use of asymmetrical integral approximations, as discussed previously in the context of the MCF method.

In practical applications one usually seeks for the solution of Eq. (5.13) in the framework of the Hartree-Fock model. By standard techniques [167, 168] the following expression can be obtained for the Fock matrix elements over the atomic basis orbitals, $\chi_\mu(r)$:

$$F_{\mu\nu} = F_{\mu\nu}^0 - \sum_{A}^{M}\int dr\,G(r,R_A)\,\chi_\mu^*(r)\,\chi_\nu(r)\,Z_A$$

$$+ \sum_{\lambda\sigma}P_{\lambda\sigma}\int\int dr\,dr'\,\chi_\mu^*(r)\,\chi_\lambda^*(r')\,G(r,r')\,\chi_\nu(r)\,\chi_\sigma(r')\,. \tag{5.16}$$

It can be seen that correction terms to the unperturbed Fock operator $F_{\mu\nu}$ involve generalized one- and two-electron integrals with the effective interaction kernel, $G(r, r')$ in place of the usual direct Coulomb interaction $|r - r'|^{-1}$.

In order to avoid the laborious evaluation of the above integrals we propose a rational approximation which consists of a multipolar expansion the elementary charge distributions around atomic centres. This goal can be attained in two steps. First a Ruedenberg-type approximation [169] is introduced to reduce all multicentre integrals to at most two-centre ones and then the $G(r, r')$ functions are developed in bipolar series with respect to atomic centres.

The elementary charge distributions involving two atomic orbitals on centres A and B can be approximated by the sum of projected monocentric distributions [170]

$$|\mu_a\rangle \langle \nu_B| \approx \tfrac{1}{2} \left\{ \sum_{\varrho\eta}^{A} S_{\nu\varrho} S_{(A)\varrho\eta}^{-1} |\mu_A\rangle \langle \eta_A| + \sum_{\tau\varepsilon}^{B} S_{\varepsilon\mu} S_{(B)\tau\varepsilon}^{-1} |\tau_B\rangle \langle \varepsilon_B| \right\}, \qquad (5.17)$$

where $|\mu_A\rangle$ and $\langle \nu_B|$ stand in the bra-ket notation for the basis functions $\chi_\mu^*(r)$ and $\chi_\nu(r)$, which are centred on atoms A and B, respectively. S is the overlap matrix, and $S_{(A)}^{-1}$ denotes the inverse of the monoatomic block of the overlap matrix, $S_{(A)}$.

By the virtue of the above approximation the intermolecular monoelectronic and bielectronic integrals can be expressed as sums of at most two-centre integrals. For example, the one-electron integral can be written as:

$$\langle \mu_A| G(r, R_C) |\nu_B\rangle \approx \tfrac{1}{2} \left\{ \sum_{\varrho\eta}^{A} S_{\nu\varrho} S_{(A)\varrho\eta}^{-1} \langle \mu_A| G(r, R_C) |\eta_A\rangle \right.$$

$$\left. + \sum_{\tau\varepsilon}^{B} S_{\varepsilon\mu} S_{(B)\tau\varepsilon}^{-1} \langle \tau_B| G(r, R_C) |\nu_B\rangle \right\}. \qquad (5.18)$$

The expression for two-electron integrals is analogous: it is the sum of four terms, each containing only two-centre integrals.

If we expand $G(r, r')$ in double Taylor series around the respective atomic centres we obtain the interaction of the multipole moments with one-centre orbital products. Note that the same kind of approximations has already been proposed for the direct intramolecular Coulomb interactions in molecular calculations with moderate success [171]. However, in the present case this approximation should work much better, since we are dealing with a sum of interactions between centres, separated by distances much larger than the intermolecular ones.

The simplest possible case is when the multipolar series is truncated at the very first term, i.e. when a point-charge approximation is used. The corresponding effective Hamiltonian makes transparent that the effective interactions are described by a net atomic charge approximation:

$$H = H^0 + \sum_{A} \sum_{B} Q_A G(R_A, R_B) \langle \Phi| Q_B |\Phi\rangle, \qquad (5.19)$$

where Q_A is the formal net atomic charge operator of atom A. This atomic charge approximation has already been discussed in detail and some test applications are presented in [140].

5.2.2 SCMP Model at the Semiempirical Level

If we start from a CNDO/INDO model Hamiltonian, instead of an ab initio one, the spherical symmetry of the CNDO electron repulsion and nuclear attraction integrals can be exploited at the very beginning. By a simple reordering and summation of the lattice series the following modified Fock-matrix elements can be obtained for the unit motif:

$$F_{\mu\nu} = F_{\mu\nu}^O - \delta_{\mu\nu} \sum_C Z_C G_{AC} + \delta_{\mu\nu} \sum_\lambda P_{\lambda\lambda} G_{AB}; \qquad \mu \in A, \quad \lambda \in B, \quad (5.20)$$

where the effective interaction kernel, G_{AB}, is the lattice sum of the usual γ integrals

$$G_{AB} = \sum_L^{lattice'} \gamma_{AB}(L) \qquad (5.21)$$

and the prime indicates that intramotif terms are to be excluded from the summation.

It is known that the γ integrals tend to R^{-1} as the distance of the centers involved increases, therefore the long-range contribution to G_{AB} can be evaluated from usual point charge lattice sums. It can be supposed that short-range corrections can be neglected for $R > R_{lim}$ ($= 7.5$ a.u.) and the effective interaction kernel can be calculated exactly in the CNDO/INDO approximation as

$$G_{AB} = \sum_L^{lattice'} R_{AB}^{-1}(L) + \sum_L^{R \leq R_{lim}} \{\gamma_{AB}(L) - R_{AB}^{-1}(L)\}. \qquad (5.22)$$

Notice that the monopolar lattice sums in the ab initio case cannot be corrected for the short-range effects in such a simple way.

5.2.3 Description of Covalent Crystals

Covalent crystals, like silicates, zeolites, etc. can be considered as giant molecules of infinite size. Nevertheless the ensemble of symmetrically unrelated strictly localized covalent bonds can be regarded as basic building blocks of these systems [172]. Therefore it is possible to describe such systems in terms of these localized bonds in an analogous manner as the individual molecules were considered in molecular crystals by the self-consistent Madelung potential method. Whereas in the case of molecular crystals we could assume that the overlap with neighbouring molecules is negligibly small, this is not true for the localized bonds of covalent crystals. This means that the non-orthogonality problem should be taken seriously. In principle we have several possibilities. One can use an explicitly Löwdin-orthogonalized basis for the construction of the SLMOs. Another possibility is to use a non-orthogonal formulation of the group function theory. The third, and obviously the simplest case is to use an NDO-type theory [172], where the basis set is supposed to be implicitly orthogonalized. Finally we mention the group function formalism of Fink et al. [173], which uses open shell MCSCF wave functions for the representation of the motifs.

5.3 Pure Molecular Liquids

A molecular liquid can be obtained from a molecular crystal by melting. In the course of melting the strongly anisotropic long-range order of the crystal disappears and a short-range, isotropically averaged structure emerges. This difference can be visualized by comparing the radial distribution functions, i.e. the probability of finding a particle of a given kind at a given distance. In crystals the radial distribution function consists of discrete peaks (eventually slightly smeared out by the lattice vibrations) which are repeated ad infinitum, while the radial distribution function in liquids shows a few peaks at short distances and then converges to a constant value, which depend on the $\varrho = N_B/V$, number density of the particles of kind B. This behaviour is illustrated on Fig. 2.

Bearing in mind the above sketched physical model it is straightforward to extend the SCMP model to the case of disordered phases, i.e. to molecular liquids. The only difference between the two cases consists of the evaluation of the effective interaction kernel functions. This can be accomplished on the basis of structural information, given in terms of the radial distribution functions of the liquid at given temperature.

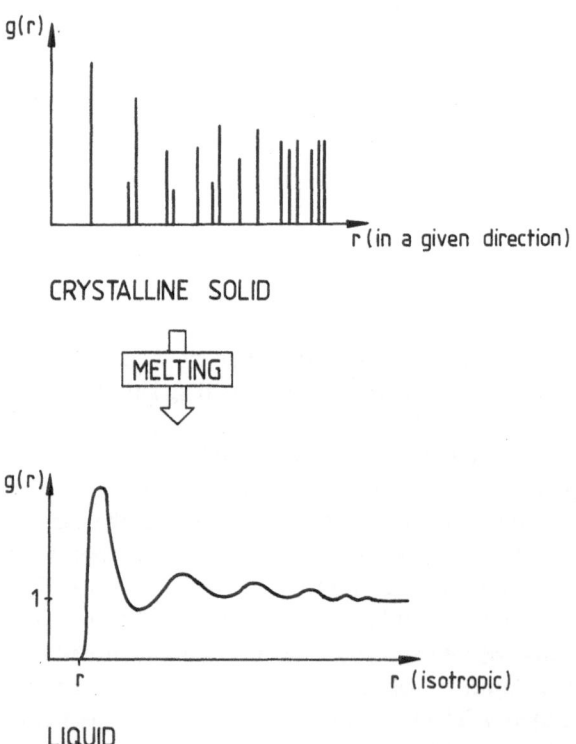

Fig. 2. Schematic illustration of the relationship between the pair correlation function in an ordered crystalline solid and the corresponding disordered liquid

Let us restrict our considerations to the simplest, point charge approximation of the effective interaction operator of the SCMP method

$$\tfrac{1}{2}\sum_A \sum_B Q_A G_{AB} \langle \Phi| Q_B |\Phi \rangle . \tag{5.23}$$

An explicit expression for the effective interaction kernel G_{AB} has already been given. Now we rewrite this infinite sum in an alternative form which emphasizes that the summation should be carried out in the order of increasing distance from atom A. Let us define the radial distribution function, $g_{AB}(r)$, in the crystal for the atom pair AB as

$$g_{AB}(r) = \frac{1}{4\pi r^2} {\sum_L}' \delta(r - |R_A - R_B(L)|) . \tag{5.24}$$

The probability that an atom of type B occurs in the volume element between r and $r + dr$ measured from an atom of type A is given by

$$dn_{AB} = g_{AB}(r)\, 4\pi r^2\, dr . \tag{5.25}$$

The effective interaction kernel can be expressed as an integral of the r^{-1} Coulomb interaction with the radial distribution function, $g_{AB}(r)$

$$G_{AB} = 4\pi \int_0^\infty dr\, r^2 g_{AB}(r)\, r^{-1} . \tag{5.26}$$

Using the G_{AB} effective interaction kernel function of Eq. (5.26) in the Hamiltonian (5.19), we can study the modification of molecular properties in disordered phases, like liquids, liquid crystals or crystals with static disorder. A similar approach has been proposed by Sesé et al. [174–176] for the study of molecular liquids at CNDO level.

6 Embedded Cluster Methods

The prototype molecule (or cluster) approach to the quantum chemical treatment of relevant portions of extended systems is an important tool in the study of localized phenomena. For example, in solid state quantum chemistry this method offers a convenient way of treating surfaces or adsorbed molecules interacting with surfaces [177], crystal impurities, [178, 179], amorphous materials [180], and so on.

Although periodic Hartree-Fock calculations may serve as starting point for the embedding problem [181–183], direct space descriptions offered e.g. by the SCMP method remain a useful alternative approach. The advantage of the prototype molecule method is that the direct space formalism is better adapted to the interpretation of localized chemical events than the reciprocal space description, available in periodic

Hartree-Fock methods. Another important field of application of embedded cluster models is the quantum chemical description of enzymatic reactions [14].

Nevertheless, there are serious difficulties in applying the prototype molecule models. The appropriate definition of the "relevant part" of an extended system may be quite ambiguous, and there remains the problem of "cutting out" the relevant set of atoms.

In the prototype molecule approach the dangling bonds are usually saturated by hydrogen atoms. This procedure, although not without controversy, offers a simple recipe for obtaining closed shell model molecules, with well-defined electronic states, easily tractable with quantum chemical methods.

The isolated prototype molecules may happen to be acceptable models, provided that the interactions with the surroundings are not too important, or one is able to minimize edge effects by increasing the size of the cluster [60, 184, 185]. Sometimes it is stated that hydrogens, used to saturate the dangling bonds, mimic in a certain way the remainder of the system. In a few cases the comparison of the electronic charge density maps of the full model (periodic HF calculations) and of the corresponding prototype molecule did not indicate important differences [64].

However in most of the cases, mainly in polar systems, the interactions with the surroundings are not negligible. There are a wide variety of theoretical methods for taking into account the effect of environment, ranging from various solvent effect models to pseudopotential-like approaches. One cannot overview all of them, but a few selected examples will be shown in the following.

6.1 Prototype Molecules in Effective Potentials

In polar systems, including dipolar subunits or ionic constituents quite important electrostatic fields may occur in the volume occupied by the relevant cluster. Various experimental and theoretical estimates suggest that the electrostatic field may attain 10 to 50 V/nm in systems like zeolites [133, 186] or protein active sites [187].

The effect of this strongly inhomogeneous external potential has to be included in the quantum chemical calculations, because its influence is probably significant on the charge distribution of the prototype molecule.

A common procedure is to include the potential of a representative set of partial charges of the environment in the Hamilton operator or the Fockian of the prototype molecule [188, 189]. This means that the core Hamiltonian is modified by a set of nuclear attraction integrals with partial point charges:

$$F_{\mu\nu} = F^0_{\mu\nu} - \sum_k q_K \langle \mu | \frac{1}{r - r_k} | \nu \rangle .$$ (6.1)

This method has been applied, for example, for the modeling of enzymatic reactions including environmental effects [190–193].

Quite recently a similar procedure has been applied to hydrogen bonded complexes in zeolites, although the form of the Fock operator is different. The interaction energy of the Q_A atomic point charges of the prototype molecule with the external potential is:

$$E_{\text{int}} = \sum_A Q_A V(r_A), \tag{6.2}$$

where Q_A is the net charge of the A-th atom in the cluster and $V(r_A)$ is the electrostatic potential at the atomic site, created by the rest of the system. The actual form of the Fock operator depends on the definition adopted for the net atomic charges in Eq. (6.1). In the case of Mulliken charges one has:

$$F_{\mu\nu} = F^0_{\mu\nu} - \tfrac{1}{2} S_{\mu\nu}\{V(r_A) + V(r_B)\}; \qquad \mu \in A, \quad \nu \in B \tag{6.3}$$

while for Löwdin charges the Fock operator is somewhat different:

$$F_{\mu\nu} = F^0_{\mu\nu} - \sum_\varrho S^{1/2}_{\mu\varrho} V(r_A) S^{1/2}_{\varrho\nu}; \qquad \varrho \in A. \tag{6.4}$$

The external potential $V(r_A)$, calculated from the Madelung potential of a crystal or from the fractional charges representing a protein charge distribution should be corrected for the frontier effects of the cluster. The presence of artificial hydrogen atoms, which saturate dangling bonds of the cluster, leads to various complications. For example, the external electrostatic potential at these hydrogens may be exceedingly large, due to the presence of closely situated charges (e.g. in zeolites or silicates one saturates an O-Si bond with a H atom, which is about 0.4 Å from the Si atom carrying a charge of about 1.2 electrons). On the other hand, charges of the H atoms themselves contribute to the total potential felt by the inner part of the cluster, and this contribution is artificial with respect to the real situation in the crystal. This particular problem necessitates appropriate corrections, but it unfortunately leaves considerable room for empirism.

6.2 Generalized Self-Consistent Reaction Field Theory

The concept of *reaction field*, originally formulated by Onsager [194], has been proved to be fruitful in the quantum chemical treatment of polar subsystems (solutes) embedded in polarizable environment (solvent) [195]. Simple cavity models, where the solvent is represented by a continuous dielectric medium and the solute is sitting in a cavity inside this dielectric, has numerous application in the framework of semiempirical [196–200] and ab initio [201–205] methods. The utility of this concept in the modelisation of biochemical processes was pointed out by Tapia and his coworkers [206].

The obvious limitations of the continuum representation of the solvent necessitated the development of microscopic models of the surroundings. Whereas for liquid phases this task is not trivial at all, for structurally well-characterized environments, like proteins [190, 207] or crystals [208] it is possible to calculate the reaction field from the polarizability distribution [209]. Assuming the existence of strongly bound solvent molecules one can attempt to represent, for example, the effect of water solvent on hydrogen bonded complexes, like $NH_3 \ldots HF$ [210].

The above-mentioned works were based on simplified representations of the solvent charge- and polarizability-distribution, and use dipolar approximation to solute-solvent interactions. A generalized version of this microscopical reaction field theory was recently proposed by Tapia [211]. In the following we present the generalized reaction field model from a slightly different aspect. We show that one can obtain the relevant equations from the coupled set of group function equations for the solute and solvent subsystems [212].

Similarly to the case of the mutually consistent field (MCF) or interaction field modified Hamiltonian (IFMH) approaches we start from the group function equations for the solute (S) and solvent (B) subsystems (c.f. Sect. 3).

$$\mathcal{H}^S_{\text{eff}} |\Psi_{Ss}\rangle = E^{Ss} |\Psi_{Ss}\rangle , \tag{6.5a}$$

$$\mathcal{H}^B_{\text{eff}} |\Psi_{Bb}\rangle = E^{Bb} |\Psi_{Bb}\rangle , \tag{6.5b}$$

where the effective Hamiltonian $\mathcal{H}^S_{\text{eff}}$ is the sum of the isolated subsystem Hamiltonian H^S_0 and of the Coulomb interaction operator (exchange interactions are neglected):

$$\mathcal{H}^S_{\text{eff}} = H^S_0 + J^B . \tag{6.6}$$

The Coulomb operator J^B describes the interaction of the charge density of S with the average charge distribution of B:

$$J^B = \int \int dr \, dr' \, \hat{\varrho}_S(r) \, T(r, r') \langle \Psi_{Bb}| \varrho^B(r') |\Psi_{Bb}\rangle . \tag{6.7}$$

Similar equation can be written for the solvent subsystem, B. The above set of coupled equations could be solved by treating subsystems S and B on an equal footing. This approach is followed in the MCF or IFMH procedures (cf. Sect. 5.1). We are interested mainly in the wave function of the solute subsystem, therefore we shall introduce the following *Ansatz* for the solvent wave function:

$$|\Psi_{Bb}\rangle = |\Phi_{Bb}\rangle - \sum_{b' \neq b} \frac{\langle \Phi_{Bb'}| \int \int dr \, dr' \, \varrho^B(r) \, T(r, r') \langle \Psi_{Ss}| \varrho^S(r') |\Psi_{Ss}\rangle |\Phi_{Bb}\rangle}{\Delta E^B(b \to b')} |\Phi_{Bb'}\rangle$$

$$= -R^B_b V |\Phi_{Bb}\rangle , \tag{6.8}$$

where Φ_{Bb} is the eigenfunction of the unperturbed solvent Hamilton operator, H^B_0. In the second part of the above equation we introduced the shorthand notation for the perturbation operator, V

$$V = \int \int dr \, dr' \, \varrho^B(r) \, T(r, r') \langle \Psi_{Ss}| \varrho^S(r') |\Psi_{Ss}\rangle \tag{6.9}$$

and the reduced resolvent R^B_b of the solvent Hamiltonian:

$$R^B_b = (1 - |\Phi_{Bb}\rangle \langle \Phi_{Bb}|) (E^B_b - H^B_0)^{-1} . \tag{6.10}$$

Substituting Eq. (6.8) into Eq. (6.7) we obtain for the effective interaction operator

$$J^B = \int\int dr\, dr'\, \varrho^S(r)\, T(r, r')\, \langle\Phi_{Bb}|\,\varrho^B(r')\,|\Phi_{Bb}\rangle$$
$$- \int\int\int\int dr\, dr'\, dr''\, dr'''\, \varrho^S(r)\, T(r, r')\, C^B(r', r'')\, T(r'', r''')\, \langle\Psi_{Ss}|\,\varrho^S(r''')\,|\Psi_{Ss}\rangle,$$
(6.11)

where $C^B(r, r')$ is the charge density response function of the solvent:

$$C^B(r, r') = \sum_{b'} \frac{\langle\Phi_{Bb}|\,\varrho^B(r)\,|\Phi_{Bb'}\rangle\,\langle\Phi_{Bb'}|\,\varrho^B(r')\,|\Phi_{Bb}\rangle}{\Delta E^B(b \to b')}$$
$$+ \sum_{b'} \frac{\langle\Phi_{Bb}|\,\varrho^B(r')\,|\Phi_{Bb'}\rangle\,\langle\Phi_{Bb'}|\,\varrho^B(r)\,|\Phi_{Bb}\rangle}{\Delta E^B(b \to b')}.$$
(6.12)

We can perform the double integral for the space variables of the solvent to get the reaction potential response function, $G^B(r, r')$ as

$$G^B(r, r') = - \int\int dr''\, dr'''\, T(r, r'')\, C^B(r'', r''')\, T(r''', r')$$
(6.13)

and we obtain for the effective solute Hamilton operator the following expression:

$$\mathcal{H}^S_{\text{eff}} = H^S_0 + \int dr\, \varrho^S(r)\, V^B(r)$$
$$+ \int\int dr\, dr'\, \varrho^S(r)\, G^B(r, r')\, \langle\Psi_{Ss}|\,\varrho^S(r')\,|\Psi_{Ss}\rangle.$$
(6.14)

This effective Hamiltonian is very similar to that of the self-consistent Madelung potential (SCMP) model, Eq. (5.14). This is a nonlinear Hamiltonian, in the sense that it depends on the wave function Ψ_{Ss}.

In order to complete the theory we have to give the total energy of the solute-solvent system. The total energy is made up of three contributions: first the energy of the perturbed solute, secondly the solute-solvent interaction energy, and thirdly the energy of the perturbed solvent system. It is easy to recognize the first two terms in the expectation value of the effective solute Hamiltonian:

$$E^S = \langle\Psi_{Ss}|\,H^S_0\,|\Psi_{Ss}\rangle,$$
(6.15a)

$$E^{SB}_{\text{int}} = \langle\Psi_{Ss}|\,\langle\Phi_{Bb}|\int\int dr\, dr'\, \varrho^S(r)\, T(r, r')\, \varrho^B(r')\,|\Phi_{Bb}\rangle\,|\Psi_{Ss}\rangle.$$
(6.15b)

The third contribution is the expectation value of the solvent Hamiltonian, H^B_0, with the first order solvent wave function, Ψ_{Bb}:

$$E^B = \langle\Psi_{Bb}|\,H^B_0\,|\Psi_{Bb}\rangle.$$
(6.15c)

Using the first-order *Ansatz* for the solvent wave function and taking into account the normalization of the perturbed wave function we can calculate the terms (6.15b) and (6.15c) to the first order

$$E_{\text{int}}^{SB} = \langle \Phi_{Bb}| \, V \, |\Phi_{Bb}\rangle - 2 \langle \Phi_{Bb}| \, V R_b^B V \, |\Phi_{Bb}\rangle, \tag{6.16}$$

$$E^B = \langle \Phi_{Bb}| \, H_0^B \, |\Phi_{Bb}\rangle + \langle \Phi_{Bb}| \, V R_b^B V \, |\Phi_{Bb}\rangle. \tag{6.17}$$

The above result of usual second-order perturbation theory is in general ignored, although its physical content is very instructive. It says that the first-order correction to the wave function raises the expectation value of the unperturbed Hamiltonian by an amount of $\frac{1}{2}$ of the expectation value of the perturbation operator. In other words, the "polarization work" usually cited to explain the factor of $\frac{1}{2}$ gains a clear interpretation in terms of the wave function: this is the energy we have to pay, when we distort (polarize) the solvent subsystem. It must be emphasised that this result is true only for a linear response (or second-order perturbation). If nonlinear polarization of the solute subsystem were allowed the above relationship would not hold.

Comparing the above equations with Eq. (6.10) we obtain the total energy of the solute-solvent system:

$$
\begin{aligned}
E_{\text{tot}} &= E^S + E^B + E^{SB} \\
&= \langle \Psi_{Ss}| \, H_0^S \, |\Psi_{Ss}\rangle + \langle \Phi_{Bb}| \, H_0^B \, |\Phi_{Bb}\rangle \\
&\quad + \int dr \, \langle \Psi_{Ss}| \, \varrho^S(r) \, |\Psi_{Ss}\rangle \, V^B(r) \\
&\quad + \tfrac{1}{2} \int\int dr \, dr' \, \langle \Psi_{Ss}| \, \varrho^S(r) \, |\Psi_{Ss}\rangle \, G^B(r, r) \, \langle \Psi_{Ss}| \, \varrho^S(r') \, |\Psi_{Ss}\rangle. \tag{6.18}
\end{aligned}
$$

These general equations are not limited by any multipolar approximation. In practical implementations, as mentioned above, the reaction field response function can be calculated from atomic or group dipolar polarizabilities, while the solvent charge density operator can be approximated by atomic multipoles.

The merit of the generalized SCRF theory is that it correctly describes polarization effects on the solute subsystem which may be important, e.g. in enzymatic reactions. The fact that the solute nonlinear Schrödinger equation is explicitly solved, allows one to have special "solvent induced" electronic states, which could appear otherwise as excited states for the isolated system.

The above formulation demonstrates the most important approximations of the generalized SCRF theory. First, we do not consider charge transfer. Fortunately, this effect can be greatly reduced through an appropriate construction of the solute system. Charge transfer effects, though small in magnitude, may be important for hydrogen-bonded systems and it is therefore advisable to include all atoms involved in H-bonds when defining the solute.

Another approximation is the neglect of intergroup antisymmetry requirement for the total wave function. This implies that important repulsion effects are missing from the above scheme. Appropriately selected non-local pseudopotentials can be useful for surmounting this difficulty, following for example the formalism developed for nonorthogonal group functions.

Finally the problem of dispersion forces can be mentioned. The so-called "direct reaction field" model has been proposed by Thole and van Duijnen [209, 213]. This is claimed to incorporate approximately dispersion effects in the self-consistent instantaneous reaction field. It should be pointed out that the intermolecular correlation, par excellence, abolishes the factorization of the wave function to solute and solvent components. If one accepts, nevertheless, the idea of an approximate separability, justified mathematically by an Unsöld-type average energy denominator approach, one can arrive at more satisfactory schemes. Another way of dealing with dispersion interaction in solvent effects, which seems to be more acceptable, was proposed recently in Ref. [214].

6.3 Protein Dipoles Langevin Dipoles Method

An alternative way to model a large molecular system (especially a protein) by partitioning it into a cage and environment was proposed by Warshel [215, 216]. The cage is embedded into its environment which is modeled by a set of polarizable atoms and permanent dipoles. The essential feature of the method is that the water surrounding the protein-cage system is also considered. The Hamiltonian is written as a sum of cage, environment and cage-environment coupling terms though this is partly formal in the present case since the energy of the environment is calculated classically.

The cage system is treated quantum mechanically. In the original version of the model all valence electrons were included and to allow a natural definition of the cage, orthogonalized atomic hybrid orbitals were used as a basis set [215]. This allows to avoid problems with the saturation of dangling bonds since all hybrids on the same atom may belong to the cage with a wave function obtained by solution of a closed-shell secular equation.

An attempt was made to overcome the difficulties associated with the reliability of current quantum mechanical approaches for the description of proton transfer reactions by introducing the Empirical Valence Bond (EVB) method [217]. For the hydrogen-bonded system, A–H ... B, the resonance structures (A–H ... B) and (A$^-$ ... H–B$^+$) are considered and the corresponding 2×2 secular equation is solved. Matrix elements are parametrized to reproduce experimental protonation energies in aqueous solution. This approach yields proton transfer energies, pK values and other quantities in reasonable agreement with experiment [218]. The effect of the environment on the cage system is considered by incorporation of the potential from the permanent and induced dipoles of the enzyme and the surrounding water molecules in the effective diagonal elements of the Hamiltonian [216]:

$$F_{ii}^c = F_{ii}^0 - \sum_j q_j r_{ij}^{-1} + \sum_j \mu_j^a r_{ij} r_{ij}^{-3} + \sum_k \mu_j^w r_{ik} r_{ik}^{-3} , \qquad (6.19)$$

where F_{ii}^c is the matrix element for the unperturbed cage system, while the second, third and fourth terms represent the effects of the protein permanent and induced dipoles and Langevin dipoles of the water molecules, respectively. q_j is the charge of the j-th ionized group or the fractional charge of the j-th atom in the protein. μ_j^a are the induced dipoles of the protein and μ_j^w are the Langevin dipoles of the water molecules. It is possible to incorporate a term into the total energy expression that accounts for the van der Waals interaction between the cage and the environment. The energy of the latter is calculated classically by molecular mechanics. In the EVB approach matrix elements are calculated similarly as in Eq. (6.19) but the effect of environment is treated using empirical quantities.

An appealing feature of the Protein Dipoles Langevin Dipoles method is that the water environment can be considered in a straightforward manner and that the dynamics of protein fluctuations can be simulated [219].

6.4 Fragment SCF Method

Following the philosophy outlined in the previous sections we proposed a model for the partitioning of very large molecular systems. This partitioning has the advantage that the different regions can be treated at a gradually decreasing sophistication allowing one to reduce the computation work drastically. We divide the system into four regions [26, 220]. In the central region, C, the actual chemical changes (geometry distortion, bond fission, etc.) take place, and this is embedded successively in a delocalization (D), induction (I) and transferable (T) region, respectively. The most important effect concerning region D is delocalization from and to C. In region I delocalization is neglected and only inductive effects are considered. Finaly, region T is composed of fully transferable SLMOs and serves as a rigid, non-polarizable environment for regions C, D and I.

The most convenient manner to describe these regions is to use a basis of strictly localized molecular orbitals (SLMOs). The SLMOs can be expressed as linear combinations of normalized atomic hybrid orbitals as in Eqs. (4.3–4.4) of Sect. 4.3.

The coefficients b_i^{ns}, b_i^{npx}, etc. determine the direction and s-character of the hybrids. The optimal calculation of these parameters is a difficult task and several procedures are known in the literature for this purpose [221, 222]. For the present purposes it is sufficient to invoke the chemical intuition and start with a set of hybrids with standard s-characters and directed along the bonds. Since such hybrids are not orthogonal on each atomic center (except in the case of some special bond angles), a consecutive Löwdin-orthogonalization allows one to readjust their s-characters and bond directions.

Once we have the proper hybrid orbitals a zeroth-order wave function can be constructed from all occupied two-centre SLMOs. To optimize their coefficients (the bond polarities) the following coupled set of 2×2 secular equations can be derived [26, 220]

$$F_i c_{mi} = \varepsilon_{mi} c_{mi}, \tag{6.20}$$

where we define the Fockian within the frame of the CNDO/2 approximation

$$F_{aa,i} = H_{aa,i}^{\text{eff}} + \sum_{m=1}^{M_i} P_{mm,i}(ai; ai \mid mi; mi) - \tfrac{1}{2}(P_{mm,i} - 1)(ai; ai \mid ai; ai) \tag{6.21}$$

$$F_{ab,i} = H_{ab,i}^{\text{eff}} - P_{ab,i}(ai; ai \mid bi; bi) \tag{6.22}$$

with the effective core Hamiltonian defined as follows

$$H_{aa,i}^{\text{eff}} = H_{aa,i} + \sum_{j=1}^{N} \sum_{m=1}^{M_j} P_{mm,j}(ai; ai \mid mj; mj)$$

$$+ \sum_{k=1}^{N_T} \sum_{m=1}^{M_k} P_{mm,k}(ai; ai \mid mk; mk), \qquad (6.23\,\text{a})$$

$$H_{ab,i}^{\text{eff}} = H_{ab,i} \qquad (6.23\,\text{b})$$

with

$$(ai; bi \mid mj; nj) = \int\int dv_1 \, dv_2 \, h_{ai}(1) \, h_{bi}(1) \, r_{12}^{-1} h_{mj}(2) \, h_{nj}(2) . \qquad (6.24)$$

The first and second term in Eq. (6.23) stand for the interaction between electrons of the i-th bond and those for other bonds within the inner (I, D and C) and outer (T) regions, respectively. The last sum comes from region T, containing N bonds and is an additive constant.

Once we have optimized parameters of the SLMOs the molecular orbitals for region C can be expanded on this basis set and a secular equation can be written for the derivation of expansion coefficients. Its dimensionality is determined by the size of region C alone. The molecular orbitals can be written as

$$\psi_m^c = \sum_{k=1}^{M_c} a_{mk}\varphi_k , \qquad (6.25)$$

where φ_k stands for the two-centre SLMO obtained from the solution of Eq. (6.20). We have to solve the following secular equation

$$F^c a_m = \varepsilon_m^c a_m \qquad (6.26)$$

with

$$F_{ij} = H_{ij}^{\text{eff}} + \sum_{k,l \in C}^{M_c} P_{kl}\{(ij \mid kl) - \tfrac{1}{2}(ik \mid jl)\} , \qquad (6.27\,\text{a})$$

$$H_{ij}^{\text{eff}} = H_{ij} + 2 \sum_{k \in D, I, T} (ij \mid kk) . \qquad (6.27\,\text{b})$$

The electron repulsion integrals $(ij \mid kl)$ are obtained formally from Eq. (6.24) by replacing h_{ai}, etc. by φ_i, etc. The Fragment SCF method has been implemented at the CNDO level of approximation and applied to the conformational study of the catalytic triad in serine proteases [223].

7 Applications

7.1 Zeolites

Some recent calculations on the $(OH)_3-Al-OH-Si-(OH)_3$ prototype molecule representing an acidic site of a zeolites interacting with an NH_3 molecule clearly demonstrated that isolated (in vacuo) clusters are in appropriate to describe correctly the ionic complex, which is formed in sodalite cages [224]. It seems that the effect of the crystalline environment plays a crucial role in the stabilization of the proton-transferred form $(OH)_3AlOSi(OH)_3^- \dots NH_4^+$. Similar qualitative conclusions were drawn from earlier studies on the hydrogen bonded subunits in the hydroxonium perchlorate crystal [208].

The Madelung potential effects were taken into account with Eqs. (6.2) and (6.3). The SCF and modified SCF calculations were carried out at the 6-31G level, using the in vacuo optimized geometries for the cluster, both in the neutral and ionic forms for the prototype system illustrated by Fig. 3.

In order to calculate the Madelung potentials we had to suppose a charge distribution for the faujasite crystal. Most of the earlier studies on the electrostatic potentials in zeolites used either fully ionic charges (4, 3 and -2 for Si, Al and O, respectively) [225, 226] or half-ionic charges (2, 1.5, -1) [227]. A more refined, but reasonably simple procedure to estimate the charge distributions in zeolites, the electronegativity equalization method (EEM) has been recently proposed by Mortier and coworkers [228–230]. The EEM parameters can be adjusted to any quantum chemically calculated atomic charges. The parameter set given in Ref. [229] is claimed to fit quite well the STO-3G ab initio charges for a series of small molecules. Slightly different parameter set was used in a systematic study of various crytalline silicate modifications and experimental trends were reasonably well reproduced [230].

The EEM charges were calculated for a faujasite model with a Si/Al ratio of 1:1. In this specific case the distribution of Al and Si atoms is unequivocally determined

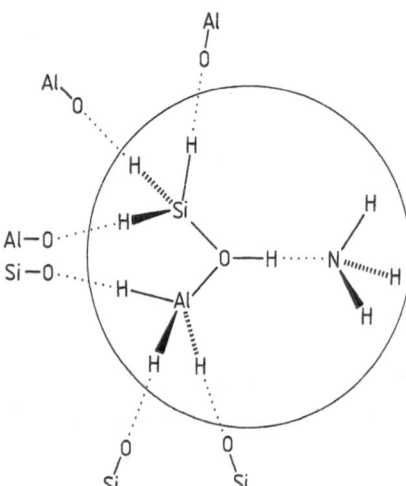

Fig. 3. Schematic illustration of the relationship between the zeolite structure and the prototype molecule representing the acidic site in a sodalite cage. Actually the covalent form of the zeolite-NH_3 complex is shown

by the Löwenstein rules [231]: each Si atom has 4 Al neighbours and conversely. The I, I' and II cationic sites were occupied by fixed point charges by $+2$, $+1$ and $+1$, respectively. One of the cationic sites (of II type) was replaced formally by the NH_4 cation in our model, therefore the contribution of this cation to the Madelung potential acting on the cluster was removed.

The Madelung potentials, corrected for edge effects, were added to the Fock operator of the prototype molecule. The geometrical parameters were kept at the values optimized for the isolated model, involving the interaction with the NH_3 molecule, in both the neutral and ionic forms, at the 6-31G level.

In the isolated model calculations the ionic form is only slightly stabilized with respect to the neutral one: by 3.13 kcal/mol. The stability of the ionic complex was greatly enhanced, when the external field has been added to the Fock operator and the difference in the energies of the ionic and neutral forms increased to about 50 kcal/mol in favour of the ionic complex. A comparison of the energies and charge distributions in the isolated and embedded clusters can be found in the Table 2 [224].

Table 2. Total energies, energy differences and proton charges in the isolated and embedded cluster in its covalent $(OH)Al-OH-Si(OH).NH_3$ and its ionic $(OH)Al-O-Si(OH).NH_4$ forms

	Isolated model	Embedded model
covalent form (a.u.)	-1115.17796	-1115.54074
ionic form (a.u.)	-1115.18295	-1115.62070
ΔE (kcal/mol)	3.13	50.0
Q (covalent)	0.65	1.02
Q (ionic)	0.61	0.73

7.2 Serine Proteases

It is known that protein dipoles induce strong electrostatic fields considerably influencing properties and function of enzymes [232–234]. Accordingly, molecular electrostatic potential maps are especially useful for the study of these giant molecules. Clearly, their size is too large for an SCF calculation therefore, if calculating electrostatic potentials, transferable fragments have to be used as described in Sect. 4. In the following we report on some results achieved in our laboratory by application of the Bond Increment method [129].

We concentrate on serine proteases catalyzing the cleavage of peptide or ester bonds since these have been studied in detail and precise information is available on their properties and action [235–237]. The catalytic reaction starts with a nucleophilic attack of the active serine hydroxyl at the carbon atom of the carboxyl group (cf. Fig. 4).

Simultaneously with a proton transfer to the neighbouring imidazole a tetrahedral intermediate is formed that breaks down to an acyl enzyme and an amine or alcohol by loss of proton. The acyl enzyme hydrolyses via the reverse route regenerating the enzyme.

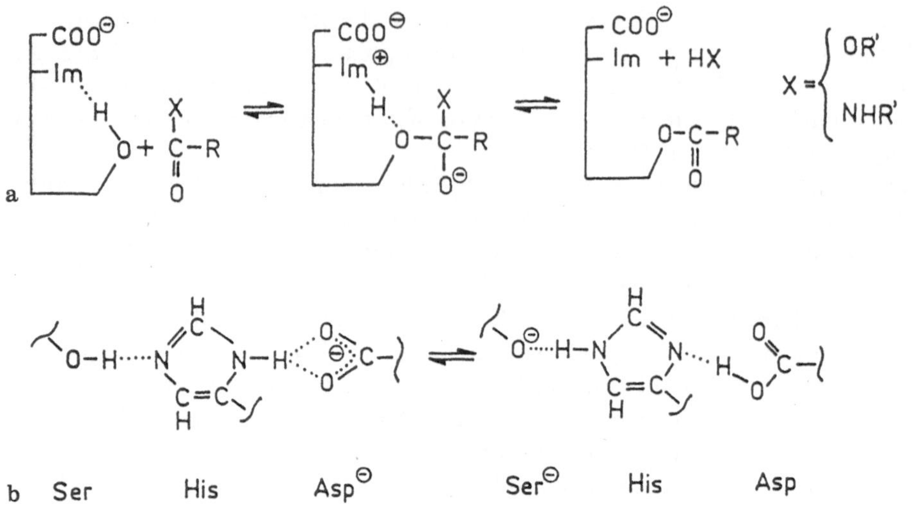

Fig. 4 a Schematic reaction sequence of hydrolysis catalyzed by serine proteases. b Hypothetical charge relay in the catalytic triad

We used Protein Data Bank coordinates [238] to model the enzymes. In earlier works we considered residues that are within a 1 nm range from the centre of the active site [239–241], while in later calculations all enzyme atoms were treated [187, 242–245].

Interpretation of early X-ray and NMR spectroscopic results lead to the hypothesis that the serine hydroxyl is activated via a "charge-relay" mechanism including proton removal from serine to the buried aspartate anion via the neighbouring histidine (cf. Fig. 4b). This double proton transfer would yield neutral aspartic acid and an alkoxide anion with enhanced nucleophilicity. Later this was questioned on the basis of more precise NMR and neutron diffraction studies (cf. [241] for references). At variance with earlier quantum chemical calculations [246, 247] predicting the aspartate as the ultimate proton acceptor, we stressed the importance of the electrostatic effect of the environment including the protein dipoles, surrounding water molecules and a counter ion, and concluded that, while the Asp-His couple exists in a neutral form in vacuo, the ion-pair form is stabilized by the environment [239, 241]. These results have been confirmed by recent sophisticated calculations [218].

On the basis of well-designed experiments and theoretical calculations it seems at present quite certain that the "charge-relay" mechanism is not operative in serine proteases. What is then the source of the rate acceleration by 6–8 orders of magnitude as compared to the uncatalyzed reaction? This qustion has been addressed by several authors [233, 235–237, 240, 248]. The problem can be best treated by theoretical calculations since these allow the consideration of hypothetical reaction intermediates, transition-state complexes and reaction routes, as well [237]. Based on simple calculations combining the CNDO/2 and BI methods we stressed the importance of the electrostatic stabilization effect of the buried aspartate on the transition state leading to a rate acceleration [240]. The same conclusion was reached independently using the ab initio method by Umeyama et al. [249]. Later on, Warshel and Russell were able to reproduce the experimental transition-state stabilization energy for the trypsin reaction almost quantitatively [218]. A very strong argument for the electrostatic rate accelerating effect of the buried aspartate in serine proteases is that we could quantitatively reproduce the rate decrease in the Asn/Asp mutant of trypsin and in the Ala/Asp mutant of subtilisin by using the Protein Dipoles Langevin Dipoles method described in Sect. 6.3 [248].

A phenomena where electrostatics is especially important is ligand binding to enzymes. On the basis of electrostatic potential calculations by the BI method we have found that the experimental Gibbs free energy of association between trypsin and subsituted benzamidine inhibitors depends linearly on the calculated electrostatic interaction energy [243]. We defined an "electrostatic lock" representing the active site of the enzyme that is complementary to the key, the charge pattern of the inhibitor. Refinement of this model lead to the definition of electrostatic potential patterns around the lock and the key as well [244, 245]. Fitting the key into the lock means a correspondence between enzyme and inhibitor potential patterns. Once the former is known simple rules can be derived to predict relative potencies of various inhibitors.

7.3 Amorphous Silicon

It is supposed that several interesting properties of amorphous silicon (a-Si) can be traced back to their non-uniform charge distribution. The prototype molecule approach where the dangling bonds of the surface are saturated by H-atoms is not appropriate for this purpose, since the electronegativity difference of Si and H is large enough to induce artificial charges in the model of the bulk a-Si. In order to avoid this artefact the "pseudoatom" approach has been implemented in the framework of the SLMO approximation [187]. The dangling bonds have been saturated by "one-legged" silicon atoms, bearing only one hybrid orbital oriented to the dangling bond direction.

The charge accumulation has been traced back to the bond angle distortion with respect to the tetrahedral angle, corresponding to a diamond structure. We proposed a simple equation for the estimation of the atomic net charges in a distorted a-Si network. Considering atom M we have the following expression

$$\Delta Q_M = A \left\{ 2 \sum_{i=1}^{6} \mathrm{d}\theta_i - \sum_{j=1}^{12} \mathrm{d}\theta_j \right\}, \tag{7.1}$$

where $\theta_i = {<}\mathrm{XMY}$ and $\theta_j = {<}\mathrm{MXZ}$ (atoms X and Y are bonded to M, Z to X or Y). Comparing charges obtained from Eq. (7.1) and from CNDO/2 calculations, it was found that our estimation is reasonable, $\Delta Q_{\mathrm{CNDO}} = (-0.69 \pm 0.024)\,\Sigma$ where Σ denotes the quantity in parentheses in Eq. (7.1).

Applying Eq. (7.1) to a continuous random-network model of 216 silicon atoms proposed by Wooten et al. [250] we obtained 0.021 electron units for the rms deviation from charge neutrality. This is in reasonable agreement with recent theoretical results [251, 252], but considerably different to experimental estimates ranging from 0.1 to 0.3 electrons [253, 254]. The discrepancy with the core-level spectroscopic estimation [253] may be explained on the basis of intra-atomic charge transfer, while the infrared spectroscopic estimation is subject to considerable simplifications as discussed by the authors themselves [254].

Analysing our results in terms of ring statistics we concluded that atomic charge distribution in a-Si is determined primarily by bond angle distortions that are dependent on the number of fivefold and perturbed by the number of sixfold rings attached to the same atom [255].

8 References

 1. Clementi E, Detrich J, Chin S, Corongiu G, Folsom D, Logan D, Caltabiano R, Carnevali A, Helin J, Russo M, Gnudi A, Palamidese P (1986) In: Clementi E, Chin S (eds) Structure and dynamics of nucleic acids, proteins, and membranes. Plenum, New York, p 403
 2. Almlöf J, Faegri Jr. K, Korsell K (1983) J Comput Chem 3: 3003
 3. Lüthi HP, Almlöf J (1987) Chem Phys Lett 135: 357
 4. Almlöf J, Lüthi HP (1988) Chem Des Automat News 2: (8): 1
 5. Häser M, Ahlrichs R (1989) J Comput Chem 10: 104
 6. Ahlrichs R, Bär M, Häser M, Horn H, Kölmel (1989) Chem Phys Lett 162: 165
 7. Pauling L (1960) The nature of the chemical bond. Cornell, Ithaca NY
 8. Náray-Szabó G, Surján PR, Ángyán JG (1987) Applied quantum chemistry. Akadémiai Kiadó Reidel, Budapest Dordrecht
 9. Surján PR (1984) Croat Chim Acta 57: 833
10. Minkin VI, Osipov OA, Zhdanov YA (1970) Dipole moments in organic chemistry. Plenum, New York
11. LeFèvre RJW (1965) Adv Phys Org Chem 3: 1
12. Claverie P (1978) In: Pullman (ed) Intermolecular interactions: From diatomics to biopolymers, Wiley. New York, p 69
13. Surján PR (1989) In: Maksić ZB (ed) Theoretical models of chemical bonding, Part 2, The concept of chemical bond, Springer, Berlin Heidelberg New York
14. Náray-Szabó G, Bleha T (1982) In: Csizmadia IG (ed) Molecular structure and conformation: Recent advances. Elsevier, Amsterdam, p 267 (Progress in theoretical organic chemistry, vol 3)
15. Jaffé HH, Orchin M (1962) Theory and application of ultraviolet spectroscopy. Wiley, New York
16. Davydov AS (1973) Theory of molecular excitons. McGraw Hill, New York
17. Snatzke G (1979) Angew Chem Int Ed Engl 18: 363
18. Patai S (ed) (1964) The chemistry of functional groups. Wiley, Chichester
19. Colthup NB, Daly LH, Wiberley SE (1975) Introduction to infared and Raman spectroscopy, 2nd edn Academic, New York
20. Tapia O (1982) In: Ratajczak H, Orville-Thomas WJ (eds) Intermolecular Interactions. vol 2, chap 2, Wiley, Chichester
21. Stoneham AM (1975) Theory of defects in solids. Oxford University Press, Oxford
22. Shustorovich E (1984) J Am Chem Soc 106: 6479
23. Siegbahn PEM, Blomberg MRA, Bauschlicher CW jr (1984) J Chem Phys 81: 4
24. Whangbo MH, Schlegel HB, Wolfe S (1977) J Am Chem Soc 99: 1296
25. Bernardi F, Bottoni A (1982) In: Csizmadia IG (ed) Molecular structure and conformation: Recent advances. Elsevier, Amsterdam, p 65 (Progress in theoretical organic chemistry, vol 3)
26. Náray-Szabó G (1984) Croat Chem Acta 57: 901 and references therein
27. Bader RFW, Nguyen-Dang TT (1981) Adv Quantum Chem 14: 63
28. Bader RFW (1985) Acc Chem Res 18: 9

29. Bader RFW, Tal Y, Anderson SG, Nguyen-Dang TT (1980) Isr J Chem 19: 8
30. Bader RFW, Nguyen-Dang TT, Tal Y (1981) Rep Prog Phys 44: 893
31. Gatti C, Fantucci P, Pacchioni G (1987) Theoret Chim Acta 72: 433
32. Epiotis ND (1988) Pure Appl Chem 60: 157; Nouv J Chim, to be published
33. Bader RFW (1988) Pure Appl Chem 60: 145
34. Srebrenik S, Bader RFW (1975) J Chem Phys 63: 3945
35. Daudel R, Leroy G, Peeters D, Sana M (1983) Quantum chemistry, Wiley-Interscience, New York
36. Becker P (1977) Phys Scripta 15: 119
37. Wiberg K (1989) In: Maksić ZB (ed) Theoretical models of chemical bonding, vol 1 Springer, Berlin Heidelberg New York
38. Malrieu JP (1977) In: Segal GA (ed) Semiempirical methods of electronic structure calculation. Part A: techniques. Plenum, New York, p 69 (Modern theoretical chemistry, vol 7)
39. Boća R (1982) Theoret Chim Acta 61: 179
40. Wolfe S, Mitchell DJ, Whangbo MH (1978) J Am Chem Soc 100: 1936, 3698
41. Kost D, Schlegel HB, Mitchell DJ, Wolfe S (1979) Can J Chem 57: 729
42. Bernardi F, Bottoni A (1981) Theoret Chim Acta 58: 245
43. Moffitt W (1954) Rep Prog Phys 17: 173
44. Bálint-Kürti GG, Karplus M (1974) In: March NH (ed) Orbital theories of molecules and solids, Clarendon, Oxford, p 250
45. Schipper PE (1987) Austrian J Chem 40: 635
46. Mayer I (1983) Int J Quantum Chem 23: 323
47. Maksić ZB, Eckert-Maksić M, Rupnik K (1984) Croat Chem Acta 57: 1295
48. Maksić ZB (1986) Comp Maths with Appls 12B: 697
49. Maksić ZB (1988) J Mol Struct (Theochem) 170: 39
50. Maksić ZB (1989) In: Maruani J (ed) Molecules in physics, chemistry and biology, vol 3 Kluwer Academic, Dordrecht, p 49
51. Hall GG (1951) Proc Roy Soc (London) Ser A 205: 541
52. Sándorfy C (1955) Can J Chem 33: 1337
53. Del Re G (1958) J Chem Soc 4031
54. Hoyland JR (1968) J Am Chem Soc 90: 2227
55. Diner S, Malrieu JP, Claverie P (1969) Theoret Chim Acta 13: 1
56. Náray-Szabó G (1976) Acta Phys Acad Sci Hung 40: 261
57. Surján PR, Révész M, Mayer I (1981) J Chem Soc Faraday Trans 2 77: 1129
58. Gibbs GV, Meagher EP, Newton MD, Swanson DK (1981) In: O'Keefe M, Navrotsky A (eds) Structure and bonding in crystals, Academic, New York, vol 1, p 195
59. Sauer J, Zahradnik R (1984) Int J Quantum Chem 26: 793
60. Messmer RP (1977) In: Segal GA (ed) Semiempirical methods of electronic structure calculation. Part B: Applications. Plenum, New York, p 215 (Modern theoretical chemistry, vol 8)
61. László I (1982) Int J Quantum Chem 21: 813
62. Náray-Szabó G, Kramer G, Nagy P, Kugler S (1987) J Comput Chem 8: 555
63. Révész M, Bertóti I, Mink G, Mayer I (1988) J Mol Struct (Theochem) 181: 335
64. Dovesi R, Pisani C, Roetti C, Silvi B (1987) J Chem Phys 86: 6967
65. McWeeny R (1959) Proc Roy Soc (London) Ser A 253: 242
66. McWeeny R (1960) Rev Mod Phys 32: 335
67. McWeeny R, Sutcliffe BT (1969) Methods of mecular quantum mechanics, Academic, London
68. Hoffman DK, Ruedenberg K, Verkade JG (1977) Structure and Bonding 33: 57
69. Bishop DM (1967) Adv Quantum Chem 3: 25
70. Lykos PG, Parr RG (1956) J Chem Phys 24: 1166
71. Parr RG, Ellison FO, Lykos PG (1956) J Chem Phys 24: 1106
72. Szász L (1985) Pseudopotential theory of atoms and molecules, J. Wiley, New York
73. McWeeny R, Ohno K (1960) Proc Roy Soc (London) Ser A 255: 367
74. Huzinaga S, Cantu AA (1971) J Chem Phys 55: 5543
75. Adams WH (1961) J Chem Phys 34: 89

76. Gilbert TL (1964) In: Löwdin PO, Pullman B (eds) Molecular orbitals in chemistry, physics and biology. Academic, New York, p 409
77. Kunz AB (1973) J Phys B 6: L47
78. Matsuoka O (1977) J Chem Phys 66: 1245
79. McWeeny R, Steiner E (1965) Adv Quantum Chem 2: 93
80. Kutzelnigg W (1966) J Chem Phys 40: 3640
81. Surján PR (1984) Phys Rev A 30: 43
82. Klessinger M, McWeeny R (1965) J Chem Phys 67: 2728
83. Mehler EL (1977) J Chem Phys 67: 2728
84. Mehler EL (1981) J Chem Phys 74: 6298
85. Mehler EL (1978) Int J Quantum Chem Quantum Chem Symp 12: 407
86. Fülscher MP, Mehler EL (1986) Int J Quantum Chem 29: 627
87. Fülscher MP, Mehler EL (1988) J Mol Struct (Theochem) 165: 319
88. Kirtman B, de Melo CP (1981) J Chem Phys 75: 4592
89. Kirtman B (1982) J Phys Chem 86: 1059
90. Kirtman B (1983) J Chem Phys 79: 835
91. Kirtman B, de Melo CP (1986) Int J Quantum Chem 89: 1209
92. Kirtman B, Dykstra CE (1986) J Chem Phys 85: 2791
93. Dykstra CE (1988) Ab initio calculation of structures and properties of molecules. Elsevier, Amsterdam. chap 5
94. McWeeny R (1962) Rev Mod Phys 32: 335
95. Gordon MS, England W (1972) J Am Chem Soc 94: 5168
96. Claverie P (1978) in Ref [12], pp 180–182
97. Mulliken RS (1955) J Chem Phys 23: 1833
98. Mayer I (1983) Chem Phys Lett 97: 270; (1983) Int J Quantum Chem 23: 341
99. Jug K (1973) Theoret Chim Acta 31: 63
100. Thole BT, van Duijnen PT (1983) Theoret Chim Acta 63: 209
101. Brobjer JT, Murrell JN (1981) Chem Phys Lett 77: 601; JCS Faraday Trans 2 78: 1853
102. Cox SR, Williams DE (1981) J Comput Chem 2: 304
103. Ray NK, Shibata M, Bolis R, Rein R (1985) Int J Quantum Chem 27: 427
104. Rullman JAC (1988) Ph D Thesis, University of Groningen, Netherlands
105. Némethy G, Pottle MS, Scheraga H (1983) J Phys Chem 87: 1883
106. Pettitt BM, Karplus M (1985) J Am Chem Soc 107: 1166
107. Weiner SJ, Kollman PA, Nguyen DT, Case DA (1986) J Comput Chem 7: 230
108. Jorgensen WL, Swenson CJ (1985) J Am Chem Soc 107: 569
109. Rullman JAC, van Duijnen PT (1988) Mol Phys 63: 451
110. Mehler EL, Paul CH (1979) Chem Phys Lett 63: 145
111. Pullman A, Pullman B (1981) Quart Rev Biophys 14: 283
112. Pullman A, Perahia D (1978) Theoret Chim Acta 48: 29
113. Bonaccorsi R, Scrocco E, Tomasi J (1976) J Am Chem Soc 98: 4049; ibid 99: 4545
114. Agresti A, Bonaccorsi R, Tomasi J (1979) Theoret Chim Acta 53: 215
115. Lavery R, Etchebest C, Pullman A (1982) Chem Phys Lett 85: 266
116. Etchebest C, Lavery R, Pullman A (1982) Theoret Chim Acta 62: 17
117. Hall GG (1973) Chem Phys Lett 20: 501
118. Stone AJ (1981) Chem Phys Lett 83: 233
119. Stone AJ, Alderton M (1985) Mol Phys 56: 1047
120. Vigné-Maeder F, Claverie P (1988) J Chem Phys 88: 4934
121. Stone AJ, Price SL (1988) J Phys Chem 92: 3325
122. Stone AJ (1989) In: Maksić ZB (ed) Theoretical models of chemical bonding, chap 6. Classical electrostatics in intermolecular interactions vol 4, Springer, Berlin Heidelberg New York
123. Faerman CH, Price SL (1990) Am Chem Soc 112: 4915
124. Náray-Szabó G (1979) Int J Quantum Chem 16: 265
125. Náray-Szabó G, Grofcsik A, Kósa K, Kubinyi M, Martin A (1981) 2: 58
126. Nagy P, Ángyán J, Náray-Szabó G, Peinel G (1987) Int J Quantum Chem 31: 927
127. Seres J, Náray-Szabó G, Simon K, Daróczi-Csuka K, Szilágyi I, Párkányi L (1981) Tetrahedron 37: 1565

128. Ösapay K, Náray-Szabó G (1983) J Mol Struct (Theochem) 92: 57
129. Náray-Szabó G (1987) In: Maksić ZB (ed) Modelling of structures and properties of molecules. Ellis Horwood, Chichester, England, p 299
130. Ewald P (1921) Ann Phys 64: 253
131. Catti M (1978) Acta Cryst 34A: 974
132. Cummings PG, Dunmur A, Munn RW, Newham RJ (1976) Acta Cryst 32A: 847
133. Ángyán JG, Ferenczy G, Nagy P, Náray-Szabó G (1988) Coll Czech Chem Commun 53: 2308
134. Ferenczy G, Ángyán JG, (1990) J Chem Soc Faraday Trans 86: 3461
135. Otto P, Ladik J (1975) Chem Phys 8: 192; (1977) ibid 19: 209
136. Otto P (1978) Chem Phys 33: 407
137. Otto P (1979) Chem Phys Lett 62: 538
138. Förner W, Otto P, Bernhardt J, Ladik J (1981) Theoret Chim Acta 60: 269
139. Otto P (1985) Int J Quantum Chem 28: 895; (1986) ibid 30: 275
140. Ángyán JG, Silvi B (1987) J Chem Phys 86: 6957
141. Weinstein H, Eilers JE, Chang SY (1977) Chem Phys Lett 51: 534
142. Klessinger M (1978) Theoret Chim Acta 49: 77
143. Klessinger M (1988) Int J Quantum Chem 23: 535
144. Longuet-Higgins HC, Murrell JN (1955) Proc Phys Soc (London) Sect A 68: 601
145. Heilbronner E, Weber JP, Michl J, Zahradnik R (1965) Theoret Chim Acta 6: 141
146. Favini G, Gamba A, Simonetta M (1969) Theoret Chim Acta 13: 175
147. Germer HA jr, Becker RS (1972) Theoret Chim Acta 28: 1
148. Fabian J, Scholtz M (1981) Theoret Chim Acta 59: 117
149. von Niessen W (1971) J Chem Phys 55: 1948; (1973) Theoret Chim Acta 31: 111; (1973) ibid 32: 13; (1974) ibid 33: 7
150. Christoffersen RE (1972) Adv Quantum Chem 6: 333
151. Frost AA (1967) J Chem Phys 47: 3707, 3714
152. Gáspár R jr, Gáspár R (1979) Int J Quantum Chem 15: 567; (1979) ibid 16: 57, (1980) ibid 19: 501
153. Gáspár R, Gáspár R jr (1979) Int J Quantum Chem 15: 559
154. O'Shea SF, Santry DF (1975) Theoret Chim Acta 37: 1
155. Santry DP (1975) Theoret Chim Acta 42: 67
156. Pastori Parravicini GP, Resca L (1973) Phys Rev B 8: 3009
157. Ghio C, Scrocco E, Tomasi J (1976) In: Pullman B (ed) Environmental effects of molecular structure and properties, Reidel, Dordrecht, p 329
158. Tsukada M (1980) J Phys Soc Jpn 49: 1183
159. Barandiarán Z, Pueyo L, Gomez-Beltrán F (1983) J Chem Phys 78: 4612 and (1983) ibid 79: 1926
160. Zyss J, Berthier G (1982) J Chem Phys 77: 3635
161. Böhm MC (1982) Chem Phys Lett 89: 126
162. Winter NW, Pitzer RM, Temple DK (1987) J Chem Phys 86: 3549; ibid 87: 2947
163. Vail JM, Pandey R (1986) Mater Res Soc Symp Proc 63: 247
164. Barandiarán Z, Seijo L (1988) J Chem Phys 89: 5739
165. Sanhueza JE, Tapia O, Laidlaw WG, Trsic M (1979) J Chem Phys 70: 3096
166. Surján PR, Ángyán J (1983) Phys Rev A 28: 45
167. Mayer I (1971) Acta Phys Acad Sci Hung 30: 373
168. Harris FE (1968) J Chem Phys 48: 4027
169. Ruedenberg K (1957) J Chem Phys 19: 1433
170. Carbó R, Arnau C (1978) Gazz Chim Ital 108: 71
171. Steinhauser O, Schuster P (1977) Theoret Chim Acta 45: 147; ibid 46: 157
172. Hashimoto M, Santry DP (1978) Theoret Chim Acta 50: 39
173. Fink WH, Banerjee A, Simons J (1983) J Chem Phys 79: 6104
174. Sesé LM, Banon A, Fernández M (1983) J Mol Struct (Theochem) 92: 231
175. Sesé LM, Fernández M (1983) J Mol Struct (Theochem) 93: 261
176. Sesé LM (1985) J Mol Liquids 30: 185
177. Whitten JL, Pakkanen TA (1980) Phys Rev B 21: 4357

178. Harding JH, Harker AH, Keegstra PB, Pandey R, Vail JM, Woodward C (1985) Physica 131B: 151
179. Schluger AL, Kotomin EA, Kantorovich LN (1986) J Phys C: Solid State Phys 19: 4183
180. Kugler S, Surján PR, Náray-Szabó G (1988) Phys Rev B 37: 9069
181. Pisani C, Dovesi R (1987) Theoret Chim Acta 72: 277
182. Fisher AJ (1987) Theoret Chim Acta 72: 319
183. Baraff GA, Schlüchter M (1986) J Phys C: Solid State Phys 19: 4383
184. Bridet J, Fliszár S, Odiot PS, Pick R (1983) Int J Quantum Chem 24: 687
185. Mix H, Sauer J, Schröder V, Merkel A (1988) Coll Czech Chem Commun 53: 2191
186. Zahradnik R, Hobza P, Sauer J (1982) In: Náray-Szabó G (ed) Steric effects in biomolecules. Akadémiai Kiadó, Elsevier, Budapest, p 327
187. Náray-Szabó G, Kramer G, Nagy P, Kugler S (1987) J Comp Chem 8: 555
188. Noell JO, Morokuma K (1975) Chem Phys Lett 36: 465
189. Noell JO, Morokuma K (1976) J Phys Chem 80: 2675
190. Tapia O, Johannin G (1981) J Chem Phys 75: 3624
191. Kollman PA, Hayes DM (1981) J Am Chem Soc 103: 2955
192. van Duijnen PT, Thole BT, Hol WGJ (1979) Biophys Chem 9: 273
193. Allen LC (1981) Ann NY Acad Sci 367: 383
194. Onsager L (1936) J Am Chem Soc 58: 1486
195. Tapia O, Goscinski O (1975) Mol Phys 29: 1653
196. Klopman G (1968) Chem Phys Lett 1: 200
197. Miertus S, Kysel O (1977) Chem Phys 21: 27
198. Rivail JL, Rinaldi D (1976) Chem Phys 18: 233
199. Constanciel R, Tapia O (1978) Theoret Chim Acta 48: 75
200. Tapia O, Lamborelle C (1979) Chem Phys
201. Newton MD (1975) J Phys Chem 79: 2795
202. Hylton J, Christoffersen RE, Hall GG (1974) Chem Phys Lett 24: 501
203. Rinaldi D, Ruiz-Lopez MF, Rivail JL (1983) J Chem Phys 78: 834
204. Miertus, Scrocco E, Tomasi J (1981) Chem Phys 55: 117
205. Mikkelsen KV, Ágren H, Jensenand HJA, Helgaker T (1988) J Chem Phys 89: 3086
206. Tapia O, Sussman F, Poulain E (1978) J Theor Biol 71: 49
207. Longo E, Stamato F, Ferreira R, Tapia O (1985) J Theor Biol 112: 783
208. Ángyán J, Allavena M, Picard M, Potier A, Tapia O (1982) J Chem Phys 77: 4723
209. Thole BT (1981) Chem Phys 59: 341
210. Ángyán J, Náray-Szabó G (1983) Theoret Chim Acta 64: 27; (1984) Acta Chim Hung 116: 141
211. Tapia O (1990) In: H Weinstein, G Náray-Szabó (eds) Reports in Molecular Theory, CRC Press, Boca Raton (in press)
212. Ángyán JG, in preparation
213. Thole BT, van Duijnen PT (1982) Chem Phys 71: 211
214. Rinaldi D, Costa Cabral BJ, Rivail JL (1986) Chem Phys Lett 125: 495
215. Warshel A, Levitt M (1976) J Mol Biol 103: 227
216. Russell ST, Warshel A (1985) J Mol Biol 185: 389
217. Warshel A, Weiss RM (1980) J Am Chem Soc 102: 6218
218. Warshel A, Russell S (1986) J Am Chem Soc 108: 6569
219. Warshel A, Sussman F (1986) Proc Natl Acad Sci USA 83: 3806
220. Náray-Szabó G, Surján PR (1983) Chem Phys Lett 96: 449
221. Del Re G (1963) Theoret Chim Acta 1: 188
222. Bálint I, Bán MI (1983) Computers and Chem 7: 199
223. Náray-Szabó G, Surján PR, Kiss AI (1985) J Mol Struct (Theochem) 123: 85
224. Allavena M, Seiti K, Kassab E, Ferenczy G, Ángyán JG (1990) Chem Phys Lett 172: 55
225. Dempsey E (1969) J Phys Chem 73: 3660
226. Preuss E, Linden G, Peuckert M (1985) J Phys Chem 89: 2955
227. Lievens J (1987) personal comm
228. Mortier WJ (1987) Structure and Bonding 66: 125
229. van Genechten KA, Mortier WJ, Geerlings J (1987) J Chem Phys 86: 5063

230. Uytterhoven L, Lievens J, van Genecheten K, Mortier WJ (1987) Preprints of Conference Proceedings in Eberswalde (GDR)
231. Löwenstein W (1954) Am Mineral 39: 92
232. Johannin G, Kellersohn N (1972) Biochem Biophys Res Commun 49: 321
233. Warshel A (1981) Acc Chem Res 14: 281
234. Náray-Szabó G (1988) J Mol Catal 47: 281
235. Kraut J (1977) Annu Rev Biochem 46: 331
236. Polgár L, Halász P (1982) Biochem J 207: 1
237. Schowen RL (1987) In: Liebman JF, Greenberg A (eds) Principles of enzyme activity. VCH Publishers, Deerfield Beach FL USA, p 1
238. Bernstein FC, Koetzle TF, Williams GJB, Meyer jr EF, Brice MD, Rodgers JR, Kennard O, Shimanouchi T, Tasumi M (1977) J Mol Biol 112: 535
239. Náray-Szabó G, Polgár L (1981) Int J Quantum Chem Quantum Biol Symp 7: 397
240. Náray-Szabó G (1982) Int J Quantum Chem 22: 575
241. Náray-Szabó G, Kapur A, Mezey PG, Polgár L (1982) J Mol Struct (Theochem) 90: 137
242. Ángyán J, Náray-Szabó G (1983) J Theor Biol 103: 349
243. Náray-Szabó G (1984) J Am Chem Soc 106: 4584
244. Nagy P, Náray-Szabó G (1985) J Mol Struct (Theochem) 123: 413
245. Náray-Szabó G (1986) J Mol Struct (Theochem) 134: 401
246. Umeyama H, Imamura A, Nagata I, Hanano M (1973) J Theor Biol 41: 485
247. Scheiner S, Kleier DA, Lipscomb WN (1975) Proc Natl Acad Sci USA 72: 2606
248. Náray-Szabó G, Warshel A (1989) In: Kotyk A (ed) Proceedings of the 14th International Congress of Biochemistry. VSP Intl Sci Publ, Zeist
249. Umeyama H, Nakagawa S, Kudo T (1981) J Mol Biol 150: 409
250. Wooten F, Winter K, Weaire D (1985) Phys Rev Lett 54: 1392
251. Kramer B, King H, Mackinnon A (1983) J Non-Cryst Solids 59–60: 73
252. Brey L, Tejedor C, Verges A (1984) Phys Rev Lett 52: 1840
253. Ley L, Reichardt J, Johnson RL (1982) Phys Rev Lett 49: 1664
254. Klug DD, Whalley E (1982) Phys Rev B 25: 5543
255. Kugler S, Náray-Szabó G (1987) J Non-Cryst Solids 97–98: 503

Semiclassical Methods for Large Molecules of Biological Importance

Charles L. Brooks, III
Department of Chemistry, Carnegie Mellon University, Pittsburgh, PA 15213, USA

This chapter provides an overview of mixed quantum/classical methods applied to biopolymers. It begins with a brief introduction providing a historical context to the methods of empirical potential energy functions, molecular dynamics, free energy simulations and the scope for application of quantum chemical methods. The following sections provide a review of these methods as they have been developed for, and applied to, biopolymer systems. Section 2 deals with the classical potentials and simulation methodologies. In section 3 a general formula for implementation of mixed quantum/classical simulation models is discussed. Finally, a concluding section provides a perspective for the future development and application of mixed model.

1 Introduction

Protein and nucleic acids are the instruments by which chemical and physical processing takes place in living systems. The twenty amino acids and four nucleotides of which proteins and nucleic acids, respectively, are composed provide the range of chemical functionality necessary to carry out this processing. Through the structural arrangement and motional characteristics of chemically labile groups, these bio-polymers control and conduct transport, chemical synthesis and degradation processes which are necessary for their functioning. Thus, the detailed investigation of motion, structure and reactivity of proteins and nucleic acids is the key to understanding biological systems at a molecular level, and presents a major challenge to biologists, chemists and physicists.

Over the past 10–15 years computational methods have been developed which permit the study of protein and nucleic acid motions and structure, as well as some aspects of their reactivity. These techniques, known as biopolymer dynamics and mechanics [1, 2], evolved from pioneering work by Alder and Wainwright [3] and Rahman [4] on the classical simulation of condensed phase systems. They were solidified by the first application of classical molecular dynamics to proteins by McCammon, Gelin and Karplus in 1977 [5]. Today a broad range of biophysical processes are explored using molecular simulation methods [1, 2].

In the classical molecular dynamics simulation method, atoms in the biopolymer interact through empirical force laws and their motions are propagated through the use of classical equations of motions. From such simulations both time-dependent and time-independent processes can be studied. Examples include the calculation of reactive trajectories for determination of rate constants, examination of the extent of atomic fluctuations and the computation of free energy differences for the binding of ligand molecules [1, 2].

The potential energy functions used in such calculations are usually of the molecular mechanics type [6]. This means they do not explicitly include electronic degrees of freedom and hence atomic or molecular polarization is not explicitly treated, and bond breaking/making processes are not possible to simulate. In addition, inherently quantum phenomena such as nuclear or electron tunneling are not included in such classical based calculations.

These problems are of great importance if chemical processes are to be studied. However, they may be overcome within the context of classical simulation methods using a variety of strategies. The simplest approach, which permits some aspects of electronic polarization to be included, involves exploiting the classical formulation of polarizable materials [7]. Using classical electrostatics and related classically based methods, e.g. the Drude oscillator model of electronic polarization, atomic and molecular polarization of electron densities can be achieved [8–9]. Even with these approaches, however, the extreme case of polarization, i.e. bond breading/making, cannot be treated due to the underlying form of the molecular mechanics potential (see Sect. 2 below).

In these situations the possibility often exists to develop an accurate quantum mechanically based potential energy surface for the "reaction path" and use this surface, fit to an appropriate analytical form, to carry out classical dynamics and examine the reactive process. These methods have long been used to study the dynamics of simple reaction systems [10–11]. The extension to reactions in solution or biopolymers has only recently been developed [12–13].

The reaction path approach is useful but suffers from the limitation that a reaction path has to be pre-determined. Thus, assumptions regarding the mechanism of the reaction process must be made. This is often a difficult task for reactions occurring in the interior of a protein. An alternative approach is to make no such assumption and to derive the energy and associated forces for a given nuclear configuration quantum mechanically during the course of a classical simulation. The demands of such a technique require that the quantum mechanical evaluation of energy and forces not be prohibitive. This currently limits the practical application of such a method to 10–15 atoms for ab initio quantum chemical methods and ~ 50–100 atoms for semi-empirical approaches. These numbers fall far short of the size of interesting biological molecules (1000–10 000 atoms). Thus, another possibility is the development of a mixed quantum/classical methodology, wherein only those atoms directly involved in the chemical process are treated quantum mechanically and the remainder are treated as classical particles. Such a scheme would partition the system into a quantum motif embedded in a classical "bath". The classical bath in such a picture provides the environment for evolution of the quantum motif. A division like that illustrated in Fig. 1 permits large many-body systems in which reactive processes occur to be efficiently modeled.

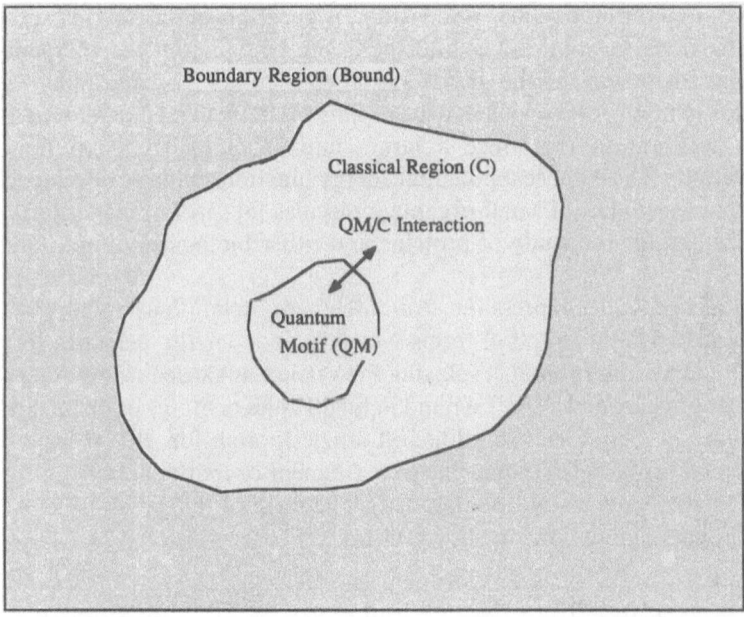

Fig. 1. A Schematic illustration of the partitioning of a many-body system in mixed quantum/-classical calculations. The partitioning shows the division of the system into three regions: the quantum motif, the classical region and the boundary region. This partitioning describes the division of the effective Hamiltonian given in Eq. (12)

Finally, we note that quantum mechanical tunneling effects may also be incorporated through the use of path integral techniques [14]. These effects are believed to be important in electron transfer processes in condensed phases and proteins [15–16].

In this chapter, we will focus on the development and application of the combined quantum/classical methods. To accomplish this we first provide background on the classical methods used in protein and nucleic acid simulations. In Sect. 2 we review the form and origin of empirical potentials used in biopolymer dynamics, the classical simulation methods, and techniques for evaluating thermodynamic averages as might be important in computing barrier heights for chemical rate processes. Next we describe the basic formalism for mixed quantum/classical simulation methods as well as some of the practical considerations in their development and implementation. This is done in Sect. 3. We conclude in Sect. 4 with an overview of these methods and their potential for chemical studies.

2 Models and Methods in Biopolymer Dynamics

In this section we briefly review the methods of classical biopolymer dynamics.

2.1 The Molecular Mechanics Potential for Biopolymers

The development of the structure-energy relation, generally referred to as the potential energy function, required for biopolymer simulations, is guided by a balance between computational restrictions (or limitations) and the desire to represent the energetic penalty of molecular distortions at the greatest level of accuracy possible. What is required is a simple functional form which may be rapidly evaluated and differentiated, yet is sufficiently accurate to reproduce a large number of experimental (and theoretical) observations. The empirical potential energy functions initially developed for molecular mechanics studies of small organic molecules [6] satisfy these criteria. They have been adapted for the study of proteins and other biologically important molecules [17–18].

Empirical potentials generally express the structure-energy relation as an additive sum of "physically intuitive" interaction terms which include energy penalties for: distorting pairs of bonded atoms (bonds); splaying the valence angle defined by triples of atoms (angles); rotating around a central bond in quadruples of atoms or distorting the chiraltity around a chiral center (dihedral angles); and for the pairwise, through-space interactions between atoms due to permanent or induced electrostatic effects, nuclear or exchange repulsion (nonbonded). The form of this potential most often used in biopolymer simulations in shown in Eq. (1) below.

$$U = \frac{1}{2} \sum_{\text{bonds}} k_b (b - b_0)^2 + \frac{1}{2} \sum_{\text{angles}} k_\theta (\theta - \theta_0)^2$$

$$+ \frac{1}{2} \sum_{\text{dihedrals}} k_\phi [1 + \cos(n\varphi - \delta)] + \sum_{\substack{\text{nonbonded} \\ \text{pairs}}} \left[\frac{A}{r^{12}} - \frac{B}{r^6} + \frac{q_1 q_2}{\varepsilon r} \right] \quad (1)$$

In this expression the total potential energy of a molecular system is related to its molecular configuration through the introduction of localized internal coordinates and nonbonded van der Waals and electrostatic terms.

The parameters which describe the "chemical interactions", i.e. collectively the bond, angle and dihedral terms, are most often derived by fitting individual terms from a target potential function such as Eq. (1) to experimental data from vibrational and NMR spectroscopy on small subunits contained in the biological macromolecules. For example, the carbonyl bond force constant (k_b) could be derived from the vibrational spectrum of a small molecule, or the dihedral potential parameters for rotation around the C–C bond in a valine sidechain would come from consideration of the gauche/trans equilibrium measured via NMR. The empirical nature of the potential thus guarantees that certain experimental properties are reproduced quite accurately; provided the chosen form of the function fits the experimental data well, and that the experiments on small fragments are transferable to larger systems. In general, both of these caveats must be kept in mind when the results from a given calculation are being interpreted.

Parameterization and evaluation of the through-space nonbonded interactions in Eq. (1) presents the greatest difficulty in the development of empirical potentials for biological macromolecules. There are a number of reasons for this difficulty. First, the A and B parameters, which represent the repulsive "van der Waals" radius and attractive dispersion interactions, are not directly observable by experimental measurement. In fact, they represent a complicated and non-uniquely decomposable set of high level quantum chemical interactions (Pauli exclusion, nuclear repulsion, induced electronic effects, etc.) which are not readily amenable to theoretical analysis. A second problem arises in attempts to represent the electrostatic interactions due to a permanent charge distribution on particular atomic centers i.e. the q's in Eq. (1).

To derive the van der Waals portion of the nonbonded parameters (the A's and B's for all pairs of atoms) one approach often used is to choose a basis set of atomic radii and van der Waals self-interaction strengths, based on optimization by comparison with experiment [19] or theoretical calculations [20], and then to generate the pairs of interaction parameters using combination rules.

To find an appropriate set of atomic charges ab initio quantum chemical calculations are often carried out on pairs of interacting molecules to generate an interaction surface [21]. If van der Waals parameters are assumed to have been chosen previously for all atoms in the pair of interacting molecules, the van der Waals interaction surface is substracted from the ab initio surface and the residual interaction is assumed to be due to electrostatic charges permanently distributed on atomic (or other) centers. These charges are adjusted to best fit the residual surface. An alternative approach is to derive the electrostatic potential due to a single molecule of interest from experimental or ab initio data and to fit an atom centered charge distribution to best reproduce this potential [22].

2.2 Algorithms for Biopolymer Dynamics

Given an explicit form for the potential and the corresponding parameters, the forces on individual atomic centers may be computed. The forces give rise to a classical description of the equations of motion. The numerical solution of these equations yields the desired molecular trajectory. We outline some of these methods below.

Before we sketch the details of the algorithms used in protein dynamics, several considerations involved in choosing the algorithm which is most appopriate should be noted. If explicit account is taken of the solvent surrounding and periodic replicas of the solute-solvent system are used to mimic a bulk phase, standard molecular dynamics algorithms using periodic boundary conditions in the NVE, NVT or NPT ensembles are appropriate. The techniques of periodic boundary conditions are presented elsewhere and the reader is referred to the literature for a detailed discussion of these methods [1, 2, 23]. If pseudo vacuum calculations are to be carried out, then it is often most appropriate to couple the motion of the protein atoms to a heatbath and Langevin dynamics algorithm is best used [24]. When localized enzyme chemistry; or thermodynamic changes associated with modification of specific amino acid sites on a protein; or chemical modification of inhibitors (drugs) bound in localized binding sites are of primary interest, the stochastic boundary molecular dynamics method is the best choice [25]. This technique combines standard molecular dynamics and Langevin dynamics algorithms to give a realistic description of local dynamic and thermodynamic properties.

A very useful algorithm for the numerical integration of the large number of coupled differential equations associated with proteins was introduced by L. Verlet in 1964 for dynamical simulations of Lennard-Jones fluids [26]. The Verlet algorithm may be derived from the simple central difference of the Taylor series expansion for the atomic position at time t. For the vector position of atom i, r_i, the algorithm is

$$r_i(t + \Delta t) = 2r_i(t) - r_i(t - \Delta t) + \frac{\Delta t^2}{m_i} F_i(t) , \qquad (2)$$

where F_i is the total force on atom i at time t and Δt is the timestep. The corresponding velocity propagation equation is

$$v_i(t) = [r_i(t + \Delta t) - r_i(t - \Delta t)]/2\Delta t . \qquad (3)$$

We note that the position propagation is of order Δt^3 and the velocity propagation is of order Δt^2, and lags a step behind the position. The Verlet algorithm has been found to provide a very stable numerical procedure for the integration of many coupled differential equations and appears to meet an appropriate balance between accuracy and timestep. The usual timestep used in solving protein equations of motion with this and the other algorithms presented below ranges from 0.5 fs to 4 fs.

The Verlet algorithm as embodied in Eqs. (2) and (3) above lends itself most readily to simulations in the NVE ensemble, since there is no simple and direct mechanism for coupling the system to a constant temperature reservoir. An interest in constant T ensembles has provided the impetus for alternate propagation algorithms to be introduced. Among these are the leap frog [27] and Beeman algorithms [28], which identically reduce to the Verlet algorithm for position propagation, Eq. (2), but differ in the velocity propagation scheme. These "summed forms" of the Verlet algorithm, allow direct coupling of the system to simple linearly (in the velocity) dissipative forces for the control of temperature. This adaptation permits simulations to be carried out in the NVT ensemble. Procedures for carrying out constant temperature simulations have been developed by Anderson [29], Nose and Hoover [30], Evans et al. [31] and Berendsen et al. [32].

2.3 Free Energy Simulation Methods

The general formalism for thermodynamic simulation methods follows from early work by Zwanzig [33], and provides a tool for the computation of thermodynamic properties, ΔA, ΔE and ΔS, as well as barriers for chemical processes occurring on long timescales [34]. These methods take on several guises in present implementations. The two approaches which we will describe are termed thermodynamic cycle perturbation theory (TP) [35] and thermodynamic integration (TI) [36]. Both of the methods are based on the definition of a hybrid Hamiltonian which represents some mixture of the initial state (1) and final state (2) of the system [37]. If λ represents the "coordinate" describing the pathway used to interconvert the two systems, then the hybrid Hamiltonian may be defined by [35, 37].

$$H(\lambda) = (1 - \lambda)^n H_1 + \lambda^n H_2 + H_{env},$$

(4)

where n describes the degree of non-linearity for this thermodynamic pathway. In this expression H_1 and H_2 represent the initial and final states of the thermodynamic process. H_{env} represents the "environment" Hamiltonian which includes all solvent-solvent interactions as well as iterations between the solvent and those portions of the solute common to both states 1 and 2. For a fixed value of λ not equal to 1 or 0 this system is unphysical but specifies a well defined thermodynamic state, with a Hamiltonian given in Eq. (4) above.

To make connections with thermodynamic properties a dynamics trajectory generated at a specific value of λ is used to compute the average quantity

$$\Delta A(\lambda, \lambda + \Delta\lambda) = -k_B T \ln \left\langle \exp\left[\frac{-(H(\lambda + \Delta\lambda) - H(\lambda))}{k_B T} \right] \right\rangle_\lambda,$$

(5)

which is the Helmholtz free energy difference between thermodynamic states at λ and $\lambda + \Delta\lambda$. Calculations at several values of λ, spanning $\lambda = 0$ to $\lambda = 1$, are carried out and the averages from these "windows" are then summed to compute the free energy change,

$$\Delta A = \sum_{\lambda=0}^{1-\Delta\lambda} \Delta A(\lambda, \lambda + \Delta\lambda).$$

(6)

Alternatively, one may use the methods of thermodynamic integration to compute the free energy change. This technique relies on the connection formula

$$\Delta A(\lambda) = \int_0^\lambda \left\langle \frac{\partial H(\lambda')}{\partial \lambda'} \right\rangle_{\lambda'} d\lambda',$$

(7)

where, again, molecular dynamics may be used to compute the average quantities within the broken brackets, $\langle...\rangle_\lambda$. In most cases this average is calculated for a number of windows spanning $\lambda = 0$ to $\lambda = 1$ and numerical quadrature is used to accumulate the free energy change [37].

The thermal decomposition of the free energy into energetic and entropic contributions may be considered as well [39]. These properties are computed from the same set of thermodynamic simulations used to derive the free energy with expressions for the connection formulas developed by considering the finite difference approximant to the temperature derivate of $\Delta A(\lambda, \lambda + \Delta \lambda)$ [36, 39]. The total energy and entropy changes are computed by summing incremental changes between each window as done for ΔA in Eq. (6).

Thermodynamic simulation methods may be used to calculate free energy surfaces using molecular dynamics [40]. This is of particular importance when one wishes to compute free energy barrier for chemical reactions [34]. A recent approach to this calculation uses a fusion of molecular dynamics and Monte Carlo techniques. The basic connection formula follows from Eq. (5) above, and consideration of the free energy difference between a particular configuration specified at ξ and that at $\xi + \Delta\xi$. If we let $W_i^\pm$ represent the weight (Monte Carlo Boltzmann factor) for displacing the pair of particles by $\pm\Delta\xi$ respectively during the ith molecular dynamics step,

$$W_i^\pm = \exp\left\{\frac{-(U(\xi \pm \Delta\xi) - U(\xi))}{k_B T}\right\}, \tag{8}$$

the free energy difference between the configuration at ξ and those at $\xi \pm \Delta\xi$ is simply related to the average of the weights for each step. [40].

$$\Delta A(\xi, \zeta \pm \Delta\xi) = \frac{1}{N}\sum_i W_i^\pm = -k_B T \ln\left\langle\exp\left\{\frac{-(U(\xi \pm \Delta\xi) - U(\xi))}{k_B T}\right\}\right\rangle_\zeta. \tag{9}$$

In Eq. (9), N is the number of dynamics steps used in computing the average and the subscript ζ on the angle brackets indicates this average is over an ensemble constrained at a fixed ξ.

This formalism is implemented by attempting (but never carrying out) Monte Carlo steps $\pm\Delta\xi$ away from the position ζ and storing the weights for each step in a conventional molecular dynamics simulation. A series of simulations with different ξ spanning the range of reaction coordinate of interest may then be used to compute the reaction free energy surface [40]. Temperature and spatial derivative approximants of $\Delta A(\xi)$ may also be computed.

3 A General Method for Mixed Quantum/Classical Simulations

In this section we describe a general approach to the development of a mixed quantum/classical simulation protocol suitable for the study of chemical processes in protein and nucleic acid systems. This approach follows closely that recently published by Field et al. [47]. The main elements of such a procedure involve system partitioning, definition of the appropriate mixed, or effective, hamiltonian and description of the quantum chemical model/method. Each is laid out in the subsections below. These developments build on the classical empirical potential function and simulation methods described in the previous section.

3.1 Partitioning the System

The system may be partitioned as illustrated in Fig. 1 into three parts; a quantum-motif, a classical or molecular mechanics region, and a boundary region. The quantum and classical regions contain the atoms that are explicitly treated in the calculation. The boundary region accounts for the surroundings that are neglected. As such either periodic boundaries or the stochastic boundary method can be used [23, 25].

Atoms in the quantum motif are represented as nuclei and electrons. The potential surface for these atoms is determined within the Born-Oppenheimer approximation [41] and energy derivatives (forces) as a function of the nuclear positions are obtained. This region includes those atoms directly involved in the reaction process.

The classical region contains the all of the remaining atoms of the system. Their interactions are determined from an empirical potential energy function of the form described above. They provide the immediate environment for the quantum mechanical atoms and are included because their interactions and dynamics will have a significant influence on the quantum motif.

3.2 Development of the Mixed Hamiltonian

The solution of the time-independent Schrödinger equation for the electronic wavefunction of the quantum motif, Ψ, is constructed from the effective Hamiltonian of the system, H_{eff}.

$$\hat{H}_{Eff}\Psi(\mathbf{r}, \mathbf{R}_\alpha, \mathbf{R}_C) = E(\mathbf{R}_\alpha, \mathbf{R}_C)\,\Psi(\mathbf{r}, \mathbf{R}_\alpha, \mathbf{R}_C). \tag{10}$$

The wavefunction Ψ in Eq. (10) depends explicitly on the coordinates of the electrons, $\mathbf{r}$, on the positions of the quantum mechanical nuclei, $\mathbf{R}_\alpha$, and the classical region atoms, $\mathbf{R}_c$, in a parametric manner. For each geometry of the nuclei and the classical region atoms the energy is given as the expectation value of H_{eff}.

$$E = \frac{\langle\Psi|\,\hat{H}_{Eff}\,|\Psi\rangle}{\langle\Psi\,|\,\Psi\rangle}. \tag{11}$$

The forces on the nuclei, $\mathbf{F}_\alpha$, and the classical atoms, $\mathbf{F}_c$, are obtained by differentiating Eq. (11) with respect to the nuclear or atomic Cartesian coordinates.

To achieve the partitioning illustrated in Fig. 1 an effective Hamiltonian for the system is written as the sum of four terms,

$$\hat{H}_{Eff} = \hat{H}_{QM} + \hat{H}_C + \hat{H}_{QM/C} + \hat{H}_{Bound}, \tag{12}$$

and the total energy is the sum of the expectation values of each term in the Hamiltonian, i.e.,

$$E_{Eff} = E_{QM} + E_C + E_{QM/C} + E_{Bound}. \tag{13}$$

The first term in Eq. (12) is the Hamiltonian describing the quantum mechanical particles (electrons and nuclei) of the system and their interactions with each other. It is most often a standard nonrelativistic electronic Hamiltonian. The classical part of the Hamiltonian, $\hat{H}_C$, describes the classical region and is analogous to that described in Eq. (1). The interaction Hamiltonian, $\hat{H}_{QM/C}$, describes the interaction between the quantum motif and classical atoms and the interaction of the quantum motif and classical regions with the environment is determined by the components of the boundary Hamiltonian, $\hat{H}_{Bound}$. Since the wavefunction depends only parametrically on the $\mathbf{R}_C$, the terms E_C and $E_{Bound(C)}$ can be taken outside the integral in Eq. (11) and we have

$$E = \frac{\langle \Psi | \hat{H}_{QM} + \hat{H}_{QM/C} + \hat{H}_{Bound(QM)} | \Psi \rangle}{\langle \Psi | \Psi \rangle} + E_C + E_{Bound(C)}, \tag{14}$$

where the boundary term has been separated into quantum motif and classical parts. The forces on the nuclei, $\mathbf{F}_\alpha$, and the classical atoms, $\mathbf{F}_c$, are obtained by differentiating Eq. (14) with respect to the nuclear or atomic Cartesian coordinates.

3.3 Considerations in the Quantum Mechanical Model

The system paritioning and Hamiltonian specifications as described above present a mathematical/computational problem which must be addressed to carry out mixed quantum/classical dynamics simulations. This involves the quantum mechanical method used to compute the expectation value of the energy and forces in the quantum motif as well as the treatment of the quantum motif/classical region interfacial atoms. In this section we provide a review of some of the techniques used in doing such calculations.

Much like the choice ot the empirical potential functional form discussed in Sect. 2 above, the choice of the quantum mechanical method and model is a compromise between speed of evaluation and accuracy. The most rigorous approach to evaluating these energy and force terms would be to use ab initio quantum chemical methods with large basis sets and correlation corrections beyond the Hartree Fock level. Clearly, this is currently not a feasible approach because of the computional demands such as model places on a single energy evaluation, not to mention the iterative evaluation over thousands of structures (timesteps) of a dynamics simulation.

Despite the demands presented by such a calculation, a number of researchers have used ab initio models to treat the electronic and nuclear degrees of freedom for the quantum motif in molecular mechanics, energy minimization studies. Examples of this include the self-consistant reaction field methods developed by Tapia and coworkers [42–44], which represent only the quantum motif explicitly and use continuum models for the environmental effects (classical and boundary regions), and the methods implemented by Kollman and coworkers [45] in their studies of condensed phase (chemical and biochemical) reaction mechanisms. In both of these implementations the expectation value of the quantum motif Hamiltonian, defined in Eqs. (11) and (14) above, is treated at the Hartree Fock level with relatively small basis sets.

In order to extend these methods to make them feasible for the study dynamical chemical processes in biopolymers, simplifying assumptions are necessary. The most obvious choice is the use of semi-empirical techniques within the Hartree Fock, linear combination of atomic orbitals framework. These methods can achieve speedups on the order of 1000 over typical ab initio calculations using split valence basis sets within the Hartree Fock approximation. Often greater accuracy can be achieved as well because of the parameterization inherent in the semi-empirical approaches. One semi-empirical approach which has proven successful in representing many chemically interesting processes is the AM1 and MNDO Hartree Fock Self-Consistent Field methods developed and paramerterized by Dewar and coworkers [46]. These methods have recently been implemented in a mixed quantum/ classical methodology for the study of chemical and biochemical processes by Field et al. [47].

Even greater speed enhancement can be achieved through the use of empirical quantum mechanically based schemes. A good example of this in the mixed quantum/classical methodology developed by Warshel and coworkers. This method uses an empirical valence bond approach to treat the quantum motif and hence is extremely inexpensive (commutationally) to include. Warshel has illustrated a number of qualitative ideas using these methods [48]. However, a key aspect to such an approach is the ability to adequately specify the appropriate resonance forms for the "ionic" and "covalent" species.

Finally, we note that consideration of the interactions between the quantum motif atoms and classical region atoms must be given. This is true not only for the electrostatic interactions but also the shorter range repulsive and dispersive forces. In the case of the electrostatic effects a very good, and widely used, approximation is to assume that the classical charge centers within the classical region provide an external electric field to the quantum motif and this is simply coupled into the quantum self-consistent field calculation at every step using standard methods [45].

The situation is somewhat more complicated for the shorter range repulsive and dispersion forces. It is essential that the classical atoms "see" the quantum atoms to avoid the possibility of a classical charge center collapsing onto a quantum nucleus. In order to circumvent this problem the quantum atoms are given empirically chosen van der Waals parameters and these are used in calculations of the energy and forces of interaction between the two regions. The specific choice of parameters must be carefully considered in such cases since it must properly maintain the balance of forces between the different interacting regions. Thus calibration and testing are necessary [47].

4 Concluding Remarks

In the previous sections we have reviewed and outlined the methods utilized in carrying out studies of chemical processes in condensed phase systems, specifically those of biological importance. The application of these mixed quantum/classical methods have only begun to be prevalent in the last few years. During this period new and differing techniques have been utilized to study basic chemical phenomena occurring in ionic solids, and solution [43, 45, 49]. Perhaps more relevant to our understanding of chemical reactivity in biological molecules have been calculations of chemical processes in liver alcohol dehydrogenase [42], triosphosphate isomerase [50], dihydrofolate reductase [51], the photosynthetic reaction center [52], and other biopolymers. As computational power increases so increases our ability to treat reactions in systems of increasing size and complexity at levels of greater accuracy. These studies will grow in number as time proceeds, and a deeper understanding of environmental influences on chemical reactivity can be expected to increase as well.

In closing this chapter we note that new and developing techniques can also be expected to play a role. Most current methods employ the traditional quantum chemical approaches, e.g., Hartree Fock, MPn and related LCAO based techniques. These methodologies require computational effort which scales with fairly high powers of the number of nuclei in the quantum motif (generally as N^3 or greater). However, density functional based approaches [53] show much more favorable scaling behavior and can be anticipated to play a role in the future. Also, varients of the more traditional methods, such as the pseudo-spectral technique under development by Freisner [55], provide a real potential for extending the range of mixed quantum/classical simulations. Lastly, we note that path integral techniques are beginning to show promise in the study of one electron phenomena such as electron transfer and processes which involve the quantum aspects of nuclear or atomic motion [14–16].

Acknowledgements: Support for this work came from the National Institutes of Health. Charles L. Brooks, III is an Alfred P. Sloan Fellow.

5 References

1. Brooks CL, III, Karplus M, Pettitt BM (1988) Adv Chem Phys 71: complete volume
2. McCammon JA, Harvey S (1987) Dynamics of proteins and nucleic acids, (Cambridge University Press, Combridge or NY
3. Alder BJ, Wainright TE (1959) J Chem Phys 31: 459
4. Rahman A (1964) Phys Rev A136: 405
5. McCammon JA, Gelin BR, Karplus M, (1977) Nature 267: 585
6. Burkhart U, Allinger NL, Molecular Mechanics (1982) Am Chem Soc Washington, D. C
7. See, for example Jackson JD, Classical and Electrodynamics, (John Wiley, New York 1975)
8. Sprik M, Klein M (1988) J Chem Phys, 89: 7556
9. Lybrand T, Kollman P (1985) J Chem Phys, 83: 2923; Cieplak P, Lybrand T Kollman P (1985) J Chem Phys 86: 6393; Cieplak P, Kollmann P, Lybrand T (1990) J Chem Phys 92: 6755

10. Karplus M, Porter RN, Sharma RD (1964) J Chem Phys, 40: 2033
11. Miller WH, Handy NC, Adams JE (1980) J Chem Phys, 72: 99
12. See, for example: Chandrasekhar J, Smith SF, Jorgensen WL (1985) J Am Chem Soc, 107: 154; Jorgensen WL, Buckner JK (1986) J Phys Chem, 90: 4651; Blake JF and Jorgensen WL (1987) J Am Chem Soc, 109: 3856; Madura JD, Jorgensen WL, (1986) J Am Chem Soc, 108: 2517
13. Bergsma JP, Reimers JR, Wilson KR, Hynes JT (1986) J Chem Phys, 85: 5625; Bergsma JP, Gertner BJ, Wilson KR, Hynes JT (1987) J Chem Phys, 86: 1356; Gertner BJ, Bergsma JP, Wilson KR, Lee S, Hynes JT (1987) J Chem Phys, 86: 1377
14. Chandler D, Wolynes PG (1981) J Chem Phys, 74: 4078
15. Kuki A, Wolynes PG (1987) Science, 236: 1647; Zheng C, Wong CF, McCammon J, Wolynes PG (1989) Chim Scripta, 29A: 171; Zheng C, McCammon JA, Wolynes PG, (1989) Proc Natl Acad Sci, 86: 6441
16. Bader JS, Kuharski RA, Chandler D (1990) J Chem Phys, 93: 230
17. Brooks BR, Bruccoleri R, Olafson B, States D, Swaminathan S, Karplus M (1983) J Comp - Chem, 4: 187
18. Weiner SJ, Kollman P, Nuguyen DT, Case D (1986) J Comp Chem, 7: 230
19. Gibson KD, Scheraga HA (1967) Proc Nat Acad Sci, USA 58: 420; Dunfield L, Burgess A Scheraga H (1978) J Phys Chem, 82: 2609
20. Stone AJ, Price SL (1988) J Phys Chem, 92: 3325
21. Jorgensen W (1981) J Am Chem Soc, 103: 335; ibid. 103, 341 (1981)
22. (a) Singh, Kollman PA (1984) J Comp Chem, 5: 129. (b) Chirlian L, Francal M, (1987) J Comp Chem, 8: 894
23. Allen MP, Tildesley DJ, Computer Simulations of Liquids, (Oxford University Press, London, 1987)
24. Brünger A, Brooks CL, III, Karplus M (1984) Chem Phys Lett, 105: 495
25. Brooks CL III, Brünger A, Karplus M (1985) Biopolymers 24: 434; Brooks CL III, Karplus M (1989) J. Mol Biol., 208: 159
26. Verlet L (1967) Phys Rev 159: 98
27. Hockney RW Easfwasel JW, Computer Simulations Using Particles, (McGraw Hill, New York, 1981)
28. Beeman D (1976) J Comp Phys 20: 130
29. Andersen HC (1980) J Chem Phys 72: 2384
30. Nose S (1984) J Chem Phys 81: 511; Hoover WG (1985) Phys Rev A, 31: 1695; Evans DJ Hollihan BL (1985) J Chem Phys, 83: 4069
31. Evans DJ, Morris GP (1983) Chem Phys, 77: 63
32. Berendsen HJC, Postma JPM, van Gusteren WF, DiNola A, Haak JR, (1984) J Chem - Phys 81: 3684
33. Zwanzig RW (1954) J Chem Phys, 22: 1420
34. Chandler D (1978) J Chem Phys, 68: 2959; McCammon JA, Karplus M, (1979) Proc Natl A- cad Sci, USA, 76: 3585
35. Tembe BL, McCammon JA (1984) Comput Chem, 8: 281
36. Fleischman SH, Brooks CL III, (1987) J Chem Phys, 87: 3029
37. Mezei M, Beveridge DL (1986) Annals NY Acad Sci, 483: 1
38. Jorgensen WL, Ravimohan C (1985) J Chem Phys, 83: 3050
39. Brooks CL III (1986) J Phys Chem, 90: 6680
40. (a) Tobias DJ, Brooks CL III (1987) Chem Phys Lett, 142: 472; (b) ibid. (1988) J Chem Phys, 89: 5115; (c) Tobias DJ, Brooks CL, III, Fleischman SH (1989) Chem Phys Lett 156: 256
41. See, for example: Born M, Huang K, Dynamical Theory of Crystal Lattices, (Oxford University Press, London, 1965); Sutcliffe BT, "Fundamentals of Computational Qantum Mechanics" in Computational Techniques in Quantum Chemistry, Dierchsen GHF, Sutcliffe BT, Veillard A, Eds, (Reidel, Boston, 1975)
42. Topia O Johannia G (1981) J Chem Phys 75: 3624
43. Tapia O (1982) I: Ratjczak H Orville-Thomas WJ (eds) Molecular interactions, Wiley, Chichester, vol 3 p 47
44. Tapia O, Goscinski O (1975) Mol Phys, 6: 1653
45. Weiner SJ, Singh UC, Kollman PA (1985) J Am Chem Soc, 107: 2219

46. Dewar MJS et al, (1985) J Am Chem Soc 107: 3902
47. Field M, Bash P, Karplus M (1990) J Comp Chem, 11: 700
48. Warshel A, Weiss RM (1980) J Am Chem Soc, 102: 6218
49. Bash PA, Field MJ, Karplus M (1987) J Am Chem Soc, 109: 8092
50. Bash PA, Field MJ, Karplus M, work in progress
51. Singh UC (1988) Proc Natl Acad Sci, 85: 4280
52. Croighton S, Hwang J-K, Warshel S, Parson WW, Norris J (1988) Biochemistry, 27: 774
53. Parr RG, Yang W, Density-functional theory of atoms, molecules, (Oxford University Press, New York, 1989)
54. Ringnaldu M, Won Y, Freisner R (1990) J Chem Phys 92: 1163.

Electronic Excited States of Biomolecular Systems: Ab Initio FSGO-based Quantum Mechanical Methods with Applications to Photosynthetic and Related Systems

Gerald M. Maggiora, James D. Petke and Ralph E. Christoffersen

Upjohn Laboratories 301 Henrietta Street, Kalamazoo, MI 49001, USA

Floating spherical Gaussian orbitals (FSGOs) possess a number of properties which make them quite useful in molecular electronic structure calculations of large molecular systems. The molecular fragment procedure provides a means for developing basis sets which mimic electronic environments in large molecules with a minimal number of FSGO basis functions. This parsimonious representation of electronic structure makes it possible to extend ab initio quantum mechanical calculations to molecular systems containing 200 to 500 electrons. The present chapter provides a review of the molecular fragment, configuration interaction, and molecular exciton procedures as they apply to studies of the electronic excited states of individual and aggregates of biomolecules. Numerous applications to photosynthetic and related systems are presented and discussed, including the singlet states of neutral chlorophylls and bacteriochlorophylls, the doublet states of their cation and anion radicals, charge-transfer states of porphyrin heterodimers, and singlet exciton states of bacteriochlorin dimers.

1 Introduction

The accelerating rate with which new computer hardware is being introduced has spurred development of a wide variety of sophisticated scientific software for the simulation of chemical and biological systems. Molecular mechanics and dynamics methods, which are based on potential energy functions derived from empirical and quantum mechanically calculated data, are now widely employed in investigations of the conformations and interactions of molecules in their ground electronic states. Applications to systems as complex as proteins and nucleic acids have made significant contributions to our understanding, especially as regards the molecular basis of structure and function. Direct application of quantum mechanical methods to such studies of large biomolecules have been sparse, and have not, in general had the impact that molecular mechanical and dynamical methods have had in shaping our understanding of these systems. When dealing with electronic excited states, however, quantum mechanical methods are required. And knowledge of the electronic structure and properties of these states is important for a detailed understanding of a number of photobiological processes (such as photosynthesis and vision), and for analyzing and interpreting the results obtained from applications of absorption, emission, and circular dichroism spectroscopy to the study of individual and inter-acting biomolecules.

Thus, development of quantum mechanical methods suitable for treating electronic excited states of large biomolecular systems (e.g., molecules containing 200 to 500 electrons) is important if we are to gain the level of "molecular understanding" that is needed for such systems. At the present time, the basis of many quantum mechanical methods is the Hartree-Fock self-consistent-field molecular orbital (HF-SCF-MO) approach, each MO being represented as a linear combination of appropriate basis functions. Electronic excited states are generally treated subsequently by configuration interaction (CI) methods [1], although multi-configurational SCF [2], equations-of-motion [3], and coupled cluster [4] methods are beginning to gain in importance.

In most instances it is the number of basis functions, N, that represents the most serious impediment to the application of these methods to biomolecular systems, the major factor being the number of two-electron integrals which must be evaluated and stored — on the order of $N^4/8$. This problem is usually addressed in one of two ways, namely through the use of semi-empirical methods or through the use of basis sets designed to describe the electronic structure of the systems with as small a number of functions as possible. An example of the latter approach is given in the work described in the present chapter, and shows that minimal basis sets, if properly designed and implemented, can mimic many of the results obtained with larger basis sets on smaller systems as well as providing a reasonably reliable means for studying much larger systems.

Specifically, floating spherical Gaussian orbitals (FSGO), developed from an extension of Frost's simple "electron pair model" of molecular electronic structure, are used as basis functions in SCF-MO and CI calculations. As will be described in the following section, these functions, unlike atomic basis orbitals, generally are not confined to atoms, but are allowed to occupy electron-rich regions corresponding to chemical bonds and "lone-pairs" as well. In addition, evaluation of the two-electron integrals is very fast for FSGOs compared to cartesian Gaussian-type orbital (CGTO) atomic basis sets. Furthermore, due to the relatively "localized" nature of FSGOs, the effective number of two-electron integrals that must be calculated and stored is often closer to N^3 than to N^4 for large molecules. As the number of FSGOs generally lies between the number of electron pairs and the number of electrons in a given system, it is possible to obtain very parsimonious representations of molecular electronic structure. Moreover, it will be shown that these elegantly simple functions possess other properties which make them useful for studies of large, interacting molecular systems such as are encountered in the exciton studies discussed in Sect. 6.

In Sect. 2 we present a brief description of FSGOs and of the molecular fragment (MF) approach which employs these functions in an economical and efficient manner for describing molecular electronic structure. Section 3 presents a description of the CI methodology used in the calculation of monomer excited states. A general discussion of the capabilities and limitations of the FSGO-based CI method, of the relationship of the present methods to other ab initio approaches, and of the computational requirements for application of the present methods to large molecules is presented in Sect. 4. Section 5 includes a review of a number of applications to the excited states and electronic spectra of porphyrins, chlorophylls (Chl), bacterio-chlorophylls (BChl), and related photosynthetic pigments. Section 6 provides a discussion of the exciton methodology used to study collective excited state properties of multimeric systems. The section also includes a discussion of the use of spherical Gaussian-based distributed point-charge models (PCMs) in evaluations of the intermolecular interactions required in exciton calculations. Applications of the exciton methodology to the investigation of spectral shifts in dimeric bacterio-chlorin (BC) systems follows in Sect. 7. Section 8 summarizes the material and presents a number of conclusions regarding the methods used for studying excited states of large systems.

2 Floating Spherical Gaussian Orbitals — FSGOs

2.1 FSGO Basis Sets

The use of Gaussian-type orbitals (GTOs) as basis functions for quantum mechanical calculations was first proposed by Boys [5] almost forty years ago. Today essentially all ab initio electronic structure calculations are carried out with these functions. Generally, cartesian Gaussian-type orbitals are employed in the construction of atomic basis sets. The present discussion, however, describes the construction of small basis sets containing only spherical Gaussian functions (FSGO), which in their normalized form are given by

$$g_m(r) = \left(\frac{2a_m}{\pi}\right)^{3/4} \exp\left[-a_m(r - R_m)^2\right], \tag{1}$$

where a_m is its "orbital exponent" and R_m describes its position relative to an arbitrary fixed origin.

2.2 Molecular Fragment Procedure

The simplest application of these functions to electronic structure calculations
was developed by Frost [6]. In his approach each FSGO describes a two-electron
spatial orbital in a "Frost model" wave function given by the normalized 2n-electron
Slater determinant

$$\Psi(1, 2, \ldots, 2n) = [(2n)!]^{-1/2} \det \{g_1(1)\, \bar{g}_1(2) \ldots g_n(2n-1)\, \bar{g}_n(2n)\} \qquad (2)$$

where a "bar" over a spatial orbital indicates a β-spin function, and the numbers
in parentheses label the electrons. Because the number of electron pairs is equal
to the number of FSGO, the Frost approach represents the case of an absolute minimum basis set. The values of the non-linear parameters, a_m and $\boldsymbol{R}_m$, are determined
variationally through minimization of an appropriate energy functional. In this
process the optimum "orbital radius" $\varrho_m = a_m^{-1/2}$ of each FSGO is determined
by variation of the orbital exponents, while each FSGO is simultaneously allowed
to "float" to an optimum position within the molecule, the final positions generally
corresponding to inner-shell, bonding, or lone-pair regions.

Frost [7] and Linnett [8] have applied this simple and intuitively appealing
model of electronic structure to the ground states of both closed- and open-shell
systems. Applications to excited electronic states have not been forthcoming due
to the overly restrictive nature of the model. Several approaches [9] have, however,
been developed to deal with this problem while maintaining the simplicity of
Frost's original model. The molecular fragment (MF) procedure [10, 11] represents
an approach aimed at addressing the problem by combining the Frost model with
HF-SCF-MO calculations.

In this procedure, FSGO parameters are obtained from Frost model calculations
on "molecular fragments" chosen to mimic various "molecular environments".
For example, a planar methyl radical would be used to mimic the environment of an
sp^2 carbon atom Fig. 1a, where the locations of FSGOs containing the inner-shell
and bonding electron pairs are indicated by the filled circles on the figure. The pair
of spherical Gaussians, g_m^U and g_m^D, located above and below the plane of the molecule
are combined as

$$g_m^\pi = [2(1 - \langle g_m^U | g_m^D \rangle]^{-1/2} (g_m^U - g_m^D) \qquad (3)$$

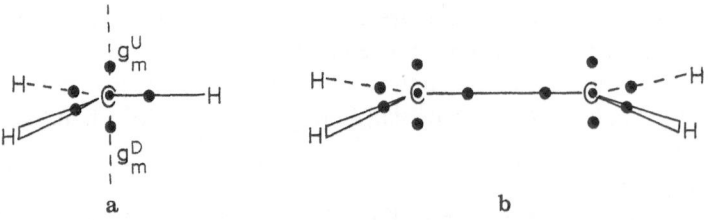

a   b

Fig. 1a, b. Positions of spherical Gaussians in "molecular fragment" FSGO basis sets for (**a**) the methyl
radical, and (**b**) ethylene. Pairs of spherical Gaussians labelled g_m^U and g_m^D are combined as in Eq. (3)
to form p_π-type basis functions

to represent a normalized p_π-type FSGO basis function, which contains the sole unpaired electron. More complex systems can then be assembled by "merging" the FSGOs of appropriate fragments as illustrated in Fig. 1 b for the case of ethylene. Note that the protons of the two methyl fragments that would lie in the carbon-carbon bond region of ethylene are removed, leaving a pair of FSGOs as primary contributors to the description of the carbon-carbon sigma bond. The combined sets of FSGOs of the two methyl radicals, with fixed orbital exponents and positions, are then used as a basis set for ethylene. This basis is larger than the absolute minimum basis of the Frost model. Thus, ethylene molecular orbitals constructed as linear combinations of fragment-based FSGOs, i.e.,

$$\varphi_1 = \sum_{s=1}^{N} c_{1s} \chi_s , \qquad (4)$$

where $\{\chi_s\}$ includes both spherical and p_π-type FSGO basis functions, can be determined using standard HF-SCF-MO methods. By initially determining the nonlinear FSGO parameters in molecular fragment environments, it is anticipated that moderate deviations from these environments can be adequately accounted for by the linear MO coefficients. Such an approach is well suited to large molecular systems including those of biological interest, in which highly accurate descriptions of the electron distribution in specific chemical bonding regions is often not required. A tabulation of FSGO data for a number of molecular fragments involving first-row atoms may be found in a previous publication [11].

3 Calculation of Electronic Excited States: Configuration Interaction (CI) Procedure

Electronic excited state calculations, whether semi-empirical or ab initio, are often carried out by some form of CI procedure [1]. Calculations described in the current work employ a multi-reference CI procedure (MRCI) [12, 13], with single and double excitations being performed from a set of "parent" configurations, which are those configurations determined in exploratory studies to be the major contributors to the electronic states of interest. Selection of configurations is determined by a second-order perturbation criteria, as originally proposed by Whitten and Hackmeyer [12]. The resultant CI expansions contain multiply-excited configurations (single, double, triple, quadruple ...) with respect to the ground state SCF determinant. The method is designed to give moderately large CI expansions (i.e. 1000 to 10,000 configurations) for a relatively large number (i.e., 10 to 40) electronic states. The molecular wave function for the Kth electronic state is given by

$$\Xi_K = \sum_L C_{KL} \Phi_L , \qquad (5)$$

where each configuration, Φ_L, is constructed from appropriate symmetry-adapted combinations of Slater determinants made up of molecular spin orbitals (MSO) obtained from molecular fragment HF-SCF-MO calculations. The expansion coefficients, C_{KL}, are obtained variationally from the usual secular equations

$$\sum_L (H_{JL} - \delta_{JL} W_K) C_{KL} = 0, \tag{6}$$

where W_K is the state energy and the Hamiltonian matrix elements, H_{JL}, are given by

$$H_{JL} = \langle \Phi_J | H | \Phi_L \rangle, \tag{7}$$

and H is the electronic Hamiltonian operator. The resulting wave functions are then used to calculate Franck-Condon vertical transition energies, oscillator strengths, and polarizations in addition to ground and excited state properties and electron distributions.

Analysis of the resulting data aims at a qualitative or semi-quantitative rather than quantitative interpretation of experimental results, consistent with the fact that small FSGO basis sets are used in the calculations. Due to the relatively restricted nature of these basis sets, and the fact that ground-state MOs are used in the CI expansions of both ground and excited states, calculated transition energies are generally too large in comparison with experimental values. To improve agreement with experiment, "estimated" transition energies, ΔE^{est}, may be obtained from calculated ones, ΔE^{calc}, by a simple scaling procedure based on the following linear relationship

$$\Delta E^{est} = m \cdot \Delta E^{calc} + b, \tag{8}$$

in which the parameters m and b must be determined for each specific class of molecule, usually by a least squares fit of calculated and experimental transition energies of a prototype molecule. Specific examples of implementation of this scaling procedure are found in the following sections.

4 Practical Considerations

4.1 Applicability of the Methods

The methods described above possess a number of advantages as well as drawbacks which should always be considered before proceeding with a particular problem. Due to the somewhat restrictive nature and small size of FSGO molecular fragment basis sets, the present methods are not appropriate for providing highly accurate descriptions of excited-state electronic structure, or for accounting for significant percentages of the correlation energy of an electronic state. Rather, they provide a means for obtaining a qualitative picture of a manifold of excited electronic states and their relative transition energies and intensities, and therefore are appropriate for applications in which such data are likely to suffice. Examples include the qualitative resolution of a spectral band into individual electronic transitions, or the determination of the polarization of a transition. Often it is possible to obtain a useful qualitative understanding of the factors responsible for spectral shifts or other spectral changes, particularly when such changes are large. It should be emphasized that, because the absorption or CD spectrum of a large molecule is typically not well resolved, the type of information provided by the present methods is precisely what is needed in interpreting experimental results. A number of examples illustrating this point are reviewed in the next section.

Because small FSGO basis sets are used in the present methods, it is important to establish the extent to which these methods are capable of describing electronic states and spectra, and one approach is a direct comparison with CI calculations performed with an extended basis set. The results of such a comparison for the pyrazine molecule [14] are summarized in Table 1. In this study, SCF and CI calculations performed by Hackmeyer and Whitten (HW) [15] using an "extended" split-valence Gaussian-lobe atomic basis set were essentially repeated with an FSGO basis. In Table 1 data for five excited states with known transition energies are given, and it is seen that both calculations give the same relative ordering of excited states, and that both calculations overestimate the transition energies.

With the HW extended basis, the calculated transition energies obtained for the lowest three states in Table 1 are quite accurate, being within about 0.3 eV of the experimental values. However, much larger errors are obtained for the two higher-lying (π, π^*) states, 1^1B_{1u} and 2^1B_{2u}. By analogy to the benzene spectrum [16, 17], these two states are expected to contain a diffuse Rydberg component, and thus basis sets containing additional diffuse functions are required to properly describe them. Therefore, while the extended basis set calculations indicate the existence of

Table 1. Calculated spectral data for pyrazine

State	Transition Energy (ev)			Oscillator Strength		CI Composition[d]		Configuration
	FSGO[a]	HW[b]	exptl.	FSGO[c]	HW	FSGO	HW	
1^3B_{3u}	4.49	3.56	3.2–3.3	0.0	0.0	0.95	0.91	$(6a_g \rightarrow 2b_{3u})$
(n, π^*)						0.12	—	$(1b_{2g}, 5b_{1u} \rightarrow 2b_{3u}^2)$
1^1B_{3u}	5.47	4.22	3.8–3.9	0.017	0.01	0.95	0.90	$(6a_g \rightarrow 2b_{3u})$
(n, π^*)						0.21	0.27	$(1b_{2g}, 5b_{1u} \rightarrow 2b_{3u}^2)$
1^1B_{2u}	6.88	5.29	4.8–4.9	0.027	0.01	0.77	0.70	$(1b_{1g} \rightarrow 2b_{3u})$
(π, π^*)						0.51	0.54	$(1b_{2g} \rightarrow 1a_u)$
1^1B_{1u}	9.03	9.10	6.5–6.6	0.221	0.12	0.91	0.87	$(1b_{1g} \rightarrow 1a_u)$
(π, π^*)						0.36	0.42	$(1b_{2g} \rightarrow 2b_{3u})$
2^1B_{2u}	11.07	9.95	7.6–7.7	1.27	0.41	0.54	—	$(1b_{1g} \rightarrow 2b_{3u})$
(π, π^*)						0.78	—	$(1b_{2g} \rightarrow 1a_u)$

[a] FSGO basis, [14].
[b] Hackmeyer and Whitten extended basis, [15].
[c] Corrected values. Oscillator strengths given in [14] are in error.
[d] Coefficients C_{KL} (Eq. 4) with magnitudes >0.1 are listed along with the corresponding configuration Φ_L.

G. M. Maggiora, J. D. Petke and R. E. Christoffersen

these two states, it is not expected that a particularly accurate characterization will be obtained. This is not of paramount importance for the present discussion, however, which is concerned primarily with the comparison of FSGO and extended-basis results.

In contrast to the extended-basis results, transition energies from the FSGO-based studies are uniformly higher than the experimental values, Fig. 2, and a linear relation similar to Eq. (8) may be obtained:

$$\Delta E^{est} \approx \Delta E^{exptl} = 0.6875\ \Delta E^{calc} + 0.1504\ (eV) \tag{9}$$

where ΔE^{exptl} and ΔE^{calc} are the experimental and calculated transition energies, respectively. This equation may be used to obtain estimates of transition energies, ΔE^{est}, for the numerous other optically-forbidden excited states of pyrazine [14]. Moreover, it is also expected that the equation may be applied in estimating transition energies calculated for related molecules, e.g., substituted pyrazines or other diazines. The fact that calculated transition energies are approximately linearly related to experimental values appears to be characteristic of the FSGO-based approach, as will be demonstrated in the forthcoming examples.

The remaining data in Table 1 show that both the intensities and compositions of excited states obtained from the FSGO-based studies mimic those of the extended-basis calculations. In particular, calculated oscillator strengths are sufficiently accurate to permit a qualitative identification and differentiation of weak and intense transitions, which may be of considerable aid in resolving the spectrum of a large molecule.

In practice, FSGO-based methods would not be applied to "small" molecules such as pyrazine, both because of the possible existence of diffuse Rydberg states in important

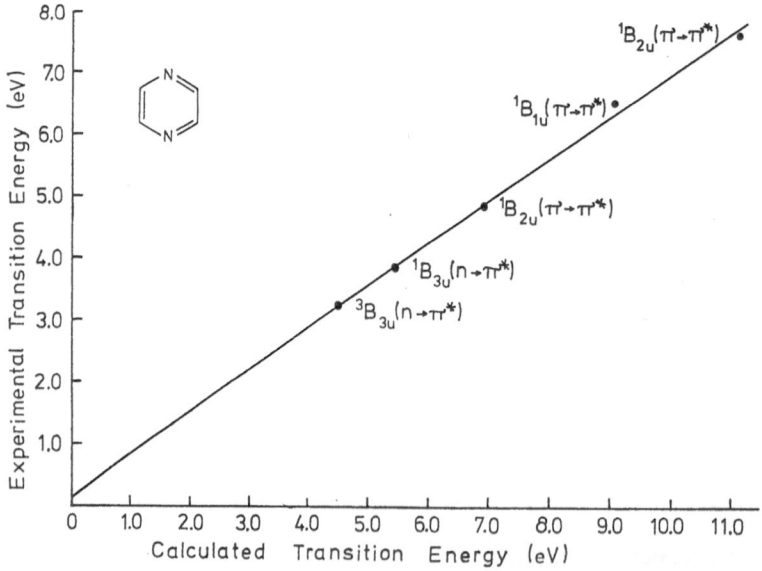

Fig. 2. Calculated versus experimental transition energies for selected excited states of pyrazine (Reprinted with permission)

regions of the spectrum, and because more accurate basis sets may be used to study such molecules. However, the present comparison demonstrates the type and quality of information which may be expected from applications of the present methods to larger molecules.

An alternative to the present methods of investigating the electronic spectra of large molecules is provided by various semi-empirical SCF-MO-CI methodologies which have been developed for excited states [18, 19]. A review of semi-empirical methods for large molecules has been given previously [20], and will not be repeated here. The semi-empirical and present ab initio techniques provide independent methods for the study of large molecules, and it is anticipated that both approaches will continue to provide useful insight into the spectra and properties of the largest systems to which quantum mechanical methods are currently being applied.

4.2 Computational Considerations

To provide an indication of the speed with which FSGO-based calculations may be carried out, examples of execution times for evaluation of the 2-electron integrals

$$\left(ij \left| \frac{1}{r_{12}} \right| kl \right) = \int \int g_i(1)\, g_j(1) \left(\frac{1}{r_{12}} \right) g_k(2)\, g_l(2)\, dV_1\, dV_2 \qquad (10)$$

are presented in Table 2. In the table, integral evaluation data for the adenine molecule obtained from calculations on an IBM 3081 computer is given for both the present FSGO and a representative CGTO minimal atomic basis set. Clearly, integral evaluation presents an insignificant computational problem for this molecule with the FSGO basis, while the equivalent calculation with the minimal atomic basis is significantly longer though not unreasonable. However, application of even minimal CGTO atomic basis sets to larger molecules such as those considered below can rapidly lead to excessive demands on computer resources.

An example more representative of the type of molecule to which the FSGO basis is generally applied is ethyl chlorophyllide a (Et-Chl-a) [21], for which the FSGO basis set contains 239 basis functions (see Fig. 3 for molecular structure). With our current integral codes, thirty-four minutes of computer time are needed for integral

Table 2. Two-electron integral evaluation data

Molecule	Adenine	Adenine	Et-Chl-a
Basis Set	FSGO	CGTO[a]	FSGO
Basis Functions	50	55	239
Integrals[b]	813,450	1,186,570	411,285,540
Calculated Integrals[c]	284,439	559,483	20,225,509
CPU Time[d] (s)	13	911	2044

[a] Minimal atomic basis. C, N, O: (6s, 3p) primitive gaussians contracted to (2s, 1p) basis functions; H: (3s) contracted to (1s).
[b] Maximum number possible for basis set.
[c] Number of integrals with magnitude $> 10^{-9}$, calculated and stored.
[d] IBM 3081 computer

evaluation, a quite reasonable value in view of the possibility of having to evaluate up to 411 million integrals. In this regard, Table 2 shows that only about five percent of the the maximum number of possible integrals were of a magnitude greater than 10^{-9}. Hence, efficient algorithms for detecting and avoiding the evaluation of small integrals are particularly essential with the FSGO basis if fast integral evaluation is to be realized.

The only other steps in the present SCF-MO-CI procedure which are potentially time consuming are the Roothaan/Hall SCF iterative procedure [22] and the 4-index transformation of 2-electron integrals from the FSGO to an MO basis [23]. For Et-Chl-a, each of these steps generally requires approximately the same ammount of time as does 2-electron integral evaluation. The number of SCF iterations required for convergence may be substantially reduced by implementation of the procedure for generating an initial guess to the density matrix developed for FSGOs by Shipman and Christoffersen [24]. Once this is employed, it is generally found that FSGO-based SCF calculations converge much faster than do those which utilize extended-atomic basis sets.

The time required for the 4-index transformation depends on the number of 2-electron FSGO integrals, and on the number of molecular orbitals M utilized in the CI excitation process. Generally, M has been chosen to be between 20 to 50 molecular orbitals in most previous studies. It should be noted that the time for a 4-index transformation may be reduced by at least an order of magnitude using a vector processor, relative to the time required for scalar execution with the same machine.

5 Applications to Selected Biomolecular Systems

The present FSGO-based SCF-MO-CI methods have been applied in calculations of the electronic spectra as well as ground and excited state properties of a variety of molecules, including carbazoles [25], anthraquinones [26], porphyrins [13, 27–29], chlorins [30], and various neutral, anionic, and cationic chlorophyllides [31–34] and bacteriochlorophyllides [34–37]. In the following paragraphs we present highlights of a number of calculations on selected porphyrins, chlorins, chlorophyllides, and bacteriochlorophyllides. The results presented below focus primarily on the calculation and resolution of absorption spectra. Additional material, including tabulated values of calculated transition energies, oscillator strengths, CI compositions of excited states, charge and spin densities, as well as calculated data on triplet states, may be found in the original publications.

5.1 Molecular Structures

The molecular structure of free-base porphine, a prototypical porphyrin, is shown in Fig. 3. The molecule is found in x-ray studies [38] to be nearly planar with D_{2h} symmetry; therefore, in most theoretical studies a planar geometry with the NH hydrogen atoms positioned as shown in assumed. A variety of metalloporphyrins may be formed by replacing the two NH hydrogen atoms by a central metal atom, which is 4-coordinated by the nitrogen atoms and lies slightly above the porphyrin plane. We shall confine our attention to the magnesium analog.

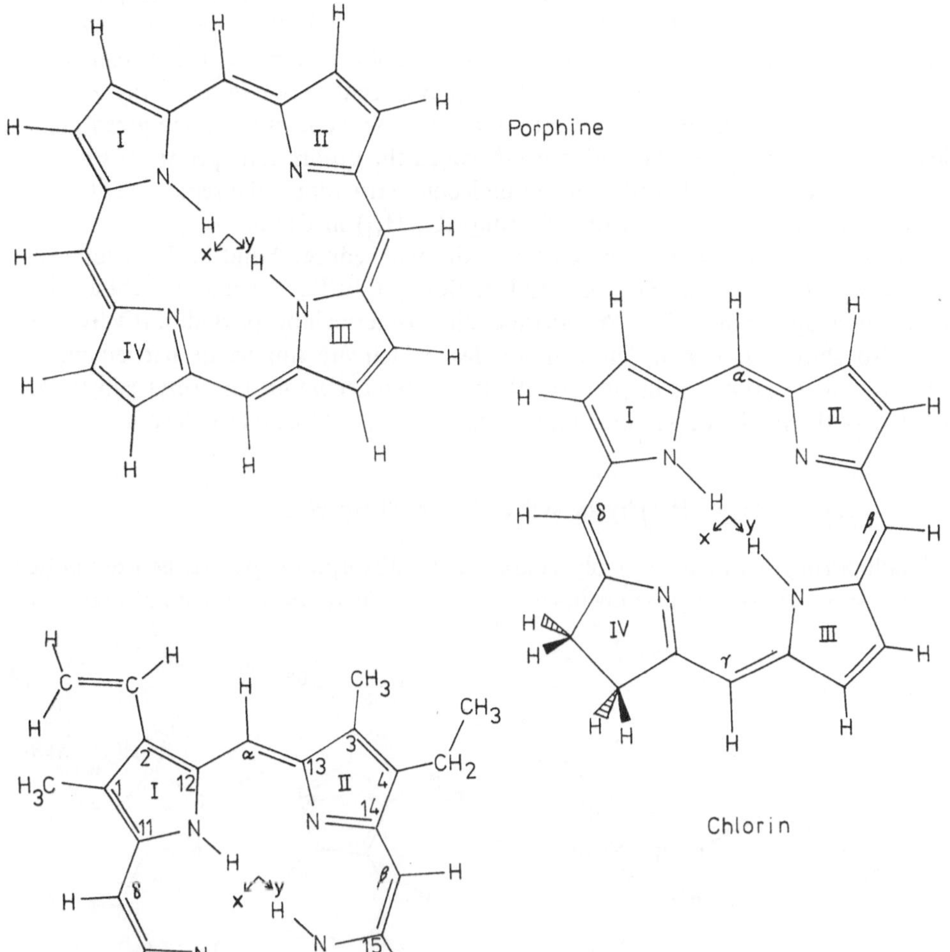

Porphine

Chlorin

Ethyl Pheophorbide *a*

Fig. 3. Molecular structures of free-base porphine, free-base chlorin, and Et-Pheo-a. Structures of the respective magnesium analogs, magnesium porphine, magnesium chlorin, and Et-Chl-a, are obtained by replacement of the NH hydrogen atoms by a magnesium atom at the center of the macrocycle (Reprinted with permission)

Magnesium and free-base chlorin, shown in Fig. 3, are derivatives of the respective porphyrins obtained by reduction of the C_7—C_8 bond. The basic chlorin macrocyclic structure is also found in ethyl pheophorbide a (Et-Pheo-a) and its magnesium analog Et-Chl-a. As shown in Fig. 3, however, the latter two molecules have much more complex molecular structures than do the chlorins, the most significant differences being the presence of isocyclic ring V and two ester side chains. Despite these differences, x-ray crystallographic studies [39, 40] of both molecules show that the macrocyclic rings remain essentially planar. Et-Chl-a is important due to its close relationship to chlorophyll a (Chl-a), which plays a central role in the primary events of light harvesting and charge separation in photosynthetic processes in green plants [41]. There is no discernable difference between the absorption spectra of Et-Chl-a and Chl-a, while the only difference in molecular structure is the replacement of the ethyl group in Et-Chl-a by a phytyl group ($C_{20}H_{39}$) in Chl-a.

In bacteriochlorin, Fig. 4, there is an additional reduced bond in the macrocycle at the C_3—C_4 position. The bacterial analogs of Et-Pheo-a and Et-Chl-a, ethyl bacteriopheophorbide a (Et-BPheo-a) and ethyl bacteriochlorophyllide a (Et-BChl-a), are also shown in Fig. 4. These molecules, which are important participants in photosynthetic processes in photosynthetic bacteria, contain an acetyl substituent in place of the vinyl group found at the 2 position in Et-Chl-a and Et-Pheo-a.

5.2 Characteristics of Porphyrin and Chlorophyll Spectra

Metalloporphyrins possess highly characteristic absorption spectra, as exemplified by the spectrum of magnesium porphine, Fig. 5. There are two distinct regions of

Fig. 4. Molecular structures of bacteriochlorin (**left**) and Et-BPheo-a (**right**, $R = C_2H_5$). The structure of Et-BChl-a is identical to that of Et-BPheo-a except for replacement of the NH hydrogen atoms by a magnesium at the center of the macrocycle (Reprinted with permission)

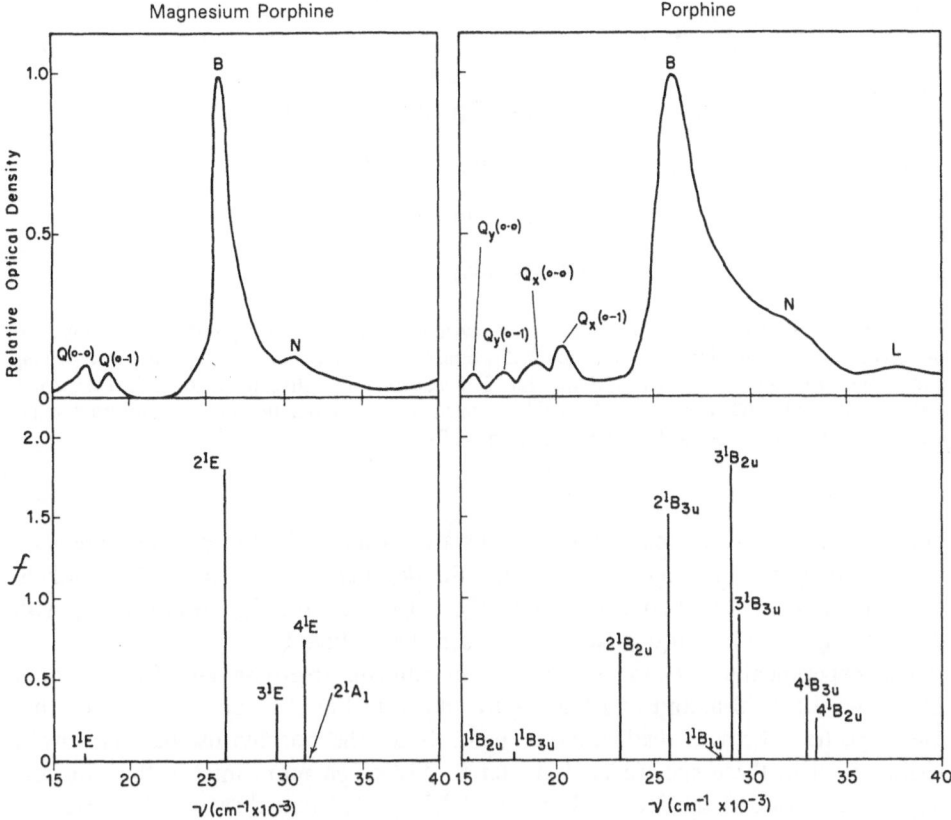

Fig. 5. Calculated and experimental absorption spectra of magnesium porphine (**left**) and free-base porphine (**right**). Locations of calculated excited singlet states are given in terms of estimated transition energies from Eq. (11), and f is the calculated oscillator strength. In free-base porphine, transitions to B_{2u} and B_{3u} states are polarized y and x, respectively, relative to the axes shown in Fig. 3 (Reprinted with permission)

absorption: one or more low-intensity bands in the visible region (15,000 to 20,000 cm^{-1}), and an intense band in the near-UV spectral region (24,000 to 30,000 cm^{-1}) called the Soret band. In the visible region, the absorption band labelled (0–1) in Fig. 5 is not due to an individual electronic transition, but rather is the envelope of a number of vibronic bands of the lowest electronic transition. The lowest vibronic band lies within the (0–0) band envelope, as confirmed by fluorescence spectroscopy [42, 43].

The overall spectral characteristics are often explained phenomenologically by the "4-orbital model" of porphyrin spectra developed by Gouterman [44], in which both the visible band (the Q band) and the Soret band (the B band) of a metallo-porphyrin arise from transitions to a pair of doubly-degenerate excited states. The general features of the 4-orbital model are illustrated in Fig. 6. In the language of CI, each excited state is composed of a pair of single excitations from the two highest-occupied molecular orbitals, 1 and 2, to the two lowest-unoccupied orbitals, 1* and 2*, which are degenerate in a square metalloporphyrin. The low intensity of the Q band is explained by a partial cancellation of the transition dipole contributions

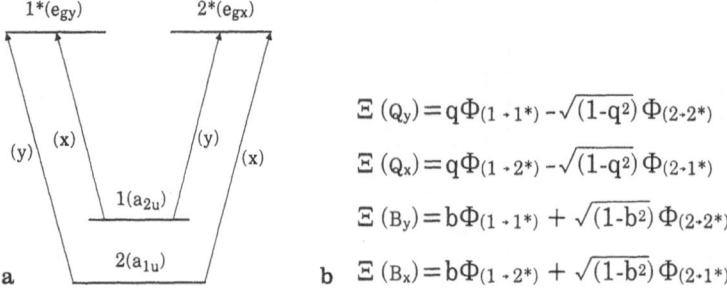

$$\Xi\,(Q_y) = q\Phi_{(1\to 1^*)} - \sqrt{(1-q^2)}\,\Phi_{(2\to 2^*)}$$

$$\Xi\,(Q_x) = q\Phi_{(1\to 2^*)} - \sqrt{(1-q^2)}\,\Phi_{(2\to 1^*)}$$

$$\Xi\,(B_y) = b\Phi_{(1\to 1^*)} + \sqrt{(1-b^2)}\,\Phi_{(2\to 2^*)}$$

$$\text{b}\quad \Xi\,(B_x) = b\Phi_{(1\to 2^*)} + \sqrt{(1-b^2)}\,\Phi_{(2\to 1^*)}$$

Fig. 6a, b. The "4-orbital model" of a metalloporphyrin with D_{4h} symmetry: (**a**) constituent molecular orbitals include the two highest-lying occupied orbitals (1 and 2) and two (degenerate) lowest-lying unoccupied orbitals (1* and 2*) of a ground state SCF wavefunction; (**b**) constituent excited electronic states of the visible (*Q*) and Soret (*B*) bands given as linear combinations of the indicated single excitations. Parameters q and b are CI expansion coefficients

of the two single excitations in the excited states, whereas for the B band there is an addition of terms. In free-base porphine, the degeneracy of 1* and 2* is broken, resulting in pairs of Q and B bands, (Q_x, Q_y) and (B_x, B_y), respectively, with distinct x and y polarizations (see the axis systems in Fig. 3).

The experimental spectra of magnesium chlorin, free-base chlorin, Et-Chl-a, Et-Pheo-a, Et-BChl-a, and Et-BPheo-a are shown in Fig. 7, 8, and 9. It is seen that these spectra possess a distinct Soret band as do the porphyrins, but the lowest visible band in these spectra is characteristically much more intense than in porphyrins. In terms of the 4-orbital model, it is expected that there will be increased splittings between both orbitals 1 and 2, and 1* and 2*. This in turn is expected to increase significantly the contribution of the $(1 \to 1^*)$ relative to the $(2 \to 2^*)$ configuration in the composition of the Q_y excited state (Fig. 6). Consequently, there will be less cancellation between the contributions of the two configurations to the transition dipole, and the resultant transition dipole will be larger than those in the porphyrins.

5.3 Calculated Spectra

5.3.1 Porphyrins

Calculated absorption spectra for magnesium and free-base porphine [13], obtained using the methods described above, are compared with experimental spectra in Fig. 5. The transition energies used are estimates, ΔE^{est}, obtained from a least-squares fit of calculated and gas-phase experimental transition energies for the two molecules [45]. The resultant linear relation

$$\Delta E^{est} = 0.61\,\Delta E^{calc} - 441.0\,(cm^{-1})\,, \tag{11}$$

has been used to estimate transition energies for other neutral porphyrins, chlorins, Chls, and related molecules, including those considered below.

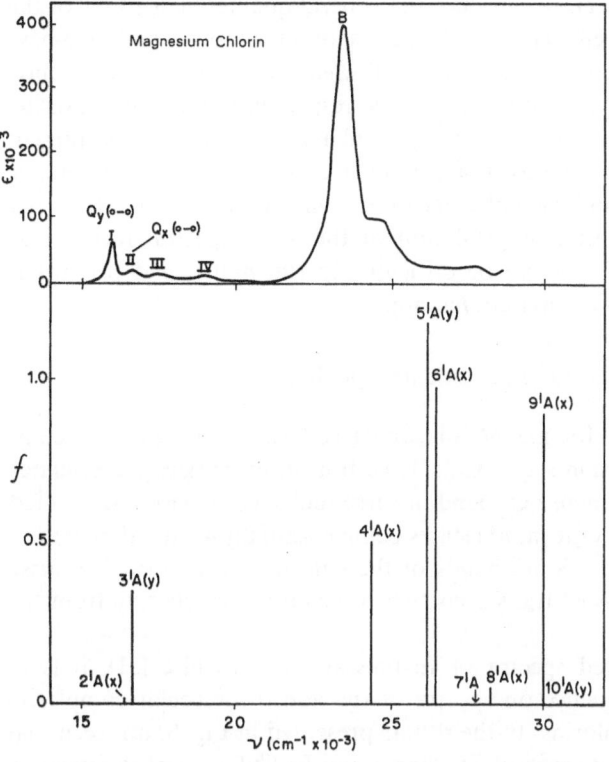

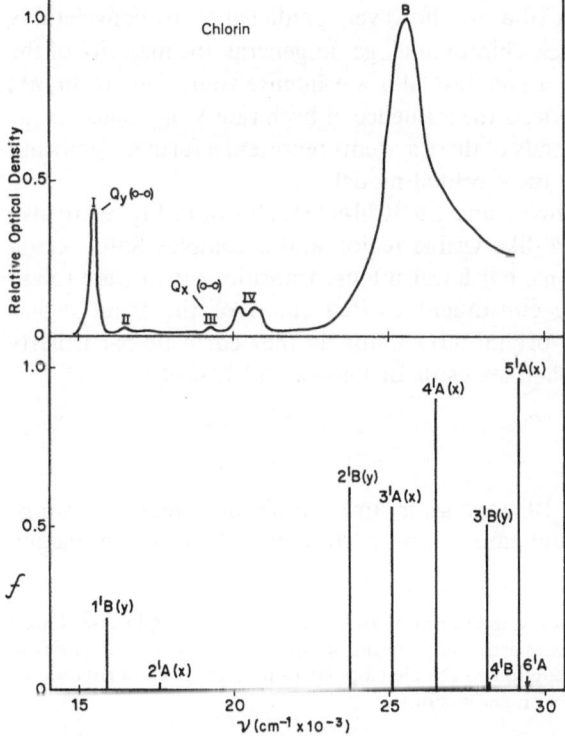

Fig. 7. Calculated and experimental absorption spectra of magnesium chlorin (**upper**) and free-base chlorin (**lower**). Locations of calculated excited singlet states are given in terms of estimated transition energies from Eq. (11), f is the calculated oscillator strength, and calculated polarizations are given in parentheses, relative to the axes in Fig. 3 (Reprinted with permission)

Generally, the calculated spectra reproduce the general qualitative features of the experimental spectra quite well. The visible bands of magnesium and free-base porphine are predicted to arise from weakly-allowed transitions to a doubly-degenerate (π, π^*) state and a pair of (π, π^*) states, respectively, a picture basically similar to the 4-orbital model. In the Soret spectral region, however, the present studies reveal that this band is composed of a number of intense $\pi \rightarrow \pi^*$ transitions, a result which represents a fundamental departure from the basic 4-orbital model [13]. The results further suggest that resolution of the Soret spectral region into individual electronic transitions would be difficult experimentally, and perhaps is best achieved currently with a computational approach.

5.3.2 Chlorins, Chlorophyllides, and Bacteriochlorophyllides

Calculated absorption spectra for magnesium and free-base chlorin [30] are compared with experimental spectra in Fig. 7. Calculated transition energies are estimates from Eq. (11). The relatively intense Q_y band in these molecules is nicely accounted for by the calculations, the relevant (π, π^*) states being essentially 4-orbital in nature. As in the porphyrin studies, the Soret bands of the chlorins are predicted to arise from a number of intense $\pi \rightarrow \pi^*$ transitions, which again differs sharply from the 4-orbital model.

In considering the calculated spectra of Et-Pheo-a and Et-Chl-a [31], it is of interest to investigate the influence on spectra of the added molecular complexity of these systems, relative to chlorins. In the results presented in Fig. 8, it is seen that the composition of the visible spectra of Et-Pheo-a and Et-Chl-a is not significantly different from those of chlorin and magnesium chlorin, respectively. The Soret bands of both Et-Pheo-a and Et-Chl-a are, however, predicted to be considerably more complex than are those of their chlorin analogs. In general, the majority of the Soret band transitions in Et-Pheo-a and Et-Chl-a are intense transitions to (π, π^*) states, the composition of which reflects the influence of both ring V and macrocyclic substituents [31]. Thus, the Soret bands of these systems represent a further significant deviation from the simple ideas of the 4-orbital model.

The calculated spectra of Et-BPheo-a and Et-BChl-a [35] shown in Fig. 9 are also characterized by a simple, 4-orbital-like visible region and a complex Soret region with a number of $\pi \rightarrow \pi^*$ transitions, but fewer intense transitions than were found for Et-Pheo-a and Et-Chl-a. The constituent excited states of the Soret region generally do not conform to the 4-orbital model, nor do they correlate particularly closely with the respective Soret band states of Et-Pheo-a and Et-Chl-a.

5.3.3 Ethyl Chlorophyllide a Enol

The enol form of Et-Chl-a, Fig. 10, represents an example of a molecule whose seemingly conflicting absorption and emission properties may be easily and unequi-

---

Fig. 8. Calculated and experimental absorption spectra of Et-Chl-a (**upper**) and Et-Pheo-a (**lower**). Locations of calculated excited singlet states are given in terms of estimated transition energies from Eq. (11), f is the calculated oscillator strength, and calculated polarizations are given in parentheses, relative to the axes in Fig. 3 (Reprinted with permission)

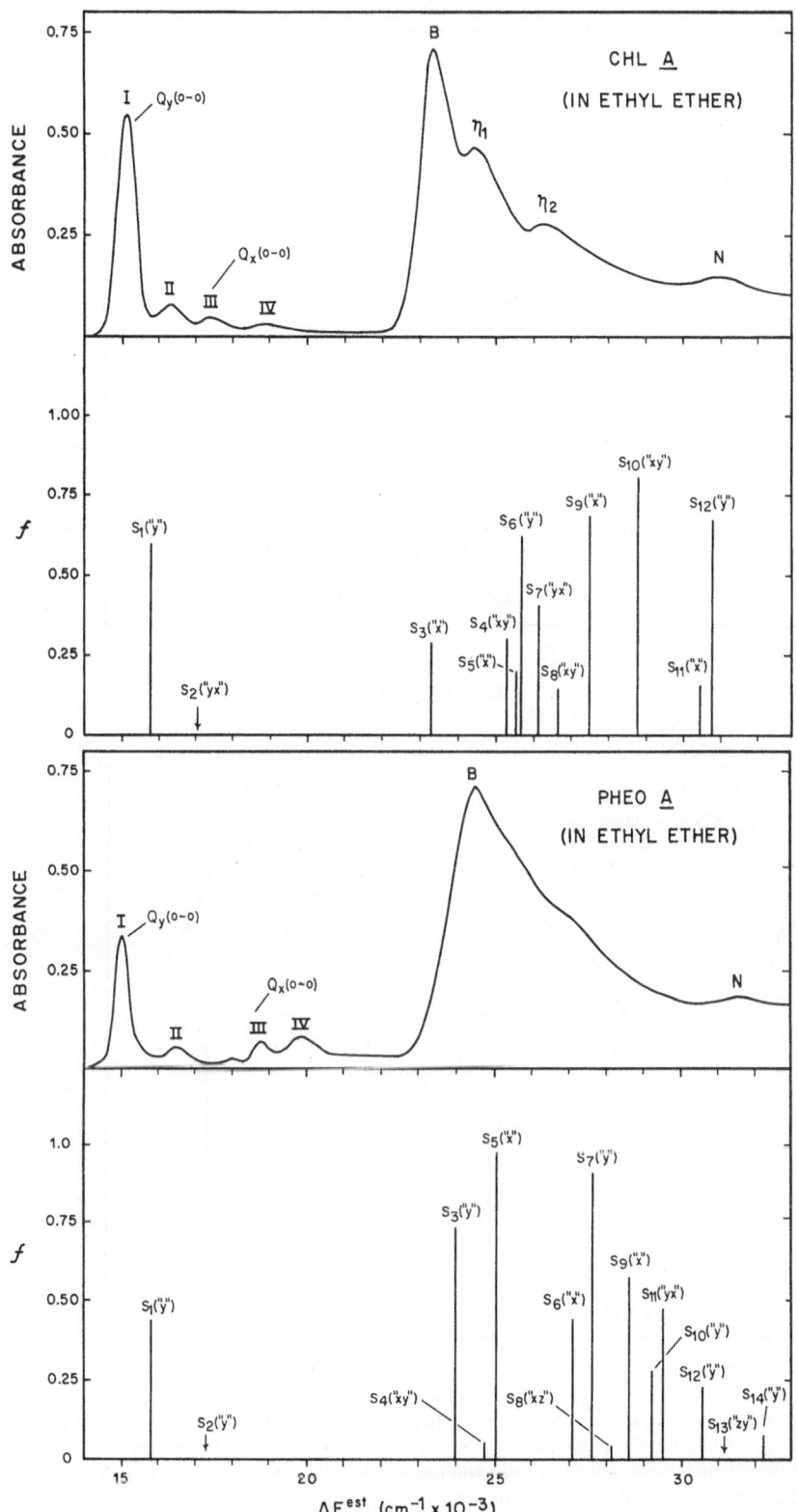

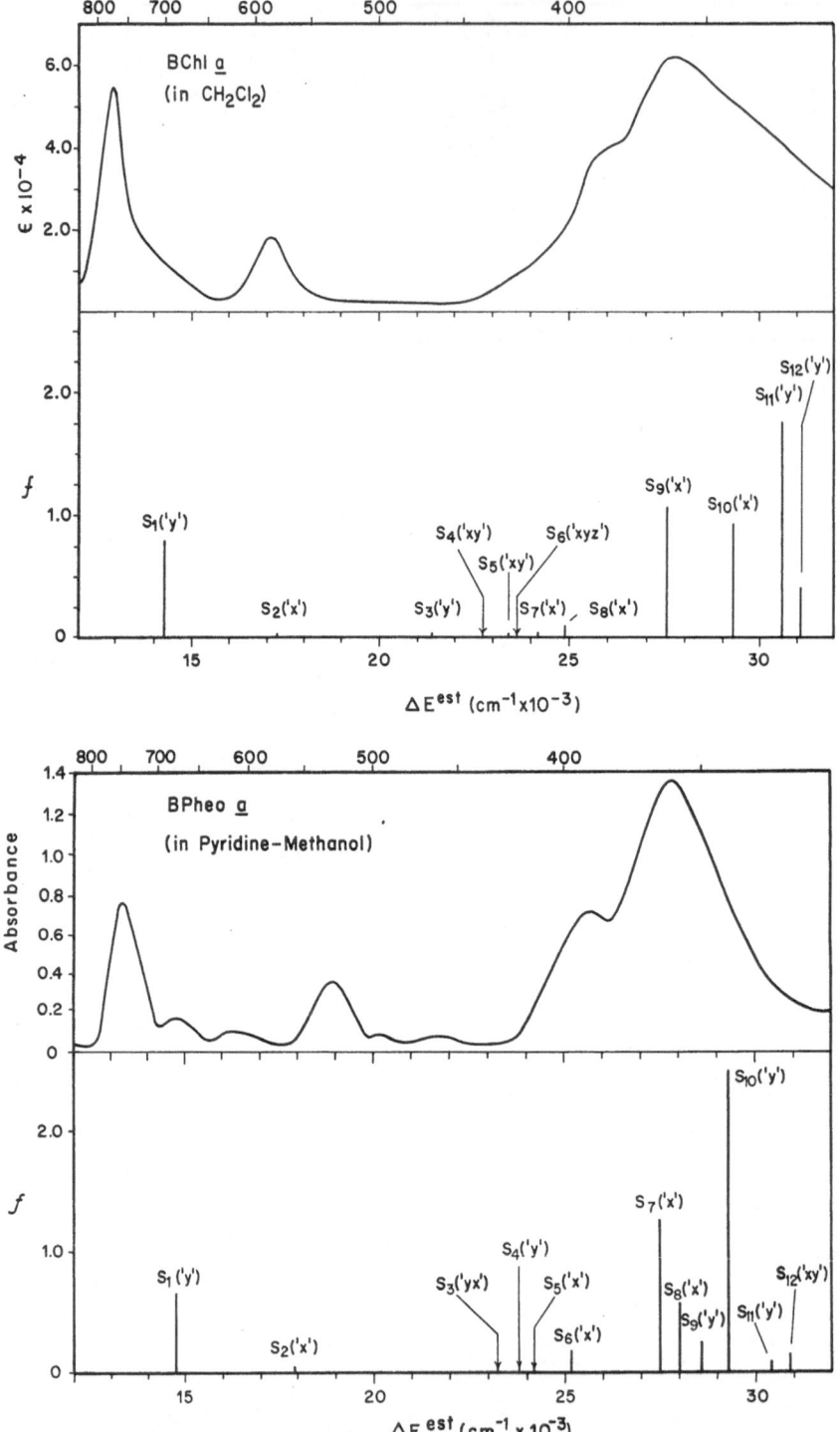

Fig. 9. Calculated and experimental absorption spectra of Et-BChl-a (**upper**) and Et-BPheo-a (**lower**). Locations of calculated excited singlet states are given in terms of estimated transition energies from Eq. (11), f is the calculated oscillator strength, and calculated polarizations are given in parentheses, relative to the axes in Fig. 3 (Reprinted with permission)

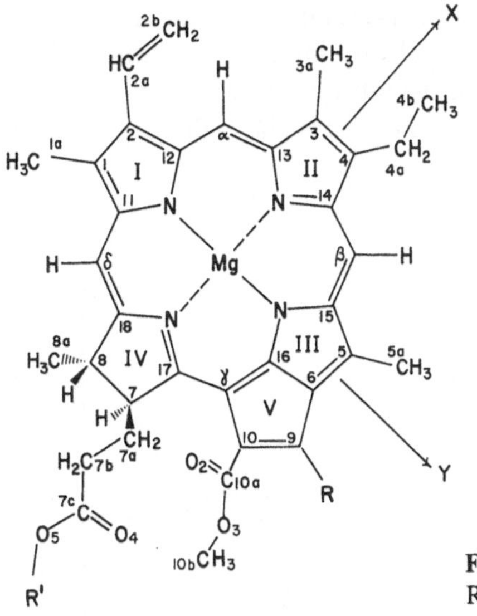

Fig. 10. Molecular structure of Et-Chl-a enol: R=OH; R′=C$_2$H$_5$ (Reprinted with permission)

vocally explained by application of the present methods. In the spectrum of Et-Chl-a enol, Fig. 11, the Soret region of the spectrum possesses a characteristic splitting which differs in appearance from that of Et-Chl-a, while in the visible region there is strong absorption analogous to the Q$_y$ band in the keto form of Et-Chl-a (Fig. 8). However, the enol form is found to have a marked absence of fluoressence, unlike the characteristically strong fluorescence from the lowest excited singlet state of the keto form and most other chlorophylls.

The reasons for the apparent inconsistencies in visible absorption and emission properties of Et-Chl-a enol are revealed by the computational results [32], Fig. 11. The visible band at approximately 15,200 cm^{-1} is not due to a transition to the lowest excited singlet state, S$_1$, but rather to the second-lowest excited singlet state, S$_2$. Excitation to S$_1$ occurs with a transition energy of approximately 9,900 cm^{-1} and an oscillator strength, f, of only 0.04. Both absorption and fluorescence emission involving S$_1$ are essentially forbidden processes, and are not observed experimentally. Thus, although both the keto and enol forms of Et-Chl-a appear to have identical visible absorption spectra, the calculations clearly show that the excited states responsible for the Q$_y$ bands in the two forms are unrelated, and that there is an unexpected dipole-forbidden low-lying excited single state, S$_1$, in the enol form.

5.3.4 Et-Chl-a and Et-BChl-a Cation Radicals

Stable chlorophyll π-cation radicals may be generated by either chemical or electrolytic one-electron oxidation of the neutral species, and their properties have been extensively studied [46]. These cation radicals are of particular significance, as there is evidence that radicals of Chl or BChl dimers are formed during the primary charge separation processes in green plant and bacterial photosynthesis, respectively [47, 48].

The experimental and calculated spectra of the π-cation radicals of Et-Chl-a and Et-BChl-a are shown in Fig. 12. The observed spectra show a qualitative resemblence to those of the corresponding neutral species in that there is an intense Soret band, while on the other hand the intense Q_y absorption in the neutral species is replaced by weak absorption which extends into the infrared region.

Perhaps the most striking feature of the calculated spectra [34] is the large number

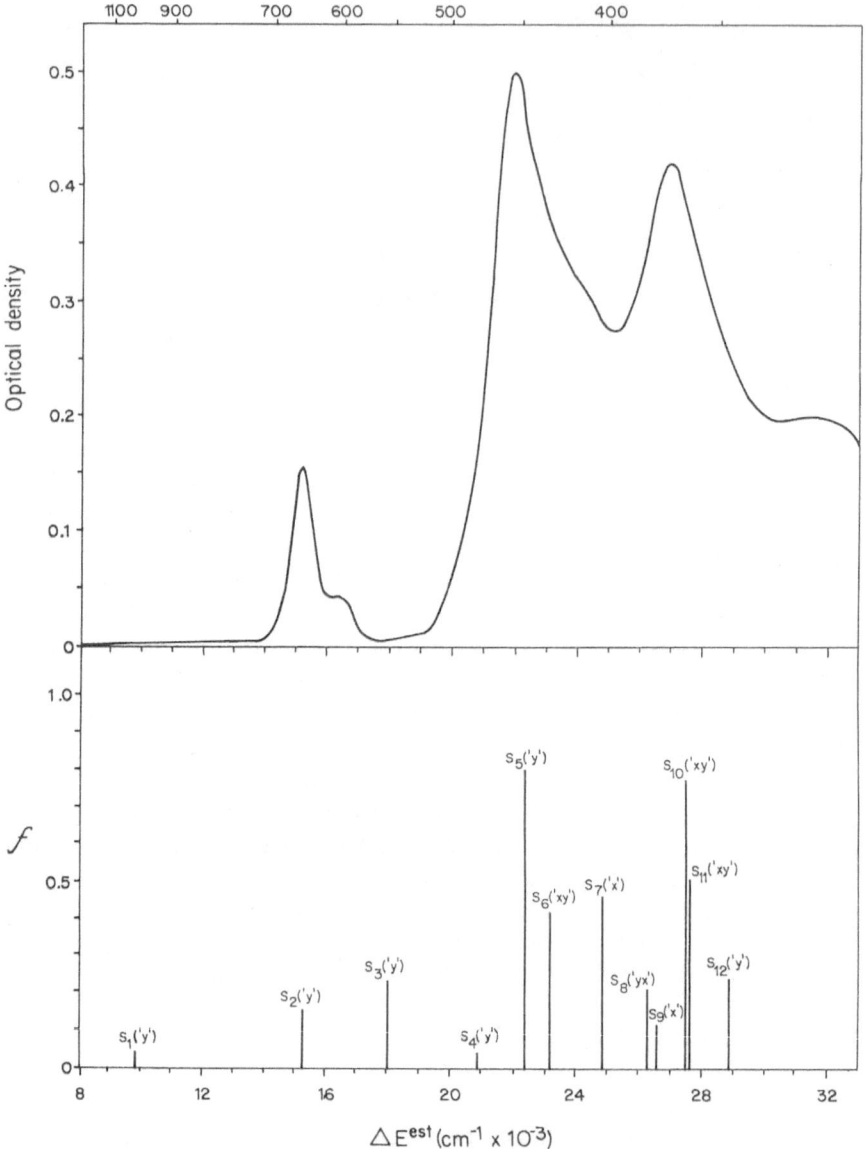

Fig. 11. Calculated and experimental absorption spectra of Et-Chl-a enol. Locations of calculated excited singlet states are given in terms of estimated transition energies from Eq. (11), f is the calculated oscillator strength, and calculated polarizations are given in parentheses, relative to the axes in Fig. 10 (Reprinted with permission)

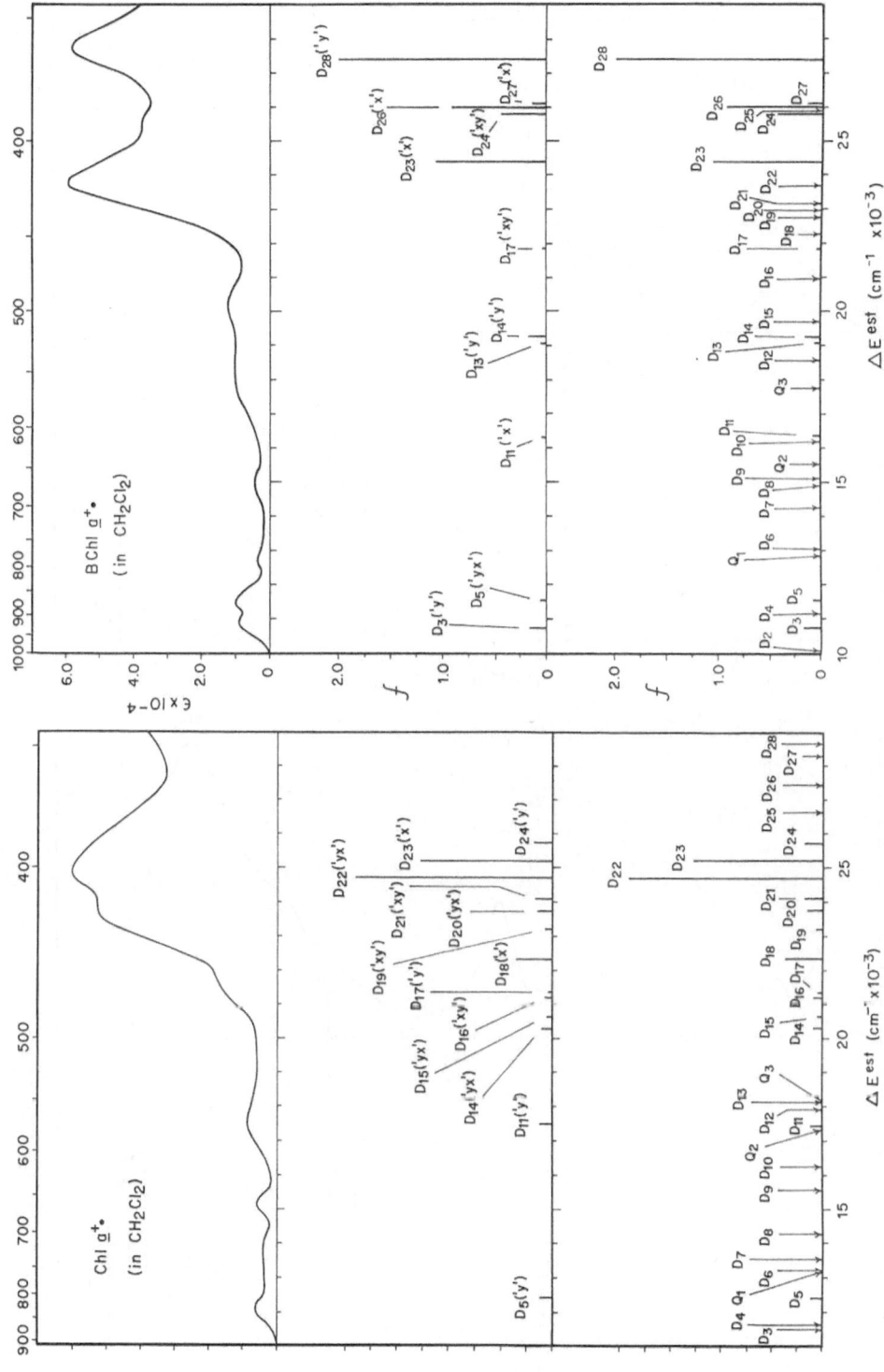

Fig. 12. Calculated and experimental absorption spectra of the cation radicals of Et-Chl-a (**left**) and Et-BChl-a (**right**). Locations of calculated excited doublet states are estimates from the equation $\Delta E^{est} = 0.532 \, \Delta E^{calc} + 762.6 \, cm^{-1}$ (see Ref. 34), and f represents the calculated oscillator strength. The bottom portion of each plot shows all computed doublet states with estimated energies above $11{,}000 \, cm^{-1}$, while the middle portion shows states with $f > 0.025$, with calculated polarizations relative to the axes in Figs. 3 and 4 given in parentheses (Reprinted with permission)

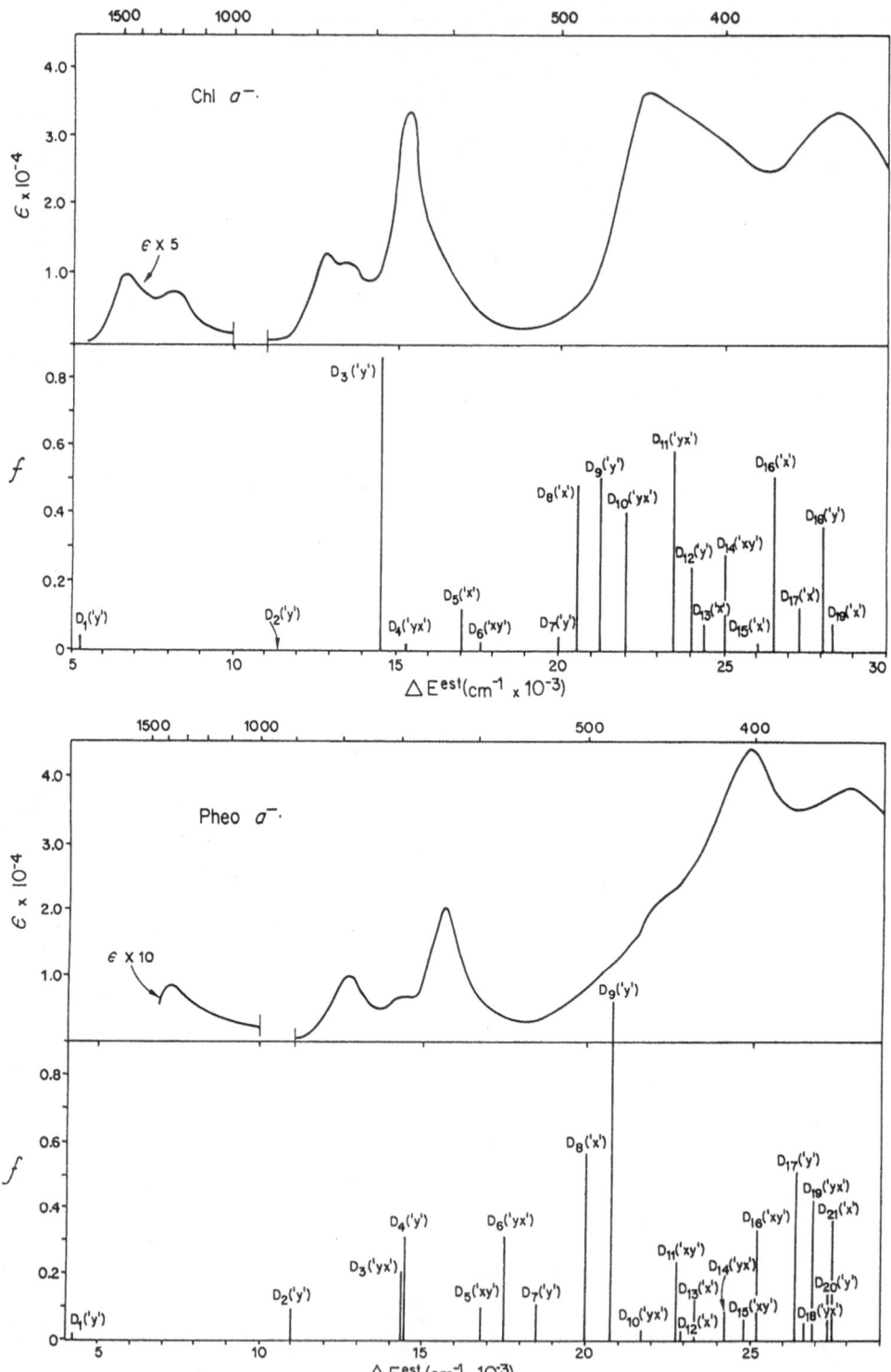

Fig. 13. Calculated and experimental absorption spectra of the anion radicals of Et-Pheo-a, Et-Chl-a, Et-BPheo-a, and Et-BChl-a (see figure labels). Locations of calculated excited doublet states ▶

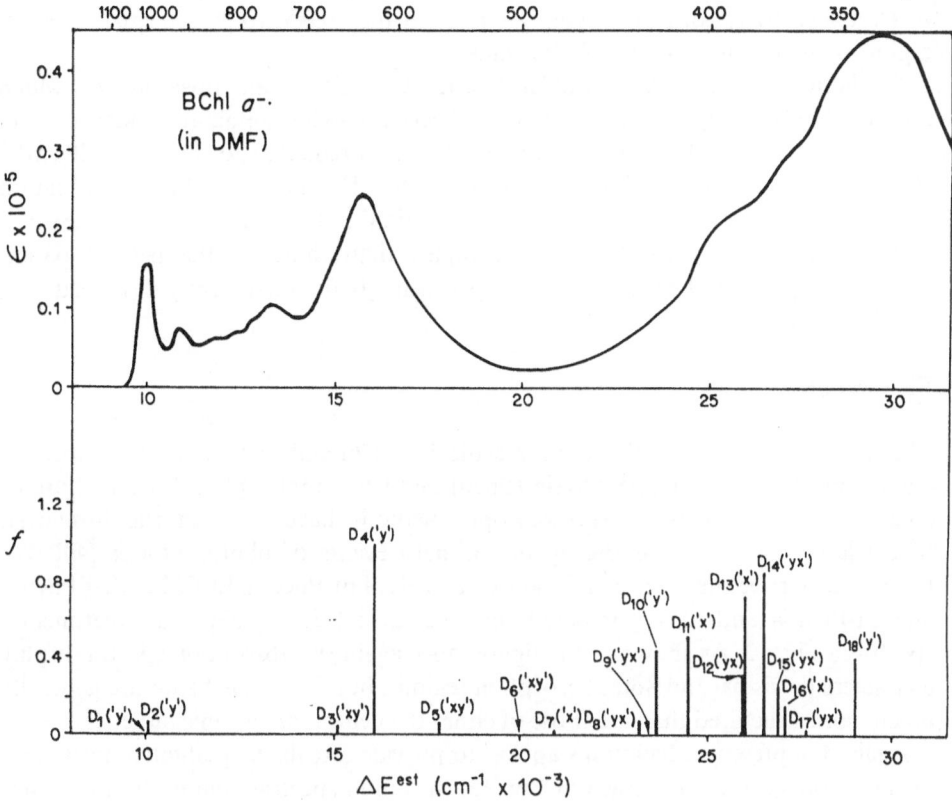

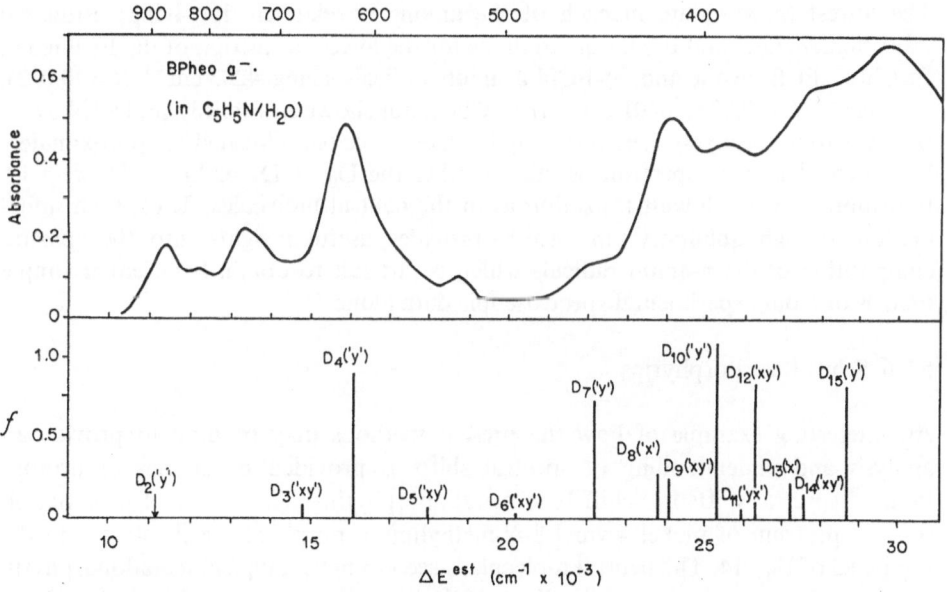

are estimates from the equation $\Delta E^{est} = 0.532 \Delta E^{calc} + 762.6\ cm^{-1}$ (see Ref. [34]), f represents the calculated oscillator strength, and calculated polarizations relative to the axes in Fig. 3 and 4 given in parentheses (Reprinted with permission)

of electronic transitions of varying intensity in the 10,000 to 30,000 cm^{-1} spectral region, many of which are dipole-forbidden.

Furthermore, the lowest doublet-doublet transitions, the positions of which are not shown in Fig. 12, are examples of very low-lying dipole-forbidden $\pi \rightarrow \pi^*$ transitions. Specifically, the lowest bands are predicted to be at 5220 cm^{-1} ($f = 0.001$) and 7112 cm^{-1} ($f = 0.0007$) for the Et-Chl-a and Et-BChl-a cation radicals, respectively. Thus, the present π-cation radicals appear to possess absorption spectra considerably more complex than those of the neutral parent molecules; spectra not easily resolved experimentally or conveniently described by a simple model.

5.3.5 Anion Radicals

The spectral properties of the π-anion radicals of Chl and BChl and their respective magnesium-free analogs pheophytin (Pheo) and bacteriopheophytin (BPheo) are of considerable interest, since spectroscopic methods have detected the formation of the latter three radicals during the primary events of photosynthesis [49]. We have characterized the spectra of the anion radicals of Et-Pheo-a, Et-Chl-a, Et-BPheo-a and Et-BChl-a and have compared the calculated [33, 36, 37] and experimental spectra in Fig. 13. As shown in the figure, once again the absorption spectra display characteristic visible and Soret absorption regions, but the visible bands are generally much more structured than are those of either the neutral or cationic moieties.

Again, the present calculations appear to provide a realistic qualitative resolution of the anion spectra. As shown in Fig. 13, the anion spectral composition does not resemble that of either the neutral or cationic species; in particular, the large number of forbidden "background" transitions found in the cation radical spectra are absent. The lowest (π, π^*) state in each of the anions is relatively low-lying; estimated transition energies and oscillator strengths for the lowest transition of the Et-Pheo-a, Et-Chl-a, Et-BPheo-a and Et-BChl-a anion radicals being 4203 cm^{-1} ($f = 0.022$), 5238 cm^{-1} ($f = 0.035$), 8601 cm^{-1} ($f = 0.002$, not shown in Fig. 13), and 9924 cm^{-1} ($f = 0.030$), respectively. The relatively intense visible band located at approximately 16,000 cm^{-1} in each spectrum is due to either the $D_0 \rightarrow D_3$ or $D_0 \rightarrow D_4 \; \pi \rightarrow \pi^*$ transitions, not the lowest transition as in the neutral molecules. Thus, such information, though qualitative in nature, provides useful insights into the spectral composition of the π-anion radicals which is difficult to obtain from either simple models or from experimental spectroscopic data alone.

5.3.6 Schiff Base Porphyrins

An interesting example of how the present methods may be used to provide an analysis and understanding of spectral shifts is provided by studies of proton-induced spectral shifts in Schiff base metalloporphyrins. The effects of protonation on the spectrum of nickel 4-vinyl-8-N-methylimino porphine are illustrated in the top panel of Fig. 14. The neutral molecule is seen to have a typical metalloporphyrin spectrum, but protonation of the N-methylimino substituent results in both a large (1200 cm^{-1}) red shift of the Q_y band and a large change in the appearance and decrease in intensity of the Soret band. To investigate these changes, calculations

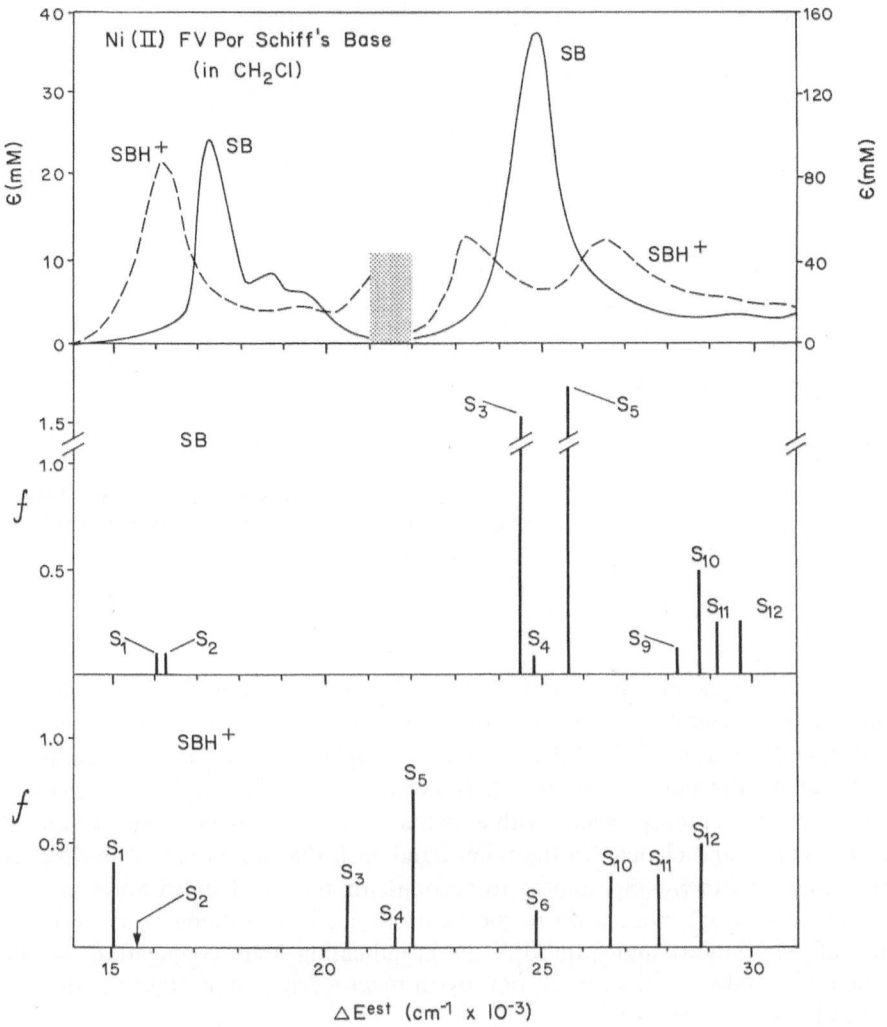

Fig. 14. The experimental spectra of the neutral (SB) and protonated (SBH$^+$) nickel 4-vinyl-8-N-methylimino porphine, compared with calculated spectra of their magnesium analogs (see Fig. 15). Locations of calculated excited singlet states are estimates from Eq. (11), and f is the calculated oscillator strength (Reprinted with permission)

[27] were performed on the analogous magnesium analog, MgPSB, Fig. 15. In these calculations, two orientations of the N-methylimino group were considered: a thirty degree rotation of the substituent from the macrocyclic plane about the C_3—C_{8a} axis, MgPSB(30), and a ninety degree orientation, MgPSB(90), to eliminate macrocycle-substituent conjugation. The vinyl substituent was rotated thirty degrees out of the macrocyclic plane in each case.

The calculated spectra of MgPSB(30) and the protonated analog, MgPSBH$^+$(30), are compared with the experimental spectra in Fig. 14. The spectrum of MgPSB(30) is only slightly perturbed from that of magnesium porphine: the visible band is due

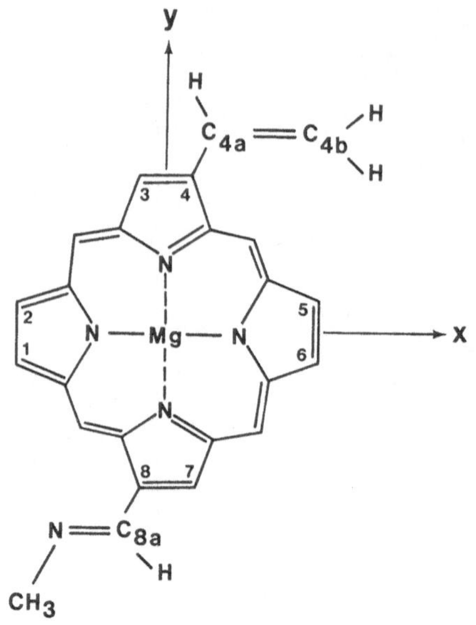

Fig. 15. Molecular structure of magnesium 4-vinyl-8-N-methylimino porphine (Reprinted with permission)

to two nearly degenerate transitions to states S_1 and S_2, while the Soret band is dominated by a near-degenerate pair of intense transitions to states S_3 and S_5. The calculated spectrum of MgPSB(90) is qualitatively the same as that of MgPSB(30). The calculated spectrum of MgPSBH$^+$(30), however, generally displays the changes seen in the experimental spectrum, with a distinct red-shift of the Q_y band estimated to be 1000 cm^{-1}, and changes in the Soret band such that a number of moderately intense transitions to S_3–S_{12} appear to account for the two-banded Soret profile. Significantly, although not shown in the figure, there is no difference between the spectra of MgPSB(90) and MgPSBH$^+$(90), indicating that conjugation of the N-methylimino substituent with the porphyrin macrocycle is important in effecting proton-induced spectral shifts.

By examining the molecular orbital correlation diagrams shown in Fig. 16, an interesting explanation of the spectral shift in MgPSB(30) may be obtained. In considering the effect of orbital perturbations on Q_y spectral shifts, it is useful to remember that only perturbations of orbitals 1, 2, 1*, and 2* (orbitals of the 4-orbital model, Fig. 6) are important in influencing the Q_y band. In the figure, orbital energies from FSGO-based SCF calculations on magnesium porphine, the indicated substituents, and the Schiff base porphyrins are shown, and orbital interactions between substituent and macrocyclic π and π^* orbitals are indicated. It is seen in both non-protonated molecules that only a weak perturbation of the highest occupied orbital of magnesium porphine primarily by the π orbital of the vinyl substituent occurs, and thus the spectra of the non-protonated Schiff bases would be expected to resemble those of unsubstituted magnesium porphine closely, as is observed. In sharp contrast, the correlation diagram for MgPSBH$^+$(30) shows a large perturbation of orbital 1* by the π^* orbital of the N-methylimino substituent, effectively lowering the orbital energy of 1* and initiating the observed red shift. In the non-protonated system, the imino π^* orbital is seen to interact with much higher-lying virtual orbitals, namely 4*–7*. In the protonated system the energy of the imino π^* orbital is differentially lowered relative to the macrocyclic π^* orbitals due to the protonic charge which influences the imino group more strongly than the macrocycle as a whole. The lowering of the imino π^* orbital is seen for both MgPSBH$^+$(30) and MgPSPH$^+$ (90). However, only in the former case are the substituent and macrocyclic orbitals able to interact through conjugation, thus resulting in a significant lowering of the 1* orbital of the protonated Schiff base, which in turn results in a red shift of the Q_y band.

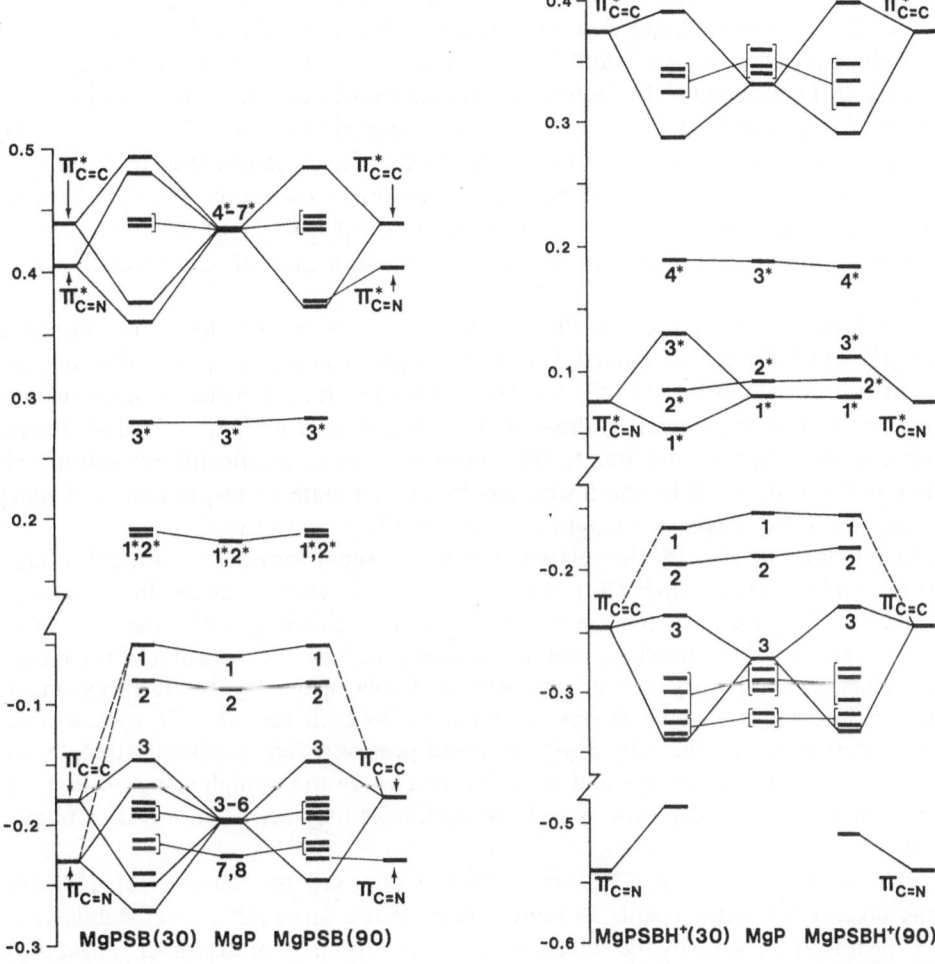

Fig. 16. Molecular orbital energy level diagrams showing orbitals of MgPSB (**left**) and MgPSBH$^+$ (**right**) and their correlations with magnesium porphine and substituent orbitals. Porphyrin occupied orbitals are numbered in the order of decreasing orbital energy, beginning with the highest occupied orbital, *1*; unoccupied orbitals are numbered in the order of increasing energy, *1** being the lowest unoccupied orbital. *Solid* and *dotted* lines connecting orbitals indicate strong and moderate orbital correlations, repectively (Reprinted with permission)

5.3.7 Charge-Transfer States in Magnesium Porphine-Porphine Dimers

In the light-driven primary charge separation processes of photosynthesis, the initial step involves irreversible electron transfer from the primary electron donor to an acceptor within 10 picoseconds [50]. In bacterial photosynthetic systems, it has been established that a BChl "special pair" acts as the primary electron donor, and that BChl and BPheo monomers act as acceptors [51]. Moreover, electron transfer in green plant photosynthetic systems is in generally known to involve Chl and Pheo moieties [41].

An important feature of photosynthetic electron transfer is the high quantum efficiency of the process. Thus, there have been a number of attempts to create chlorophyll- or porphyrin-based biomimetic systems which might retain this efficiency in a model system. A series of cofacial, double covalently-linked magnesium porphine-porphine dimers (MgP-P) synthesized by Chang [52] and studied spectroscopically by Netzel et al. [53] represent a simple model with a reasonably well-defined molecular geometry. Using picosecond spectroscopic methods, Netzel et al. [53] have reported the formation of an intramolecular singlet charge-transfer (CT) state of the type MgP^+-P^- within 6 picoseconds following excitation to a (π, π^*) state, in studies performed with a mixed solvent, $CH_2Cl_2/(C_2H_5)_4NCl$. However, in tetrahydrofuran, a less polarizable solvent, no evidence of CT formation was found.

We have used the present methods to investigate the relative locations of excited (π, π^*) and CT states of unlinked MgP-P complexes in a number of different geometries as shown in Fig. 17 [28, 29]. The geometries studied include several face-to-face conformations similar to those studied by Netzel et al. [53], as well as others. Among the properties required for the complex to act as an effective phototrap, the lowest CT state must be the lowest singlet excited state of the system, and must carry negligible oscillator strength to insure stability against radiative decay to the ground state. In Fig. 18, the relative energies of the 4 lowest lying (π, π^*) excited states and the lowest singlet CT state are shown for each complex. In every case, the lowest four excited states are (π, π^*) states which closely resemble the lowest two excited states in both free-base and magnesium porphine. The location of the lowest CT state is seen to vary considerably with molecular geometry, but is always much higher in energy than the lowest (π, π^*) state. In each case the CT state can be characterized as an ion pair, MgP^+-P^-, and possesses zero oscillator strength for radiative emission to the ground state. However, due to the high energy of the CT state, none of the complexes as isolated molecules may act as an effective phototrap.

The solvent effects on CT state formation observed by Netzel et al. [53] and the present calculated results strongly suggest that a large differential stabilization of the lowest CT state by the solvent or micro-environment of the dimer is necessary for stable CT formation in these complexes. Using dielectric continuum theory, an estimate of the stabilization of the lowest CT state in several face-to-face complexes is shown by the data in Table 3. This table gives the energy of stabilization of the CT state, represented as a point-dipole, due to a medium with a dielectric constant of 9.08, the value for CH_2Cl_2 [28, 29]. Since the calculated dipole moments of the ion-pair CT states are large, such stabilization is substantial. Furthermore, solvent stabili-

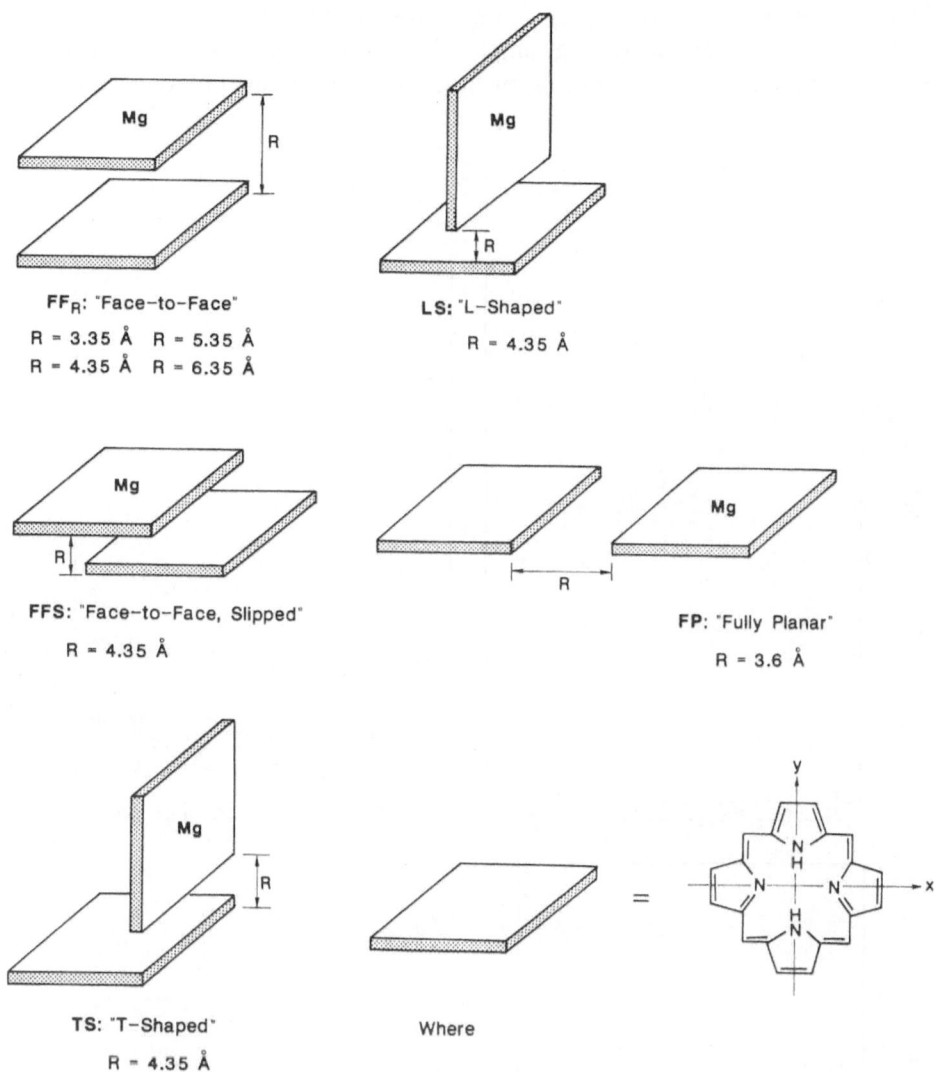

Fig. 17. Molecular geometries of MgP—P complexes. The magnesium atom is positioned 0.5 Å above the macrocyclic plane in each case (Reprinted with permission)

zation of both the ground and (π, π^*) excited states, which have calculated dipole moments of only about 3.5 Debye, is not expected to be more than approximately 100 cm^{-1}. Therefore, large differential stabilization of the lowest CT state by a polarizable medium is to be expected in these systems.

Another way of achieving a differential stabilization of the CT state is provided by coordination of a negative ion to the magnesium atom of the complex. In fact, in the studies performed by Netzel et al. [53] using $CH_2Cl_2/(C_2H_5)_4NCl$ media, it was assumed that such coordination of Mg by the chloride ion was realized, ostensibly providing stabilization of the CT state. To investigate the effect of Cl^-

coordination on the CT state, calculations were performed on Cl⁻-MgP-P complexes in the geometries depicted in Fig. 17. In Fig. 18 the computed location of the lowest CT state in Cl⁻-MgP-P complexes is given, showing a dramatic differential lowering such that the CT state is found near or below the lowest (π, π^*) state in several complexes.

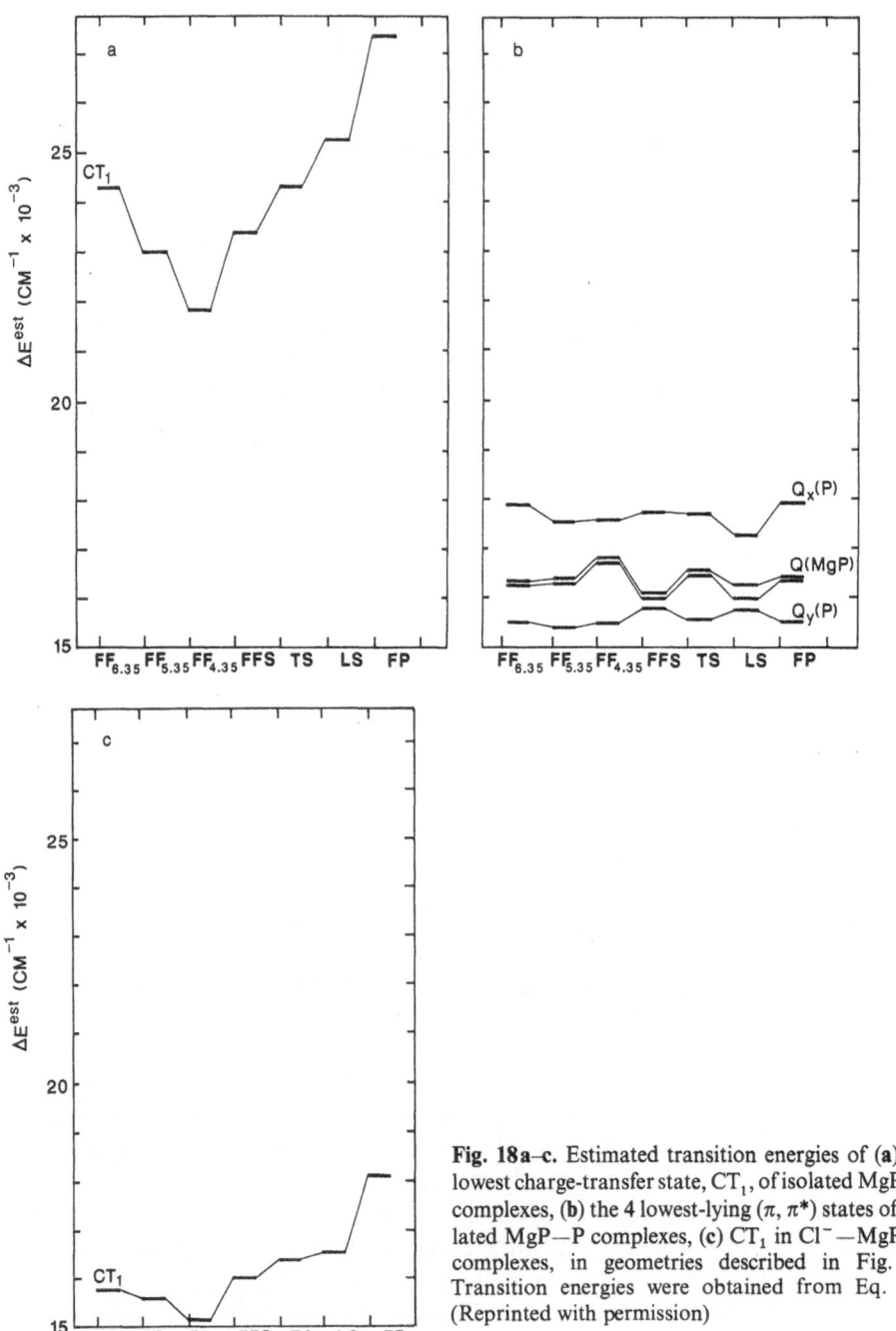

Fig. 18a–c. Estimated transition energies of (**a**) the lowest charge-transfer state, CT_1, of isolated MgP—P complexes, (**b**) the 4 lowest-lying (π, π^*) states of isolated MgP—P complexes, (**c**) CT_1 in Cl⁻—MgP—P complexes, in geometries described in Fig. 17. Transition energies were obtained from Eq. (11) (Reprinted with permission)

Table 3. Solvent Stabilization Energies, U, of the Lowest CT State of Face-to-Face MgP—P Dimers

R^a (Å)	U^b (cm^{-1})	μ^c (Debye)	a (Å)
4.35	6,660	24.1	5.7
5.35	8,820	29.2	5.9
6.35	10,240	33.9	6.2

[a] Intermacrocyclic distance, see Fig. 17.
[b] Computed using dielectric continuium theory:

$$U = [\mu^2/2a^3] \{[2(D-1)]/[2D+1]\},$$

where μ is the dipole moment, a is the radius of a cavity enclosing the molecule, and D is the solvent dielectric constant. D = 9.08, the value for CH_2Cl_2, was used.
[c] Calculated dipole moment of the lowest CT state.

The effect of Cl$^-$ binding on the location of the lowest (π, π^*) state is minimal, since this state is essentially localized on the free-base porphine ring. However, the negative ion coordination facilities loss of an electron by the MgP moiety, resulting in a CT state of lower energy than in the uncoordinated complex. Thus, the present studies have shown that environmental effects, such as stabilization by a polarizable micro-environment and/or negative ion coordination, are essential in order to produce a low-lying excited CT state in the MgP-P dimer system.

6 Electronic Excited States of Aggregated Systems: Exciton Based CI Procedures

6.1 Molecular Exciton Theory

Exciton theory provides a theoretical model for investigating the collective excitation properties of multimeric or multi-chromophoric systems. It can be thought of as a subtype of the typical CI method described in Sect. 3, where configurations of generalized Hartree products replace the usual configurations of symmetry-adapted Slater determinants. In the exciton method, the Hartree-product configurations

$$\Phi_L = \psi_\alpha(1)\,\psi_\beta(2)\dots\psi_\mu(M)\,, \tag{12}$$

are made up of monomer (chromophore) wave functions, where $\psi_\alpha(i)$ is the electronic wavefunction for the αth state of the ith molecule. Electron exchange among molecules (chromophores) is assumed to be negligible, and all wave functions are taken to be orthogonal, standard assumptions in molecular exciton theory [54, 55]. Introduction of antisymmetry among molecular (chromophore) wave functions, which is tantamount to assuming that electron exchange is no longer negligible, leads to a "triplet" exciton theory [56]. In the present work, however, we will confine our discussion to the more familiar "singlet" form of the theory.

Generally, only limited excitations are considered — primarily "singly-excited" exciton configurations in which a single molecule (chromophore) i is excited, e.g.,

$$\Phi_L = \psi_0(1) \ldots \psi_\alpha(i) \ldots \psi_0(M), \tag{13}$$

and to a limited extent "doubly-excited" configurations, where two molecules (chromophores) i and j are simultaneously excited,

$$\Phi_L = \psi_0(1) \ldots \psi_\alpha(i) \ldots \psi_\beta(j) \ldots \psi_0(M). \tag{14}$$

Doubly-excited configurations are generally neglected due to their expected small effect on the wave function.

The wave function for the Kth exciton state is of the same form as that given by Eq. (5) for CI wave functions, the expansion coefficients again being obtained as solutions of an appropriate set of secular equations, (see Eq. (6)). The Hamiltonian matrix elements, H_{JL}, can be written in terms of the appropriate charge and transition density distributions, $\varrho(i \mid \alpha, \alpha)$ and $\varrho(i \mid \alpha, \beta)$, respectively, as [57]

$$H_{JJ} = \sum_{i=1}^{M} E_\alpha(i) + \sum_{i=1}^{M} \sum_{j>i}^{M} \iint \varrho(i \mid \alpha, \alpha)\, v(i, j)\, \varrho(j \mid \beta, \beta)\, d\mathbf{r}(i)\, d\mathbf{r}(j), \quad \alpha, \beta \in \{J\} \tag{15}$$

and

$$H_{JL} = \sum_{i=1}^{M} \sum_{j>i}^{M} \iint \varrho(i \mid \alpha, \gamma)\, v(i, j)\, \varrho(j \mid \beta, \delta)\, d\mathbf{r}(i)\, d\mathbf{r}(j) \prod_{k \neq i, j}^{M} \delta_{\mu\nu}(k) \tag{16}$$

$$\alpha, \beta, \mu \in \{J\}; \qquad \gamma, \delta, \nu \in \{L\},$$

where $E_\alpha(i)$ is the energy of the "α^{th}" state of the ith molecule, and $v(i, j)$ is the inter-molecular interaction energy operator between the ith and j^{th} molecules, the integrations, $d\mathbf{r}(i)$ and $d\mathbf{r}(j)$, being performed over single electron coordinates of both molecules. The electron density distribution function is defined as

$$\varrho(i \mid \alpha, \beta) = N(i) \iint \ldots \int \psi_\alpha^*(i)\, \psi_\beta(i)\, d\mathbf{r}_2 \ldots d\mathbf{r}_{N(i)}\, d\sigma_1 \ldots d\sigma_{N(i)}, \tag{17}$$

where $N(i)$ is the number of electrons in molecule i, and the integration is performed over $N(i)-1$ spatial variables, $d\mathbf{r}_2 \ldots d\mathbf{r}_{N(i)}$, and $N(i)$ spin variables, $d\sigma_1 \ldots d\sigma_{N(i)}$. Methods designed to provide an accurate and computationally efficient approach to the approximation of these matrix elements will be presented in Sect. 6.2.

Exciton theory is essentially a "state-interaction" theory, and as such tacitly assumes that the electronic structure of an individual molecular state does not change appreciably due to intermolecular interactions. While in principle the inclusion of higher excitations (e.g. triple, quadrupole), ...), as well as excitations to higher states within a given molecule's excited state manifold can account for large changes, in practice this is rarely done due to the added computational burden it entails. Lastly, it should be noted that while we have emphasized local (i.e. molecular or chromophoric) excitations, the theory in no way prohibits inclusion of charge-transfer (CT) states. In fact, their inclusion only requires data on the state energies and wavefunctions of appropriate radical cation and anion species [58].

6.2 Intermolecular Interaction Integral Applications: Application of FSGO-based Distributed Point-Charge Models

Due to the size of the multimeric systems investigated, approximations of the inter-molecular interaction integrals appearing in Eq. (15) and Eq. (16) are essential if reasonable computation times are to be obtained. In this regard, FSGOs possess a number of properties which lead to computationally practical approximations to the charge and transition densities through the generation of distributed point-charge models (PCMs) which conserve total charge and dipole moment. This stands in sharp contrast to PCMs derived by other methods based on Mulliken population analysis [59] or electrostatic potential fitting [60], which generally do not conserve dipole moment. FSGO-based PCMs, originally developed by Hall [61], have also been shown to provide accurate representations of electrostatic potentials and other electronic properties compared to values calculated from the wave function directly [62].

A number of FSGO-based PCMs have been developed including the original work of Hall [61], the simplified PCM of Shipman [63], and the generalized PCMs developed recently by Petke [64]. All FSGO-based PCMs arise from various ways of parti-tioning the charge and transition density distribution functions. The general form of these distribution functions for the ith molecule, which is valid for wavefunctions defined in terms of an orbital basis [65] is given in Eq. (18).

$$\varrho(i \mid \alpha, \beta) = \sum_{s=1}^{N} \sum_{t=1}^{N} D_{st}(i \mid \alpha, \beta) \, g_s g_t \,, \tag{18}$$

where α and β label the wavefunctions $\psi_\alpha(i)$ and $\psi_\beta(i)$, respectively, and $D_{st}(i \mid \alpha, \beta)$ is an element of the matrix representation of $\varrho(i \mid \alpha, \beta)$ in the basis set $\{g\}$. For the present discussion the basis consists of FSGOs, and it should be pointed out that each spherical gaussian function in the FSGO basis, including the components g_m^U and g_m^D of the p_π-type FSGOs in Eq. (3), are counted individually in the sum-mations in Eq. (18).

Utilizing the fact that the product of two FSGOs is itself an FSGO, Hall [61] derived a charge and dipole moment conserving PCM of the following form:

$$\varrho(i \mid \alpha, \beta) = \sum_{s=1}^{N} Q_{ss}^{H}(i \mid \alpha, \beta) \, \delta(r - R_s) + \sum_{s=1}^{N} \sum_{t>s}^{N} Q_{st}^{H}(i \mid \alpha, \beta) \, \delta(r - R_{st}) ; \tag{19}$$

where

$$Q_{ss}^H(i \mid \alpha, \beta) = D_{ss}(i \mid \alpha, \beta) \,, \tag{20}$$

$$Q_{st}^H(i \mid \alpha, \beta) = [D_{st}(i \mid \alpha, \beta) + D_{ts}(i \mid \alpha, \beta)] \langle g_s \mid g_t \rangle \,; \tag{21}$$

$$R_{st} = \frac{a_s R_s + a_t R_t}{a_s + a_t} \,; \tag{22}$$

and $\delta(r - R_{st})$ is the Dirac delta function. While Hall's PCM produces excellent agreement with values obtained from explicit evaluation of the wave function for a number of molecular electrostatic properties, it suffers from the excessive number of effective point charges generated: $N(N + 1)/2$, N being the number of spherical gaussians in the basis.

Shipman [63] avoided this difficulty through the development of an alternative PCM that required only N effective point charges, while retaining the advantages of charge and dipole moment conservation. The Shipman-PCM is of the form

$$\varrho^S(i \mid \alpha, \beta) = \sum_{s=1}^{N} Q_s^S(i \mid \alpha, \beta) \, \delta(r - R_s) \,, \tag{23}$$

where

$$Q_s^S = Q_{ss}^H + \sum_{t \neq s}^{N} \frac{a_s}{a_s + a_t} Q_{st}^H \,. \tag{24}$$

From Eq. (24) it is clear that the reduction in the number of point charges is obtained by "reallocating" the Hall point charges located off FSGO centers, Q_{st}^H, $s \neq t$, onto these centers such that each FSGO center R_t now has an effective charge given by Eq. (24). Not unexpectedly, the Shipman-PCM does not perform as well as the Hall-PCM in approximating electrostatic potentials [62]. Recently, Petke [64] has developed a family of charge and dipole moment conserving PCMs intermediate to those of Hall and Shipman in both number of effective point charges and accuracy of calculated electrostatic properties. The form of the Petke-PCM is similar to that of the other PCMs, the effective charges now being given by

$$\varrho^P(i \mid \alpha, \beta) = \sum_{\substack{s=1 \\ (s,\,t) \in \{\lambda\}}}^{N} \left(Q_{ss}^H + \sum_{t \neq s}^{N} \frac{a_s}{a_s + a_t} Q_{st}^H \right) \delta(r - R_s)$$

$$+ \sum_{\substack{s=1 \\ (s,\,t) \notin \{\lambda\}}}^{N} \sum_{t > s}^{N} Q_{st}^H \delta(r - R_{st}) \,. \tag{25}$$

In these models, only a selected set $\{\lambda\}$ of off-center Hall point charges, Q_{st}^H, $(s, t) \in \{\lambda\}$, are divided and placed on the FSGO centers, in contrast to the Shipman-PCM in which this is done for all off-center Hall point charges. Thus, the resulting PCM contains a mixture of Hall and Shipman point charges, and the set $\{\lambda\}$ may be chosen to construct a PCM of minimum size consistent with reproduction of electrostatic properties to a desired level of accuracy. For example, the electrostatic potential of the porphyrin molecule may be reproduced to four significant figures with a Petke-PCM containing fifteen percent of the number of charges of the Hall model [64].

Implementation of any of the above PCMs in exciton matrix element evaluation involves straightforward substitution of the desired point-charge-based expression for the density distribution (Eq. (19), Eq. (23), or Eq. (25)) into Eq. (15) and Eq. (16), followed by the indicated integrations. This results in matrix element expressions given as sums of simple electrostatic interactions [57].

7 Application to Bacteriochlorin Dimers

To conclude the discussion of exciton theory, we present a brief account of the application of PCM-based exiton methods to studies of Q_y spectral shifts in bacterio-chlorin (BC) dimers [57, 66]. This work was motivated by earlier exciton studies of a model methyl bacteriopheophorbide a (Me-BPheo-a) dimer reported by Scherz and Parson [67]. Using the point-dipole model to approximate monomeric charge and transition densities [54–56], they obtained a large (1330 cm^{-1}) red shift of the Q_y band in a dimer whose geometry was defined as a translation of one monomer relative to the other by $\Delta x = 0.5$ Å, $\Delta y = 3.5$ Å, and $\Delta z = 3.6$ Å along the axes in Fig. 4.

This result appears inconsistent with the 300 cm^{-1} blue shifts determined experimentally in double covalently-linked Chl dimers with approximately the same geometry [68]. Moreover, it is questionable from a theoretical viewpoint whether the density distributions of a spatially-extended molecule such as Me-BPheo-a can be adequately represented by point-dipoles. To investigate these points, exciton calculations using the point-dipole, Hall-PCM, and Shipman-PCM electron density approximations were performed on BC dimers. Calculations with singly-excited configurations as given in Eq. (13) were performed for each approximation, and a "full" calculation in which all singly- and doubly-excited configurations were included was carried out using the Shipman-PCM. Monomer excitations to both visible- and Soret-band states were allowed.

In Fig. 19 calculated Q_y spectral shifts are presented for dimers with geometries in which the relative offsets of the two monomers were fixed along the x- and z-axes at $\Delta x = 0.0$ Å and $\Delta z = 3.5$ Å, respectively (see Fig. 4), while Δy was allowed to vary. Clearly, spectral shifts obtained with the point-dipole approximation exhibit little resemblence to those obtained using either the Hall-PCM or Shipman-PCM. Thus, the point-dipole model does not appear to be a viable approximation for the present systems.

In contrast, spectral shifts obtained with the Shipman-PCM are in excellent agreement with those obtained using the more rigorous Hall-PCM at each geometry studied. Furthermore, as seen in Fig. 19, addition of doubly-excited configurations does not significantly change spectral shifts.

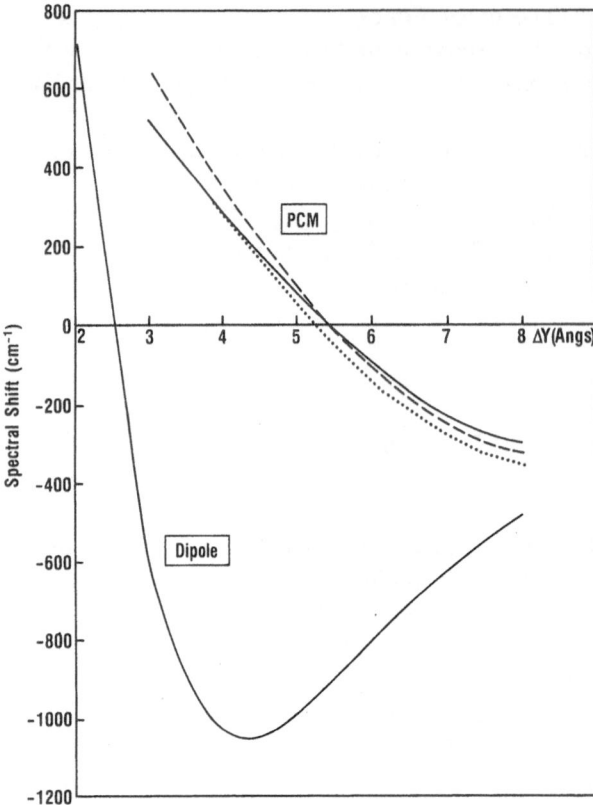

Fig. 19. Q_y spectral shifts of bacteriochlorin dimers as a function of Δy, calculated with the point-dipole [DIPOLE] and selected PCM [PCM] approximations. Negative shifts are red shifts relative to the momomer band. The PCM-based calculations are described in the text and are given as:
——— Hall-PCM, single excitations;
...... Shipman-PCM, single excitations;
——— Shipman-PCM, single and double excitations. (Reprinted with permission)

Both the Hall- and Shipman-PCM-based studies appear to predict spectral shifts as a function of Δy consistent with general experimental trends observed in Chl dimer spectra: blue shifts are observed in dimers with $\Delta y = 3$–4 Å while red shifts occur only when Δy is much larger, i.e., 7–8 Å [68, 69]. The Shipman-PCM appears to be a satisfactory as well as economical approximation for use in exciton calculations, and has been employed in further exciton studies of bacteriochlorin and Me-BPheo-a oligomers [66, 70].

8 Concluding Remarks

The examples presented in the preceeding sections have illustrated that the present FSGO-based CI methods are very effective in providing a qualitative or semi-quantitative resolution of the electronic spectra of large molecules. Calculations on molecules containing 200 to 500 or more electrons may be performed with reasonable demands on computer resources. The methods are widely applicable, and the results provide for a useful interplay between theory and experiment in interpreting electronic spectra and understanding spectra-structure correlations.

Distributed point-charge models may be constructed from FSGO-based wave-functions to provide accurate representations of electrostatic potentials and properties, while naturally preserving both total charge and dipole moment. Incorporation of such PCM's within molecular exciton theory provides an attractive and economical method for the study of spectral features of large molecular aggregates.

9 References

1. Shavitt I (1977) In: Schaefer HF (ed) Methods of electronic structure theory. Plenum, New York, p 189
2. Shepard R (1987) Adv. Chem. Phys. 59: 63
3. Hansen AE, Bouman TD (1980) Adv. Chem. Phys. 44: 545
4. Bartlett RJ, Dykstra CE, Paldus J (1984) In: Dykstra CE (ed) Advanced theories and computational approaches to the electronic structure of molecules. D. Reidel, Dordrecht, p 127
5. Boys SF (1950) Proc. Roy. Soc. A200: 542
6. Frost AA, Prentice BH, Rouse RA (1967) J. Am. Chem. Soc. 89: 3064; Frost AA (1967) J. Chem. Phys. 47: 3707; Frost AA (1977) J. Chem. Phys. 47: 3714
7. Frost AA (1977) In: Schaefer HF (ed) Methods of electronic structure theory. Plenum, New York, p 29
8. Pakiari AH, Linnett JW (1980) Int. J. Quantum Chem. 18: 661
9. Maggiora GM, Christoffersen RE, Yoffe JA, Petke JD (1981) Ann. N.Y. Acad. Sci. 367: 1
10. Christoffersen RE, Maggiora GM (1969) Chem. Phys. Lett. 3: 419; Christoffersen RE, Genson DW, Maggiora GM (1971) J. Chem. Phys. 54: 239; Christoffersen RE (1972) Adv. Quantum Chem. 6: 333
11. Christoffersen RE, Spangler D, Hall GG, Maggiora GM (1973) J. Am. Chem. Soc. 95: 8526
12. Whitten JL, Hackmeyer M (1969) J. Chem. Phys. 51: 5584
13. Petke JD, Maggiora GM, Shipman LL, Christoffersen RE (1978) J. Mol. Spectrosc. 71: 64
14. Petke JD, Christoffersen RE, Maggiora GM, Shipman LL (1977) Int. J. Quantum Chem.: Quantum Biol. Symp. 4: 343
15. Hackmeyer M, Whitten JL (1971) J. Chem. Phys. 54: 3739
16. Hay PJ, Shavitt I (1974) J. Chem. Phys. 60: 2865
17. Matos JMO, Roos BO, Malmqvist P-A (1988) J. Chem. Phys. 86: 1458
18. Del Bene J, Jaffe HH (1968) J. Chem. Phys. 48: 1807
19. Zerner MC, Loew G, Kirchner R, Mueller-Westerhoff UT (1980) J. Am. Chem. Soc. 102: 589
20. Zerner MC (1981) Ann. N.Y. Acad. Sci. 367: 35
21. Spangler D, McKinney R, Christoffersen RE, Maggiora GM, Shipman LL (1975) Chem. Phys. Lett. 36: 427; Spangler D, Maggiora GM, Shipman LL, Christoffersen RE (1977) J. Am. Chem. Soc. 99: 7478
22. Roothaan CCJ (1951) Rev. Mod. Phys. 23: 69; Hall GG (1951) Proc. Roy. Soc. A205: 541
23. Bender CF (1972) J. Comput. Phys. 9: 547
24. Shipman LL, Christoffersen RE (1972) Chem. Phys. Lett. 15: 469
25. Nitzsche LE, Chabalowski C, Christoffersen RE (1976) J. Am. Chem. Soc. 98: 4794; Murk D, Nitzsche LE, Christoffersen RE (1978) J. Am. Chem. Soc. 100: 1371
26. Petke JD, Butler P, Maggiora GM (1984) Int. J. Quantum Chem. 27: 71
27. Petke JD, Maggiora GM (1984) J. Am. Chem. Soc. 106: 3129
28. Petke JD, Maggiora GM (1983) Chem. Phys. Lett. 105: 31; Petke JD, Maggiora GM (1986) J. Chem. Phys. 84: 1640
29. Petke JD, Maggiora GM In: Gouterman M, Rentzepis PM, Straub KD (eds) Porphyrins: excited states and dynamics. ACS symposium series 321, p 20
30. Petke JD, Maggiora GM, Shipman LL, Christoffersen RE (1978) J. Mol. Spectrosc. 73: 311
31. Petke JD, Maggiora GM, Shipman LL, Christoffersen RE (1979) Photochem. Photobiol. 30: 203

32. Petke JD, Shipman LL, Maggiora GM, Christoffersen RE (1981) J. Am. Chem. Soc. 103: 4622
33. Petke JD, Maggiora GM, Shipman LL, Christoffersen RE (1982) Photochem. Photobiol. 36: 383
34. Petke JD, Maggiora GM, Shipman LL, Christoffersen RE (1980) Photochem. Photobiol. 31: 243
35. Petke JD, Maggiora GM, Shipman LL, Christoffersen RE (1980) Photochem. Photobiol. 32: 399
36. Petke JD, Maggiora GM, Shipman LL, Christoffersen RE (1980) Photochem. Photobiol. 32: 661
37. Petke JD, Maggiora GM, Shipman LL, Christoffersen RE (1981) Photochem. Photobiol. 33: 663
38. Webb LE, Fleischer EB (1975) J. Am. Chem. Soc. 87: 667
39. Kratky C, Dunitz JD (1977) J. Mol. Biol. 113: 431
40. Chow HC, Serlin R, Strouse CE (1975) J. Am. Chem. Soc. 97: 7230
41. Katz JJ, Norris JR, Shipman LL, Thurnauer MC, Wasielewski MR (1978) Ann. Rev. Biophys. Bioeng. 7: 393
42. Gurinovich GP, Sevchenko AN, Solovev KN (1961) Opt. Spectrosc. 10: 396
43. Gradyushko AT, Solovev KN, Starukhin AS (1976) Opt. Spectrosc. 40: 267
44. Gouterman M, Wagniere GH, Snyder LC (1963) J. Mol. Spectrosc. 11: 108
45. Edwards L, Dolphin DH, Gouterman M (1970) J. Mol. Spectrosc. 35: 90
46. Fajer J, Borg DC, Forman A, Felton RH, Dolphin D, Vegh L (1974) Proc. Natl. Acad. Sci. USA 71: 994
47. McElroy JD, Feher G, Mauzerall DC (1972) Biochim. Biophys. Acta 267: 363
48. Norris JR, Scheer H, Druyan ME, Katz JJ (1974) Proc. Natl. Acad. Sci. USA 71: 4897
49. Fajer J, Davis MS, Forman A, Klimov VV, Dolan E, Ke B (1980) J. Am. Chem. Soc. 102: 7143; Okamura MT, Isaacson RA, Feher G (1979) Biochim. Biophys. Acta 546: 394; Shuvalov VA, Klevanik AV, Sharkov AV, Matveetz JA, Krukov PG (1978) FEBS Lett. 91: 135; Davis MS, Forman A, Hanson LK, Thornber JP, Fajer J (1979) J. Phys. Chem. 83: 3325
50. Jortner J (1980) Biochim. Biophys. Acta 594: 193
51. Deisenhofer J, Epp O, Miki K, Huber R, Michel H (1984) J. Mol. Biol. 180: 385; Parson WW (1982) Ann. Rev. Biophys. Bioeng. 11: 57
52. Chang CK (1977) J. Heterocyclic Chem. 14: 1285
53. Netzel TL, Kroger P, Chang CK, Fujita I, Fajer J (1979) Chem. Phys. Lett. 67: 223; Netzel TL, Bergkamp MA, Chang CK (1982) J. Am. Chem. Soc. 104: 1952
54. Kasha M, Rawls HR, Aahraf El-Bayoumi M (1965) Pure Appl. Chem. 11: 371
55. Davydov AS (1962) Theory of molecular excitons. McGraw-Hill, New York
56. Craig DP, Walmsley SH (1968) Excitons in molecular crystals. Benjamin, New York
57. LaLonde DE, Petke JD, Maggiora GM (1988) J. Phys. Chem. 92: 4746
58. Warshel A, Parson WW (1987) J. Am. Chem. Soc. 109: 6143
59. Mulliken RS (1955) J. Chem. Phys. 23: 1833
60. Momany FA (1978) J. Phys. Chem. 82: 592; Singh UC, Kollman PA (1984) J. Comp. Chem. 5: 129
61. Hall GG (1973) Chem. Phys. Lett. 20: 501; Tait AD, Hall GG (1973) Theoret. Chim. Acta 31: 311
62. Martin D, Hall GG (1981) Theoret. Chim. Acta 59: 281
63. Shipman LL (1975) Chem. Phys. Lett. 31: 361
64. Petke JD (1986) Chem. Phys. Lett. 126: 26
65. McWeeny R, Sutcliffe BT (1969) Methods of molecular quantum mechanics. Academic, New York
66. LaLonde DE (1985) Evaluation of approximations in molecular exciton theory: applications to photosynthetic systems. Thesis, University of Kansas, Lawrence, KS
67. Scherz A, Parson WW (1984) Biochim. Biophys. Acta 766: 666
68. Bucks RR, Boxer SG (1982) J. Am. Chem. Soc. 104: 340
69. Boxer SG (1983) Biochim. Biophys. Acta 726: 265
70. LaLonde DE, Petke JD, Maggiora GM (1989) J. Phys. Chem. 93: 608

Classical Electrostatics in Molecular Interactions

Anthony J. Stone

University Chemical Laboratory, Lensfield Road, Cambridge CB2 1EW, England

It has recently become clear that classical electrostatics is much more useful in the description of intermolecular interactions than was previously thought. The key is the use of distributed multipoles, which provide a compact and accurate picture of the charge distribution but do not suffer from the convergence problems associated with the conventional one-centre multipole expansion. The article describes how the electrostatic interaction can be formulated efficiently and simply, by using the best features of both the Cartesian tensor and the spherical tensor formalisms, without the need for inconvenient transformations between molecular and space-fixed coordinate systems, and how related phenomena such as induction and dispersion interactions can be incorporated within the same framework. The formalism also provides a very simple route for the evaluation of electric fields and field gradients. The article shows how the forces and torques needed for molecular dynamics calculations can be evaluated efficiently. The formulae needed for these applications are tabulated.

1 Introduction

For many years it has been known that interactions between molecules at long range can be understood in terms of perturbation theory in which the perturbation is the electrostatic interaction between the particles comprising the molecules [1]. The first-order energy is the classical electrostatic interaction between the molecular charge distributions, and the second-order energy can be separated into the induction energy, arising from the distortion of each molecule in the field of its neighbours, and the dispersion energy, which describes the electrostatic interaction arising through correlated fluctuations in the molecular charge distributions.

Now it is a simple matter to write down this electrostatic perturbation in terms of the charges e_i and e_j of particle i of molecule A and particle j of molecule B and the distance r_{ij} between them:

$$V = \sum_{i \in A, j \in B} \frac{e_i e_j}{4\pi\varepsilon_0 r_{ij}}, \tag{1}$$

and it is also simple to write down an expression for the first-order electrostatic interaction in terms of the charge densities $\varrho^A(\mathbf{r})$ and $\varrho^B(\mathbf{r}')$ at position $\mathbf{r}$ in molecule A and position $\mathbf{r}'$ in molecule B:

$$U_{es} = \int \frac{\varrho^A(\mathbf{r}) \, \varrho^B(\mathbf{r}')}{4\pi\varepsilon_0 \, |\mathbf{r} - \mathbf{r}'|} \, d^3\mathbf{r} \, d^3\mathbf{r}' . \tag{2}$$

However these formulae are not useful as they stand. Eq. (1) is an adequate formal expression (except at short range, where the overlap of the wavefunctions of the two molecules makes it impossible to distinguish between identical particles and to assign some of them to molecule A and some to molecule B), but it does not lead to practical expressions for the interaction energy. Eq. (2) is an exact expression, but is very cumbersome to evaluate.

2 The Cartesian Multipole Expansion

Consequently it has become usual to use the *multipole expansion* of the interaction operator. This can be obtained in its most basic form from a Taylor expansion of $1/r_{ij}$ about suitable origins $\mathbf{A}$ and $\mathbf{B}$ in each molecule. We write the position of particle i in molecule A as $\mathbf{A} + \mathbf{r}_i$, and the position of particle j in molecule B as $\mathbf{B} + \mathbf{r}_j$, and then, writing $\mathbf{R} = \mathbf{B} - \mathbf{A}$, and using the usual repeated-suffix summation convention, we have

$$1/r_{ij} = 1/(\mathbf{B} + \mathbf{r}_j - \mathbf{A} - \mathbf{r}_i) = 1/(\mathbf{R} + \mathbf{r}_j - \mathbf{r}_i)$$

$$= 1/R + (\mathbf{r}_j - \mathbf{r}_i)_\alpha \nabla_\alpha(1/R) + \frac{1}{2!}(\mathbf{r}_j - \mathbf{r}_i)_\alpha \, (\mathbf{r}_j - \mathbf{r}_i)_\beta \, \nabla_\alpha\nabla_\beta(1/R) + \dots$$

and if we substitute this into Eq. (1) the electrostatic interaction becomes

$$V_{es} = \sum_{i \in A, j \in B} \frac{e_i e_j}{4\pi\varepsilon_0 r_{ij}}$$

$$= q^A q^B T - (q^A \mu_\alpha^B - \mu_\alpha^A q^B) T_\alpha + \frac{1}{2} (q^A Q_{\alpha\beta}^B - 2\mu_\alpha^A \mu_\beta^B + Q_{\alpha\beta}^A q^B) T_{\alpha\beta} + \cdots$$

(3)

where the charge q^A, dipole moment μ^A and second moment Q^A of molecule A are defined by

$$q^A = \sum_{i \in A} e_i \,,$$

$$\mu_\alpha^A = \sum_{i \in A} e_i r_{i\alpha} \,,$$

$$Q_{\alpha\beta}^A = \sum_{i \in A} e_i r_{i\alpha} r_{i\beta} \,,$$

and the interaction tensors T, T_α and $T_{\alpha\beta}$ are

$$T = \frac{1}{4\pi\varepsilon_0 R} \,,$$

$$T_\alpha = -\nabla_\alpha T = \frac{R_\alpha}{4\pi\varepsilon_0 R^3} \,,$$

$$T_{\alpha\beta} = \nabla_\alpha \nabla_\beta T = \frac{3R_\alpha R_\beta - R^2 \delta_{\alpha\beta}}{4\pi\varepsilon_0 R^5} \,,$$

and in general

$$T_{\alpha\beta\ldots v}^{(n)} = (-1)^n \nabla_\alpha \nabla_\beta \ldots \nabla_v \frac{1}{4\pi\varepsilon_0 R} \,;$$

the superscript (n) is used where necessary to indicate the number of suffixes.

The interaction tensors are symmetric with respect to permutation of their suffixes, as is clear from the definition. Moreover $1/R$ satisfies Laplace's equation: $\nabla^2(1/R) = 0$. Consequently $T_{\alpha\beta\gamma\ldots v}$ is traceless with respect to any pair of suffixes: $T_{\alpha\alpha\gamma\ldots v} = 0$. Because of this, it is customary to define the quadrupole moment $\Theta_{\alpha\beta}$ as the second moment with the trace removed:

$$\Theta_{\alpha\beta} = \frac{3}{2} Q_{\alpha\beta} - \frac{1}{2} Q_{\sigma\sigma} \delta_{\alpha\beta} \,,$$

and then it follows that

$$\Theta_{\alpha\beta} T_{\alpha\beta\gamma\ldots v} = \left(\frac{3}{2} Q_{\alpha\beta} T_{\alpha\beta\gamma\ldots v} - \frac{1}{2} Q_{\sigma\sigma} T_{\alpha\alpha\gamma\ldots v} \right),$$

$$- \frac{3}{2} Q_{\alpha\beta} T_{\alpha\beta\gamma\ldots v} \,.$$

This shows that the trace of $Q_{\alpha\beta}$ has no effect on the electrostatic interaction, and consequently the traceless form $\Theta_{\alpha\beta}$ is always used. The factor 3/2 is needed for consistency with other formulations.

The expansion given in Eq. (3) can be continued to higher order, to include the octopole moment, the hexadecapole moment, and so on. In general the moment of rank n (the '2^n-pole moment' in the conventional but rather unsatisfactory terminology) is defined by [1]

$$\xi_{\alpha\beta\ldots\nu}^{(n)} = (-1)^n \, (n!)^{-1} \sum_{i \in A} e_i r_i^{2n+1} \, \nabla_\alpha \nabla_\beta \ldots \nabla_\nu (1/r_i) , \tag{4}$$

and the general term in the multipole expansion is

$$\sum_{n,\,n'} (-1)^n \, \frac{1}{1.3 \ldots (2n-1)} \, \frac{1}{1.3 \ldots (2n'-1)} \, \xi_{\alpha\beta\ldots\nu}^{(n)\,(A)} \xi_{\alpha'\beta'\ldots\nu'}^{(n')\,(B)} \, T_{\alpha\beta\ldots\nu\alpha'\beta'\ldots\nu'}^{(n+n')} \tag{5}$$

This formulation is reasonably simple to understand, but it has a number of disadvantages which make it cumbersome to use. Firstly, the cartesian tensor for rank n is an object with n suffixes. This in itself introduces some awkwardness of notation; while it is possible to use the same symbol (ξ above) for all ranks, distinguishing between the ranks by the number of suffixes, it is necessary in general expressions such as (5) to include the superscript (n) to indicate the rank. Moreover it has become conventional to use different symbols (q, μ, Θ, Ω, Φ) for the moments of ranks 0 to 4. Both of these features lead to cumbersome algebra and inhibit the development of general expressions. A more troublesome feature is that the tensor of rank n has 3^n components; but the definition (4) shows that $\xi_{\alpha\beta\ldots\nu}$ is symmetric with respect to all permutations of suffixes, and traceless in all pairs of suffixes. These conditions mean that only $2n + 1$ of the components are independent, so that there is considerable redundancy in the higher moments. The same is true of the interaction tensors $T_{\alpha\beta\ldots\nu}$.

A less obvious disadvantage of the formula (5) is that it is expressed in terms of a global coordinate system, so that the components μ_x, μ_y and μ_z are the dipole moments of the molecule in the global or 'space-fixed' coordinate system. This is a great inconvenience, since the dipole moment is a property of the molecule and is much more conveniently expressed in terms of a 'molecule-fixed' coordinate system. In the case of the ammonia molecule, for instance, the dipole moment is directed along the symmetry axis, the symmetry ensuring that the components at right angles to the symmetry axis are zero. In order to calculate the electrostatic energy it therefore becomes necessary to transform from the local to the global axis system. This is not a particularly difficult procedure, but it has been a considerable disincentive to the use of higher-rank multipole moments, and there is a widespread preference for point-charge models, where the problem of transformation does not arise. Part of the purpose of this paper is to show that it is in fact possible to express the electrostatic interactions entirely in terms of molecule-fixed multipole moments and to avoid the need for transformation altogether, or rather to perform the transformation once and for all and to incorporate the transformation coefficients into the interaction functions.

3 The Spherical Tensor Formulation

In spite of the disadvantages of the Cartesian formulation, it is preferred by many workers because the alternative, the spherical tensor formulation, is perceived as mathematically difficult. There is undoubtedly some truth in this view. Moreover the spherical-tensor formulation deals in complex quantities which are more difficult to comprehend than the cartesian-tensor components. However the power and versatility of the spherical tensor approach should not be abandoned lightly, and the main purpose of the present paper is to show that it is possible to combine the best features of the cartesian and spherical-tensor methods. We will show that this hybrid approach leads to very compact expressions for the electrostatic energy and related quantities such as the induction and dispersion energies, and that these can be expressed entirely in terms of real multipole moments referred to molecule-fixed coordinate systems. The transformation between molecule-fixed and space-fixed coordinates can be carried out once and for all, and the analogues in this method of the interaction tensors $T_{\alpha\beta\ldots\nu}$ contain the necessary orientational information.

A more practical objection to the spherical tensor formulation has been that it is inefficient to use in computationally-intensive applications such as molecular dynamics, because of the supposed need to describe the molecular orientations in terms of Euler angles and to evaluate trigonometrical functions of these angles. This objection is completely unfounded. The Euler angles have traditionally been used to describe orientation, but they are by no means necessary, and the formulae to be presented below will not mention them at all. It is possible, within the spherical tensor formalism, to describe orientation in terms of the same direction cosines $l_{\alpha\alpha'}$ that are used to transform from one cartesian axis system to another in the cartesian tensor formalism.

I now give, for completeness, a brief account of the derivation of the spherical tensor formulation of the multipole expansion. This does require an understanding of spherical tensor methods and of Racah algebra, but it may be omitted by readers who are unfamiliar with these techniques, who should skip to the beginning of the next section.

The derivation [2] starts with an expansion of $1/r_{ij}$, just as the Cartesian formulation does, but this time we use the expansion in terms of spherical harmonics (the spherical harmonic addition theorem) [3]. As before, we write $\mathbf{R} = \mathbf{B} - \mathbf{A}$, and obtain

$$\frac{1}{|\mathbf{R} + \mathbf{r}_j - \mathbf{r}_i|} = \sum_{k=0}^{\infty} \sum_{m=-k}^{k} (-1)^m R_{k,-m}(\mathbf{r}_j - \mathbf{r}_i)\, I_{km}(\mathbf{R}) , \tag{6}$$

where $R_{km}(\mathbf{r})$ and $I_{km}(\mathbf{r})$ are the regular and irregular spherical harmonics respectively:

$$R_{km}(\mathbf{r}) = r^k C_{km}(\theta, \phi), \qquad I_{km}(\mathbf{r}) = r^{-k-1} C_{km}(\theta, \phi).$$

Here C_{lm} is the modified spherical harmonic $\sqrt{4\pi/(2l + 1)}\, Y_{lm}$. Eq. (6) is valid only if $|\mathbf{r}_j - \mathbf{r}_i| < R$. Another standard addition theorem [3] states that

$$R_{k,\,-m}(\mathbf{r}_1 + \mathbf{r}_2) = \sum_{l_1 l_2} \sum_{m_1 m_2} \delta_{l_1 + l_2,\, k}(-1)^{k-m} \left(\frac{(2k + 1)!}{(2l_1)!\,(2l_2)!}\right)^{1/2}$$

$$\times R_{l_1 m_1}(\mathbf{r}_1)\, R_{l_2 m_2}(\mathbf{r}_2) \begin{pmatrix} l_1 & l_2 & k \\ m_1 & m_2 & m \end{pmatrix},$$

so that, remembering that $R_{lm}(-\mathbf{r}) = (-1)^l\, R_{lm}(\mathbf{r})$,

$$V = \frac{1}{4\pi\varepsilon_0} \sum_{i \in A,\; j \in B} \frac{e_i e_j}{|\mathbf{R} + \mathbf{r}_j - \mathbf{r}_i|}$$

$$= \frac{1}{4\pi\varepsilon_0} \sum_{l_1, l_2} \sum_{m_1 m_2 m} (-1)^{l_2} \left(\frac{(2l_1 + 2l_2 + 1)!}{(2l_1)!\,(2l_2)!}\right)^{1/2}$$

$$\times \sum_{i \in A} e_i R_{l_1 m_1}(\mathbf{r}_i) \sum_{j \in B} e_j R_{l_2 m_2}(\mathbf{r}_j)\, I_{l_1 + l_2,\, m}(\mathbf{R}) \begin{pmatrix} l_1 & l_2 & l_1 + l_2 \\ m_1 & m_2 & m \end{pmatrix}$$

$$= \frac{1}{4\pi\varepsilon_0} \sum_{l_1, l_2} \sum_{m_1 m_2 m} (-1)^{l_2} \left(\frac{(2l_1 + 2l_2 + 1)!}{(2l_1)!\,(2l_2)!}\right)^{1/2}$$

$$\times \hat{Q}^A_{l_1 m_1} \hat{Q}^B_{l_2 m_2} I_{l_1 + l_2,\, m}(\mathbf{R}) \begin{pmatrix} l_1 & l_2 & l_1 + l_2 \\ m_1 & m_2 & m \end{pmatrix}. \tag{7}$$

where in the last line we have introduced the multipole moment operators:

$$\hat{Q}^A_{lm} = \sum_{i \in A} e_i R_{lm}(\mathbf{r}_i)\,.$$

The multipole moment operators in Eq. (7) are still referred to the global coordinate frame. We now transform them to the local or molecule-fixed frame:

$$\hat{Q}^A_{lm} = \sum_k \tilde{Q}^A_{lk} D^l_{mk}(\Omega_A)^*\,,$$

where $D^l_{mk}(\Omega_A)$ is a Wigner rotation matrix whose argument Ω_A describes the orientation of molecule A. In this way we reach the following expression for the interaction operator:

$$V = \frac{1}{4\pi\varepsilon_0} \sum_{l_1, l_2} \sum_{k_1 k_2} \begin{pmatrix} l_1 + l_2 \\ l_1 \end{pmatrix} R^{-l_1 - l_2 - 1} \tilde{Q}^A_{l_1 k_1} \tilde{Q}^B_{l_2 k_2} \bar{S}^{k_1 k_2}_{l_1 l_2 l_1 + l_2}\,. \tag{8}$$

Note that the similar expression given in Ref. 2 is in error; it contains a spurious factor of $(-1)^{l_1+l_2}$. In (8) $\begin{pmatrix} l_1 + l_2 \\ l_1 \end{pmatrix}$ is a binomial coefficient, and the S-function $\bar{S}^{k_1 k_2}_{l_1 l_2 j}$ is defined by

$$\bar{S}^{k_1 k_2}_{l_1 l_2 j} = i^{l_1 - l_2 - j} \begin{pmatrix} l_1 & l_2 & j \\ 0 & 0 & 0 \end{pmatrix}^{-1} \sum_{m_1 m_2 m} \begin{pmatrix} l_1 & l_2 & j \\ m_1 & m_2 & m \end{pmatrix}$$

$$\times D^{l_1}_{m_1 k_1}(\Omega_A)^* \, D^{l_2}_{m_2 k_2}(\Omega_B)^* \, C_{jm}(\theta, \phi). \tag{9}$$

In (9), D^l_{mk} is a Wigner rotation matrix, and the arguments Ω_A and Ω_B describe the orientations of the two molecules. The angles θ and ϕ describe the direction of the intermolecular vector $\mathbf{R}$.

It is convenient to define a new quantity $T^A{}_{l_1 k_1}{}^B{}_{l_2 k_2}$ as follows:

$$T^A{}_{l_1 k_1}{}^B{}_{l_2 k_2} = \frac{1}{4\pi\varepsilon_0} \begin{pmatrix} l_1 + l_2 \\ l_1 \end{pmatrix} R^{-l_1 - l_2 - 1} \bar{S}^{k_1 k_2}_{l_1 l_2 l_1 + l_2}, \tag{10}$$

and in terms of this the interaction becomes

$$V = \sum_{l_1, l_2} \sum_{k_1 k_2} \tilde{Q}^A_{l_1 k_1} T^A{}_{l_1 k_1}{}^B{}_{l_2 k_2} \tilde{Q}^B_{l_2 k_2},$$

in which the $T^A{}_{l_1 k_1}{}^B{}_{l_2 k_2}$ take the place of the interaction tensors $T_{\alpha\beta\ldots\nu}$ of the cartesian formulation.

Eq. (8) is the starting-point for further manipulations using angular momentum methods. For our present purposes, however, we now transform from the standard spherical tensor components $\tilde{Q}_{lk}$ to closely-related real quantities, since this gives an expression which is easier to comprehend and more suited to computation. Following Griffith [4], we define the quantities Q_{lkc} and Q_{lks}, for $k > 0$, by

$$Q_{lkc} = \sqrt{\frac{1}{2}} \left[(-1)^k Q_{lk} + Q_{l,-k}' \right],$$

$$iQ_{lks} = \sqrt{\frac{1}{2}} \left[(-1)^k Q_{lk} - Q_{l,-k} \right]. \tag{11}$$

No transformation is needed for Q_{l0}, which is always real. The notation reflects the fact that Q_{lkc} transforms like $\cos k\phi$ and Q_{lks} like $\sin k\phi$. The factors of $\sqrt{1/2}$ ensure that a rotation of axes induces an orthogonal transformation of the moments. The first few of these moments coincide precisely with the Cartesian charge and dipole moment, and later ones describe the quadrupole, octopole and so on. A complete list is given in Table 1 for moments up to hexadecapole.

Some authors use a different notation for the new components, writing, for example, $Z_{l,-k}$ for Q_{lks} and Z_{lk} for Q_{lkc}, with $k > 0$. This has the disadvantage that it becomes necessary to introduce a new symbol (here Z) for the new components, somewhat

Table 1. Definitions of the multipole moments to rank 4, and equivalent expressions in Cartesian tensor notation

---

$$Q_{00} = \sum_i e_i \qquad\qquad = q$$

$$Q_{10} = \sum_i e_i z_i \qquad\qquad = \mu_z$$

$$Q_{11c} = \sum_i e_i x_i \qquad\qquad = \mu_x$$

$$Q_{11s} = \sum_i e_i y_i \qquad\qquad = \mu_y$$

$$Q_{20} = \sum_i e_i \frac{1}{2}(3z_i^2 - r_i^2) \qquad\qquad = \Theta_{zz}$$

$$Q_{21c} = \sum_i e_i \sqrt{3}\, x_i z_i \qquad\qquad = \sqrt{\frac{4}{3}}\, \Theta_{xz}$$

$$Q_{21s} = \sum_i e_i \sqrt{3}\, y_i z_i \qquad\qquad = \sqrt{\frac{4}{3}}\, \Theta_{yz}$$

$$Q_{22c} = \sum_i e_i \frac{1}{2}\sqrt{3}(x_i^2 - y_i^2) \qquad\qquad = \sqrt{\frac{1}{3}}(\Theta_{xx} - \Theta_{yy})$$

$$Q_{22s} = \sum_i e_i \sqrt{3}\, x_i y_i \qquad\qquad = \sqrt{\frac{4}{3}}\, \Theta_{xy}$$

$$Q_{30} = \sum_i e_i \frac{1}{2}(5z_i^3 - 3z_i r_i^2) \qquad\qquad = \Omega_{zzz}$$

$$Q_{31c} = \sum_i e_i \sqrt{\frac{3}{8}}\, x_i(5z_i^2 - r_i^2) \qquad\qquad = \sqrt{\frac{3}{2}}\, \Omega_{xzz}$$

$$Q_{31s} = \sum_i e_i \sqrt{\frac{3}{8}}\, y_i(5z_i^2 - r_i^2) \qquad\qquad = \sqrt{\frac{3}{2}}\, \Omega_{yzz}$$

$$Q_{32c} = \sum_i e_i \frac{1}{2}\sqrt{15}\, z_i(x_i^2 - y_i^2) \qquad\qquad = \sqrt{\frac{3}{5}}(\Omega_{xxz} - \Omega_{yyz})$$

$$Q_{32s} = \sum_i e_i \sqrt{15}\, x_i y_i z_i \qquad\qquad = 2\sqrt{\frac{3}{5}}\, \Omega_{xyz}$$

$$Q_{33c} = \sum_i e_i \sqrt{\frac{5}{8}}(x_i^3 - 3x_i y_i^2) \qquad\qquad = \sqrt{\frac{1}{10}}(\Omega_{xxx} - 3\Omega_{xyy})$$

$$Q_{33s} = \sum_i e_i \sqrt{\frac{5}{8}}(3x_i^2 y_i - y_i^3) \qquad\qquad = \sqrt{\frac{1}{10}}(3\Omega_{xxy} - \Omega_{yyy})$$

$$Q_{40} = \sum_i e_i \frac{1}{8}(35z_i^4 - 30z_i^2 r_i^2 + 3r_i^4) \quad = \Phi_{zzzz}$$

$$Q_{41c} = \sum_i e_i \sqrt{\frac{5}{8}}(7x_i z_i^3 - 3x_i z_i r_i^2) \qquad = \sqrt{\frac{8}{5}}\, \Phi_{xzzz}$$

$$Q_{41s} = \sum_i e_i \sqrt{\frac{5}{8}}(7y_i z_i^3 - 3y_i z_i r_i^2) \qquad = \sqrt{\frac{8}{5}}\, \Phi_{yzzz}$$

$$Q_{42c} = \sum_i e_i \frac{1}{4}\sqrt{5}(x_i^2 - y_i^2)(7z_i^2 - r_i^2) = \sqrt{\frac{4}{5}}(\Phi_{xxzz} - \Phi_{yyzz})$$

Table 1. (Continued)

$$Q_{42s} = \sum_i e_i \frac{1}{2} \sqrt{5} \, x_i y_i (7z_i^2 - r_i^2) \qquad = \sqrt{\frac{16}{5}} \, \Phi_{xyzz}$$

$$Q_{43c} = \sum_i e_i \sqrt{\frac{35}{8}} \, z_i (x_i^3 - 3x_i y_i^2) \qquad = \sqrt{\frac{8}{35}} \, (\Phi_{xxxz} - 3\Phi_{xyyz})$$

$$Q_{43s} = \sum_i e_i \sqrt{\frac{35}{8}} \, z_i (3x_i^2 y_i - y_i^3) \qquad = \sqrt{\frac{8}{35}} \, (3\Phi_{xxyz} - \Phi_{yyyz})$$

$$Q_{44c} = \sum_i e_i \frac{1}{8} \sqrt{35} (x_i^4 - 6x_i^2 y_i^2 + y_i^4) = \sqrt{\frac{1}{35}} \, (\Phi_{xxxx} - 6\Phi_{xxyy} + \Phi_{yyyy})$$

$$Q_{44s} = \sum_i e_i \frac{1}{2} \sqrt{35} (x_i^3 y_i - x_i y_i^3) \qquad = \sqrt{\frac{16}{35}} \, (\Phi_{xxxy} - \Phi_{xyyy})$$

obscuring the fact that they are so closely related to the old ones. However this is a matter of personal preference, and does not in any way affect what follows.

It is convenient to arrange the moments into a vector **Q** whose components are $Q_{00}, Q_{10}, Q_{11c}, Q_{11s}, Q_{20}, Q_{21c}, Q_{21s}, Q_{22c}, Q_{22s}, Q_{30}, \ldots$ We adopt here a convention that the elements are listed in order of increasing l, for given l in order of increasing $|k|$, and for given nonzero $|k|$ with the c component before the s. This too is a matter of personal preference, and the order in which the components are listed is arbitrary.

We now abbreviate the notation further by using the single subscript t to label these components. A different convention for the ordering of the components now affects only the mapping from the subscript t on to the explicit labels lkc and lks, or $l, -k$ and lk, so the theory that follows can be used with either convention.

Since we have transformed the Q_{lk} to real form, we must do the same with the $\bar{S}$ functions. The transformation (11) can be written in the form

$$Q_{lk} = \sum_\varkappa Q_{l\varkappa} X_{\varkappa k}, \qquad (12)$$

where $\varkappa$ denotes one of the labels $0, 1c, 1s$, etc., and the matrix X has the form

$$
X =
\begin{array}{c}
\\
0 \\
1c \\
1s \\
2c \\
2s \\
\vdots
\end{array}
\begin{array}{cccccc}
\cdots & -2 & -1 & 0 & 1 & 2 \quad \cdots \\
\left(\cdots\right. & 0 & 0 & 1 & 0 & 0 \left.\cdots\right) \\
\cdots & 0 & \sqrt{\tfrac{1}{2}} & 0 & -\sqrt{\tfrac{1}{2}} & 0 \quad \cdots \\
\cdots & 0 & -i\sqrt{\tfrac{1}{2}} & 0 & -i\sqrt{\tfrac{1}{2}} & 0 \quad \cdots \\
\cdots & \sqrt{\tfrac{1}{2}} & 0 & 0 & 0 & \sqrt{\tfrac{1}{2}} \quad \cdots \\
\cdots & -i\sqrt{\tfrac{1}{2}} & 0 & 0 & 0 & i\sqrt{\tfrac{1}{2}} \quad \cdots \\
\vdots & \vdots & \vdots & \vdots & \vdots
\end{array}
$$

Substituting Eq. (12) into Eq. (8) then gives

$$V = \frac{1}{4\pi\varepsilon_0} \sum_{l_1 l_2} \sum_{k_1 k_2} \sum_{\varkappa_1 \varkappa_2} \binom{l_1 + l_2}{l_1}$$

$$\times R^{-l_1-l_2-1} \tilde{Q}^A_{l_1\varkappa_1} \tilde{Q}^B_{l_2\varkappa_2} X_{\varkappa_1 k_1} X_{\varkappa_2 k_2} \bar{S}^{k_1 k_2}_{l_1 l_2 l_1 + l_2},$$

$$\equiv \frac{1}{4\pi\varepsilon_0} \sum_{l_1 l_2} \sum_{\varkappa_1 \varkappa_2} \binom{l_1 + l_2}{l_1} R^{-l_1-l_2-1} \tilde{Q}^A_{l_1\varkappa_1} \tilde{Q}^B_{l_2\varkappa_2} \bar{S}^{\varkappa_1 \varkappa_2}_{l_1 l_2 l_1 + l_2},$$

and we see that the $\bar{S}$ function corresponding to the real multipole components labelled by $\varkappa_1$ and $\varkappa_2$ is

$$\bar{S}^{\varkappa_1 \varkappa_2}_{l_1 l_2 l_1 + l_2} = \sum_{k_1 k_2} X_{\varkappa_1 k_1} X_{\varkappa_2 k_2} \bar{S}^{k_1 k_2}_{l_1 l_2 l_1 + l_2}.$$

4 The Electrostatic Interaction

We now have an expression for the electrostatic interaction operator between two charge distributions A and B. It takes the form

$$V^{AB} = \sum_{t, u} \tilde{Q}^A_t T^{AB}_{tu} \tilde{Q}^B_u, \tag{13}$$

where the sum runs over all moments t and u for A and B respectively.

Now it has been known for many years that the multipole expansion of the exact electrostatic interaction between two molecules is divergent at short range. Indeed it is formally divergent at any separation, since the molecular charge distribution extends to infinity, so that points $\mathbf{r}_i$ and $\mathbf{r}_j$ can be found within the two charge distributions that do not satisfy the condition $|\mathbf{r}_j - \mathbf{r}_i| < R$ required for Eq. (6). However it is possible to separate the electrostatic interaction into two parts. One of these is the long-range limit of the multipole expansion, and takes the form already described. The other is the correction that is needed at short range because of the overlap of the charge distributions, and decays exponentially with increasing distance. It is impossible to represent the latter part in terms of an expansion in powers of $1/R$, and it is much more satisfactory to regard it as part of the short-range interaction. If this is done, then the multipole expansion proper can be shown to converge at all separations for which non-overlapping spheres can be drawn to enclose the nuclei of each molecule [5, 6].

Unfortunately this is still too severe a requirement in many cases. It is often possible for molecules to approach to distances at which such spheres would overlap, without encountering the repulsive part of the potential; and even when they do not overlap, they may approach so closely that convergence becomes very slow. For this reason, it has become common to adopt a distributed multipole description, in which each molecule is divided into a number of regions, each described by its own multipole moments. There are many ways of determining these distributed multipole moments [6–11]; many authors have used distributed charges alone, but it is now widely accepted that an accurate and efficient description requires multipoles up to at least quadrupole.

Accordingly we allow for the possibility of such a distributed multipole description, in which a molecule, A say, may be divided into a number of regions, labelled by a superscript a. Thus Q_t^a is a moment for region a of molecule A. Each region requires an origin, since the multipole moments depend in general on the choice of origin, and the origin for region a will be called 'site a', and will be at position $\mathbf{a}$ relative to the molecular origin at $\mathbf{A}$. All of the previous working still applies, except that the vector $\mathbf{R}$ is now $\mathbf{B} + \mathbf{b} - \mathbf{A} - \mathbf{a}$. The electrostatic interaction between two molecules now takes a slightly more general form than Eq. (13), in that we must sum over the regions of each molecule:

$$V^{AB} = \sum_{a,\,b} \sum_{t,\,u} \tilde{Q}_t^a T_{tu}^{ab} \tilde{Q}_u^b, \tag{14}$$

The conventional multipole description is the special case where each molecule is treated as a single region. The terms in this sum can be viewed diagrammatically as in Fig. 1a, where each small circle represents a site and the line joining sites represents the interaction function T.

The first-order electrostatic energy is now the expectation value of the operator (14) over the molecular wavefunction, i.e.

$$U^{AB} = Q_t^a T_{tu}^{ab} Q_u^b.$$

Here and subsequently we adopt a repeated-index summation convention: if a suffix t appears twice in the same term of an expression, summation over all moments on a particular site is implied, while a repeated site index a implies summation over all sites a in the molecule concerned. Sums over molecules will always be indicated explicitly.

It is apparent that the T_{tu}^{ab} take the place in this formulation of the interaction tensors $T_{\alpha\beta\gamma...}$ of the conventional Cartesian formulation, but it should be emphasized once again that all the formulae given here refer to multipole moment components in the local, molecule-fixed frame of each molecule, whereas the corresponding Cartesian formulae deal in space-fixed components throughout and require a separate transformation between molecule-fixed and space-fixed frames. ('Space-fixed' is perhaps a misleading term here, since the calculation is commonly carried out in a coordinate system with one of its axes along the intermolecular vector. However, the point is that in the Cartesian tensor notation there has to be a common set of axes for the system as a whole, and this can be the molecule-fixed frame for at most one of the molecules involved.)

The T_{tu}^{ab} are functions of relative orientation, and we should now consider how the relative orientation is to be described. It is common to describe relative orientation by means of Euler angles, but this is cumbersome and would entail the evaluation of trigonometric functions. Instead, we can describe the relative orientations that we need in terms of direction cosines, just as in Cartesian tensor theory. If the local axes of region a are described by unit vectors $\mathbf{e}_\alpha^a$ and those of region b by unit vectors $\mathbf{e}_\beta^b$, then the relative orientation of the two regions is described by the direction cosines

Table 2. Electrostatic energy formulae. The interaction function T_{tu}^{ab} for the electrostatic interaction between a multipole moment Q_t on site a and a moment Q_u on site b, both referred to local axes, is given in terms of the direction cosines r_α^a, r_β^b and $c_{\alpha\beta}$. If $\mathbf{e}_x^a$, $\mathbf{e}_y^a$ and $\mathbf{e}_z^a$ are the unit vectors defining the local axis system for site a, and $\mathbf{e}_x^b$, $\mathbf{e}_y^b$ and $\mathbf{e}_z^b$ similarly for site b, and $\mathbf{e}_{ab}$ is a unit vector in the direction from a to b, then $r_\alpha^a = \mathbf{e}_\alpha^a \cdot \mathbf{e}_{ab}$, $r_\beta^b = \mathbf{e}_\beta^b \cdot \mathbf{e}_{ba} = -\mathbf{e}_\beta^b \cdot \mathbf{e}_{ab}$ (note the minus sign) and $c_{\alpha\beta} = \mathbf{e}_\alpha^a \cdot \mathbf{e}_\beta^b$. The components of a dipole moment may be written as Q_{10}, Q_{11c} and Q_{11s}, or as Q_z, Q_x and Q_y

t	u	$4\pi\varepsilon_0 T_{tu}^{ab}$
00	00	R^{-1}
1α	00	$R^{-2} r_\alpha^a$
20	00	$R^{-3}(3r_z^{a2} - 1)/2$
21c	00	$R^{-3}\sqrt{3}r_x^a r_z^a$
21s	00	$R^{-3}\sqrt{3}r_y^a r_z^a$
22c	00	$R^{-3} \cdot \frac{1}{2}\sqrt{3}(r_x^{a2} - r_y^{a2})$
22s	00	$R^{-3}\sqrt{3}r_x^a r_y^a$
1α	1β	$R^{-3}(3r_\alpha^a r_\beta^b + c_{\alpha\beta})$
30	00	$R^{-4}(5r_z^{a3} - 3r_z^a)/2$
31c	00	$R^{-4} \cdot \frac{1}{4}\sqrt{6}r_x^a(5r_z^{a2} - 1)$
31s	00	$R^{-4} \cdot \frac{1}{4}\sqrt{6}r_y^a(5r_z^{a2} - 1)$
32c	00	$R^{-4} \cdot \frac{1}{2}\sqrt{15}r_z^a(r_x^{a2} - r_y^{a2})$
32s	00	$R^{-4}\sqrt{15}r_x^a r_y^a r_z^a$
33c	00	$R^{-4} \cdot \frac{1}{4}\sqrt{10}r_x^a(r_x^{a2} - 3r_y^{a2})$
33s	00	$R^{-4} \cdot \frac{1}{4}\sqrt{10}\,r_y^a(3r_x^{a2} - r_y^{a2})$
20	1β	$R^{-4} \cdot \frac{1}{2}(15r_z^{a2}r_\beta^b + 6r_z^a c_{z\beta} - 3r_\beta^b)$
21c	1β	$R^{-4}\sqrt{3}(r_x^a c_{z\beta} + c_{x\beta}r_z^a + 5r_x^a r_z^a r_\beta^b)$
21s	1β	$R^{-4}\sqrt{3}(r_y^a c_{z\beta} + c_{y\beta}r_z^a + 5r_y^a r_z^a r_\beta^b)$
22c	1β	$R^{-4} \cdot \frac{1}{2}\sqrt{3}(5(r_x^{a2} - r_y^{a2})\,r_\beta^b + 2r_x^a c_{x\beta} - 2r_y^a c_{y\beta})$
22s	1β	$R^{-4}\sqrt{3}(5r_x^a r_y^a r_\beta^b + r_x^a c_{y\beta} + r_y^a c_{x\beta})$
40	00	$R^{-5} \cdot \frac{1}{8}(35r_z^{a4} - 30r_z^{a2} + 3)$
41c	00	$R^{-5} \cdot \frac{1}{4}\sqrt{10}(7r_x^a r_z^{a3} - 3r_x^a r_z^a)$
41s	00	$R^{-5} \cdot \frac{1}{4}\sqrt{10}(7r_y^a r_z^{a3} - 3r_y^a r_z^a)$
42c	00	$R^{-5} \cdot \frac{1}{4}\sqrt{5}(7r_z^{a2} - 1)\,(r_x^{a2} - r_y^{a2})$

Table 2. (continued)

t	u	$4\pi\varepsilon_0 T_{tu}^{ab}$
$42s$	00	$R^{-5} \cdot \dfrac{1}{2}\sqrt{5}(7r_z^{a2} - 1)\, r_x^a r_y^a$
$43c$	00	$R^{-5} \cdot \dfrac{1}{4}\sqrt{70}\, r_x^a r_z^a (r_x^{a2} - 3r_y^{a2})$
$43s$	00	$R^{-5} \cdot \dfrac{1}{4}\sqrt{70}\, r_y^a r_z^a (3r_x^{a2} - r_y^{a2})$
$44c$	00	$R^{-5} \cdot \dfrac{1}{8}\sqrt{35}\ (r_x^{a4} - 6r_x^{a2}r_y^{a2} + r_y^{a4})$
$44s$	00	$R^{-5} \cdot \dfrac{1}{2}\sqrt{35}\, r_x^a r_y^a (r_x^{a2} - r_y^{a2})$
30	1β	$R^{-5} \cdot \dfrac{1}{2}(35 r_z^{a3} r_\beta^b + 15 r_z^{a2} c_{z\beta} - 15 r_z^a r_\beta^b - 3c_{z\beta})$
$31c$	1β	$R^{-5} \cdot \dfrac{1}{4}\sqrt{6}(35 r_x^a r_z^{a2} r_\beta^b + 5r_z^{a2} c_{x\beta} + 10 r_x^a r_z^a c_{z\beta} - 5r_x^a r_\beta^b - c_{x\beta})$
$31s$	1β	$R^{-5} \cdot \dfrac{1}{4}\sqrt{6}(35 r_y^a r_z^{a2} r_\beta^b + 5r_z^{a2} c_{y\beta} + 10 r_y^a r_z^a c_{z\beta} - 5r_y^a r_\beta^b - c_{y\beta})$
$32c$	1β	$R^{-5} \cdot \dfrac{1}{2}\sqrt{15}((r_x^{a2} - r_y^{a2})(7r_z^a r_\beta^b + c_{z\beta}) + 2r_z^a(r_x^a c_{x\beta} - r_y^a c_{y\beta}))$
$32s$	1β	$R^{-5}\sqrt{15}(r_x^a r_y^a(7r_z^a r_\beta^b + c_{z\beta}) + r_z^a(r_x^a c_{y\beta} + r_y^a c_{x\beta}))$
$33c$	1β	$R^{-5} \cdot \dfrac{1}{4}\sqrt{10}(7r_x^{a3} r_\beta^b + 3(r_x^{a2} - r_y^{a2})\, c_{x\beta} - 21 r_x^a r_y^{a2} r_\beta^b - 6r_x^a r_y^a c_{y\beta})$
$33s$	1β	$R^{-5} \cdot \dfrac{1}{4}\sqrt{10}(-7r_y^{a3} r_\beta^b + 3(r_x^{a2} - r_y^{a2})\, c_{y\beta} + 21 r_x^{a2} r_y^a r_\beta^b + 6r_x^a r_y^a c_{x\beta})$
20	20	$R^{-5} \cdot \dfrac{3}{4}\ (35 r_z^{a2} r_z^{b2} - 5(r_z^{a2} + r_z^{b2}) + 20 r_z^a r_z^b c_{zz} + 2c_{zz}^2 + 1)$
20	$21c$	$R^{-5} \cdot \dfrac{1}{2}\sqrt{3}\ (35 r_z^{a2} r_x^b r_z^b - 5r_x^b r_z^b + 2c_{zx}c_{zz} + 10 r_z^a r_z^b c_{zx} + 10 r_z^a r_x^b c_{zz})$
20	$21s$	$R^{-5} \cdot \dfrac{1}{2}\sqrt{3}(35 r_z^{a2} r_y^b r_z^b - 5r_y^b r_z^b + 2c_{zy}c_{zz} + 10 r_z^a r_z^b c_{zy} + 10 r_z^a r_y^b c_{zz})$
20	$22c$	$R^{-5} \cdot \dfrac{1}{4}\sqrt{3}(5(7r_z^{a2} - 1)(r_x^{b2} - r_y^{b2}) + 20 r_z^a r_x^b c_{zx} - 20 r_z^a r_y^b c_{zy} + 2(c_{zx}^2 - c_{zy}^2))$
20	$22s$	$R^{-5} \cdot \dfrac{1}{2}\sqrt{3}(5(7r_z^{a2} - 1)\, r_x^b r_y^b + 10 r_z^a r_x^b c_{zy} + 10 r_z^a r_y^b c_{zx} + 2c_{zx}c_{zy})$
$21c$	$21c$	$R^{-5}(35 r_x^a r_z^a r_x^b r_z^b + 10 r_z^a r_x^b c_{xz} + 10 r_x^a r_z^b c_{zx} + 2c_{xx}c_{zz} - 5r_y^a r_y^b + 6c_{yy})$
$21c$	$21s$	$R^{-5}(35 r_x^a r_z^a r_y^b r_z^b + 10 r_z^a r_y^b c_{zy} + 10 r_x^a r_z^b c_{xz} + 2c_{xy}c_{zz} + 5r_x^p r_y^{"} + 6c_{yx})$
$21c$	$22c$	$R^{-5} \cdot \dfrac{1}{2}(35 r_x^a r_z^a(r_x^{b2} - r_y^{b2}) + 10 r_z^a r_x^b c_{zx} + 10 r_z^a r_x^b c_{xx} - 10 r_z^a r_y^b c_{xy} - 10 r_x^a r_y^b c_{zy}$ $+ 2c_{xx}c_{zx} - 2c_{xy}c_{zy})$
$21c$	$22s$	$R^{-5}(35 r_x^a r_z^a r_x^b r_y^b + 5r_z^a r_x^b c_{xy} + 5r_z^a r_y^b c_{xx} + 5r_z^a r_y^b c_{zy} + 5r_x^a r_y^b c_{zx} + c_{xx}c_{xy} + c_{xy}c_{zx})$
$21s$	$21s$	$R^{-5}(35 r_y^a r_z^a r_y^b r_z^b + 10 r_z^a r_y^b c_{yz} + 10 r_y^a r_z^b c_{zy} + 2c_{yy}c_{zz} - 5r_x^a r_x^b + 6c_{xx})$

Table 2. (continued)

t	u	$4\pi\varepsilon_0 T_{tu}^{ab}$
$21s$	$22c$	$R^{-5} \cdot \frac{1}{2}(35 r_y^a r_z^a (r_x^{b2} - r_y^{b2}) + 10 r_z^a r_x^b c_{yx} - 10 r_z^a r_y^b c_{yy} + 10 r_y^a r_x^b c_{zx} - 10 r_y^a r_y^b c_{zy}$ $+ 2c_{yx}c_{zx} - 2c_{yy}c_{zy})$
$21s$	$22s$	$R^{-5}(35 r_y^a r_z^a r_x^b r_y^b + 5 r_z^a r_x^b c_{yy} + 5 r_z^a r_y^b c_{yx} + 5 r_y^a r_x^b c_{zy} + 5 r_y^a r_y^b c_{zx} + c_{yx}c_{zy} + c_{yy}c_{zx})$
$22c$	$22c$	$R^{-5} \cdot \frac{1}{4}(35 (r_x^{a2} - r_y^{a2})(r_x^{b2} - r_y^{b2}) + 20 r_x^a r_x^b c_{xx} - 20 r_x^a r_y^b c_{xy} - 20 r_y^a r_x^b c_{yx}$ $+ 20 r_y^a r_y^b c_{yy} + 2(c_{xx} - c_{xy}^2 - c_{yx}^2 + c_{yy}^2))$
$22c$	$22s$	$R^{-5} \cdot \frac{1}{2}(35 (r_x^{a2} - r_y^{a2}) r_x^b r_y^b + 10 r_x^a r_x^b c_{xy} + 10 r_x^a r_y^b c_{xx} - 10 r_y^a r_y^b c_{yx} - 10 r_y^a r_x^b c_{yy}$ $+ 2c_{xx}c_{xy} - 2c_{yx}c_{yy})$
$22s$	$22s$	$R^{-5}(35 r_x^a r_y^a r_x^b r_y^b + 5 r_x^a r_x^b c_{yy} + 5 r_x^a r_y^b c_{yx} + 5 r_y^a r_x^b c_{xy} + 5 r_y^a r_y^b c_{xx}$ $+ c_{xx}c_{yy} + c_{xy}c_{yx})$

$c_{\alpha\beta} = \mathbf{e}_\alpha^a \cdot \mathbf{e}_\beta^b$. The direction of the inter-site vector $\mathbf{r}^{ab} = \mathbf{B} + \mathbf{b} - \mathbf{A} - \mathbf{a}$ relative to region a is given by the direction cosines $r_\alpha^a = \mathbf{e}_\alpha^a \cdot \mathbf{e}_{ab}$, where $\mathbf{e}_{ab}$ is a unit vector in the direction of $\mathbf{B} + \mathbf{b} - \mathbf{A} - \mathbf{a}$, while its direction relative to region b is given by the direction cosines $r_\beta^b = \mathbf{e}_\beta^b \cdot \mathbf{e}_{ba}$, where $\mathbf{e}_{ba} = -\mathbf{e}_{ab}$. This definition of the direction cosines makes the formulae more symmetrical.

The reader will realise that this description of the relative orientations is somewhat redundant. Only five parameters are needed to describe the relative orientation of two molecules [1, 12], whereas we have nine $c_{\alpha\beta}$, three r_α^a and three r_β^b. However, it is in practice more convenient to use this redundant description of the orientation than to attempt a more economical one, because the direction cosines are so easily calculated.

The T_{tu}^{ab} are invariant under a rotation of the entire system, since they depend only on a length (the separation R) and on scalar products (the direction cosines), and are also unchanged by simultaneous exchange of the subscripts and superscripts: $T_{tu}^{ab} = T_{ut}^{ba}$. All the T_{tu}^{ab} to terms in R^{-5} have been calculated [13], and are tabulated in Table 2. Notice that the orientations Ω^a and Ω^b of the two sites do not appear explicitly in these expressions, but only implicitly through the direction cosines r_α^a etc. In fact it is never necessary to evaluate any trigonometric functions to obtain the T_{tu}^{ab}. A computer program in FORTRAN 77 is available from the author to evaluate the T_{tu}^{ab} and the electrostatic energy for a system of multipoles.

4.1 Manipulation of the Multipole Description

There is always an element of arbitrariness in the multipole description of a charge distribution. Even when the conventional single-site description is used, the position of the origin can be chosen freely. However the values of all multipole moments of higher rank than the first nonvanishing moment depend on the choice of origin. If we have a set of multipoles $Q_{lk}(\mathbf{a})$ referred to an origin at $\mathbf{a}$, then the same charge distribution may be represented by multipoles $Q_{l'k'}(\mathbf{a}')$ referred to an origin at $\mathbf{a}'$ [7]:

$$Q_{l'k'}(\mathbf{a}') = \sum_{l=0}^{l'} \sum_{k=-l}^{l} \left[\binom{l'+k'}{l+k} \binom{l'-k'}{l-k} \right]^{1/2} Q_{lk}(\mathbf{a}) R_{l'-l,\,k'-k}(\mathbf{a} - \mathbf{a}'). \quad (15)$$

This equation can be used to express the overall multipole moments Q_{LK} of a molecule in terms of the distributed multipoles Q_{lk}^a, by referring all the moments to a common origin:

$$Q_{LK}(0) = \sum_{a} \sum_{l=0}^{L} \sum_{k=-l}^{l} \left[\binom{L+K}{l+k} \binom{L-K}{l-k} \right]^{1/2} Q_{lk}^a R_{L-l,\,K-k}(\mathbf{a}). \quad (16)$$

This equation also applies to the *operators* for the multipole moments of the individual regions. There is a degree of arbitrariness in the way that the molecule is divided into regions [7]; but however the regions are defined, the multipole operators for the individual regions must be related to the multipole operators for the molecule as a whole by Eq. (16).

Notice that the sum over l in Eq. (15) runs over values up to l', so that the multipole moment of rank l' at the new origin depends on the values of moments of ranks less than or equal to l' at the old origin. An alternative way of expressing this is to say that a moment of rank l referred to the old origin is equivalent to a set of moments of ranks l and above at the new origin. This means that we can modify a multipole representation of a charge distribution by suppressing a particular moment at $\mathbf{a}$ and replacing it by a set of moments at $\mathbf{a}'$. If the suppressed moment at $\mathbf{a}$ is $Q_{lk}(\mathbf{a})$, then the moments at $\mathbf{a}'$ that replace it are $Q_{l'k'}(\mathbf{a}') = Q_{lk}(\mathbf{a}) W_{lk,\,l'k'}(\mathbf{a} - \mathbf{a}')$, where the coefficient is given by [14]

$$W_{lkl'k'}(\mathbf{x}) = \left[\binom{l'+k'}{l+k} \binom{l'-k'}{l-k} \right]^{1/2} R_{l'-l,\,k'-k}(\mathbf{x}).$$

Similar formulae can be used to describe changes of origin for the real components. It turns out [14] that when a moment $Q_{lx}(\mathbf{a})$ at site $\mathbf{a}$ is suppressed and replaced by the set of moments $Q_{l'x'}(\mathbf{a}') = Q_{lx}(\mathbf{a}) W_{lx,\,l'x'}(\mathbf{a} - \mathbf{a}')$ at the new origin $\mathbf{a}'$, the coefficients are given by

$$W_{lkc,\,l'k'c}(\mathbf{x}) = W_{lks,\,l'k's}(\mathbf{x})$$
$$= \left[\binom{l'+k'}{l+k} \binom{l'-k'}{l-k} \right]^{1/2} \frac{b_{k'}}{2 b_k b_{k'-k}} R_{l'-l,\,k'-k,\,c}(\mathbf{x}),$$

$$W_{lkc,\,l'k's}(\mathbf{x}) = -W_{lks,\,l'k'c}(\mathbf{x})$$
$$= \left[\binom{l'+k'}{l+k} \binom{l'-k'}{l-k} \right]^{1/2} \frac{b_{k'}}{2 b_k b_{k'-k}} R_{l'-l,\,k'-k,\,s}(\mathbf{x}), \quad (17)$$

where

$$
b_m = \begin{cases}
(-1)^m \sqrt{\dfrac{1}{2}}\,, & m > 0\,, \\[2mm]
\dfrac{1}{2}\,, & m = 0\,, \\[2mm]
\sqrt{\dfrac{1}{2}}\,, & m < 0\,.
\end{cases}
$$

When a distributed multipole description is used, the same principles apply. However, when there are several multipole sites in the same molecule, the moments can be moved from site to site using these formulae, so that for instance moments of ranks above dipole at a hydrogen site, which are usually small, can be moved to the neighbouring heavy-atom site, giving a description in which the hydrogen atoms carry no quadrupole, octopole and higher moments, but the effects of the small high-rank moments originally on the hydrogen atoms are correctly described by adjustments of the moments on the neighbouring atom. This gives a more economical description, with fewer parameters, but with no loss of accuracy. It is important however that moments are not moved too far; if they are, then the resulting multipole expansion will not converge sufficiently well.

5 Potential, Electric Field and Field Gradient

The formulae given above describe the interaction energy of a pair of molecules. However the electrostatic potential due to a charge distribution is the energy of a unit test charge. Consequently we obtain the potential at $\mathbf{a}$ due to molecule B as the interaction energy between it and a 'molecule' A at $\mathbf{a}$ whose only multipole moment is a unit charge. In this way we find

$$
V(\mathbf{a}) = T_{00u}^{a\,b}Q_u^b\,. \tag{18}
$$

(Remember that a sum over the repeated suffixes b and u is implied.)

This simple idea can be generalised to give other useful formulae. The electric field $\mathbf{E}$ is minus the derivative of the potential, but it is not most efficiently obtained by differentiating Eq. (18). Instead we recall that the energy of a dipole μ in an electric field is $-\mu \cdot \mathbf{E}$, so the component E_α of the field is minus the coefficient of the dipole moment μ_α in the interaction energy:

$$
E_\alpha(\mathbf{a}) = -T_{1\alpha u}^{a\,b}Q_u^b\,.
$$

Similarly, the energy of a quadrupole Q_{2k} in an electric field gradient with components V_{2k} is $\Sigma_k(-1)^k V_{2k}Q_{2,-k} = \Sigma_\varkappa V_{2\varkappa}Q_{2\varkappa}$ (remember that $\varkappa$ denotes one of the labels

0, 1c, 1s, etc.) and the field gradient is therefore obtained from the coefficient of the quadrupole moment in the interaction energy:

$$V_{2\varkappa}(\mathbf{a}) = T^{a\ b}_{2\varkappa u} Q^b_u .$$

The quantities $V_{2\varkappa}(\mathbf{a})$ that appear here are second derivatives of the electrostatic potential, evaluated at the point $\mathbf{a}$ of interest. They take the form

$$V_{2\varkappa}(\mathbf{a}) = \frac{1}{3} R_{2\varkappa}(\nabla)\ V(\mathbf{r})|_{\mathbf{r}=\mathbf{a}},$$

where $R_{2\varkappa}(\nabla)$ is a regular spherical harmonic in which the components x, y, z of $\mathbf{r}$ are replaced by the components ∇_x, ∇_y, ∇_z of the gradient operator ∇. Specifically, in terms of the cartesian derivatives $V_{\alpha\beta} = \partial^2 V / \partial x_\alpha \partial x_\beta$,

$$V_{20} = \frac{1}{6}\ (3V_{zz} - \nabla^2 V) = \frac{1}{2}\ V_{zz} ,$$

$$V_{21c} = \frac{1}{3}\ \sqrt{3} V_{xz} ,$$

$$V_{21s} = \frac{1}{3}\ \sqrt{3} V_{yz} ,$$

$$V_{22c} = \frac{1}{6}\ \sqrt{3}(V_{xx} - V_{yy}) ,$$

$$V_{21c} = \frac{1}{3}\ \sqrt{3} V_{xy} .$$

Higher derivatives of the field are rarely needed, but the formulae are very similar. In general [15], the derivative $V_{l\varkappa}$ is defined by

$$V_{l\varkappa}(\mathbf{a}) = [1.3\ldots(2l-1)]^{-1} R_{l\varkappa}(\nabla)\ V(\mathbf{r})|_{\mathbf{r}=\mathbf{a}}, \qquad (19)$$

and can be evaluated from the expression

$$V_{l\varkappa}(\mathbf{a}) = T^{a\ b}_{l\varkappa u} Q^b_u . \qquad (20)$$

We shall find this formula useful in discussing induction.

6 Forces and Torques

The formulae for the electrostatic energy in Table 2 are expressed in terms of the distance R between the sites and the direction cosines r_α^a, r_β^b and $c_{\alpha\beta}$. The force on molecule B is minus the derivative of the energy with respect to the position $\mathbf{B}$ of its centre of mass, and is found by using the chain rule and remembering that $\mathbf{R} = \mathbf{B} + \mathbf{b} - \mathbf{A} - \mathbf{a}$:

$$\mathbf{F}^B = -\nabla_B U$$

$$= -\sum_\alpha \frac{\partial U}{\partial r_\alpha^a} \nabla_B r_\alpha^a - \sum_\beta \frac{\partial U}{\partial r_\beta^b} \nabla_B r_\beta^b - \frac{\partial U}{\partial R} \nabla_B R , \tag{21}$$

Accordingly we need the derivatives $\nabla_B r_\alpha^a$, $\nabla_B r_\beta^b$ and $\nabla_B R$:

$$\nabla_B R = \nabla_B (\mathbf{R} \cdot \mathbf{R})^{1/2} = \frac{1}{2} (\mathbf{R} \cdot \mathbf{R})^{-1/2} 2\mathbf{R} \nabla_B \cdot \mathbf{R} = R^{-1} \mathbf{R} ,$$

$$\nabla_B r_\alpha^a = \nabla_B (R^{-1} \mathbf{R} \cdot \mathbf{e}_\alpha^a) = R^{-1} \mathbf{e}_\alpha^a - R^{-3} \mathbf{R} (\mathbf{R} \cdot \mathbf{e}_\alpha^a) = R^{-1} \hat{Q}_\mathbf{R} \mathbf{e}_\alpha^a ,$$

$$\nabla_B r_\beta^b = \nabla_B (-R^{-1} \mathbf{R} \cdot \mathbf{e}_\beta^b) = -R^{-1} \mathbf{e}_\beta^b + R^{-3} \mathbf{R} (\mathbf{R} \cdot \mathbf{e}_\beta^b) = -R^{-1} \hat{Q}_\mathbf{R} \mathbf{e}_\beta^b ,$$

In these equations $\hat{Q}_\mathbf{R} = (1 - R^{-2} \mathbf{R}\mathbf{R} \cdot)$ is a projection operator that removes from its vector operand the component parallel to $\mathbf{R}$. (The symbol $\hat{Q}_\mathbf{R}$ is used because $\hat{Q}$ is the conventional notation for this type of projection operator; it should not be confused with the multipole moment operators.)

While these expressions are somewhat complicated, and the formula (21) is certainly very tedious to evaluate by hand, the important features to notice are first that the differentiation can be carried out very straightforwardly by an algebra program such as REDUCE [16], and secondly that the resulting formulae are expressed entirely in terms of the vectors $\mathbf{e}_\alpha^a$ and $\mathbf{e}_\beta^b$ describing the orientation of the sites, the inter-site vector $\mathbf{R}$, and the direction cosines describing their relative orientation. This means that it is no more difficult conceptually to calculate forces involving quadrupoles, octopoles and hexadecapoles than those between dipoles, and because the higher-rank interactions involve all the same basic quantities, they can be included in the calculation for little additional computational cost.

Torques are calculated in a similar fashion. If the energy contains a term of the form $\mathbf{s} \cdot \mathbf{t} = st \cos \theta$, where the vector $\mathbf{t}$ rotates with one of the molecules but $\mathbf{s}$ does not, then the torque $\mathbf{G}$ on that molecule is minus the derivative of the energy with respect to angular variables, here θ, and is a vector quantity with magnitude $st \sin \theta$ and direction perpendicular to both $\mathbf{s}$ and $\mathbf{t}$. That is, the torque arising from the energy $\mathbf{s} \cdot \mathbf{t}$ is the vector product $\mathbf{s} \times \mathbf{t}$. Any term in the energy involving a scalar product of two vectors both fixed in space clearly makes no contribution to the torque; less obviously, the same is true of a term involving the scalar product of two vectors that both rotate with the molecule, because the scalar product is then independent of orientation. A more rigorous derivation of these results is given in Ref. 13. Since there is a differentiation involved, the chain rule can be invoked in the same way as for the force:

$$\mathbf{G}_{B.} = -\hat{\mathbf{G}}_B U$$

$$= -\sum_\alpha \frac{\partial U}{\partial r_\alpha^a} \hat{\mathbf{G}}_B r_\alpha^a - \sum_\beta \frac{\partial U}{\partial r_\beta^b} \hat{\mathbf{G}}_B r_\beta^b - \sum_{\alpha\beta} \frac{\partial U}{\partial c_{\alpha\beta}} \hat{\mathbf{G}}_B c_{\alpha\beta} - \frac{\partial U}{\partial R} \hat{\mathbf{G}}_B R, \quad (22)$$

In evaluating this expression, we are looking for the torque about the centre of mass of molecule B, and if we take the origin $\mathbf{B}$ to be at the centre of mass, then $\mathbf{B}$ stays fixed as the molecule rotates, but the position vector $\mathbf{b}$ giving the displacement of the interaction sites rotates with the molecule, as do the vectors $\mathbf{e}_\beta^b$ describing its orientation. Accordingly, the torques are calculated as follows:

$$\hat{\mathbf{G}}_B(\mathbf{R} \cdot \mathbf{R}) = \hat{\mathbf{G}}_B[(\mathbf{B} + \mathbf{b} - \mathbf{A} - \mathbf{a}) \cdot (\mathbf{B} + \mathbf{b} - \mathbf{A} - \mathbf{a})]$$

$$= 2(\mathbf{B} - \mathbf{A} - \mathbf{a}) \times \mathbf{b})$$

$$= 2\mathbf{R} \times \mathbf{b},$$

(remembering that $\mathbf{b} \times \mathbf{b} = 0$) so

$$\hat{\mathbf{G}}_B R^n = \hat{\mathbf{G}}_B(\mathbf{R} \cdot \mathbf{R})^{n/2}$$

$$= \frac{1}{2} n (\mathbf{R} \cdot \mathbf{R})^{n/2-1} \cdot 2(\mathbf{R} \times \mathbf{b})$$

$$= n R^{n-2}(\mathbf{R} \times \mathbf{b}).$$

r_α^a involves the scalar product of the vector $\mathbf{e}_\alpha^a$, which remains fixed as molecule B rotates, with the vector $\mathbf{R} = \mathbf{B} + \mathbf{b} - \mathbf{A} - \mathbf{a}$, of which only $\mathbf{b}$ rotates with molecule B:

$$\hat{\mathbf{G}}_B r_\alpha^a = \hat{\mathbf{G}}_B(R^{-1} \mathbf{e}_\alpha^a \cdot (\mathbf{B} + \mathbf{b} - \mathbf{A} - \mathbf{a}))$$

$$= -R^{-3}(\mathbf{R} \times \mathbf{b}) \, \mathbf{e}_\alpha^a \cdot \mathbf{R} + R^{-1}\mathbf{e}_\alpha^a \times \mathbf{b}$$

$$= R^{-1}(\hat{Q}_R \mathbf{e}_\alpha^a) \times \mathbf{b}.$$

r_β^b, on the other hand, involves the scalar product of $\mathbf{e}_\beta^b$, which rotates with molecule B, and $\mathbf{R} = \mathbf{B} + \mathbf{b} - \mathbf{A} - \mathbf{a}$, of which $\mathbf{B} - \mathbf{A} - \mathbf{a}$ is the part that remains fixed:

$$\hat{\mathbf{G}}_B r_\beta^b = -\hat{\mathbf{G}}_B(R^{-1} \mathbf{e}_\beta^b \cdot (\mathbf{B} + \mathbf{b} - \mathbf{A} - \mathbf{a}))$$

$$= R^{-3}(\mathbf{R} \times \mathbf{b}) \, \mathbf{e}_\beta^b \cdot \mathbf{R} - R^{-1}(\mathbf{B} - \mathbf{A} - \mathbf{a}) \times \mathbf{e}_\beta^b$$

$$= R^{-1}[\mathbf{e}_\beta^b \times \mathbf{R} - (\hat{Q}_R \mathbf{e}_\beta^b) \times \mathbf{b}].$$

Finally, $c_{\alpha\beta}$ is the scalar product of $\mathbf{e}_\alpha^a$, which remains fixed, and $\mathbf{e}_\beta^b$, which rotates:

$$\hat{\mathbf{G}}_B c_{\alpha\beta} = \mathbf{e}_\alpha^a \times \mathbf{e}_\beta^b.$$

The torque on molecule A is derived in a similar fashion, using the basic results

$$
\begin{aligned}
\hat{\mathbf{G}}_A R^n &= -nR^{n-2}(\mathbf{R} \times \mathbf{a}) , \\
\hat{\mathbf{G}}_A r_\alpha^a &= R^{-1}[\mathbf{R} \times \mathbf{e}_\alpha^a - (\hat{Q}_R \mathbf{e}_\alpha^a) \times \mathbf{a}] , \\
\hat{\mathbf{G}}_A r_\beta^b &= R^{-1}(\hat{Q}_R \mathbf{e}_\beta^b) \times \mathbf{a} , \\
\hat{\mathbf{G}}_A c_{\alpha\beta} &= \mathbf{e}_\beta^b \times \mathbf{e}_\alpha^a .
\end{aligned}
\tag{23}
$$

As in the case of the forces, the fact that these are rather complicated expressions should not discourage their use in computations. The algebra required to evaluate the formulae can be carried out with the help of an algebra program. The formulae given here for the torques are complete; that is, they give the total torque about the molecular origin, including any terms that might arise from forces on sites not at the centre of mass. The implementation in a molecular dynamics or molecular mechanics program requires only the computation of some vector products, and all the torque formulae can then be evaluated for interactions up to any rank required.

7 Induction

We have seen that we cannot describe the electrostatic interaction between molecules at short range adequately in terms of multipole moments for each molecule as a whole, and that a distributed-multipole treatment is needed. Similar refinements are required for a proper description of the distortion of the charge distribution in response to the fields due to neighbouring molecules.

The ordinary polarizability $\alpha_{\alpha\beta}$ describes the dipole induced in a molecule by a uniform external field $E = -\nabla V$. In conventional cartesian form:

$$
\begin{aligned}
\Delta\mu_\alpha &= \alpha_{\alpha\beta} E_\beta \\
&= -\alpha_{\alpha\beta} V_\beta , \quad -
\end{aligned}
\tag{24}
$$

The external field is often non-uniform. The potential may be expanded in a Taylor series about some suitable origin:

$$
V(\mathbf{r}) = V(0) + r_\alpha V_\alpha(0) + \frac{1}{2!} r_\alpha r_\beta V_{\alpha\beta}(0) + \dots ,
$$

where $V_{\alpha\beta\dots} = \nabla_\alpha \nabla_\beta \dots V$ are the derivatives of the potential. If this expansion converges over a region large enough to contain a molecule, the induced dipole of the molecule may be expressed in terms of these potential derivatives or 'field gradients' [17]:

$$
\Delta\mu_\alpha = -\alpha_{\alpha\beta} V_\beta - \frac{1}{3} A_{\alpha, \beta\gamma} V_{\beta\gamma} - \frac{1}{15} E_{\alpha, \beta\gamma\delta} V_{\beta\gamma\delta} - \dots
$$

Here $A_{\alpha,\beta\gamma}$ is the dipole-quadrupole polarizability, describing the dipole induced by the field gradient (and also the quadrupole induced by a uniform field) and $E_{\alpha,\beta\gamma\delta}$ is the dipole-octopole polarizability, describing the dipole induced by the second derivative of the field (the third derivative of the potential), and again we note in passing the cumbersome nature of the cartesian notation, which requires a new symbol for each of these effects. If we are interested in higher moments induced by external fields, further polarizabilities arise; for instance the quadrupole induced by an external field takes the form

$$\Delta\Theta_{\alpha\beta} = -A_{\gamma,\alpha\beta}V_{\gamma} - C_{\alpha\beta,\gamma\delta}V_{\gamma\delta} - \dots,$$

so that $C_{\alpha\beta,\gamma\delta}$ describes the quadrupole induced by an external field gradient. In the spherical-tensor notation all these equations take the same general form:

$$\Delta Q_t = -a_{tt'}V_{t'},$$

where $V_t = V_{lx}$ is a derivative of the potential, as defined by Eq. (19), and $\alpha_{tt'}$ is a polarizability:

$$\alpha_{tt'} = \sum_n{}' \frac{(\langle 0|\,\hat{Q}_t\,|n\rangle\,\langle n|\,\hat{Q}_{t'}\,|0\rangle + \langle 0|\,\hat{Q}_{t'}\,|n\rangle\,\langle n|\,\hat{Q}_t\,|0\rangle)}{W_n - W_0},$$

Unfortunately, when the external field at a molecule is due to its neighbours, it is usually very non-uniform, so that the Taylor expansion of the potential may not converge over the whole molecule. It then becomes necessary to divide the molecule into regions and to describe the response of each region to the electric field, field gradient, and so forth, experienced by that region. Further complications then arise. Firstly, the field experienced by any one region of a molecule is not just the external field but includes a contribution arising from the distortion of other regions of the same molecule. Applequist [18] attempted to take this into account by adding to the external field in each region the multipole fields due to the moments induced in other regions. This recipe is unsound, however, because adjacent regions of a molecule are very intimately associated with each other, and cannot possibly be regarded as regions of charge density contained in non-overlapping spheres, as is necessary if the multipole description is to be valid. It is necessary to perform a self-consistent treatment of the distortion of charge density in each region in the field of all other regions without using a multipole expansion. This leads to the author's Distributed Polarizability description, in which the induced moment in region a of molecule A depends on the field and field gradients not only in that region but in all other regions also [15]:

$$\Delta Q_t^a = -\alpha_{tt'}^{aa'}V_{t'}^{a'}, \tag{25}$$

where now $\alpha_{tt'}^{aa'}$ is a distributed polarizability:

$$\alpha_{tt'}^{aa'} = \sum_n{}' \frac{(\langle 0|\,\hat{Q}_t^a\,|n\rangle\,\langle n|\,\hat{Q}_{t'}^{a'}\,|0\rangle + \langle 0|\,\hat{Q}_{t'}^{a'}\,|n\rangle\,\langle n|\,\hat{Q}_t^a\,|0\rangle)}{W_n - W_0}. \tag{26}$$

The second complication is that the boundaries between regions are arbitrary, and there is no need for the charge contained in any region to be conserved. Electron density may flow from one region to another, in response not only to electric fields within regions but to differences in electrostatic potential between regions. In fact this is less of a complication than might appear at first sight; it means merely that the values of t and t' in (25) and (26) may be 00 as well as 10, 11c, etc. There is a sum rule for the distributed polarizabilities $\alpha_{tt'}^{aa'}$ which ensures that the total charge of a molecule is conserved and that changing the potential in every region by the same amount has no effect on the induced moments [15]:

$$\sum_{a'} \alpha_{t00}^{aa'} = 0 \,. \tag{27}$$

To summarise, the polarizabilities $\alpha_{tt'}^{aa'}$ specify the change in moment t at site a resulting from the field $V_{t'}^{a'}$ at site a'. The moment t can be the charge, dipole or quadrupole, etc., labelled as before by the symbols 00, 10, 11c, 11s, ... The V_t^a are the potential V_{00}^a and its derivatives, evaluated at the origin of site a:

$$V_{lx}^a = V_{lx}(\mathbf{a}) = [1.3 \dots (2l-1)]^{-1} R_{lx}(\nabla) V(\mathbf{r})|_{\mathbf{r}=\mathbf{a}} \,, \tag{28}$$

as given in Eq. (19) above. The polarizabilities can be calculated by coupled Hartree-Fock perturbation theory [19], and the procedure has been implemented by the author as part of the CADPAC ab initio package [20].

7.1 Induced Moments

If the external field arises from the neighbouring molecules, we can obtain V_{lx}^a from Eq. (20). However, the neighbouring molecules are also polarizable, and the moments that appear in Eq. (20) are the polarized moments $Q_u^b + \Delta Q_u^b$. Consequently the induced moments in molecule A are

$$\begin{aligned} \Delta Q_{t'}^{a'} &= -\alpha_{t't}^{a'a} V_t^a \\ &= -\sum_{B \neq A} \alpha_{t't}^{a'a} T_{tu}^{ab} (Q_u^b + \Delta Q_u^b) \,, \end{aligned} \tag{29}$$

so that the induced moments are the solutions of a set of coupled equations. Equation (29) can be solved iteratively, and convergence is usually reached in about six to eight iterations. Indeed the first iteration (represented diagrammatically by Fig. 1 b) gives induced moments that are usually within about 10% of the converged solution. Use of distributed multipoles to describe the charge distribution and distributed polarizabilities to describe the polarization has given very accurate estimates of induced moments in a number of small van der Waals molecules [21, 22]. More recent calculations have given induced moments in good agreement with experiment for HCCH $\cdots$ HCCH and HCCH $\cdots$ CO_2 [23].

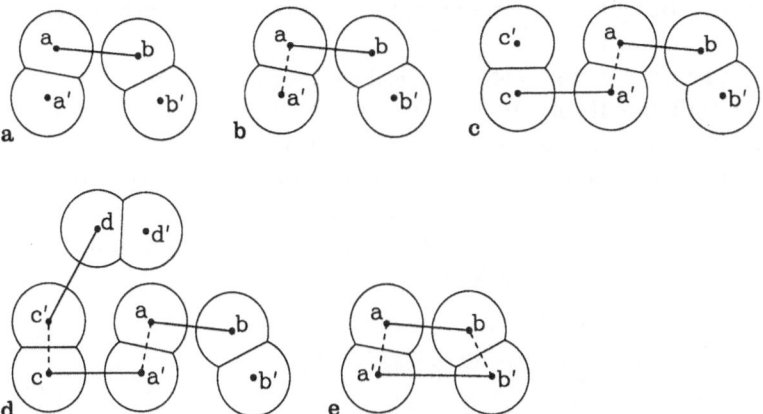

Fig. 1a–e. Diagrammatic representation of some electrostatic effects. The *solid lines* represent electrostatic interactions, and the *dashed lines* represent polarizabilities. **a)** Electrostatic energy: $Q_t^a T_{tu}^{ab} Q_u^b$. **b)** Induced moments: $\Delta Q_{t'}^{a'} = \alpha_{t't}^{a'a} T_{tu}^{ab} Q_u^b$. **c)** Induction energy in lowest order: $Q_u^b T_{ut}^{ba} \alpha_{tt'}^{aa'} T_{t'v}^{a'c} Q_v^c$. **d)** Higher-order term in the induction energy: $Q_u^b T_{ut}^{ba} \alpha_{tt'}^{aa'} T_{t'v}^{b'c} \alpha_{vv'}^{cc'} T_{v'w}^{c'd} Q_w^v$. **e)** Dispersion energy: $T_{tu}^{ab} T_{t'u'}^{a'b'} X_{tt'uu'}^{aa'bb'}$

7.2 Induction Energies

The calculation of the induction energy is more troublesome. It can be separated formally into a sum of terms, one for each molecule in the system [24]:

$$E_{ind}^A = \frac{1}{2} \sum_{B \neq A} \Delta Q_t^a \, T_{tu}^{ab} Q_u^b , \qquad (30)$$

but this is only a formal separation, since the induced moments ΔQ_t^a themselves depend on the total field due to the neighbouring molecules, according to Eq. (29). By using this expression, it is possible to develop a series expansion of the induction energy in powers of the polarizabilities:

$$E_{ind}^A = \frac{1}{2} \sum_{B \neq A} \Delta Q_t^a \, T_{tu}^{ab} Q_u^b$$

$$= -\frac{1}{2} \sum_{B \neq A} \sum_{C \neq A} (Q_v^c + \Delta Q_v^c) \, T_{vt'}^{ca'} \alpha_{t't}^{a'a} T_{tu}^{ab} Q_u^b$$

$$= -\frac{1}{2} \sum_{B \neq A} \sum_{C \neq A} Q_v^c T_{vt'}^{ca'} \alpha_{t't}^{a'a} T_{tu}^{ab} Q_u^b$$

$$+ \frac{1}{2} \sum_{B \neq A} \sum_{C \neq A} \sum_{D \neq C} (Q_w^d + \Delta Q_w^d) \, T_{wv'}^{dc} \alpha_{v'v}^{c'c} T_{vt'}^{ca'} \alpha_{t't}^{a'a} T_{tu}^{ab} Q_u^b . \qquad (31)$$

This procedure can be continued to arbitrary order in the polarizabilities to give a generalisation of the expression derived by Barker for the case where only dipole polarizabilities are included [25]. The terms in the series can be represented by diagrams such as those in Figs. 1c and 1d.

This formula can be derived either from classical considerations or by using quantum mechanical perturbation theory. In either case it is clear that only linear polarizabilities are included; that is, the induced moments described by Eq. (29) are a linear response to the fields and field gradients of the neighbouring molecules. For a complete description we should also allow for contributions to the induced moments that depend on quadratic and higher powers of the electric fields. For a uniform field, we should use, instead of Eq. (24), the form

$$\Delta\mu_\alpha = -\alpha_{\alpha\beta}V_\beta + \frac{1}{2}\beta_{\alpha\beta\gamma}V_\beta V_\gamma - \frac{1}{6}\gamma_{\alpha\beta\gamma\delta}V_\beta V_\gamma V_\delta + \dots,$$

where $\beta_{\alpha\beta\gamma}$ and $\gamma_{\alpha\beta\gamma\delta}$ are the first and second hyperpolarizabilities. Corresponding quantities can be defined in the spherical tensor formalism [26], but their use in calculations of induction energies has not been attempted within the formalism described here. Piecuch [27] has given expressions for some of the hyperpolarizability contributions to the interaction energy, but has used a recoupled formulation which, although elegant, is probably less suited to practical calculations.

In fact there are reasons for believing that inclusion of these higher-order effects may be counter-productive. The calculation of the induction energy via Eq. (30) using the induced moments calculated self-consistently from Eq. (29), or alternatively via Eq. (31), is equivalent to carrying out quantum mechanical perturbation theory to infinite order, but including only some of the terms. The neglected terms include not only the dispersion energy, an omission which can be remedied, and the hyperpolarizability terms just mentioned, but a variety of more complicated terms. Some of the terms which arise at low order have been discussed by Piecuch in some detail [27]. The omission of these other terms is questionable because the quantum-mechanical perturbation expansion is known to be an asymptotic series [28, 29], and the inclusion of only some of the terms of such a series is a very dubious procedure. In fact, numerical investigations [30] suggest that better results can be obtained by truncating the series at the second order of perturbation theory, provided that polarizabilities up to octopole or hexadecapole rank are included; this corresponds in the case of the induction energy to taking only the first term in the expansion (31) in powers of the polarizabilities, so that terms illustrated by Fig. 1c are included, but higher-order contributions like Fig. 1d are not.

8 Dispersion Energy

It may seem surprising that the dispersion energy is included in a survey of the applications of classical electrostatics to molecular interactions, since it is well known that the dispersion energy has no classical analogue. The same techniques and limitations apply, however, and the connection with the classical approach is made through the

Casimir-Polder formulation of the dispersion energy [31], which for dipole polarizabilities in the cartesian notation takes the form:

$$E_{\text{disp}} = -\frac{\hbar}{2\pi} \, T_{\alpha\beta} T_{\gamma\delta} \int_0^\infty \alpha_{\alpha\gamma}^A(iu) \, \alpha_{\beta\delta}^B(iu) \, du \, , \tag{32}$$

so that the dispersion energy is described by an integral over the product of the individual molecular polarizabilities evaluated at the imaginary frequency iu. The polarizability $\alpha_{tt'}(\omega)$ describing the magnitude of the oscillating moment Q_t, induced by an oscillating field $V_t \cos \omega t$ is

$$\alpha_{tt'}(\omega) = \sum_n{}' \frac{\omega_{0n}(\langle 0| \hat{Q}_t |n\rangle \langle n| \hat{Q}_{t'} |0\rangle + \langle 0| \hat{Q}_{t'} |n\rangle \langle n| \hat{Q}_t |0\rangle)}{\hbar(\omega_{0n}^2 - \omega^2)} \, ,$$

where $\hbar\omega_{0n} = W_n - W_0$ is the energy difference between states 0 and n. The polarizability at imaginary frequency is obtained by setting $\omega = iu$. It is difficult to attach a satisfactory physical interpretation to such polarizabilities, though they can be regarded as describing the response to an exponentially increasing electric field. However it is apparent from the formula that they are well defined mathematically, and that they are much more well-behaved than polarizabilities at real frequencies, which exhibit singularities when $\omega = \omega_{0n}$. (The singularities disappear if proper account is taken of the finite lifetime of the excited states [17], but the polarizability is still anomalous near such resonances.) In contrast, the polarizability $\alpha_{tt'}(iu)$ is a monotonically decreasing function of u, which makes it relatively easy to calculate accurately using the same Coupled Hartree-Fock procedure that is used for the static polarizability.

In terms of these polarizabilities, the dispersion energy takes the more general form

$$E_{\text{disp}} = T_{tu}^{AB} T_{t'u'}^{AB} X_{tt'uu'}^{AB} \, , \tag{33}$$

where $X_{tt'uu'}^{AB}$ is a dispersion integral:

$$X_{tt'uu'}^{AB} = -\frac{\hbar}{2\pi} \int_0^\infty \alpha_{tt'}^A(iu) \, \alpha_{uu'}^B(iu) \, du \, ,$$

The dispersion integrals need only be evaluated once for each pair of molecules, since all information about the relative positions of the molecules is contained in the interaction functions T_{tu}^{AB}, not in the dispersion integrals. This is therefore an efficient way to calculate the dispersion energy.

However the polarizability at imaginary frequency, like the static polarizability, rests on the multipole expansion of the electrostatic perturbation. At short distances this must, as before, be replaced by the distributed multipole expansion, and the dispersion energy then becomes

$$E_{\text{disp}} = T_{tu}^{ab} T_{t'u'}^{a'b'} X_{tt'uu'}^{aa'bb'} \, , \tag{34}$$

where the dispersion integral $X^{aa'bb'}_{t\,t'uu'}$ now involves distributed polarizabilities:

$$X^{aa'bb'}_{tt'uu'} = -(4\pi\varepsilon_0)^{-2}\,\frac{\hbar}{2\pi}\int \alpha^{aa'}_{tt'}(iu)\,\alpha^{bb'}_{uu'}(iu)\,\mathrm{d}u\,,$$

where

$$\alpha^{aa'}_{tt'}(\omega) = \sum_n{}'\frac{\omega_{0n}(\langle 0|\,\hat{Q}^a_t\,|n\rangle\,\langle n|\,\hat{Q}^{a'}_{t'}\,|0\rangle + \langle 0|\,\hat{Q}^{a'}_{t'}\,|n\rangle\,\langle n|\,\hat{Q}^a_t\,|0\rangle)}{\hbar(\omega^2_{0n} - \omega^2)}\,. \quad (35)$$

There is an implicit sum in Eq. (34), as usual, over the repeated suffixes a, a', b, b', t, t', u and u'. A typical term in this sum can be represented by Fig. 1e, and can be interpreted in terms of correlated fluctuations. A fluctuation in moment t at site a produces a change in the field component u at site b in the other molecule, and because molecule B is polarizable this induces a change in moment u' at site b'. Site b' may in fact be the same as b, in which case we have a *local* polarization, or it may be a different one, in which case the polarization is *non-local*. The induced moment u' produces a field component t' at site a' in the first molecule, which interacts with the original fluctuation via the polarizability. Again a' may be the same as a or different, corresponding to a local or non-local polarization of molecule A.

If a distributed description is used, there is a very large number of terms in Eq. (34), even for small molecules. If each molecule has two sites, and if polarizabilities are included up to dipole, there are four moments (charge and three dipoles) for each site, giving a total of $2^4 4^4 = 4096$ terms. Fortunately there are symmetry relationships that greatly reduce the number of independent dispersion coefficients. For linear molecules, the polarizability $\alpha^a_{l\varkappa l'\varkappa'}$ vanishes unless $\varkappa = \varkappa'$, and moreover $\alpha^a_{lmc,\,l'mc} = \alpha^a_{lms,\,l'ms}$. For a heteronuclear diatomic, with one site for each atom, there are then nine independent non-vanishing polarizabilities up to dipole rank, and $9^2 = 81$ independent X coefficients for a pair of dissimilar diatomics. If the molecules are identical, then this number is reduced to 45. If the molecules are homonuclear, then there are only six independent polarizabilities up to dipole, and then the number of independent X coefficients for identical molecules is 21.

In accurate models of the dispersion interaction, it is necessary to include dipole-quadrupole and quadrupole-quadrupole polarizabilities too. The number of independent dispersion coefficients then increases dramatically, as Table 3 shows.

The main reason for the large number of dispersion coefficients is the presence of non-local polarizabilities in the distributed polarizability description. If there were

Table 3. The number of independent non-vanishing dispersion coefficients for a pair of diatomic molecules

Molecules	Number of sites:	1	2	2
	Rank:	2	1	2
Dissimilar Heteronuclear		49	81	784
Identical Heteronuclear		28	45	406
Identical Homonuclear		15	21	153

no dispersion integrals involving non-local polarizabilities, Eq. (34) would reduce immediately to the conventional, and quite successful, model of the dispersion energy, which is commonly assumed to be a sum of atom-atom interactions, in which the leading terms take the form

$$E_{\text{disp}} = \sum_{ab} C_6^{ab}(\Omega)\, R_{ab}^{-6},$$

and the coefficient C_6 depends on the relative orientation Ω of the molecules. However the leading terms in Eq. (34) are proportional to $R_{ab}^{-1} R_{a'b'}^{-1}$, and the corresponding dispersion integrals involve charge-flow polarizabilities, which are inherently non-local. It appears, then, that these two forms for the dispersion energy are incompatible. In fact this is not the case. Although at short range there are terms in the dispersion energy that behave like R^{-2}, the sum rule for the polarizabilities (Eq. (27)) guarantees that at long range the terms cancel in such a way as to give R^{-6} behaviour.

Furthermore it appears to be possible to transform Eq. (34) into an equivalent expression that involves only local dispersion terms, though this has so far been done successfully only for diatomics [14]. It is done by using Eq. (17) in the following way. Suppose that the local axes of the diatomic are set up with the z axis along the molecule and the origin at the bond centre, so that atom 1 is at $(0, 0, -a)$ in local axes and atom 2 at $(0, 0, a)$. Then if, for example, an external potential causes a movement of charge from atom 1 to atom 2, then the change $-\delta q$ at atom 1 may be represented by multipoles at the origin: $Q_{l'\varkappa'}(0) = -\delta q\, W_{00,\, l'\varkappa'}(0, 0, -a)$. Inserting the appropriate W coefficients, we find that the new multipoles are a charge $-\delta q$, a dipole $+a\delta q$, a quadrupole $-a^2\delta q$, and so on. Similarly, the change δq in the charge at atom 2 may be represented at the origin by a charge δq, a dipole $a\delta q$, a quadrupole $a^2\delta q$, and so on. The even moments cancel, and the charge flow is represented by a dipole $2a\delta q$, an octopole $2a^3\delta q$, and so on. In this way the inherently non-local charge flows are representing by local induced moments.

A similar procedure can be applied to the non-local polarizabilities $\alpha_{tt'}^{aa'}$. Here we are dealing (in Eq. (35)) with multipole moment operators Q_t^a and $Q_{t'}^{a'}$ for sites a and a'. By representing these, via eq. (17), in terms of moments at a convenient common site a'', the non-local polarizability $\alpha_{tt'}^{aa'}$ is replaced by local polarizabilities of the form $\alpha_{t\,t'}^{a''a''}$, though at the cost of introducing higher-rank polarizabilities.

The main uncertainty at present about such manipulations is the convergence of the resulting power series in reciprocal distance for the dispersion energy. The great merit of the distributed polarizability picture is that the distortion of individual atoms is described in terms of multipole moments induced on that atom, which can be done efficiently and accurately without going to very high ranks of multipole. If it becomes necessary to describe such distortions in terms of moments referred to a different origin, then the convergence of the new multipole description will inevitably be less satisfactory. It is not yet clear whether the higher accuracy of the non-local dispersion model can justify the considerable computational cost that it entails, or whether the local dispersion model can be made sufficiently accurate for present purposes.

9 Summary

Evidence is accumulating that the interactions responsible for the bonding in van der Waals molecules are primarily electrostatic, in the sense that they can be adequately described in terms of classical electrostatics and that it is unnecessary to invoke chemical bonding or electron exchange. It has been shown conclusively by the work of Buckingham and Fowler [32] that the structures of hydrogen-bonded van der Waals complexes can be well understood, and usually predicted quantitatively, by a model that includes an accurate description of the electrostatic interaction (using distributed multipoles) and only the crudest representation, via a hard-sphere model, of the other terms in the interaction. In the few cases where the model fails, the explanation is usually either that the hard-sphere repulsion model is inadequate or that the electrostatic potential surface is unusually flat [33, 34]. The induced dipole moments in a number of van der Waals complexes are inconsistent with simple predictions based on one-centre electrostatic models, and this has been adduced as evidence that 'chemical' bonding is present in such cases [35], but here too a more accurate treatment of the classical electrostatics, using distributed moments and polarizabilities, gives calculated dipole moments in excellent agreement with the observed ones [21, 22].

All of this is compelling evidence that accurate models of intermolecular interactions can be constructed on the basis of electrostatic models, provided that the electrostatic interactions are treated correctly. This requires a distributed description of the moments and polarizabilities. One way of handling the electrostatic interaction itself (but not induction) is to use a point-charge model of the charge distribution, and this has indeed been a very popular approach, since the calculations are then very simple. However it has also become clear that the repulsive part of the potential must treat the atoms as non-spherical [5], and the implementation of anisotropic atom-atom potentials uses the same functions of relative molecular orientation, the $\bar{S}$ functions, as are required in the calculation of electrostatic interactions involving dipoles, quadrupoles and higher moments. Although the resulting formulae are more complicated than those for isotropic atom-atom potentials, they can be, and have been, implemented efficiently in molecular dynamics programs [36]. The purpose of the present paper has been to give the formulae that are needed for such calculations, and to show that similar methods can be extended to the induction and dispersion terms in the interaction.

10 References

1. Buckingham AD (1967) Adv. Chem. Phys. 12: 107
2. Stone AJ, Tough RJA (1984) Chem. Phys. Lett. 110: 123
3. Brink DM, Satchler GR (1968) Angular momentum, Clarendon Press, Oxford, p 151
4. Griffith JS (1961) The theory of transition-metal ions, Cambridge University Press, p 200
5. Stone AJ, Price SL (1988) J. Phys. Chem. 92: 3325
6. Vigné-Maeder F, Claverie P (1988) J. Chem. Phys. 88: 4934

7. Stone AJ (1981) Chem. Phys. Lett. 83: 233; Stone AJ, Alderton M (1985) Molec. Phys. 56: 1047
8. Pullman A, Perahia D (1978) Theor. Chim. Acta 48: 29
9. Rico JF, Alvarez-Collado JR, Paniagua M (1985) Molec. Phys. 56: 1145
10. Cooper DL, Stutchbury NCJ (1985) Chem. Phys. Lett. 120: 167
11. Sokalski WA, Sawaryn A (1987) J. Chem. Phys. 87: 526
12. Stone AJ (1984) In: Orville-Thomas WJ, Yarwood J (eds) Molecular liquids. D. Reidel, NATO A.S.I. series
13. Price SL, Stone AJ, Alderton M (1985) Molec. Phys. 56: 1047
14. Tong C-S, Stone AJ (in preparation)
15. Stone AJ (1985) Molec. Phys. 56: 1065
16. REDUCE, an algebraic programming system (1984) The Rand Corporation, Santa Monica, California
17. Buckingham AD (1978) In: Pullman B (ed) Intermolecular forces: from diatomics to biopolymers, Wiley, New York, p 1
18. Applequist J (1977) Acc. Chem. Research 10: 79; (1983) J. Math. Phys. 24: 736; (1984) Chem. Phys. 85: 279; (1985) J. Chem. Phys. 83: 809
19. Stevens RM, Pitzer R, Lipscomb WN (1963) J. Chem. Phys. 38: 550
20. Amos RD, Rice JE (1987) CADPAC: The Cambridge Analytical Derivatives Package, issue 4.0, Cambridge
21. Buckingham AD, Fowler PW, Stone AJ (1986) Intern. Rev. Phys. Chem. 5: 107
22. Stone AJ, Fowler PW (1987) J. Phys. Chem. 91: 509
23. Stone AJ, Le Sueur CR, Fowler PW, Muenter JS (in preparation)
24. Stone AJ (1989) Chem. Phys. Lett. 155: 102
25. Barke JA (1953) Proc. Roy. Soc. A 219: 367
26. Gray CG, Lo BWN (1976) Chem. Phys. 14: 73
27. Piecuch P (1986) Molec. Phys. 59: 1067, 1085, 1097
28. Dalgarno A, Stewart AL (1956) Proc. Roy. Soc. A 238: 276; Dalgarno A, Lynn N (1957) Proc. Phys. Soc. (London) A70: 223
29. Kreek H, Meath WJ (1969) J. Chem. Phys. 50: 2289
30. Stone AJ (1989) Chem. Phys. Lett. 155: 111
31. Casimir HBG, Polder D (1948) Phys. Rev. 73: 360
32. Buckingham AD, Fowler PW (1983) J. Chem. Phys. 79: 6426; (1985) Canad. J. Chem. 63: 2018
33. Hurst GJB, Fowler PW, Buckingham AD, Stone AJ (1986) Int. J. Quantum Chem. 29: 1223
34. Rendell AP, Bacskay GB, Hush NS (1985) Chem. Phys. Lett. 117: 400
35. Altman RS, Marshall MD, Klemperer W (1982) Disc. Faraday Soc. 73: 116; (1983) J. Chem. Phys. 79: 57
36. Rodger PM, Stone AJ, Tildesley DJ (1988) Molec. Phys. 63: 173; (1988) Chem. Phys. Letters 145: 365

Weak Interactions Between Molecules and Their Physical Interpretation

Valerio Magnasco
Istituto di Chimica Industriale dell'Università, 16132 Genoa, Italy

Roy McWeeny
Dipartimento di Chimica e Chimica Industriale dell'Università, 56100 Pisa, Italy

Exchange perturbation theory is used to give, to second order, a direct, unequivocal, classification of the components of the weak interactions occurring between molecules in the Van der Waals region. Electron distribution functions together with one- and two-particle potentials (e.g. the molecular electrostatic potential) are used in reducing the formal many-electron expressions of the various coulombic components to space integrals. These represent the semi-classical interaction between static charge distributions localized on the isolated molecules (electrostatic energy) and their deformations. The distortion of the charge distribution of each molecule in the static electric field provided by the other (induction) and the correlation between the instantaneous fluctuations in the charge distributions of the two molecules (dispersion) occurring in second order of perturbation theory are expressed in terms of static and dynamic polarization propagators of the individual molecules. Non-classical effects, originating in quantum mechanical exchange and the Pauli repulsion between closed shells, are described in terms of suitable exchange and overlap densities of the subsystems. A generalized spherical tensor expansion of the intermolecular potential is used to relate the different components of the interaction to the electric properties, in principle observable, of the free molecules (multipole moments and polarizabilities, and their generalizations). Applications are described to the calculation of molecular interaction coefficients and to the study of the angular geometry of some Van der Waals dimers, including hydrogen-bonded systems.

1 Introduction

The forces acting between atoms or molecules (regarded as stable aggregates of nuclei and electrons) are prevalently electric in origin. Other forces, such as gravitational, nuclear or magnetic forces, are negligible either because their intensity is sensibly smaller (as for the case of magnetic or gravitational forces) or because they are extremely short-ranged (as for nuclear forces) [1]. The intermolecular forces can be derived from a potential (the intermolecular potential) which can be defined in the Born-Oppenheimer approximation [2, 3], where the motion of the electrons is separated from the slow motion of the nuclei, and the molecular energy $E(q) = E_e(q) + E_N$ is studied for a given configuration q of the nuclei. $E(q)$ defines a molecular potential surface, independent of the velocity of the nuclei, the latter moving so as to satisfy the nuclear wave equation

$$\left\{ - \sum_\alpha \frac{1}{2M_\alpha} \nabla_\alpha^2 + E(q) \right\} \Psi_n(q) = W \Psi_n(q),$$

where M_α is the mass of nucleus α expressed in units of the electron mass, and $\Psi_n(q)$ the nuclear wavefunction (wf)[1]. In the following, we shall tacitly assume that the conditions for the validity of the Born-Oppenheimer approximation are all satisfied. The intermolecular energy is then the interaction energy of two molecules for fixed positions of all nuclei, and is naturally defined as the difference between the energy of the system A + B in a given particular configuration and the energy resulting when A and B are brought to infinite separation. Assuming that the distance is such that we can safely neglect the internal flexibility of the molecules [4], the interaction energy will depend on the distance R characterizing the separation of the molecules and the Euler angles Ω specifying their orientation in space.

Even if not directly observable, intermolecular forces influence the microscopic and bulk properties of matter, being responsible for a variety of interesting phenomena such as the equilibrium and transport properties of real fluids, the structure and properties of liquids and molecular crystals, the structure and binding of Van der Waals (VdW) molecules (which can be observed under high resolution rotational spectroscopy [5–8] or molecular beam electric resonance spectroscopy [9]), the shape of reaction paths and the structure of transition states determining chemical reactions [10].

The intermolecular energies are extremely small, their ratio with respect to the energy E_o of the individul molecules being of the order of 10^{-5} (Table 1). The appropriate units are small fractions of the atomic unit of energy.

---

[1] Here and elsewhere we use atomic units in which e, m, $\hbar$ and the permittivity $4\pi\varepsilon_o$ take unit values. The other units most commonly encountered are: length (Bohr, a_o) = $4\pi\varepsilon_o \hbar^2 m^{-1} e^{-2}$ = 5.29177×10^{-11} m, energy (Hartree, E_h) = $e^2 (4\pi\varepsilon_o)^{-1} a_o^{-1}$ = 4.35944×10^{-18} J, charge (e) = 1.60218×10^{-19} C, 2^l-pole moment (μ_l) = ea_o^l, multipole polarizability $(\alpha_{ll'})$ = $e^2 a_o^{l+l'} E_h^{-1}$ = $4\pi\varepsilon_o a_o^{l+l'+1}$. Interaction energies are conveniently expressed in fractions of E_h restricted to steps of 10^{-3}: $10^{-3} E_h$ = mE_h (milliHartree) ~ 2625.5 J mol^{-1} ~ 27.21 meV ~ 219.47 cm^{-1} ~ 315.78 K, or $10^{-6} E_h$ = μE_h (microHartree).

Table 1. Well depths at the experimental geometries, total energies of the isolated molecules and their ratio for some Van der Waals dimers

Dimer	He_2	$(H_2)_2$	$(N_2)_2$	$(HF)_2$	$(H_2O)_2$
R/a_o	5.6	6.5	8.0	5.1	5.6
$E^{int}/10^{-3}\,E_h$	−0.034	−0.117	−0.392	−7.76	−8.92
E_0/E_h	−5.807	−2.349	−218.768	−200.607	−152.877
$\left\|\dfrac{E^{int}}{E_0}\right\|\,10^{-5}$	0.58	4.98	1.79	3.87	5.83

It is clear that the required interaction energies are in general so small compared with the energies of the separate molecules A and B that any attempt to calculate them at the ab initio level is bound to meet enormous difficulties. For example, a Hartree-Fock calculation on a single N_2 molecule gives an energy of about $-109\,E_h$, with a "correlation error" of roughly 1 percent; this error, which cannot be eliminated without the use of huge basis sets and a vast amount of configuration interaction, is more than a thousand times as large as the required interaction energy! Such difficulties are inescapable in "supermolecule" calculations [11, 12], where the energy of A + B is computed as a function of geometry. A more attractive approach is to treat the interaction between the two molecules as a small perturbation, extending the Rayleigh-Schroedinger (RS) perturbation theory to include exchange and overlap effects [13, 14]. The difficulties of both perturbation and supermolecule approaches have been discussed elsewhere e.g. [15, 16]. Basic difficulties of the perturbation approach are (i) the inappropriate symmetry of unperturbed wavefunctions of product type (which do not allow for electron exchange) and (ii) the unavailability of exact eigenfunctions of the unperturbed Hamiltonian (which refers to two noninteracting molecules). To avoid these difficulties it is possible to introduce a *variational* element into the calculation by using *antisymmetrized* products of free-molecule functions, with optimization of the resultant wavefunction by standard methods. Ideally, the free-molecule functions would be *exact* molecular wavefunctions (for both ground and excited states) but in practice it is necessary, at some point, to introduce orbital approximations. Approaches of this kind [15–18], which admit exchange and non-orthogonality effects, are often said to be of Epstein-Nesbet (EN) type; they differ mainly in the precise point at which the approximations are introduced. With the assumption of *exact* free-molecule functions it is possible to obtain [16, 18, 19] formal expressions for the interaction energy itself, introducing approximations only in the subsequent calculations. Since, however, the general formulation is of considerable complexity, we shall for the most part consider orbital-based approaches in which the usual approximations are introduced at the very beginning.

An up-to-day and exhaustive discussion of the computational aspects of ab initio calculations of intermolecular interactions for small to medium size molecules can be found in [12]. In this work, we shall be mostly concerned with the weak interactions occurring between closed-shell molecules in the VdW region, and with the physical interpretation of the different components of the interaction. Almost the whole totality of the experimental data in this region refers in fact to VdW molecules [8, 9, 20, 21], whose monomers are represented by stable neutral molecules having a closed-shell structure which is not lost in the dimer.

To this end, exchange-perturbation theory seems the most appropriate tool, since it gives a direct way of obtaining the small binding energies of such VdW dimers, relating the intermolecular interaction to free-molecule quantities (like electron densities, polarization propagators, frequency-dependent polarizabilities) which can be calculated *directly*, to any degree of accuracy, without reference to the total electronic energy of the molecules. The interaction energy may in this way be expressed in terms of electron distribution functions (e.g. the charge and transition densities) of the separate molecules, together with their first- and second-order properties (e.g. multipole moments and polarizabilities). This description must be supplemented by the inclusion of exchange and interpenetration effects as soon as the "charge clouds" of the two molecules begin to overlap; although such effects are non-classical they can be *interpreted* classically in terms of a deformation of the free-molecule charge densities.

A generalized spherical tensor expansion of the intermolecular potential [22] allows one to extend to the whole range of intermolecular separations the multipolar analysis of the energy components which is familiar in long range [23, 24]. The new expansion maintains all conceptual and practical advantages offered by the long-range expansion, provided the electric properties of the molecules are replaced by generalized k-dependent quantities which can still be calculated once and for all for the individual monomers. Such quantities can then be used later to evaluate the interaction energy of any pair of molecules, for whatever orientation Ω and whatever separation R, so avoiding the repeated cumbersome evaluation of a huge number of multicentre integrals characteristic of many other approaches.

Applications are described to the calculation of molecular interaction coefficients and to the study of the angular geometry of some VdW dimers, including hydrogen-bonded systems.

2 Classification of Intermolecular Forces

Since intermolecular forces are repulsive at short range and attractive at large distances, there must be at least two contributions of opposite sign to the total force and hence to the intermolecular potential. The attractive contributions are due to the coulombic energies of first and second order (electrostatic, induction, dispersion), the repulsive contribution to the requirement of antisymmetry of the total wf due to the Pauli principle, which for closed shells results in a reduction of the electron density in the region of overlap between the molecules and hence in an increased repulsion.

We shall consider the following kinds of intermolecular interaction:

(i) **Electrostatic (or coulombic) energy:** this gives the semiclassical interaction between *rigid* charge distributions of the two molecules. Strictly pairwise additive, it appears in first order of perturbation theory. For systems with permanent moments (molecules or atoms in states with $L \neq 0$) it varies at long range as

R^{-n}, while it decreases exponentially with distance for neutral atoms in S-states and for molecules which do not have permanent moments.

(ii) **Induction (or polarization) energy**: this is the energy lowering resulting from the distortion of the charge distribution of one molecule by the mean electric field provided by the other molecule, and vice versa. If one or both molecules possess permanent moments, this interaction will be of long-range. Not pairwise additive, it is described in second order of perturbation theory.

(iii) **Dispersion energy**: this describes the intermolecular electron correlation due to the instantaneous coupling of the density fluctuations mutually induced in each molecule. It is a long-range interaction, pairwise additive in second order of perturbation theory. For neutral atoms having spherical symmetry, dispersion is more important than induction at large distances, but the situation is reversed in short range.

(iv) **Exchange energy**: this is a first-order term arising from the antisymmetry requirement of the wf. Attractive for closed shells, it is a 2-electron contribution surviving even for zero-overlap.

(v) **Overlap (or penetration) energy**: this is the other first-order term arising from the antisymmetry requirement, and describes the Pauli repulsion due to interpenetration of the charge clouds of the molecules with a closed-shell structure. It comprises 1-electron and 2-electron terms, the first being larger, and vanishes for zero overlap. Like (iv), it is a non-classical contribution depending on the nature of the spin coupling of the interacting systems. Strictly non-additive, it decays exponentially ($\propto e^{-\alpha R}$) with the intermolecular separation R. For open shells, appropriate pairing of the electron spins can give an attractive contribution, yielding chemical bonding. This possibility will not be considered in this work, where we confine ourselves to weak interactions between closed shells in the VdW region of intermediate separations. (iv) and (v) should be referred to cumulatively as exchange-overlap contributions, arising respectively from exchange and overlap densities, even if, loosely speaking, we shall often use, in short, the word penetration for both of them [25] as opposed to coulombic.

(vi) **Second-order penetration energy**: a second-order quantity corresponding to (iv) and (v), it includes the exchange-overlap corrections to induction and dispersion contributions. In an orbital approximation, it implies intermolecular overlap between occupied orbitals of one molecule and vacant orbitals of the other [15, 26–28]. Since the second-order terms containing induction and dispersion are small anyway, such overlap corrections should be even smaller, and will not be considered further in this paper.

3 The Perturbation Theory of Intermolecular Forces

In the perturbation theory of intermolecular interactions at intermediate separations the majority of the physically meaningful effects can be described in second order, provided the coulombic energies (i)–(iii) are supplemented by terms (iv)–(v) arising from intermolecular electron exchange. First-order effects, which are dominant in short range, can be described by antisymmetrizing the product ψ_o of the unperturbed wfs of the separate molecules. The small second-order effects can then be calculated by seeking variational approximations to induction and dispersion energies, eventually including later second-order exchange (Murrell-Shaw-Musher-Amos or MS-MA exchange perturbation theory). The MS-MA theory is just one of the possible exchange perturbation theories [13, 29, 30], and owes its simplicity to the fact that the energy corrections can be calculated without solving perturbation equations involving the antisymmetrizer A (an operator which implies interchange of electrons between different molecules). The same is true of the more general methods [16, 18, 19] in which antisymmetrized functions are introduced at the outset.

If V is the intermolecular potential and H_o the unperturbed Hamiltonian (sum of the Hamiltonians H_o^A and H_o^B of the isolated molecules), then the unperturbed product (ψ_o) of isolated-molecule functions will satisfy

$$(H_o - E_o)\,\psi_o = 0\,. \tag{1}$$

The interaction energy in second-order of the MS-MA expansion is obtained *directly* (after antisymmetrizing ψ_o) as

$$E^{\text{int}} = E_1 + E_2\,, \tag{2}$$

where

$$E_1 = \frac{\langle A\psi_o|\,V\,|\psi_o\rangle}{\langle A\psi_o\,|\,\psi_o\rangle} = E_1^{cb} + E_1^{pn} \tag{3}$$

is the first-order interaction with its coulombic and penetration (exchange-overlap) components:

$$E_1^{cb} = \langle \psi_o|\,V|\psi_o\rangle\,, \qquad E_1^{pn} = \frac{\langle P\psi_o|\,V - E_1^{cb}\,|\psi_o\rangle}{1 + \langle P\psi_o\,|\,\psi_o\rangle} \tag{4}$$

and E_2 is the second-order interaction

$$E_2 = \frac{\langle A\psi_1^p|\,V - E_1\,|\Psi_o\rangle}{\langle A\psi_o\,|\,\psi_o\rangle} = E_2^{cb} + E_2^{pn} \tag{5}$$

with

$$E_2^{cb} = \langle \psi_1^p|\,V|\psi_o\rangle = E_2^{\text{ind}} + E_2^{\text{disp}}\,, \tag{6}$$

$$E_2^{pn} = \frac{\langle P\psi_1^p|\,V - E_1\,|\psi_o\rangle - E_2^{cb}\langle P\psi_o\,|\,\psi_o\rangle}{1 + \langle P\psi_o\,|\,\psi_o\rangle} \tag{7}$$

In these formulae A is the partial (idempotent) antisymmetrizer

$$\mathsf{A} = Q^{-1}(1 + P)\,, \qquad Q^{-1} = \frac{N_A!\,N_B!}{N!}\,, \tag{8}$$

P the operator interchanging electrons between *different* molecules

$$P = -\sum_{i}^{A}\sum_{j}^{B} P_{ij} + \sum_{i<i'}^{A}\sum_{j<j'}^{B} P_{ij}P_{i'j'} - \cdots \qquad (9)$$

and ψ_1^p the first-order polarization function satisfying the non-homogeneous differential equation

$$(H_o - E_o)\,\psi_1^p + (V - E_1^{cb})\,\psi_o = 0 \quad \text{with} \quad \langle\psi_o\,|\,\psi_1^p\rangle = 0. \qquad (10)$$

Variational approximations [31] can be given for this differential equation, seeking essentially to minimize with respect to an appropriate trial function $\tilde{\psi}_1^p$ the second-order (Hylleraas) functional

$$\tilde{E}_2^{cb}[\tilde{\psi}_1^p] = \langle\tilde{\psi}_1^p|\,H_o - E_o\,|\tilde{\psi}_1^p\rangle + \langle\tilde{\psi}_1^p|\,V\,|\psi_o\rangle + \langle\psi_o|\,V\,|\tilde{\psi}_1^p\rangle \geq E_2^{cb}. \qquad (11)$$

The value assumed by this functional at its minimum gives an upper bound to the second-order coulombic energy. The problem of obtaining E_2, Eq. (5), directly through the complementary bounds of a single bivariational functional has also been considered [32, 33].

The Coulombic components alone give what is known as the "polarization approximation" [13], and can be obtained from ordinary RS perturbation theory for stationary states with the intermolecular potential V playing the role of a small perturbation. Not reflecting the full symmetry of the Hamiltonian H, the RS expansion converges very slowly, unlike expansions using antisymmetrized products, and an accurate energy cannot be obtained in finite order. There are two essential points in the RS expansion:

a) the eigenvalue equation (1) must be exactly satisfied;
b) the second-order energy takes the "sum-over-states" expression:

$$E_2^{cb} = -\oint_{\kappa \neq 0} \frac{|\langle\psi_\kappa|\,V\,|\psi_o\rangle|^2}{E_\kappa^o - E_o} = -\oint_{\kappa \neq 0} \frac{|\langle\psi_\kappa|\,V\,|\psi_o\rangle|^2}{\varepsilon_\kappa} \qquad (12)$$

where $\{\psi_\kappa\}$ denotes the *complete* set of eigenstates of H_o with eigenvalues $\{E_\kappa^o\}$ and excitation energies $\{\varepsilon_\kappa = E_\kappa^o - E_o > 0\}$. The symbol $\oint$ means that we must sum over the *discrete* spectrum of H_o and integrate over the *continuous* spectrum as well.

Such formal properties are certainly satisfied by the *exact* wfs of each molecule, which unfortunately are not available. Henceforth we make the more realistic assumption that only approximate free-molecule functions are available and that these "pseudo-states" have been obtained by separate diagonalization of the matrix representatives of $(H_o^A - E_o^A)$ and $(H_o^B - E_o^B)$ over a *finite* set of suitable basis functions, individually antisymmetrized, $\{A_a\}$ for A and $\{B_b\}$ for B [31, 36]. The final conclusions are then

a) that the first-order term, E_1 in (3), must be corrected by the addition of the term

$$\Delta = \frac{\langle A\psi_o| \, H_o - E_o \, |\psi_o\rangle}{\langle A\psi_o \, | \, \psi_o\rangle} \tag{13}$$

which after reduction must be added to the penetration component in (4) and is at least of order $O(S^2)$, if S is an overlap integral between an occupied orbital of A and one of B, and of order $O(S^4)$ for Hartree-Fock wfs of A and B [12, 34, 35];
b) that the second-order RS term (12) is replaced by an upper bound

$$\tilde{E}_2^{cb} = - \sum_{\substack{a \quad b \\ \text{(not both zero)}}} \frac{|\langle A_a B_b| \, V |A_o B_o\rangle|^2}{\varepsilon_a + \varepsilon_b} \geq E_2^{cb} \tag{14}$$

in which ε_a, for example, is a (pseudo)excitation energy for molecule A, being an expectation value of $(H_o^A - E_o^A)$ in pseudo-state A_a. (14) has the same structure as the "sum-over-states" expression and coincides with the exact second-order energy when the bases $\{A_a\}$ and $\{B_b\}$ become complete, in which case the equality sign holds.

The essential features of variationally-based approaches of MS-MA type are that the matrix elements contain the total Hamiltonian H instead of V (i.e. there is *no separation* into $H_o + V$); that the free-molecule product functions are replaced by *antisymmetrized* products [44b]; that the expansions are *truncated*; and that the coefficients are obtained by treating the expansion as a linear variation function [31]. When the corresponding *matrix* equations are solved by partitioning or perturbation techniques [44b] the resultant first-order interaction resembles (3), while the second-order contribution resembles (12) except that the product functions are antisymmetrized and the summation is discrete and finite. The formal reduction of E_1 and E_2 for *exact* free-molecule functions is dealt with elsewhere [16, 18, 19], in the context of a second-order EN expansion. Except in cases of heavy overlap, the antisymmetrizer in the off-diagonal elements may safely be omitted, as E_2 is already of second-order.

For closed-shell molecules $(S = M_S = 0)^2$ the product $\psi_o = A_o B_o$ of the one-determinant Hartree-Fock functions for the separate molecules gives a reasonable first approximation, while the excited pseudostates $\psi_\kappa = A_a B_b$ (at least one of a, b differing from 0) can be represented in terms of single or double excitations from the occupied orbitals of A and B into corresponding vacant orbitals. The terms in which only one molecule is excited lead to the induction energy; those in which both are excited lead to the dispersion interaction. In using Hartree-Fock functions, however, H_o^A and H_o^B must be re-interpreted as *model* Hamiltonians, not containing explicit electron interaction terms, and V must consequently by supplemented by the residual "fluctuation" or "correlation" potentials for the separate molecules; these extra terms may be admitted by using a double perturbation

---

^{2} S is here the total spin quantum number and M_S the eigenvalue of the z-component of the spin operator.

theory [13, 42] but, since they are relatively large, it is frequently desirable to include them to high order. The alternative approach [16], in which H_o^A and H_o^B are regarded as true free-molecule Hamiltonians, deals with this problem in a different way by using (wherever possible) expansion functions in which correlation effects are already included i.e. free-molecule functions calculated with a precision beyond the Hartree-Fock level. In this case, of course, the required molecular *properties* (multipole moments etc.) must be calculated with a similar precision; but this can be done by methods (see for example [37]) which are now standard.

In the present work, for simplicity, we consider the evaluation of all terms *at the Hartree-Fock level* and for closed-shell molecular ground states. The admission of correlation effects changes neither the essential framework of the theory nor the physical interpretation of the intermolecular interactions.

4 Distribution Functions and Interaction Potentials

Longuet-Higgins [43] first drew attention to the fact that the dispersion interaction between two molecules could be calculated directly in terms of charge density functions, without making the usual multipole expansion of the interaction terms in the Hamiltonian. The charge density *operator* for molecule A at point $\mathbf{r}$ may be defined as (using a for particle index, α for nuclei, i for electrons)

$$\gamma^A(\mathbf{r}) = \sum_a Z_a \delta(\mathbf{r} - \mathbf{r}_a) = \underbrace{\sum_\alpha Z_\alpha \delta(\mathbf{r} - \mathbf{r}_\alpha)}_{\text{nuclei}} - \underbrace{\sum_i \delta(\mathbf{r} - \mathbf{r}_i)}_{\text{electrons}} \tag{15}$$

and the transition charge density at point $\mathbf{r}$, associated with a transition $o \to a$ in molecule A, is then given by

$$\gamma^A(oa \mid \mathbf{r}) = \langle A_a \mid \gamma^A(\mathbf{r}) \mid A_o \rangle = \sum_\alpha Z_\alpha \delta(\mathbf{r} - \mathbf{r}_\alpha)\, \delta_{ao} - P^A(oa \mid \mathbf{r}; \mathbf{r}) \tag{16}$$

where $P^A(oa \mid \mathbf{r}; \mathbf{r})$ is the *electron* transition density and is a "diagonal element" $(\mathbf{r}' = \mathbf{r})$ of the one-electron *transition density matrix*[3] [44a]

$$P^A(oa \mid \mathbf{r}; \mathbf{r}')$$

$$= N_A \int ds_1\, d\mathbf{x}_2 \ldots d\mathbf{x}_{N_A}\, A_o(s_1\mathbf{r}, \mathbf{x}_2 \ldots \mathbf{x}_{N_A})\, A_a^*(s_1\mathbf{r}', \mathbf{x}_2 \ldots \mathbf{x}_{N_A}). \tag{17}$$

For $a = o$ (17) becomes the one-electron density matrix for a single state of A, the ground state, whith a diagonal element which is simply the electron density, as

---

[3] The Dirac delta function is formally defined by the relation $\int d\mathbf{r}'\, f(\mathbf{r}')\, \delta(\mathbf{r} - \mathbf{r}') = f(\mathbf{r})$, while δ_{ao} is a Kronecker delta. The density *matrix* (17) is defined [44c] as the matrix element of a more general density operator in which the delta function in (15) is replaced by an operator kernel $\delta(\mathbf{r}_i - \mathbf{r}')\, \delta(\mathbf{r} - \mathbf{r}_i)$.

observed in X-ray crystallography; and (16) thus gives the *total* charge density, nuclei included. The normalization is such that for $a = o$ the integrals

$$\int d\mathbf{r}\, P^A(oa \,|\, \mathbf{r}; \mathbf{r}) = N_A \delta_{ao}, \qquad \int d\mathbf{r}\, \gamma^A(oa \,|\, \mathbf{r}) = \left(\sum_\alpha Z_\alpha - N_A\right) \delta_{ao} \qquad (18)$$

represent, respectively, the total number of electrons and the total electric charge carried by molecule A.

The density functions are useful both as a basis for physical interpretation and as a means of reducing matrix elements to a more transparent form. A key result is that for any one-electron operator of the form $v = \sum_i v(\mathbf{r}_i)$, $v(\mathbf{r}_i)$ being the potential energy of electron i at point $\mathbf{r}_i$ in any potential field whatever,

$$\langle A_a | v | A_0 \rangle = \int d\mathbf{r}\, v(\mathbf{r})\, P^A(oa \,|\, \mathbf{r}; \mathbf{r}). \qquad (19)$$

The integrand is clearly the classical expression for the energy of $P^A(oa \,|\, \mathbf{r}; \mathbf{r})\, d\mathbf{r}$ electrons, in volume element $d\mathbf{r}$, in the given field; and the integral gives the energy of the whole transition "charge cloud". All the preceding results are valid for any system and thus apply equally to molecule B.

For wavefunctions of product form, such as occur in (14), there is an immediate generalization of the last result to two-electron operators such as $g = \sum_{i,j} g(\mathbf{r}_i, \mathbf{r}'_j)$ where $\mathbf{r}_i$ is the position of electron i in molecule A and $\mathbf{r}'_j$ is that of electron j in B. For the coulomb interaction $g(\mathbf{r}_i, \mathbf{r}'_j) = |\mathbf{r}_i - \mathbf{r}'_j|^{-1}$ it is clear that

$$\langle A_a B_b | g | A_o B_o \rangle = \iint d\mathbf{r}_1\, d\mathbf{r}_2\, \frac{P^A(oa \,|\, \mathbf{r}_1; \mathbf{r}_1) P^B(ob \,|\, \mathbf{r}_2; \mathbf{r}_2)}{r_{12}}. \qquad (20)$$

Equation (20) also applies in cases where $a = o$ and/or $b = o$.

From the previous results it is easy to obtain matrix elements of the interaction potential

$$V = \sum_a \sum_b \frac{Z_a Z_b}{r_{ab}} = \sum_{i=1}^{N_A} V^B(\mathbf{r}_i) + \sum_{j=1}^{N_B} V^A(\mathbf{r}'_j) + \sum_{i=1}^{N_A} \sum_{j=1}^{N_B} \frac{1}{|\mathbf{r}_i - \mathbf{r}'_j|} + V^{AB}_{\text{nuc}}$$

$$(21)$$

in which, for example,

$$V^B(\mathbf{r}_i) = -\sum_\beta \frac{Z_\beta}{r_{i\beta}} \qquad (22)$$

is the potential energy of an electron at $\mathbf{r}_i$ in the field of the nuclei (with charges Z_β) of molecule B, while V^{AB}_{nuc} is the nuclear repulsion energy between A and B. We then find

$$\langle A_a B_b | V | A_o B_o \rangle = -\delta_{bo} \sum_\beta \int d\mathbf{r}_1\, \frac{Z_\beta P^A(oa \,|\, \mathbf{r}_1; \mathbf{r}_1)}{r_{1\beta}}$$

$$- \delta_{ao} \sum_\alpha \int d\mathbf{r}_2\, \frac{Z_\alpha P^B(ob \,|\, \mathbf{r}_2; \mathbf{r}_2)}{r_{2\alpha}}$$

$$+ \iint d\mathbf{r}_1 \, d\mathbf{r}_2 \, \frac{P^A(oa \mid \mathbf{r}_1; \mathbf{r}_1) \, P^B(ob \mid \mathbf{r}_2; \mathbf{r}_2)}{r_{12}}$$

$$+ \delta_{ao}\delta_{bo} \sum_{\alpha, \beta} \frac{Z_\alpha Z_\beta}{r_{\alpha\beta}} \tag{23}$$

or, in terms of the total charge densities (16),

$$\langle A_a B_b \mid V \mid A_o B_o \rangle = \iint d\mathbf{r}_1 \, d\mathbf{r}_2 \, \frac{\gamma^A(oa \mid \mathbf{r}_1) \, \gamma^B(ob \mid \mathbf{r}_2)}{r_{12}}, \tag{24}$$

where from now on we use $\mathbf{r}_1$ for a space point in molecule A and $\mathbf{r}_2$ for one in B (Fig. 1).

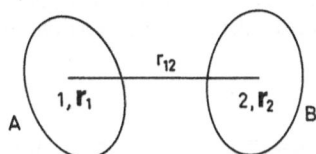

Fig. 1. Distance between two particles, 1 on A and 2 on B. $\mathbf{r}_1$ and $\mathbf{r}_2$ denote any two points in space, on A and B respectively, where we can find electrons or nuclei

The above result (24), valid also for $a = o$ and/or $b = o$, allows us to express the first- and second-order Coulombic interactions in terms of densities and transition densities for the separate molecules. Such interactions are also conveniently described using potentials: the potential energy of an electron at $\mathbf{r}_1$ in the field of the unperturbed electron distribution of molecule B is

$$J^B(oo \mid \mathbf{r}_1) = \int d\mathbf{r}_2 \, \frac{P^B(oo \mid \mathbf{r}_2; \mathbf{r}_2)}{r_{12}} \tag{25}$$

and this, together with $V^B(\mathbf{r}_1)$, allows us to define the *molecular electrostatic potential* (MEP)[4] of molecule B, namely

$$U^B(\mathbf{r}_1) = \int d\mathbf{r}_2 \, \frac{\gamma^B(oo \mid \mathbf{r}_2)}{r_{12}} = - V^B(\mathbf{r}_1) - J^B(oo \mid \mathbf{r}_1). \tag{26}$$

Such potentials can be routinely calculated for any given approximation to the unperturbed wf of the single molecule and have found a wide field of application in the interpretation of molecular properties and chemical reactivity [45, 46]. Figure 2 illustrates two sections of the equipotential map for H_2O calculated for a SCF wf with a $4-31G^*$ basis [46].

---

[4] The electric potential experienced at any point in space by a unit of *positive* charge approaching the molecule [45, 46].

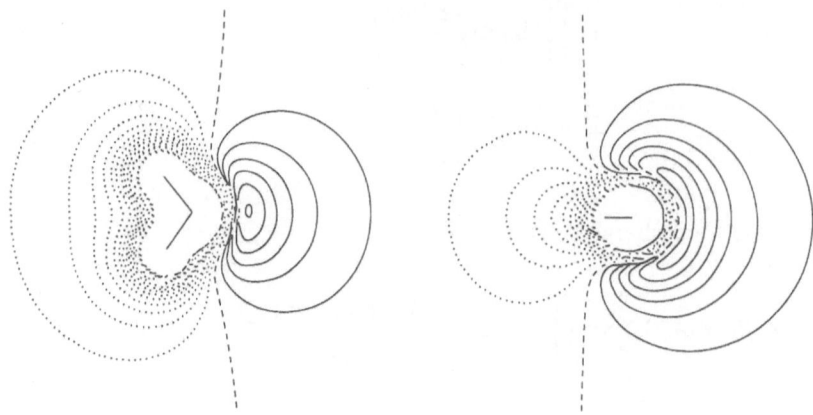

Fig. 2. Equipotential contour map of the molecular electrostatic potential (MEP) of H_2O at intervals of $15.94 \times 10^{-3}\, E_h$(SCF/4−31G*). Full lines, MEP < 0; dotted lines, MEP > 0; dashed line, MEP = 0. **Left:** section in the molecular plane (σ_h); **right:** perpendicular section in the symmetry plane (σ_v)

For describing first-order exchange-overlap effects, however, we also need the non-diagonal elements of the unperturbed electron density (in short, the *exchange* density), that for B being

$$P^B(oo \mid \mathbf{r}_1; \mathbf{r}_2)\,, \tag{27}$$

which gives rise to the exchange potential[5] [44 b]

$$\hat{K}^B(oo \mid \mathbf{r}_1) = \int d\mathbf{r}_2\, \frac{P^B(oo \mid \mathbf{r}_1; \mathbf{r}_2)}{r_{12}}\, \hat{P}_{\mathbf{r}_1\mathbf{r}_2}\,. \tag{28}$$

And in discussing the overlap-dependent term in the penetration energy E_1^{pn} in (3), we shall introduce the *overlap* density matrix [47−50]

$$P_{ov}^B(oo \mid \mathbf{r}; \mathbf{r}') = 2 \sum_j^B \sum_q^{all} \varphi_j(\mathbf{r})\, \Delta_{jq} \varphi_q^*(\mathbf{r}') \tag{29}$$

whose diagonal element has the property

$$\int d\mathbf{r}\, P_{ov}^B(oo \mid \mathbf{r}; \mathbf{r}) = 0\,, \tag{30}$$

---

[5] An integral operator with kernel $K^B(\mathbf{r}_1; \mathbf{r}_2) = \dfrac{P^B(oo \mid \mathbf{r}_1; \mathbf{r}_2)}{r_{12}}$, whose effect on function $F(\mathbf{r}_1)$ is:

$$\hat{K}^B(oo \mid \mathbf{r}_1)\, F(\mathbf{r}_1) = \int d\mathbf{r}_2\, K^B(\mathbf{r}_1; \mathbf{r}_2)\, F(\mathbf{r}_2)\,.$$

In (28), $\hat{P}_{\mathbf{r}_1\mathbf{r}_2}$ is the operator which replaces $\mathbf{r}_1$ with $\mathbf{r}_2$ in the argument of function $F(\mathbf{r}_1)$.

and similarly for A^6. In (29), j runs over the occupied molecular orbitals (MOs) of B, q over *all* occupied MOs of A and B, and Δ_{jq} is an element of the matrix

$$\Delta = -S(1 + S)^{-1} \tag{31}$$

which depends in a complicated way on the intermolecular overlap between the occupied MOs of A and B^7.

For $r' = r$, (29) gives the *overlap density* of B; this is the *change* of electron density at point r in molecule B, caused by overlap with A. Although contributing to the first-order interaction E_1 and to the error Δ (see later), the overlap density gives a vanishing contribution to the integral of the total electron density, sharing this property with the transition densities which occur in higher orders of perturbation theory.

The transition density matrix (17) can be expanded in any convenient orbital basis[8] as

$$P^A(oa \mid r; r') = \sum_{pq} p(r) \, (P^A_{oa})_{pq} \, q^*(r') , \tag{32}$$

where $(P^A_{oa})_{pq}$ is the coefficient of the orbital product $\varphi_p(r) \, \varphi_q^*(r')$, in short pq^*, in the transition $o \to a$. For closed shells, when the unperturbed reference state of A is described by a Hartree-Fock wf, (32) is zero unless p is an occupied (occ) orbital and q^* a virtual (vacant, vac) orbital of A. Henceforth we shall consistently use $p\,q\,r\,s$ to denote *atomic* orbitals of A, and $t\,u\,v\,w$ to denote those of B.

5 The Exchange-Overlap Energy

As mentioned in Sect. 2, the exchange-overlap energy depends on the nature of the spin coupling of the interacting molecules [18, 19]. For closed-shell molecules the resultant total spin is zero, and the first-order contribution to the exchange-overlap component of the interaction can be expressed in *closed form* if $A\psi_o$ is approximated as a single determinant of Hartree-Fock spin-orbitals of the individual molecules [48–50].

First-order penetration can then be expressed in terms of two contributions of opposite sign, the *attractive* one due to pure 2-electron exchange between A and B^9:

$$K = -\frac{1}{2} \int dr_1 \, \hat{K}^B(oo \mid r_1) \, P^A(oo \mid r_1; r_1')$$

$$= -\frac{1}{2} \iint dr_1 \, dr_2 \, \frac{P^B(oo \mid r_1; r_2)}{r_{12}} \, P^A(oo \mid r_2; r_1) , \tag{33}$$

---

[6] In this way, both the electron and overlap densities of the undistorted system A + B are divided into parts referring to A and B *separately*.

[7] We recall that, in general, MOs on A and B separately can be taken orthogonal, whereas MOs of A are *not* orthogonal to MOs of B. S is the *complete* overlap matrix, for all the MOs of A and B. For a generalization, see [28].

[8] Atomic bases are usually of Slater (STO) or gaussian (GTO) type.

[9] The prime on r_1' being removed *after* the action of the operator and *before* the integration.

and the one expressing the *Pauli repulsion* due to overlapping of the *static* electron distributions of the two molecules:

$$E_{ov} = \int d\mathbf{r}_1 \left[-U^B(\mathbf{r}_1) - \frac{1}{2} \hat{K}^B(oo \mid \mathbf{r}_1) \right] P_{ov}^A(oo \mid \mathbf{r}_1; \mathbf{r}_1')$$

$$+ \frac{1}{2} \iint d\mathbf{r}_1 \, d\mathbf{r}_2 \, \frac{P_{ov}^B(oo \mid \mathbf{r}_2; \mathbf{r}_2) - \frac{1}{2} P_{ov}^B(oo \mid \mathbf{r}_1; \mathbf{r}_2) \hat{P}_{\mathbf{r}_1\mathbf{r}_2}}{r_{12}} P_{ov}^A(oo \mid \mathbf{r}_1; \mathbf{r}_1')$$

$$+ (A \leftrightarrow B, 1 \leftrightarrow 2) \tag{34}$$

with

$$P_{ov}(oo \mid \mathbf{r}; \mathbf{r}') = P_{ov}^A(oo \mid \mathbf{r}; \mathbf{r}') + P_{ov}^B(oo \mid \mathbf{r}; \mathbf{r}') . \tag{35}$$

E_{ov} expresses the interaction of the overlap density of one molecule with the electrostatic-exchange potential of the other, plus the coulomb-exchange interaction of the overlap densities of the two molecules. All terms in (34) are rigorously zero for non-overlapping molecules, and have a non-classical origin arising from exchange and overlap densities. For clusters of many interacting molecules, the overlap energy is the source of first-order non-additivity observed for intermolecular forces [51, 52]. In fact, the partition (35) is only apparently additive, since it contains the effect of the *whole* intermolecular overlap which occurs through definitions (29) and (31). Nevertheless, partition (35) allows for the precise definition of (i) the additional density which must supplement the ordinary and exchange electron densities of each molecule when overlap occurs, and (ii) the error Δ occurring when A_o and B_o are not the "exact" eigenstates of H_o, but rather *approximations* satisfying the eigenvalue equation for some model Hamiltonian H_o (e.g. the one constructed in terms of the usual one-electron Fock operators of the isolated molecules).

Let us define

$$\Delta = \Delta^A + \Delta^B \tag{36}$$

where, for example,

$$\Delta^A = \frac{\langle A\psi_o \mid H_o^A - E_o^A \mid \psi_o \rangle}{\langle A\psi_o \mid \psi_o \rangle} = \frac{\langle A\psi_o \mid H_o^A - E_o^A \mid \psi_o \rangle}{1 + \langle P\psi_o \mid \psi_o \rangle} \tag{37}$$

and

$$E_o^A = \frac{\langle A_o \mid H_o^A \mid A_o \rangle}{\langle A_o \mid A_o \rangle} . \tag{38}$$

Clearly, E_o^A is the expectation value of H_o^A over the (approximate) state A_o, and similarly for B. If

$$\hat{F}^A(oo \mid \mathbf{r}) = -\frac{1}{2} \nabla^2 + V^A(\mathbf{r}) + J^A(oo \mid \mathbf{r}) - \frac{1}{2} \hat{K}^A(oo \mid \mathbf{r}) \tag{39}$$

is the *spatial* Fock operator for the closed-shell molecule A, it can be shown that [48]

$$\Delta^A = \int d\mathbf{r}_1 \, \hat{F}^A(oo \mid \mathbf{r}_1) \, P_{ov}^A(oo \mid \mathbf{r}_1; \mathbf{r}_1')$$

$$+ \frac{1}{2} \iint d\mathbf{r}_1 \, d\mathbf{r}_2 \, \frac{P_{ov}^A(oo \mid \mathbf{r}_2; \mathbf{r}_2) - \frac{1}{2} P_{ov}^A(oo \mid \mathbf{r}_1; \mathbf{r}_2) \, \hat{P}_{\mathbf{r}_1 \mathbf{r}_2}}{r_{12}}$$

$$\times P_{ov}^A(oo \mid \mathbf{r}_1; \mathbf{r}_1') \,. \tag{40}$$

Since P_{ov}^A is at least of order $O(S^2)$, it can be seen that the first term in (40) is of order $O(S^2)$ and vanishes for Hartree-Fock A_o [50], while the second term is of order $O(S^4)$ and survives in any case.

To give a simple example, consider the complex $He_2({}^1\Sigma_g^+)$ of two $He({}^1S)$ atoms whose ground states are described in term of the single determinants $A_o = \|a\bar{a}\|$ and $B_o = \|b\bar{b}\|$. If $S = \langle a \mid b \rangle$ is the interatomic overlap, then

$$P^A(oo \mid \mathbf{r}; \mathbf{r}') = 2a(\mathbf{r}) \, a^*(\mathbf{r}') \,,$$

$$P_{ov}^A(oo \mid \mathbf{r}; \mathbf{r}') = 2(1 - S^2)^{-1} \, [S^2 a(\mathbf{r}) \, a^*(\mathbf{r}') - Sa(\mathbf{r}) \, b^*(\mathbf{r}')]$$

and simple calculation shows that (with a common integral notation)

$$\Delta^A = 2(1 - S^2)^{-1} \, \{S^2[h_{aa}^A + 2(a^2 \mid a^2) - (a^2 \mid a^2)]$$

$$- S[h_{ba}^A + 2(ab \mid a^2) - (ab \mid a^2)]\}$$

$$+ (1 - S^2)^{-2} \, [S^4(a^2 \mid a^2) - 2S^3(ab \mid a^2) + S^2(ab \mid ab)] \,.$$

It can be seen that the first term is of order $O(S^2)$ and vanishes if $a(\mathbf{r})$ is the Hartree-Fock orbital of $He({}^1S)$ with orbital energy ε_o^A:

$$\hat{F}^A(oo \mid \mathbf{r}) \, a(\mathbf{r}) = \varepsilon_o^A a(\mathbf{r}) \,,$$

while the second is of order $O(S^4)$.

As a further numerical example, we give in Table 2 some accurate numerical results for "exchange" energies $(E^{exch} + \Delta = E_1^{pn})$ obtained by Murrell and Varandas [35] with STO/HF bases for the ${}^1\Sigma_g^+$ states of He_2 and Ne_2. The correction Δ, small for

Table 2. Exchange-overlap energies and correction Δ $(10^{-6} E_h)$ for the ${}^1\Sigma_g^+$ states of He_2 and NE_2 at $R = 7a_0$ calculated for Hartree-Fock wfs of separate atoms

Dimer	E^{exch}	Δ	E_1^{pn}
He_2	1.121	−0.034	1.087
	1.064 [a]	0.012 [a]	1.076 [a] (1.07) [b]
Ne_2	8.202	−0.930	7.272

[a] Ref. [53]; [b] Ref. [15].

He_2, is larger than 10% of E^{exch} for Ne_2. The numerical data for He_2 are compared with the extremely accurate results by Gutowski et al. [53] obtained with a basis of 30 "Regular Even Tempered" GTOs (RET30), and with the E_1^{pn} value of Ref. 15 calculated with a 5-term STO/HF basis of Clementi and Roetti.

Table 3 illustrates the dependence of E_1^{pn} and its exchange and overlap components on the intermolecular separation R for the linear dimer $(H_2O)_2$ in the VdW region [28], and the results at the well depth for $(HF)_2$ [17]. Although less accurate than the previous ones[10], these results are interesting for what concerns the relative value of the components. It is evident that near the VdW minimum, E_1^{pn} is dominated by the overlap component describing Pauli repulsion of the closed shells (E_{ov}).

Table 3. First-order penetration energy and its components ($10^{-3} E_h$) for some dimers in the VdW region

Dimer	R/a_o [a]	K.	E_{ov}	E_1^{pn}
$(H_2O)_2$ [b]	5.20	-29.99	51.28	21.29
$\theta_A = 128°, \theta_B = 0°$	5.67	-13.74	22.41	8.67
	6.14	-6.12	9.70	3.58
	6.61	-2.57	4.16	1.59
$(HF)_2$ [c]	5.35	-5.43	10.34	4.91
$\theta_A = 5°, \theta_B = 60°$				

[a] Distance between heavy atoms;   [b] Ref. [28];   [c] Ref. [17].

6 Non-Expanded Coulombic Interactions

The coulombic energies (i)–(iii) give the main attractive component of the inter-molecular potential. Their expression for the different orders of perturbation theory is immediately deduced from (24) according to the possible excitations occurring on A or B.

Electrostatic energy (first-order coulombic)

With $a = b = o$ there is *no* excitation on either A or B and (24) becomes

$$E_1^{cb} = \langle A_o B_o | V | A_o B_o \rangle = \iint dr_1 \, dr_2 \, \frac{\gamma^A(oo \mid r_1) \, \gamma^B(oo \mid r_2)}{r_{12}}$$

$$= \int dr_1 \, U^B(r_1) \, \gamma^A(oo \mid r_1) = \int dr_2 \, U^A(r_2) \, \gamma^B(oo \mid r_2). \tag{41}$$

---

[10] The basis functions are *not* of Hartree-Fock quality, GTO bases are an unscaled MEDIUM ($11s \, 7p \mid 8s$) contracted to $[2s \, 1p \mid 1s]$ for $(H_2O)_2$ [28] and DZ* + 2d for $(HF)_2$ [17].

The electrostatic energy gives the average value of the molecular electrostatic potential of B(A) weighted with the *static* distribution function of A(B). It also follows, as in (23), that

$$E_1^{cb} = \sum_\alpha Z_\alpha U^B(\mathbf{r}_\alpha) - \int d\mathbf{r}_1 \, U^B(\mathbf{r}_1) \, P^A(oo \,|\, \mathbf{r}_1; \mathbf{r}_1) = \sum_\alpha \sum_\beta \frac{Z_\alpha Z_\beta}{r_{\alpha\beta}}$$

$$+ \int d\mathbf{r}_1 \, V^B(\mathbf{r}_1) \, P^A(oo \,|\, \mathbf{r}_1; \mathbf{r}_1) + \int d\mathbf{r}_2 \, V^A(\mathbf{r}_2) \, P^B(oo \,|\, \mathbf{r}_2; \mathbf{r}_2)$$

$$+ \iint d\mathbf{r}_1 \, d\mathbf{r}_2 \, \frac{P^A(oo \,|\, \mathbf{r}_1; \mathbf{r}_1) \, P^B(oo \,|\, \mathbf{r}_2; \mathbf{r}_2)}{r_{12}}, \tag{42}$$

where the different terms have a clear physical meaning: the first is the repulsion between the positively charged point-like nuclei of the two molecules, the second the potential energy of the A-electron density in the field provided by the nuclei of B, the third a similar term with $(A \leftrightarrow B, 1 \leftrightarrow 2)$, the last one the mutual repulsion of the electron charge clouds.

In terms of the *atomic* basis,

$$\int d\mathbf{r}_1 U^B(\mathbf{r}_1) \, P^A(oo \,|\, \mathbf{r}_1; \mathbf{r}_1) = \sum_{pq}^A (P_{oo}^A)_{pq} \, (pq \,|\, U^B), \tag{43}$$

where we use the charge density notation

$$(pq \,|\, U^B) = \int d\mathbf{r}_1 \, U^B(\mathbf{r}_1) \, p(\mathbf{r}_1) \, q^*(\mathbf{r}_1) \tag{44}$$

and introduce the AO density matrix with elements

$$(P_{oo}^A)_{pq} = 2 \sum_i^A C_{pi} C_{iq}^*, \tag{45}$$

the summation running over all *occupied* MOs of molecule A.

Since, in the absence of intermolecular overlap, $\psi_o = A_o B_o C_o \ldots$ and the potential V is by its very nature pairwise additive [51]

$$V = \sum_{A<B} V^{AB}, \tag{46}$$

the electrostatic energy of a cluster of molecules is pairwise additive. As an example, for three molecules:

$$E_1^{cb} = E_1^{cb}(AB) + E_1^{cb}(AC) + E_1^{cb}(BC). \tag{47}$$

Induction energy (polarization)

This term, obtained from (14), involves *single* excitations either on A ($a \neq o, b = o$) or B ($a = o, b \neq o$):

$$E_2^{ind} = E_2^{ind,A} + E_2^{ind,B} \quad \text{(A polarized by B + B polarized by A)}$$

$$= -\sum_a \frac{|\langle A_a B_o | V | A_o B_o \rangle|^2}{\varepsilon_a} - \sum_b \frac{|\langle A_o B_b | V | A_o B_o \rangle|^2}{\varepsilon_b}$$

$$= -\sum_a \frac{|\int d\mathbf{r}_1 \, U^B(\mathbf{r}_1) \, P^A(oa \,|\, \mathbf{r}_1; \mathbf{r}_1)|^2}{\varepsilon_a}$$

$$- \sum_b \frac{|\int d\mathbf{r}_2 \, U^A(\mathbf{r}_2) \, P^B(ob \,|\, \mathbf{r}_2; \mathbf{r}_2)|^2}{\varepsilon_b} . \tag{48}$$

This (non-expanded) induction energy arises from the *distortion* (polarization) of the electron charge cloud of one molecule by the molecular electrostatic potential of the other.

Equation (48) shows that induction is *not* pairwise additive. In the presence of a third molecule C, the polarization of A implies, in second order, the 3-body term

$$- \sum_a \frac{\int d\mathbf{r}_1 \, U^B(\mathbf{r}_1) \, P^A(oa \,|\, \mathbf{r}_1; \mathbf{r}_1) \int d\mathbf{r}_1' \, U^C(\mathbf{r}_1') \, P^A(ao \,|\, \mathbf{r}_1'; \mathbf{r}_1')}{\varepsilon_a} + \text{c.c.} \tag{49}$$

so that the polarization of A by B + C is *not* the sum of the corresponding 2-body terms.

The induction energy of each molecule can be related to another quantity, describing the *response* of the molecule in its ground state to the perturbation provided by the other, called the *polarization propagator*. We have, in fact:

$$E_2^{ind,A} = -\sum_a \frac{\int d\mathbf{r}_1 \, U^B(\mathbf{r}_1) \, P^A(oa \,|\, \mathbf{r}_1; \mathbf{r}_1) \int d\mathbf{r}_1' \, U^B(\mathbf{r}_1') \, P^A(ao \,|\, \mathbf{r}_1'; \mathbf{r}_1')}{\varepsilon_a}$$

$$= -\frac{1}{2} \int d\mathbf{r}_1 \, U^B(\mathbf{r}_1) \int d\mathbf{r}_1' \, \Pi^A(\mathbf{r}_1; \mathbf{r}_1') \, U^B(\mathbf{r}_1')$$

$$= -\frac{1}{2} \int d\mathbf{r}_1 \, U^B(\mathbf{r}_1) \, \hat{\Pi}^A(\mathbf{r}_1) \, U^B(\mathbf{r}_1), \tag{50}$$

where

$$\Pi^A(\mathbf{r}_1; \mathbf{r}_1') = \sum_a \frac{P^A(oa \,|\, \mathbf{r}_1; \mathbf{r}_1) \, P^A(ao \,|\, \mathbf{r}_1'; \mathbf{r}_1') \, P^A(oa \,|\, \mathbf{r}_1'; \mathbf{r}_1') \, P^A(ao \,|\, \mathbf{r}_1; \mathbf{r}_1)}{\varepsilon_a} \tag{51}$$

is the symmetrized *static* polarization propagator for molecule A. Mathematically it appears as the kernel of the integral operator (see footnote 5)

$$\hat{\Pi}^A(\mathbf{r}_1) = \int d\mathbf{r}_1' \, \Pi^A(\mathbf{r}_1; \mathbf{r}_1') \, \hat{P}_{\mathbf{r}_1\mathbf{r}_1'}, \tag{52}$$

and is the limit at zero frequency of the corresponding *dynamic* propagator $\Pi^A(\mathbf{r}_1; \mathbf{r}_1' \mid iu)$ which determines the linear response of the charge density at $\mathbf{r}_1$ to an oscillating perturbation (at pure imaginary frequency iu) applied ar $\mathbf{r}_1'$ (Fig. 3) [39, 54]:

$$\Pi^A(\mathbf{r}_1; \mathbf{r}_1' \mid iu)$$
$$= \sum_a \varepsilon_a \frac{P^A(oa \mid \mathbf{r}_1; \mathbf{r}_1) \, P^A(ao \mid \mathbf{r}_1'; \mathbf{r}_1') + P^A(oa \mid \mathbf{r}_1'; \mathbf{r}_1') \, P^A(ao \mid \mathbf{r}_1; \mathbf{r}_1)}{(\varepsilon_a)^2 + u^2}, \tag{53}$$

$$\lim_{u \to o} \Pi^A(\mathbf{r}_1; \mathbf{r}_1' \mid iu) = \Pi^A(\mathbf{r}_1; \mathbf{r}_1') = \sum_a \Pi^A(\mathbf{r}_1; \mathbf{r}_1' \mid a). \tag{54}$$

The general polarization propagators, and the corresponding frequency-dependent polarizabilities (FDPs) that they determine, can be calculated via TDHF (also known as the Random Phase Approximation or RPA) or MC–TDHF methods, which are time-dependent generalizations of Hartree-Fock theory and its multiconfigurational variants [18, 37–41]. According to (54), the *static* propagators can be obtained either as the limit of dynamic propagators or, more directly, in terms of a finite number of suitable *pseudostates* $\{A_a\}$ according to the prescriptions given in Sect. 3.

In terms of the *atomic* basis, the non-expanded induction energy can be written as

$$E_2^{\text{ind}, A} = -\frac{1}{2} \sum_{pqrs}^A \Pi_{pq,rs}^A (pq \mid U^B)(rs \mid U^B), \tag{55}$$

where

$$\Pi_{pq,rs}^A = \sum_a \Pi_{pq,rs}^A(a) = \sum_a \frac{2(P_{oa}^A)_{pq} \, (P_{ao}^A)_{rs}}{\varepsilon_a} \tag{56}$$

and the *transition* density matrices in the numerator are analogous to those in (45), which refer to the ground state.

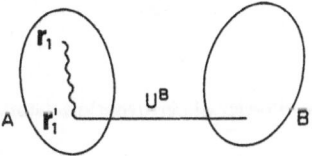

Fig. 3. Origin of induction energy for B polarizing A. The one-electron perturbation U^B acting at $\mathbf{r}_1'$ on A determines a density fluctuation propagating to $\mathbf{r}_1$

Dispersion energy

This term, also obtained from (14), involves a simultaneous *double* excitation, one on A and the other on B $(a \neq o, b \neq o)$,

$$
E_2^{\text{disp}} = - \sum_a \sum_b \frac{\left| \iint d\mathbf{r}_1\, d\mathbf{r}_2\, \dfrac{P^A(oa \mid \mathbf{r}_1; \mathbf{r}_1)\, P^B(ob \mid \mathbf{r}_2; \mathbf{r}_2)}{r_{12}} \right|^2}{\varepsilon_a + \varepsilon_b}
\tag{57}
$$

and describes the second-order contribution arising from the instantaneous interaction of the electron density fluctuations mutually induced in the two molecules. As distinct from induction, dispersion is pairwise additive in second order of perturbation theory[11].

Non-expanded dispersion can be given two fully *equivalent* formulations, in terms of either static (better, their pseudostate *components*) or dynamic propagators. In fact, we can write [39]:

$$
E_2^{\text{disp}} = - \sum_a \sum_b \frac{1}{\varepsilon_a + \varepsilon_b} \iint d\mathbf{r}_1\, d\mathbf{r}_2\, \frac{P^A(oa \mid \mathbf{r}_1; \mathbf{r}_1)\, P^B(ob \mid \mathbf{r}_2; \mathbf{r}_2)}{r_{12}}
$$

$$
\times \iint d\mathbf{r}_1'\, d\mathbf{r}_2'\, \frac{P^A(ao \mid \mathbf{r}_1'; \mathbf{r}_1')\, P^B(bo \mid \mathbf{r}_2'; \mathbf{r}_2')}{r_{1'2'}}
$$

$$
= - \iint d\mathbf{r}_1\, d\mathbf{r}_2\, \frac{1}{r_{12}} \iint d\mathbf{r}_1'\, d\mathbf{r}_2'\, \frac{1}{r_{1'2'}}
$$

$$
\times \frac{1}{4} \sum_a \sum_b \frac{\varepsilon_a \varepsilon_b}{\varepsilon_a + \varepsilon_b}\, \Pi^A(\mathbf{r}_1; \mathbf{r}_1' \mid a)\, \Pi^B(\mathbf{r}_2; \mathbf{r}_2' \mid b),
\tag{58}
$$

$$
= - \iint d\mathbf{r}_1\, d\mathbf{r}_2\, \frac{1}{r_{12}} \iint d\mathbf{r}_1'\, d\mathbf{r}_2'\, \frac{1}{r_{1'2'}}
$$

$$
\times \frac{1}{2\pi} \int_0^\infty du\, \Pi^A(\mathbf{r}_1; \mathbf{r}_1' \mid iu)\, \Pi^B(\mathbf{r}_2; \mathbf{r}_2' \mid iu),
\tag{59}
$$

where we have used the integral transform

$$
\frac{1}{\varepsilon_a + \varepsilon_b} = \frac{2}{\pi} \int_0^\infty du\, \frac{\varepsilon_a \varepsilon_b}{[(\varepsilon_a)^2 + u^2]\,[(\varepsilon_b)^2 + u^2]} \qquad (\varepsilon_a, \varepsilon_b > 0).
\tag{60}
$$

---

[11] The first non-additive contribution for dispersion occurs in third order of perturbation theory [51].

We shall henceforth refer to (58) and (59) as of generalizations of the London formula [55] and the Casimir-Polder formula [56], respectively[12]. The latter, in fact, refer to C_6 dispersion coefficients for atoms expressed in terms of static (London) or dynamic (Casimir-Polder) polarizabilities, whereas (58) and (59) describe, in a completely general way, non-expanded dispersion between atoms or molecules. (59) expresses the coupling of two electrostatic interactions ($1/r_{12}$ and $1/r_{1'2'}$) involving four space points in the two molecules, with a strength factor which depends on how readily density fluctuations propagate between $\mathbf{r}_1'$ and $\mathbf{r}_1$ on A, $\mathbf{r}_2'$ and $\mathbf{r}_2$ on B (Fig. 4).

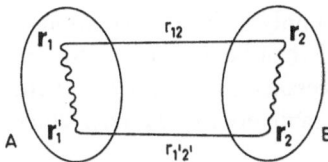

Fig. 4. Origin of intermolecular dispersion. The two electrostatic interactions $1/r_{12}$ and $1/r_{1'2'}$ are coupled by the density fluctuations simultaneously propagating from $\mathbf{r}_1'$ to $\mathbf{r}_1$ on A and from $\mathbf{r}_2'$ to $\mathbf{r}_2$ on B

The coupling between the two fluctuating systems is instead expressed in (58) by the presence of the denominator $(\varepsilon_a + \varepsilon_b)$. The two couplings are equivalent through the transform (60).

In terms of an appropriate *atomic* basis, the non-expanded dispersion can be written as

$$E_2^{\text{disp}} = - \sum_{pqrs}^{A} \sum_{tuvw}^{B} (pq \mid tu)\,(rs \mid vw)\,\frac{1}{4} \sum_a \sum_b \frac{\varepsilon_a \varepsilon_b}{\varepsilon_a + \varepsilon_b}\, \Pi^A_{pq,rs}(a)\, \Pi^B_{tu,vw}(b)\,, \tag{61}$$

$$= - \sum_{pqrs}^{A} \sum_{tuvw}^{B} (pq \mid tu)\,(rs \mid vw)\,\frac{1}{2\pi} \int_0^{\infty} du\, \Pi^A_{pq,rs}(iu)\, \Pi^B_{tu,vw}(iu)\,, \tag{62}$$

where

$$(pq \mid tu) = \iint d\mathbf{r}_1\, d\mathbf{r}_2\, \frac{t(\mathbf{r}_2)\, u^*(\mathbf{r}_2)}{r_{12}}\, p(\mathbf{r}_1)\, q^*(\mathbf{r}_1) \tag{63}$$

is the 2-electron integral[13] expressing the interaction between the elementary charge distributions pq^* on A and tu^* on B, and

$$\Pi^A_{pq,rs}(iu) = \sum_a \varepsilon_a\, \frac{2(P^A_{oa})_{pq}\,(P^A_{ao})_{rs}}{(\varepsilon_a)^2 + u^2}\,. \tag{64}$$

---

[12] An expression similar to (59) was also derived by Longuet-Higgins (apparently unpublished, but cited in ref. 39a).

[13] In general, a many-centre integral.

7 Generalized Multipole Expansion for the Intermolecular Potential

The previous expressions for the non-expanded coulombic energies are exact but imply the calculation of a large number of one- and two-electron many-centre integrals over the orbitals of the atomic basis: for small molecules this is feasible (see for example [16, 17]) but for large molecules it becomes prohibitive. Furthermore, the calculation of the integrals must be repeated for all different intermolecular separations R and the different orientations (Ω_A, Ω_B) of the two molecules. From the standpoint of computational convenience, an ideal method would imply a single evaluation of all the electric properties of each molecule, which could then be used unaltered for all different values of (R, Ω_A, Ω_B). To do this, it is necessary to make *explicit* the dependence of the intermolecular interaction on the parameters (R, Ω_A, Ω_B), which is instead implicit in all formulae given so far.

To this end we use the *generalized multipole expansion* of the intermolecular potential V[22, 57, 58], based on the Fourier transform (FT) of the reciprocal of the interparticle distance (Fig. 5)

$$\frac{1}{r_{ab}} = \frac{1}{2\pi^2} \int \frac{d\mathbf{k}}{k^2} \, e^{i\mathbf{k}\cdot\mathbf{r}_{ab}} = \frac{1}{2\pi^2} \int\limits_0^\infty dk \int\limits_0^{2\pi} d\theta_k \sin\theta_k \int\limits_0^{2\pi} d\varphi_k \, e^{i\mathbf{k}\cdot\mathbf{R}} \, e^{i\mathbf{k}\cdot\mathbf{r}_b} \, e^{-i\mathbf{k}\cdot\mathbf{r}_a} \,. \quad ^{14} \qquad (65)$$

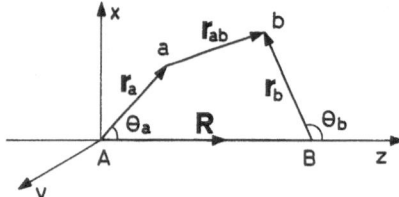

Fig. 5. The intermolecular reference system. A and B are arbitrary centres on each molecule, from which $\mathbf{r}_a$ and $\mathbf{r}_b$ are measured

In k-space, V is then expressed by the *factorized* expression:

$$V = \frac{1}{2\pi^2} \int \frac{d\mathbf{k}}{k^2} \, e^{i\mathbf{k}\cdot\mathbf{R}} \, Q^A(-\mathbf{k}) \, Q^B(\mathbf{k}) \,, \qquad (66)$$

where

$$Q^A(-\mathbf{k}) = \sum_a Z_a \, e^{i(-\mathbf{k})\cdot\mathbf{r}_a} \qquad (67)$$

is the resolution in *plane* waves of the charge density operator $\gamma^A(\mathbf{r})$[15].

---

[14] The transformation vector $\mathbf{k}$ has spherical coordinates (k, θ_k, φ_k).

[15] More precisely, the Longuet-Higgins charge density operator is the FT of $Q^A(-\mathbf{k})$:

$$\frac{1}{8\pi^3} \int d\mathbf{k} \, e^{i\mathbf{k}\cdot\mathbf{r}} \, Q^A(-\mathbf{k}) = \sum_a Z_a \delta(\mathbf{r} - \mathbf{r}_a) = \gamma^A(\mathbf{r}) \,.$$

Each plane wave can be expanded in *spherical* waves (Rayleigh)

$$e^{i\mathbf{k}\cdot\mathbf{r}} = 4\pi \sum_{l=0}^{\infty} i^l j_l(rk) \sum_{m=-l}^{+l} Y_{lm}(\theta_k, \varphi_k)\, Y_{lm}(\theta_r, \varphi_r)\,, \tag{68}$$

where the Y's are spherical harmonics in *real* form pertaining to vectors $\mathbf{k}$ and $\mathbf{r}$, and j_l is a spherical Bessel function of order l [59]. Introducing (68) in (65), and integrating over the angles (θ_k, φ_k), we obtain for the particular choice of $\mathbf{R}$ in Fig. 5:

$$\frac{1}{r_{ab}} = 4\pi \sum_{l_a=0}^{\infty} \sum_{l_b=0}^{\infty} (-1)^{l_b + l_{ab}} \sqrt{(2l_a + 1)(2l_b + 1)}$$

$$\times \sum_{m=-l_{ab}}^{+l_{ab}} (-1)^m Y_{l_a m}(\theta_a, \varphi_a)\, Y_{l_b m}(\theta_b, \varphi_b)$$

$$\times \sum_{\substack{L=|l_a - l_b| \\ (l_a + l_b + L = \text{even})}}^{l_a + l_b} (-1)^{[l_a + l_b + L - 2|l_a - l_b|]/2}\, C(l_a l_b L; 00)$$

$$\times C(l_a l_b L; |m|, -|m|) \frac{2}{\pi} \int_0^{\infty} dk\, j_L(Rk)\, j_{l_a}(r_a k)\, j_{l_b}(r_b k) \tag{69}$$

$$= \sum_{l_a} \sum_{l_b} \sum_{m} \sum_{L} (-1)^{l_b + m}\, K(l_a l_b Lm) \frac{2}{\pi} \int_0^{\infty} dk\, j_L(Rk)\, R^A_{l_a m}(\mathbf{r}_a, k)\, R^B_{l_b m}(\mathbf{r}_b, k)\,. \tag{70}$$

In these expressions, $l_{ab} = \min(l_a, l_b)$, the lesser between l_a and l_b, the C's are Clebsch-Gordan vector coupling coefficients [60], K the numerical factor

$$K(l_a l_b Lm) = (-1)^{[l_a + l_b + L - 2|l_a - l_b|]/2}\, \frac{C(l_a l_b L; 00)\, C(l_a l_b L; |m|, -|m|)}{(2l_a - 1)!!\,(2l_b - 1)!!}\,, \tag{71}$$

R_{lm} the *generalized* spherical tensor (Appendix 1)

$$R_{lm}(\mathbf{r}, k) = (2l + 1)!!\, j_l(rk) \sqrt{\frac{4\pi}{2l + 1}}\, Y_{lm}(\theta, \varphi)\,, \tag{72}$$

and $n!!$ the double factorial function

$$(2l)!! = 2^l l!, \qquad (2l-1)!! = \frac{(2l)!}{2^l l!}, \qquad 0!! = (-1)!! = 1. \tag{73}$$

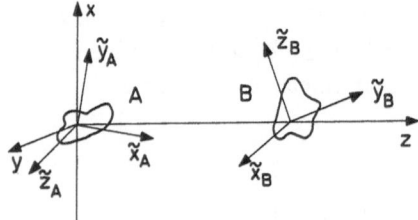

Fig. 6. The intramolecular reference system. The origin of the local axes is at the centre-of-mass of each molecule

Transforming to an *intramolecular* reference system (Fig. 6), we get:

$$R_{lm}(\mathbf{r}, k) \qquad = \sum_{q=-l}^{+l} r_{lq}^m(\Omega)\, \tilde{R}_{lq}(\mathbf{r}, k) \tag{74}$$

intermolecular
frame (space−fixed)

intramolecular
frame (body−fixed)

where the r_{lq}^m are coefficients describing the rotations that bring the intramolecular frame into self-coincidence with the original intermolecular frame. These coefficients depend on the representations of the 3-dimensional rotation group [60], and have been worked out *explicitly* (Appendix 2) for real spherical harmonics, using a convenient polynomial form [61] involving the Euler angles $\Omega = (\alpha, \beta, \gamma)$.

Using (70) in (21), we then get the *generalized multipole expansion* of V in a form where the angular variables (Ω_A, Ω_B), which define the relative orientation of the two molecules, are *separated* from R:

$$V = \sum_{l_a} \sum_{l_b} \sum_{m} \sum_{q_a} \sum_{q_b} T_{q_a q_b}^{l_a l_b m}(\Omega_A, \Omega_B)\, U_{q_a q_b}^{l_a l_b m}(R). \tag{75}$$

This 5-index expansion gives the most detailed description of the intermolecular potential V, and is valid for any value of R. In (75)

$$T_{q_a q_b}^{l_a l_b m}(\Omega_A, \Omega_B) = (-1)^{l_b + m}\, r_{l_a q_a}^m(\Omega_A)\, r_{l_b q_b}^m(\Omega_B) \tag{76}$$

is an angular factor *independent* of R, while

$$U_{q_a q_b}^{l_a l_b m}(R) = \sum_L K(l_a l_b L\, m)\, \frac{2}{\pi} \int_0^\infty dk\, j_L(Rk)\, \tilde{Q}_{l_a q_a}^A(k)\, \tilde{Q}_{l_b q_b}^B(k) \tag{77}$$

is the factor containing all the *R-dependence* through the integral over k. The $\tilde{Q}_{lq}(k)$ are *generalized* (k-dependent)[16] multipole moment operators referred to body-fixed axes, that in the limit $k \to 0$ transform into the *ordinary* multipole moment operators $\tilde{Q}_{lq} = \sum_a Z_a \tilde{R}_{lq}(\mathbf{r}_a)$.

The k-integral in (77) can be evaluated analytically by contour integration in the complex plane [22, 62] and, for $R \geq r_a + r_b$, U takes the simple form

$$
U^{l_a l_b m}_{q_a q_b}(R \geq r_a + r_b)
$$

$$
= \frac{(l_a + l_b)!}{\sqrt{(l_a + m)!\,(l_a - m)!\,(l_b + m)!\,(l_b - m)!}} \frac{\tilde{Q}^A_{l_a q_a} \tilde{Q}^B_{l_b q_b}}{R^{l_a + l_b + 1}} \tag{78}
$$

implying the complete decoupling of $\tilde{Q}^A$ and $\tilde{Q}^B$, and making explicit the R-dependence.

Introducing (76) and (78) into (75), we obtain the well-known *long-range* expansion of V in real spherical tensor form

$$
V(R \geq r_a + r_b) = \sum_{l_a}\sum_{l_b} (-1)^{l_b} \frac{(l_a + l_b)!}{R^{l_a + l_b + 1}} \sum_{q_a}\sum_{q_b} P^{l_a l_b}_{q_a q_b} \tilde{Q}^A_{l_a q_a} \tilde{Q}^B_{l_b q_b} \,, \tag{79}
$$

where

$$
P^{l_a l_b}_{q_a q_b} = \sum_m (-1)^m \frac{r^m_{l_a q_a}(\Omega_A)\, r^m_{l_b q_b}(\Omega_B)}{\sqrt{(l_a + m)!\,(l_a - m)!\,(l_b + m)!\,(l_b - m)!}} \tag{80}
$$

is a function describing the relative orientation of the two molecules [63]. Equation (79) is a classical result due to Carlson-Rushbrooke-Rose-Fontana (see Ref. 63) resolving V into the sum of elementary interactions between the multipole moment operators of the two molecules.

From the general expansion (75) we can immediately obtain the corresponding expansions for the coulombic interaction energies. As an example, we have for *expanded dispersion*:

$$
E^{\text{disp}}_2 = \sum_{\substack{l_a l_b m \\ q_a q_b}} \sum_{\substack{l'_a l'_b m' \\ q'_a q'_b}} T^{\,l_a l_b m}_{q_a q_b} T^{\,l'_a l'_b m'}_{q'_a q'_b} (U(\text{disp}))^{l_a l_b m;\, l'_a l'_b m'}_{q_a q_b;\, q'_a q'_b} \,, \tag{81}
$$

where the angular factors are defined in (76), and the factor $U(\text{disp})$, which contains all the R-dependence including the charge-overlap effects due to the damping of

---

[16] For the generalized spherical tensors see the last part of Appendix 1.

the multipoles [64], can be written in the two equivalent forms:

$$(U(\text{disp}))_{q_a q_b ; q'_a q_b}^{l_a l_b m ; l'_a l'_b m'} = - \sum_L \sum_{L'} K(l_a l_b L\, m)\, K(l'_a l'_b L'\, m')$$

$$\times \frac{2}{\pi} \int_0^\infty dk\, j_L(Rk)\, \frac{2}{\pi} \int_0^\infty dk'\, j_{L'}(Rk')$$

$$\times \frac{1}{4} \sum_a \sum_b \frac{\varepsilon_a \varepsilon_b}{\varepsilon_a + \varepsilon_b}\, \tilde{\alpha}_{l_a q_a,\, l'_a q'_a}^A(k, k' \mid a)\, \tilde{\alpha}_{l_b q_b,\, l'_b q'_b}^B(k, k' \mid b)$$

$$(82)$$

which is a PSEUDOSTATE decomposition, with $\varepsilon_a = \varepsilon_{lq,\, l'q'}^A(a)$, or

$$(U(\text{disp}))_{q_a q_b ; q'_a q_b}^{l_a l_b m ; l'_a l'_b m'} = - \sum_L \sum_{L'} K(l_a l_b L\, m)\, K(l'_a l'_b L'\, m')$$

$$\times \frac{2}{\pi} \int_0^\infty dk\, j_L(Rk)\, \frac{2}{\pi} \int_0^\infty dk'\, j_{L'}(Rk')$$

$$\times \frac{1}{2\pi} \int_0^\infty du\, \tilde{\alpha}_{l_a q_a,\, l'_a q'_a}^A(k, k' \mid iu)\, \tilde{\alpha}_{l_b q_b,\, l'_b q'_b}^B(k, k' \mid iu) \quad (83)$$

which is a CASIMIR-POLDER like expression.

In these expressions, $\tilde{\alpha}(k, k' \mid iu)$ is a *generalized* dynamic polarizability referred to intramolecular axes [22]

$$\tilde{\alpha}_{lq,\, l'q'}^A(k, k' \mid iu) = \sum_a \varepsilon_a \frac{\tilde{\mu}_{lq}^A(oa \mid k)\, \tilde{\mu}_{l'q'}^A(ao \mid k') + \tilde{\mu}_{l'q'}^A(oa \mid k')\, \tilde{\mu}_{lq}^A(ao \mid k)}{(\varepsilon_a)^2 + u^2}$$

$$= \int d\mathbf{r}_1\, \tilde{R}_{lq}^A(\mathbf{r}_1, k) \int d\mathbf{r}'_1\, \Pi^A(\mathbf{r}_1; \mathbf{r}'_1 \mid iu)\, \tilde{R}_{l'q'}^A(\mathbf{r}'_1, k'), \qquad (84)$$

with a similar expression for the static pseudostate component $\tilde{\alpha}_{lq,\, l'q'}^A(k, k' \mid a)$, and $\tilde{\mu}_{lq}^A(oa \mid k)$ is the k-dependent generalization of the q-component of a transition moment of order l. The ordinary multipole moments and polarizabilities are simply the limit for $k \to 0$ of these k-dependent quantities (Appendix 1).

Equations (81)–(83) give our final result for the dispersion energy: once the k-dependent polarizabilities have been calculated for a given series of molecules by any one of the standard methods at our disposal (pseudostates, RPA, TDHF, MC-TDHF), they can be used to evaluate the dispersion energies for *any pair* of molecules from that set, for whatever orientation Ω and whatever separation R. In this sense, the generalized expansion (75) is given a *separable* form [58]. The method is relatively new[17], and has been applied so far to the accurate calculation of the

---

[17] The method was first introduced by Koide [57] for atoms, and later extended to molecules [22, 58].

damping factors for dispersion and induction in dimers derived from He, Ne, HF [58]. The corresponding expressions for electrostatic and induction energies have also been worked out and can be found elsewhere [22].

8 Long-Range Molecular Coefficients

As a consequence of the particular form assumed by the potential V at large distances (Eq. 79), each coulombic component of the interaction energy (electrostatic, induction, dispersion) varies asymptotically as $C_n R^{-n}$ [13, 23, 61, 63]:

$$E_1^{cb}(\text{long-range}) = \sum_{l_a} \sum_{l_b} C_n^{l_a, l_b}(\text{es}) \, R^{-n}, \tag{85}$$

$$E_2^{cb}(\text{long-range}) = -\sum_{l_a} \sum_{l_b} \sum_{l_a'} \sum_{l_b'} C_{n+n'}^{l_a l_a', l_b l_b'}(\text{ind/disp}) \, R^{-n-n'}, \tag{86}$$

$$n = l_a + l_b + 1, \qquad n' = l_a' + l_b' + 1. \tag{87}$$

At intermediate to small separations, charge-overlap effects (which decrease exponentially with distance) are no longer negligible [64], and it is conventional to introduce empirical factors which *damp* the interaction in short range. These damping factors, which take any value ranging from 0 to 1 according to the intermolecular distance, depend on R for atoms and on (R, Ω) for molecules, and may be defined as the ratio between the *exact* expanded energy resulting from (75) and its corresponding *long-range* value from (79). In the present formulation there is clearly no need to introduce empirical factors.

The long-range molecular coefficients C_n are anyway very interesting by themselves: they describe exactly the long-range behaviour of the bimolecular system, embodying all dependence on the electric properties which characterize the charge distributions of the individual molecules and their relative orientation in the dimer [23, 63].

Assuming point multipoles located at the centre-of-mass of each molecule and neglecting translation of the reference frame [18], we obtain the general formulae [61]:

$$C_n^{l_a, l_b}(\text{es}) = (-1)^{l_b} (l_a + l_b)! \sum_{q_a q_b} P_{q_a q_b}^{l_a l_b} \tilde{\mu}_{l_a q_a}^A(oo) \, \tilde{\mu}_{l_b q_b}^B(oo), \tag{88}$$

$$C_{n+n'}^{l_a l_a', l_b l_b'}(\text{ind}, A) = \tfrac{1}{2} (-1)^{l_b + l_b'} (l_a + l_b)! \, (l_a' + l_b')!$$
$$\times \sum_{q_a q_b} \sum_{q_a' q_b'} P_{q_a q_b}^{l_a l_b} P_{q_a' q_b'}^{l_a' l_b'} \tilde{\alpha}_{l_a q_a, l_a' q_a'}^A \tilde{\mu}_{l_b q_b}^B(oo) \, \tilde{\mu}_{l_b' q_b'}^B(oo), \tag{89}$$

---

[18] In the most general case, the body-fixed frame differs from the space-fixed frame by rotations of the Euler angles followed by *translation* of the origin. General formulae can be found in [63].

$$C_{n+n'}^{l_a l_a', l_b l_b'}(\text{ind}, B) = \tfrac{1}{2}(-1)^{l_b+l_b'}(l_a+l_b)!\,(l_a'+l_b')!$$

$$\times \sum_{q_a q_b}\sum_{q_a' q_b} P_{q_a q_b}^{l_a l_b} P_{q_a' q_b}^{l_a' l_b} \tilde{\mu}_{l_a q_a}^A(oo)\,\tilde{\mu}_{l_a' q_a'}^A(oo)\,\tilde{a}_{l_b q_b,\,l_b q_b}^B, \qquad (90)$$

$$C_{n+n'}^{l_a l_a', l_b l_b'}(\text{disp}) = (-1)^{l_b+l_b'}(l_a+l_b)!\,(l_a'+l_b')!$$

$$\times \sum_{q_a q_b}\sum_{q_a' q_b} P_{q_a q_b}^{l_a l_b} P_{q_a' q_b}^{l_a' l_b} \frac{1}{4}\sum_a\sum_b \frac{\varepsilon_a \varepsilon_b}{\varepsilon_a+\varepsilon_b}\,\tilde{\alpha}_{l_a q_a,\,l_a' q_a'}^A(a)\,\tilde{\alpha}_{l_b q_b,\,l_b q_b}^B(b), \qquad (91)$$

or

$$\times \sum_{q_a q_b}\sum_{q_a' q_b} P_{q_a q_b}^{l_a l_b} P_{q_a' q_b}^{l_a' l_b} \frac{1}{2\pi}\int_0^\infty du\,\tilde{\alpha}_{l_a q_a,\,l_a' q_a'}^A(iu)\,\tilde{\alpha}_{l_b q_b,\,l_b q_b}^B(iu). \qquad (92)$$

In these formulae,

$$\tilde{\alpha}_{lq,\,l'q'}^A(iu) = \sum_a \varepsilon_a \frac{\tilde{\mu}_{lq}^A(oa)\,\tilde{\mu}_{l'q'}^A(ao) + \tilde{\mu}_{l'q'}^A(oa)\,\tilde{\mu}_{lq}^A(ao)}{(\varepsilon_a)^2 + u^2}$$

$$= \int d\mathbf{r}_1\,\tilde{R}_{lq}^A(\mathbf{r}_1) \int d\mathbf{r}_1'\,\Pi^A(\mathbf{r}_1;\mathbf{r}_1'\mid iu)\,\tilde{R}_{l'q'}^A(\mathbf{r}_1') \qquad (93)$$

is a *frequency-dependent* polarizability of molecule A in real spherical tensor form[19] (compare eq. 84);

$$\tilde{\alpha}_{lq,\,l'q'}^A = \lim_{u\to 0}\tilde{\alpha}_{lq,\,l'q'}^A(iu) = \sum_a \tilde{\alpha}_{lq,\,l'q'}^A(a) \qquad (94)$$

the corresponding *static* polarizability (with its pseudostate *decomposition*); and

$$\tilde{\mu}_{lq}^A(oa) = \int d\mathbf{r}_1\,\tilde{R}_{lq}^A(\mathbf{r}_1)\,\gamma^A(oa\mid\mathbf{r}_1) \qquad (95)$$

is the q-component $(-l \leq q \leq +l)$ of the 2^l-pole transition moment of molecule A (all referred to intramolecular axes).

All the above coefficients have a simple physical meaning:

- (88) leads to the first-order interaction between the *permanent* electric moments of A and B;
- (89) leads to the *polarization* of A by the permanent moments of B, and similarly for (90)[20];
- (91) and (92) lead to, in an *equivalent* (London/pseudostates or Casimir-Polder) and by now familiar way, the *coupling* of the (static/dynamic) polarizabilities of the two molecules which gives rise to the dispersion energy.

---

[19] Transformation to *cartesian* tensor form can be achieved using existing literature formulations [63, 65, 66].

[20] (90) cannot be obtained from (89) simply by interchanging A ↔ B, because of the asymmetry in the phase factor $(-1)^{l_b+l_b'}$ resulting from the choice of the reference system in Fig. 5.

Further simplifications of (88)–(92) will depend on the local symmetry of the monomers[21]. In particular, atoms in S-states have no permanent moments in ground state, $\mu_{lq}(oo) = 0$, and have isotropic polarizabilities, $\alpha_{lq,l'q'} = \alpha_l \delta_{ll'} \delta_{qq'}$; linear molecules have only axial permanent moments, $\mu_{l_0}(oo)$, and polarizabilities restricted to $\alpha_{lq,l'q'} = \alpha_{lq,l'q} \delta_{qq'}$ with $\alpha_{l\bar{q},l'\bar{q}} = \alpha_{lq,l'q}$ ($\bar{q} = -q$); for centrosymmetric molecules all odd moments vanish.

Explicit formulae for the coefficients C_n can be obtained easily by making explicit the dependence of P, eq. (80), on the Euler angles (Ω_A, Ω_B) using the polynomial formula for the rotation coefficients r_{lq}^m given in Appendix 2. Other formulae exist in the Literature, but are either incomplete to a given order [1, 23] or rather implicit [67–69], involving calculation of a number of 3j and 9j Wigner symbols [67] or polynomial functions of the direction cosines [68, 69] which give the relative orientation of the two molecules. For the particular case of atoms in S-states only the dispersion coefficient is non zero. All Euler angles are zero, and $r_{lq}^m = \delta_{mq}$ for any l. Furthermore:

$$l_a' = l_a, \qquad l_b' = l_b, \qquad n' = n,$$

$$[(l_a + l_b)!]^2 (P_{q_a q_b}^{l_a l_b})^2 = \binom{2l_a + 2l_b}{2l_a}, \tag{96}$$

where $\binom{p}{q}$ is a binomial coefficient, so that:

$$C_{2n}^{l_a, l_b}(\text{disp}) = \binom{2l_a + 2l_b}{2l_a} \frac{1}{4} \sum_a \sum_b \frac{\varepsilon_a \varepsilon_b}{\varepsilon_a + \varepsilon_b} \tilde{\alpha}_{l_a}^A(a) \tilde{\alpha}_{l_b}^B(b) \tag{97}$$

$$= \binom{2l_a + 2l_b}{2l_a} \frac{1}{2\pi} \int_0^\infty du \, \tilde{\alpha}_{l_a}^A(iu) \, \tilde{\alpha}_{l_b}^B(iu), \tag{98}$$

a result first given by Dalgarno and Davison [70]. In this case, $\varepsilon_a = \varepsilon_{l_a}^A(a)$. Even if intensively studied in the last two decades [14], accurate dispersion coefficients have been calculated so far by ab initio methods only for relatively few small atomic systems.

For atoms, an N-term linear pseudostate decomposition of the one-centre variation function determining polarizabilities, based on a rapidly convergent expansion in powers of the radial variable, is found to give extremely accurate values of dispersion coefficients even with a limited number of terms [36]. The excited pseudostates resulting from the solution of the one-centre problem are used as they stand in the calculation of the related dispersion coefficients, Eq. (97), so that additional multicentre optimization of non-linear parameters [73], iterative procedures [74], or numerical integration over dynamic polarizabilities [75] are avoided from the outset. Results for the model system $H(1s) + H(1s)$ are given in Table 4 for the first three C_{2n} coefficients [36]. They are obtained from pseudostate decomposition of dipole, quadrupole and octopole *static* polarizabilities of the H atom resulting from N-term

---

[21] Exhaustive tables giving the number of independent components of electric multipole moments and polarizabilities for molecules of different symmetry can be found in [1].

diagonalization of the matrix of the excitation energies $\{\langle A_a | H_o^A - E_o^A | A_{a'} \rangle\}$. The expansion is rapidly convergent and the results for $N = 5$ are exact to all figures given in Table 4. In principle, the *same* pseudostate technique applies also to molecules according to Eq. (91).

Table 4. Convergence of the one-centre calculation of the C_{2n} dispersion coefficients (atomic units) for the $H(1s) + H(1s)$ system according to the number N of linear pseudostates describing the static multipole polarizabilities of $H(1s)$

N	$C_6/E_h a_o^6$	$C_8/E_h a_o^8$	$C_{10}/E_h a_o^{10}$	
1	6	115.7143	2016[a]	1063.125[b]
2	6.4821	124.0932	2145.819	1132.610
3	6.4984	124.3865	2150.319	1135.107
4	6.4990_0	124.3984	2150.602	1135.209
5	6.4990_2	124.3990	2150.613_5	1135.214

[a] Dipole-octopole; [b] Quadrupole–quadrupole.

Expansion in a finite set of discrete pseudostates evidently converges much faster than expansion in true eigenstates of H_0 (which would involve also continuum states). The same behaviour is observed for the expansion in eigenstates of F_0 (the one-particle Fock Hamiltonian), so that convergence of conventional TDHF methods may prove rather slow, especially with large bases of atomic orbitals[22].

A time-dependent extension of the pseudostate method, where the true spectrum is replaced by a small number of "effective states", has been proposed by Visser and Wormer [76], and applied to the TDHF calculation of dynamic polarizabilities for He, Ne, H_2, N_2 and the related dispersion coefficients [42]. The TDHF values (without correlation) calculated in this way for He and He_2, as well as those resulting from a SDT-MBPT calculation which includes the real correlation effects in second order of the correlation potential [42], are compared in Table 5 with estremely accurate values[23] taken from [77, 78].

Table 5. Static multipole polarizabilities α_l for He, and C_{2n} dispersion coefficients for He_2 (a.u.)

Property	TDHF	SDT-MBPT	Accurate
α_1	1.322[a]	1.354[a]	1.383[b,c]
α_2	2.316	2.372	2.443
α_3	9.941	10.22	10.61
C_6	1.375	1.431	1.461
C_8	13.12	13.66	14.11
C_{10}	168.4	175.8	183.6

[a] Ref. [42]; [b] Ref. [77]; [c] Ref. [78].

---

[22] The dimensionality of the TDHF eigenvalue problem can easily be of the order of 1000 or more, even for medium size molecules [76].

[23] Based on a highly correlated ψ_o having $E_o = -2.9037243\, E_h$ and giving an overlap of 0.999999991 with the exact wf for the $1^1 S$ state of He.

TDHF values are about 5–8% smaller than the accurate ones, this error being halved for the correlated values (SDT-MBPT). For Ne, Ne_2, H_2, $(H_2)_2$, N_2, $(N_2)_2$ no such correspondingly exact comparison values are available. The same authors have recently [79] calculated, by the same many-body perturbation method, correlated frequency-dependent polarizabilities for HF, H_2O, NH_3, CO_2, computing the dispersion C_n coefficients up to $n = 10$ for all the dimers consisting of these molecules and for all their combinations with He, Ne, H_2 and N_2.

As a second example, let us consider the system formed by a spherical atom A and a linear molecule B (Fig. 7).

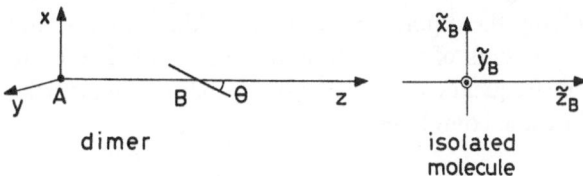

Fig. 7. The system spherical atom A — linear molecule B

In this case, a single Euler angle ($\beta_B = -\theta$) suffices to define the orientation of the molecule with respect to the intermolecular z-axis. Furthermore, we have:

$$\tilde{\mu}^A_{l_a q_a}(oo) = 0, \qquad \tilde{\alpha}^A_{l_a q_a, l'_a q'_a} = \tilde{\alpha}^A_{l_a} \delta_{l_a l'_a} \delta_{q_a q'_a}, \tag{99}$$

$$\tilde{\mu}^B_{l_b q_b}(oo) = \tilde{\mu}^B_{l_b o} \delta_{q_b o}, \qquad \tilde{\alpha}^B_{l_b q_b, l_b q_b} = \tilde{\alpha}^B_{l_b q_b, l_b q_b} \delta_{q_b q_b} \tag{100}$$

so that only $C(\text{ind}, A)$ and $C(\text{disp})$ are non zero. Using the results of Appendix 2, we find

$$C^{l_a l'_a, l_b l_b}_{n+n'}(\text{ind}, A) = \delta_{l_a l'_a} I^{l_a l_b l_b}(\theta)\, \tilde{\alpha}^A_{l_a} \tilde{\mu}^B_{l_b o} \tilde{\mu}^B_{l_b o}, \tag{101}$$

where the angular factor $I(\theta)$ is given by

$$I^{l_a l_b l_b}(\theta) = (-1)^{l_b + l_b}\, (l_a + l_b)!\, (l_a + l'_b)!$$

$$\times \sum_{m=0}^{\min(l_a, l_b,\, l_b)} \frac{P^{|m|}_{l_b}(\cos \theta)\, P^{|m|}_{l_b}(\cos \theta)}{2^{\delta_{mo}}(l_a + m)!\, (l_a - m)!\, (l_b + m)!\, (l'_b + m)!}; \tag{102}$$

and for the dispersion coefficient in the Casimir-Polder form

$$C^{l_a l'_a, l_b l_b}_{n+n'}(\text{disp}) = \delta_{l_a l'_a} D^{l_a l_b l_b}(\theta) \frac{1}{2\pi} \int_0^\infty du\, \tilde{\alpha}^A_{l_a}(iu)\, \tilde{\alpha}^R_{l_b q_b, l_b q_b}(iu) \tag{103}$$

with the angular factor

$$D^{l_a l_b l_b'}(\theta) = (-1)^{l_b + l_b'} (l_a + l_b)! (l_a + l_b')! l_b! l_b'!$$

$$\times \sum_{m=0}^{\min(l_a, l_b, l_b')} \sum_{q_b=0}^{\min(l_b, l_b')} \frac{[(l_a + m)! (l_a - m)! (l_b + m)! (l_b' + m)!]^{-1}}{\sqrt{(l_b + q_b)! (l_b - q_b)! (l_b' + q_b)! (l_b' - q_b)!}}$$

$$\times \left\{ \frac{V_{l_b q_b}^m(\theta) V_{l_b' q_b}^m(\theta)}{2^{\delta_{mo} + \delta_{q_b o}}} + (1 - \delta_{mo})(1 - \delta_{q_b o}) U_{l_b \bar{q}_b}^m(\theta) U_{l_b \bar{q}_b}^m(\theta) \right\}. \qquad (104)$$

Anisotropy in the Rg–HX system (Rg = rare gas, X = F, Cl) is largely determined by the R^{-7} induction term, involving the dipole polarizability of the rare gas atom and the dipole and quadrupole moments of the hydride molecule [80]. The corresponding dispersion term, involving frequency-dependent polarizabilities, gives further stabilization to the H-bonding in these complexes.

9 Van der Waals Complexes

Weakly bound complexes with a large-amplitude vibrational structure have been aptly called *Van der Waals molecules* [20]. Complexes of the heavier rare gases Rg_2 [81], or weak complexes between centrosymmetric molecules like $(H_2)_2$ or $(N_2)_2$, fit equally well into this definition; but complexes between proton donor and proton acceptor molecules, for example $(HF)_2$ or $(H_2O)_2$, which involve hydrogen bonding [11], may be regarded as borderline cases between VdW molecules and "good" molecules [20]. In such complexes, often classified as VdW, the bonding involved is essentially electrostatic in origin. However, all the above complexes are characterized by having monomers with a closed-shell structure which are held together, at moderately large distances, by weak forces[24]; and in this sense we may describe them as VdW type.

As has become apparent from previous Sections, VdW forces differ from those which determine "chemical" forces through the sign and magnitude of the penetration component (*attractive* for chemical bonding, *repulsive* for VdW bonding) and through the relative importance of the other components. Apparently, the stability of the complexes depends on a subtle balance between Pauli repulsion, decreasing exponentially with the intermolecular separation R, and the attractive forces due to electrostatic, induction and dispersion interactions, decreasing as R^{-n} at large distances.

Even if the interaction between higher multipoles determines the details of the potential surface near the minimum [82], VdW complexes can be conveniently classified in order of increasing stability on the basis of the dominant long-range attractive component:

- Dimers of rare gases or highly symmetrical molecules (like CH_4): very weak binding energy (of the order of 10^{-1} mE_h or less), dominated by dispersion (R^{-6});

---

[24] Binding energy comparable to $kT = 0.95$ mE_h at 300 K [20].

- Dimers of centrosymmetric molecules: weak binding energy, dominated by the interaction between quadrupoles (R^{-5});
- Heterodimers between dipolar and centrosymmetrical molecules (e.g. complexes between NH_3 and halogens): intermediate binding, dominated by dipole–quadrupole interaction (R^{-4});
- Dimers of dipolar molecules: relatively stronger binding (few mE_h), dominated by dipole–dipole interaction (R^{-3}).

Although the R^{-n} expansion is *not* strictly convergent at any R [13], (85) and (86) are still useful for studying the *angular dependence* of the interaction energy in the asymptotic region and near the well depth in the VdW region ($R \geq r_a + r_b$). For polar molecules, the dominant component is the electrostatic energy. Ab initio calculations for 20 dimers of small polar molecules at the experimental equilibrium distance R, using in (88) single-centre moments of Hartree-Fock quality placed at the centre-of-mass of each molecule, show that for $R \geq r_a + r_b$ the absolute minima of the electrostatic energy expanded up to R^{-6} (namely, including dipole–hexadeca-pole and quadrupole–octopole interactions) converge to angular structures that are very close to the ones observed from molecular beam experiments [82]. This seems to suggest that large cancellation occurs between the remaining energy components, and that polar molecules act mostly through the permanent moments of their charge distribution, these being the true "fingerprints" of the molecules. Similar results are obtained with models involving distributed multipoles (DM) [83], and for the second-order energies of the systems Rg − HX [80].

The above results have prompted the formulation of an even more elementary electrostatic model [84], that allows one to predict the angular shape of most symmetrical VdW dimers simply on the basis of the knowledge of the first two central electric moments characterizing the monomers. This model, based on physically *observable* properties of the interacting molecules, turns out to be a useful complement to other models, based on physically *unaccessible* properties, like DMA models [83] or the MO description of H-bonding in terms of HOMO/LUMO interactions [85].

Acknowledgements: We acknowledge support by the Italian Ministry of Public Education (M.P.I., 40%) under grants in Theoretical Chemistry.

10 Appendix 1

The *spherical tensors* $R_{lm}(\mathbf{r})$ are defined here in terms of (normalized) modified spherical harmonics in *real* form [63]

$$
R_{lm}(\mathbf{r}) = r^l \sqrt{\frac{4\pi}{2l+1}} \; Y_{lm}(\theta, \varphi) = r^l \begin{cases} \sqrt{2\dfrac{(l+m)!}{(l-m)!}} \; P_l^{\bar{m}}(\cos\theta)\sin(\bar{m}\varphi) & m < 0, \\[2ex] P_l(\cos\theta) & m = 0, \\[2ex] \sqrt{2\dfrac{(l-m)!}{(l+m)!}} \; P_l^m(\cos\theta)\cos(m\varphi) & m > 0, \end{cases}
$$

where $\bar{m} = -m$, and $P_l^{|m|}$ is the associated Legendre polynomial [59]

$$P_l^{|m|}(\cos\theta) = \frac{(\sin\theta)^{|m|}}{(2l)!!} \frac{d^{l+|m|}}{d(\cos\theta)^{l+|m|}} (\cos^2\theta - 1)^l.$$

Spherical tensors in real form are eigenfunctions of the operator $\mathscr{L}^2$ alone, but still form a *complete* orthogonal set with respect to integration over the angular variables. The first few of them ($l = 0, 1, 2$) are easily written in terms of cartesian coordinates:

$$R_{00} = 1; \quad R_{1\bar{1}} = y, \quad R_{10} = z, \quad R_{11} = x;$$

$$R_{2\bar{2}} = \sqrt{3}\, xy, \quad R_{2\bar{1}} = \sqrt{3}\, yz, \quad R_{20} = \tfrac{1}{2}(3z^2 - r^2),$$

$$R_{21} = \sqrt{3}\, zx, \quad R_{22} = \frac{\sqrt{3}}{2}(x^2 - y^2).$$

We note that the spherical tensors appear in the definition of the multipole moments of a system: thus

$$\mu_{lm}(oo) = \int d\mathbf{r}\, R_{lm}(\mathbf{r})\, \gamma(oo \mid \mathbf{r})$$

is the m component of an l-th order multipole, for a system in state o with charge density function $\gamma(oo \mid \mathbf{r})$, while the components of a *transition* multipole ($o \to a$) are determined similarly using $\gamma(oa \mid \mathbf{r})$.

The *generalized* spherical tensors in k-space [22] are defined by replacing r^l by $(2l + 1)!!\, j_l(rk)$

$$R_{lm}(\mathbf{r}, k) = (2l + 1)!!\, j_l(rk) \sqrt{\frac{4\pi}{2l + 1}}\, Y_{lm}(\theta, \varphi)$$

where j_l is a spherical Bessel function of order l [59]. They are related to the *ordinary* spherical tensors by the relation

$$R_{lm}(\mathbf{r}) = \lim_{k \to 0} \frac{R_{lm}(\mathbf{r}, k)}{k^l} = \frac{1}{l!} \left[\frac{\partial^l}{\partial k^l} R_{lm}(\mathbf{r}, k) \right]_{k=0}$$

which can be obtained by representing j_l as the Taylor series

$$j_l(x) = \sum_{p=0}^{\infty} \frac{(-1)^p}{(2p)!!\,(2p + 2l + 1)!!} x^{2p+l}.$$

A *finite* representation in inverse powers of the variable is particularly useful for integrals involving j_l [22, 62]

$$j_l(x) = \frac{1}{2}\{W_l(x)\,e^{ix} + W_l^*(x)\,e^{-ix}\}, \qquad W_l(x) = (-i)^{l+1} \sum_{p=0}^{l} \frac{i^p(l+p)!}{(2p)!!\,(l-p)!}\,x^{-p-1}.$$

11 Appendix 2

The rotation coefficients r_{lq}^m occurring in (74) have been recently worked out in a convenient polynomial form [61] involving the Euler angles $\Omega = (\alpha, \beta, \gamma)$ [60]

$$r_{lq}^m = (-1)^m\,l!\,\sqrt{\frac{(l-|m|)!}{(l+|m|)!\,(l+q)!\,(l-q)!}}$$

$$\times \left\{ H(m) \left[H(q)\,(-1)^q\,\frac{U_{lq}^m(\beta)\sin m\gamma\sin q\alpha + V_{lq}^m(\beta)\cos m\gamma\cos q\alpha}{(\sqrt{2})^{\delta_{mo}+\delta_{qo}}} \right.\right.$$

$$\left. + H(-q-1)\,\frac{U_{lq}^m(\beta)\sin m\gamma\cos q\alpha - V_{lq}^m(\beta)\cos m\gamma\sin q\alpha}{(\sqrt{2})^{\delta_{mo}}} \right]$$

$$+ H(-m-1)\left[H(q)\,(-1)^q\,\frac{U_{lq}^m(\beta)\cos m\gamma\sin q\alpha + V_{lq}^m(\beta)\sin m\gamma\cos q\alpha}{(\sqrt{2})^{\delta_{qo}}} \right.$$

$$\left.\left. + H(-q-1)\,(U_{lq}^m(\beta)\cos m\gamma\cos q\alpha - V_{lq}^m(\beta)\sin m\gamma\sin q\alpha) \right] \right\},$$

where the δ's are Kronecker deltas and $H(t)$ the Heaviside step-function ($= 1$ for $t \geqq 0$, $= 0$ for $t < 0$). The coefficients U and V are given by

$$U_{lq}^m(\beta) = (-1)^q\,P_{lq}^{|m|}(\cos\beta) - P_{l\bar{q}}^{|m|}(\cos\beta),$$

$$V_{lq}^m(\beta) = (-1)^q\,P_{lq}^{|m|}(\cos\beta) + P_{l\bar{q}}^{|m|}(\cos\beta),$$

where

$$P_{lq}^{|m|}(\cos\beta) = \left(\frac{\sin\beta}{\cos\beta + 1}\right)^q \frac{(\sin\beta)^{|m|}}{(2l)!!}$$

$$\times \frac{d^{l+|m|}}{d(\cos\beta)^{l+|m|}}\left[(\cos\beta+1)^{l+q}\,(\cos\beta-1)^{l-q}\right]$$

is a generalization of the associated Legendre polynomials [59], to which it reduces for $q = 0$.

12 References

1. Buckingham AD (1978) In: Pullman B (ed) Intermolecular interactions: from diatomics to biopolymers. Wiley, New York, p 1
2. Longuet-Higgins HC (1961) Adv. Spectry 2: 429
3. Varandas AJC (1988) Adv. Chem. Phys. 74: 255
4. Buckingham AD (1980) In: Bratos S, Pick RM (eds) Vibrational spectroscopy of molecular liquids and solids. Plenum, New York, p 1
5. Le Roy RJ, Carley JS (1980) Adv. Chem. Phys. 42: 353
6. Legon AC, Millen DJ, Rogers SC (1980) Proc. Roy. Soc. Lond. A 370: 213
7. Legon AC (1983) Ann. Rev. Phys. Chem. 34: 275
8. Legon AC, Millen DJ (1986) Chem. Rev. 86: 635
9. Dyke TR (1984) Topics Curr. Chem. 120: 85
10. Bernardi F, Robb MA (1987) Adv. Chem. Phys. 67:155
11. Pople JA (1982) Faraday Discuss. Chem. Soc. 73: 7
12. Van Lenthe JH, Van Duijneveldt-Van de Rijdt JGCM, Van Duijneveldt FB (1987) Adv. Chem. Phys. 69: 521
13. Jeziorski B, Kołos W (1982) In: Ratajczak H, Orville-Thomas WJ (eds) Molecular interactions, vol 3, Wiley, New York, p 1
14. Buckingham AD, Fowler PW, Hutson JM (1988) Chem. Rev. 88: 963
15. Magnasco V, Figari G (1986) Mol. Phys. 59: 689
16a. McWeeny R (1988) Pure & Applied Chem. 60: 223
16b. Amovilli C, McWeeny R (1990) Chem. Phys. 140: 343
17. Amovilli C, McWeeny R (1986) Chem. Phys. Lett. 128: 11
18. McWeeny R (1989) Methods of molecular quantum mechanics, 2nd ed, Academic, London
19. Dacre PD, McWeeny R (1970) Proc. Roy. Soc. Lond. A 317: 435
20. Buckingham Ad (1982) Faraday Discuss. Chem. Soc. 73: 421
21. Novick SE (1987) In: Weber A (ed) Structure and dynamics of weakly bound molecular complexes. Reidel, Dordrecht, p 201
22. Magnasco V, Figari G (1989) Mol. Phys. 67: 1261
23. Buckingham AD (1967) Adv. Chem. Phys. 12: 107
24. Stone AJ, Tough RJA (1984) Chem. Phys. Lett. 110: 123
25. Magnasco V, Musso GF, McWeeny R (1967) J. Chem. Phys. 47: 4617
26. Hayes IC, Stone AJ (1984) Mol. Phys. 53: 69, 83
27. Figari G, Magnasco V (1985) Mol. Phys. 55: 319
28. Magnasco V, Musso GF, Costa C, Figari G (1985) Mol. Phys. 56: 1249
29. Chipman DM, Bowman JD, Hirschfelder JO (1973) J. Chem. Phys. 59: 2830
30. Chałasinski G, Gutowski M (1988) Chem. Rev. 88: 943
31. Epstein ST (1974) The variation method in quantum chemistry. Academic, New York, p 187
32. Battezzati M, Magnasco V (1977) Chem. Phys. Lett. 48: 190
33. Battezzati M (1989) Rivista Nuovo Cimento, N. 5 12: 1
34. Conway A, Murrell JN (1972) Mol. Phys. 23: 1143; (1974) 27: 873
35. Murrell JN, Varandas AJC (1975) Mol. Phys. 30: 223
36. Magnasco V, Figari G (1987) Mol. Phys. 62: 1419
37. Jaszunski M, McWeeny R (1982) Mol. Phys. 46: 483
38. McWeeny R (1983) J. Mol. Struct. (Theochem) 93: 1
39a. Longuet-Higgins HC (1965) Faraday Discuss. Chem. Soc. 40: 7
39b. McWeeny R (1984) Croatica Chem. Acta 57: 865
39c. Magnasco V (1987) In: Atti Seminario Nazionale Chimica Fisica Sistemi Biologici, 24 Sept–3 Oct 1987. Vico Equense (Naples), p 149
40. McWeeny R (1985) J. Mol. Struct. (Theochem) 123: 231
41. Jaszunski M, McWeeny R (1985) Mol. Phys. 55: 1275
42. Rijks W, Wormer PES (1988) J. Chem. Phys. 88: 5704
43. Longuet-Higgins HC (1956) Proc. Roy. Soc. Lond. A 235: 537
44a. McWeeny R (1955) Proc. Roy. Soc. Lond. A 232: 114

44b. McWeeny R (1960) Rev. Mod. Phys. 32: 335
44c. McWeeny R, Mizuno Y (1961) Proc. Roy. Soc. Lond. A 259: 554
45a. Scrocco E, Tomasi J (1973) Topics Curr. Chem. 42: 95
45b. Tomasi J (1982) In: Ratajczak H, Orville-Thomas WJ (eds) Molecular interactions, vol. 3. Wiley, New York, p 119
46. Hall GG (1989) J. Chem. Soc. Faraday Trans. 2 85: 251
47. Magnasco V, Musso GF (1974) J. Chem. Phys. 60: 3744
48. Jeziorski B, Bulski M, Piela L. (1976) Int. J. Quantum Chem. 10: 281
49. Magnasco V (1982) Faraday Discuss. Chem. Soc. 73: 109
50. Magnasco V, Figari G, Costa C (1990) Chem. Phys. Lett. 168: 84
51. Margenau H, Stamper J (1967) Adv. Quantum Chem. 3: 129
52. Musso GF, Magnasco V, Giardina MP (1974) J. Chem. Phys. 60: 3749
53. Gutowski M, Chałasinski G, Van Duijneveldt-Van de Rijdt JGCM (1984) Int. J. Quantum Chem. 26: 971
54. Stone AJ (1985) Mol. Phys. 56: 1065
55. London F (1930) Z. Phys. 63: 245
56. Casimir HBG, Polder P (1948) Phys. Rev. 73: 360
57. Koide A (1976) J. Phys. B: Atom. Mol. Phys. 9: 3173
58. Knowles PJ, Meath WJ (1987) Mol. Phys. 60: 1143
59. Abramowitz M, Stegun IA (1965) Handbook of mathematical functions. Dover, New York
60. Rose ME (1957) Elementary Theory of Angular Momentum. Wiley, New York
61. Magnasco V, Costa C, Figari G (1990) J. Mol. Struct. (Theochem) 206: 235
62. Figari G, Magnasco V (1989) Chem Phys. Lett. 161: 539
63. Magnasco V, Figari G, Costa C (1988) J. Mol. Struct. (Theochem) 164: 49
64. Kreek H, Meath WJ (1969) J. Chem. Phys. 50: 2289
65. Stone AJ (1975) Mol. Phys. 29: 1461; (1976) J. Phys. A: Math. Gen. 9: 485
66. Gray CG, Lo BWN (1976) Chem. Phys. 14: 73
67. Mulder F, Van der Avoird A, Wormer PES (1979) Mol. Phys. 37: 159
68. Isnard P, Robert D, Galatry L (1976) Mol. Phys. 31: 1789
69. Price SL (1986) Acta Cryst. B 42: 388
70. Dalgarno A, Davison W (1966) Adv. Atom. Mol. Phys. 2: 1
71. Koga T, Matsumoto S (1985) J. Chem. Phys. 82: 5127
72. Koga T, Ujiie M (1986) J. Chem. Phys. 84: 335
73. Figari G, Musso GF, Magnasco V (1985) Mol. Phys. 54: 689
74. Kutzelnigg W (1969) Chem. Phys. Lett. 4: 435
75. Deal WJ (1972) Int. J. Quantum. Chem. 6: 593
76. Visser F, Wormer PES (1984) Mol. Phys. 52: 923
77. Thakkar AJ (1981) J. Chem. Phys. 75: 4496
78. Koide A, Meath WJ, Allnatt AR (1982) J. Phys. Chem. 86: 1222
79. Rijks W, Wormer PES (1989) J. Chem. Phys. 90: 6507
80. Magnasco V, Costa C, Figari G (1989) Chem. Phys. Lett. 156: 585
81. Huxley P, Knowles DB, Murrell JN, Watts JD (1984) J. Chem. Soc. Faraday Trans. 2 80: 1349
82. Magnasco V, Costa C, Figari G (1989) Chem. Phys. Lett. 160: 469
83. Buckingham AD, Fowler PW (1985) Can. J. Chem. 63: 2018
84. Magnasco V, Costa C, Figari G (1990) J. Mol. Struct. (Theochem) 204: 229
85. Janda KC, Steed JM, Novick SE, Klemperer W (1977) J. Chem. Phys. 67: 5162

Ab Initio Studies of Hydrogen Bonding

Steve Scheiner

Department of Chemistry & Biochemistry, Southern Illinois University, Carbondale, IL 62901-4409, USA

Ab initio calculations carried out over the last few years of H-bonded complexes are reviewed. The systems of interest are composed of simple hydride molecules, e.g. H_2O, NH_3, HCl. The methods are shown to be adequate for accurate reproduction of the equilibrium geometries of binary complexes. Although high-level calculations are required to obtain the details of the structure such as the correct intermolecular distance, a simple analysis based on electrostatic considerations can provide excellent predictions of the intermolecular orientations. A breakdown of the interaction energies indicates that correlation plays a major role in the stabilization of these complexes, becoming progressively more important as one moves down a column of the periodic table. Recent years have witnessed attempts to reproduce various features of the vibrational spectra of H-bonded complexes. It appears possible at this point to calculate the frequencies to good accuracy; the intensities require considerably higher levels of theory for quantitative reliability. In either case, however, the perturbations of the intramolecular vibrations induced by formation of the H-bond can be studied with only modest basis sets in many cases. A comparison with the structural, energetic, and spectroscopic features of Li bonds highlights those features which are unique to H-bonding as well as those characteristic of molecular interactions in general. The cooperativity associated with formation of a sequence of H-bonds is discussed in the context of vibrational spectra as well as structure and energetics.

1 Introduction

Rapid technological advances in computing over the last decade have opened up new opportunities for the application of molecular orbital procedures to chemical problems. Coupled to the increase in raw speed and power of the hardware, incorporation of new methodology into the programs has brought dramatic improvements in the level of theory which may be applied to a given system.

For example, identification of the equilibrium structure of a molecule is one of the most important steps in any theoretical study. Geometry optimizations formerly required a large number of tedious single point calculations, the results of which were fit to assumed functional forms of the energy with respect to the particular geometrical parameter. The same procedure was then applied to the next parameter, and so on. It was usually necessary to complete the entire cycle of all parameters several times until the geometry was completely optimized. As a consequence, many of the earlier calculations included only "partial" geometry optimizations in which several parameters were arbitrarily assigned constant values based on grounds such as experimental measurements of similar systems or considerations of expected symmetry. Such an approach was necessary to limit the number of parameters to be optimized, thereby bringing the entire project within the framework of available computational resources.

This procedure has been greatly simplified by the use of energy gradient techniques which allow the program to efficiently sample the multidimensional potential energy surface and choose a path which will lead towards the minimum [1–3]. After taking a small step along this path, the surface is again scanned for any necessary modifications of the path direction. Full optimization of *all* parameters is usually achieved in a relatively small number of steps. This procedure is much simpler for the user, frequently requiring a single run. Furthermore, since the method generates second derivatives of the energy with respect to all geometrical parameters, the Hessian may be diagonalized to ensure that a true minimum has been located, as opposed to a higher-order stationary point which would be characterized by one or more negative eigenvalues. Indeed, for this very reason, these gradient procedures are an invaluable aid if one is searching for a transition state structure rather than the minimum. Calculation of the second derivatives allows for a valuable spin-off result; the vibrational frequencies may be obtained in a straightforward manner by normal mode analysis.

A second principal shortcoming that formerly plagued ab initio work was the frequent limitation to remain within the Hartree-Fock approximation. Electron correlation plays a major role in a wide range of chemical processes but this phenomenon is completely ignored unless one can go beyond the SCF approximation in some manner. Most means of incorporating electron correlation [4, 5] were formerly very demanding in terms of computer resources and, moreover, the available programs required a good deal of expertise on the part of the user. Recent years have seen giant strides in a solution to this problem. For example, Møller-Plesset perturbation theory [6, 7] has found wide use in a number of straightforward applications and has proven itself to be quite accurate for systems which can be well described by a single determinant. It is also possible to carry out configuration interaction calculations up through single and double excitations without exceeding computer resources for systems which were previously much too large.

Due to these major advances which have enabled researchers to probe the hydrogen bonding phenomenon in greater detail, our understanding has improved substantially since 1982 when this author completed a review on molecular orbital treatments of H-bonding [8]. The present report will hence concentrate on work that has been accomplished in the last six years or so. Prior to 1982, many of the forces contributing to H-bonds had already been established. However, the limitations discussed above had hindered a quantitative assessment of their relative magnitude. This difficulty has been largely circumvented now and we present below an up-to-date summary as well as a discussion of where current and future research is headed.

The theoretical work on H-bonding has been spurred on by concomitant advances in experimental techniques which have supplied new information with unprecedented accuracy concerning the equilibrium geometries of H-bonded complexes in the gas phase [9–14]. Also emerging is a large data base of vibrational spectra of these systems, new grist for the theoretician's mill. Ab initio techniques have responded well to the challenge; spectra have been calculated for a number of systems for direct comparison with experiment. The field has been aided greatly by the more recent ability of these techniques to supply not only the vibrational frequencies but the intensities as well. Analysis of the calculated intensities has provided an open window into the electronic redistributions that accompany the formation of a H-bond.

The term "hydrogen bond" is frequently used to denote interactions of varying strength, ranging up to the very strong ionic complexes, e.g. $(HO-H-OH)^-$ or $(H_2O-H-OH_2)^+$. A number of very informative reviews on this broad-ranging topic, of both a theoretical and experimental nature, have recently appeared in the literature [9–17] to which the interested reader is enthusiastically referred. We will restrict our attention here to neutral complexes. Rather than attempt to cover the entire vast literature of ab initio calculations of systems involving a H-bond, the scope will be further restricted to complexes involving simple hydrides such as H_2O, NH_3, or HCl. The small size of these systems has generally permitted the use of fairly high levels of theory and therefore a more rigorous treatment. Further, the equilibrium geometries and vibrational spectra of many such complexes have been determined experimentally. As a convention to be used throughout this review, hydrogen halides are referred to in a general sense as HX, hydrides of the oxygen group as H_2Y, while H_3Z represents molecules of the ammonia family.

It should be emphasized at the outset that experimental and theoretical data are generally not directly comparable as they frequently refer to different quantities. For example, the theoretically optimized equilibrium geometry corresponds to the bottom of the potential energy well whereas the analogous experimental structure includes the effects of vibrational averaging (R_e vs R_0). Likewise, most of the calculated binding energies reported below are pure electronic contributions, containing neither zero-point vibrational energies nor thermal corrections.

Section 2 is devoted to a systematic discussion of the equilibrium geometries of various complexes and the forces that hold them together. Various facets of the geometries are related to the properties of the constituent subunits and to the overall binding energies. The calculated vibrational frequencies and intensities associated with hydrogen bonding are analyzed in Sect. 3. The level of theory required to treat different aspects of the spectra is discussed at length and conclusions are drawn concerning the root of certain outstanding experimental observations. The comparison of H and Li bonding in Sect. 4 provides insights into the unique properties of the former and points out features common to other molecular interactions. Section 5 considers the situation when more than one H-bond is present in a given system and demonstrates that the cooperativity extends to the spectra as well as to aspects of the geometry and energetics. The Concluding Remarks include a set of recommendations as to fruitful means of treating the H-bond at various levels of sophistication.

2 Structure and Energetics

2.1 $H_3Z\cdots HX$

Let us begin our discussion with H-bonded complexes formed between molecules of the type H_3Z (Z = N, P, As) and a hydrogen halide HX (X = F, Cl, Br, I). Experimental and theoretical work [18–33] concur that the equilibrium geometries of these $H_3Z\cdots HX$ complexes belong to the C_{3v} point group; the HX molecule is collinear with the single lone pair of the Z atom. This aspect of the geometry greatly reduces the complexity of the potential energy surface which must be searched to locate the absolute minimum and thereby simplifies our discussion.

Considering first the case of the first-row atoms, Pople's calculations of $H_3N\cdots HF$ [18], had yielded a total binding energy at the SCF level of 51 kJ/mol using a basis set of the 6-31G* type. Later calculations by Hinchliffe [19, 20] and by Kurnig et al. [21] demonstrated that other basis sets of the general double-ζ plus polarization function type (DZP) lead to a very similar result. Pople's calculations [18] further suggested that electron correlation makes a minor contribution to the total binding energy, adding only 5 kJ/mol at the MP4SDQ level. However, when Latajka and Scheiner [22] made use of a somewhat expanded basis set, they found a substantially larger contribution of 13 kJ/mol arising from correlation. This result suggested that the full measure of correlation's importance may have been underestimated in prior calculations due to the use of inflexible basis sets.

A common approximation in the past, and one which persists in more current work even in some cases where it is inappropriate, has been to calculate the correlated energies using a geometry optimized at the SCF level. Such a procedure assumes that the SCF and correlated equilibrium geometries are quite similar. Latajka and Scheiner [22] checked this assumption and found that (MP2) correlation leads to a contraction of the $N\cdots F$ distance by some 4 pm while the HF bond is stretched by 3 pm. Nonetheless, the binding energy is little affected by these relatively minor geometry changes in this case.

However, the situation is quite different when the N or F centers are replaced with atoms below them in the periodic table. First considering substitution of one of these atoms by its second-row analogue, the MP2 value of $R(N\cdots Cl)$ in $H_3N\cdots HCl$

is shorter than the SCF H-bond length by 16 pm [22] whereas a similar correlation-induced contraction of 17 pm occurs in $H_3P \cdots HF$ [23]. The effects of correlation are even larger when both atoms are substituted: $R(P \cdots Cl)$ is reduced from 417 to 380 pm, a shrinkage of 37 pm [23].

Another aspect of the geometry which is directly related to H-bonding is the stretch incurred in the HX molecule. Like the $R(Z \cdots X)$ distance, this stretch, denoted $\Delta r(HX)$ is quite sensitive to inclusion of correlation. Also similar is the progressively larger impact of correlation as N and F are replaced by atoms lower in the periodic table. For example, while SCF calculations make up for 80% of the total HX stretch in $H_3N \cdots HF$, this contribution is reduced to only 36% in $H_3P \cdots HCl$.

In addition to a significant effect on the geometry, correlation makes progressively larger contributions to the binding energy as well. In $H_3N \cdots HCl$ and in $H_3P \cdots HF$, correlation contributes 19 and 9 kJ/mol, respectively, accounting for about 40% of the total in either case. This percentage contribution rises to 66% in the case of $H_3P \cdots HCl$ [24]. Since correlation produces larger perturbations on both the optimized geometry and the binding energy, spurious results are likely to ensue if one attempts to calculate correlated energies at an SCF-optimized geometry.

Recent calculations by Alabart and Caballol [25] have extended the systems examined up to the third-row complex $H_3As \cdots HBr$, and have employed a multideterminantal zeroth-order wavefunction in the second-order perturbation theory approach to

Table 1. Computed binding energies D_e of $H_3Z \cdots HX$ (kJ/mol)

SCF

H_3Z	Basis set	HF	HCl	HBr	HI	Ref.
H_3N	DZP	49.4	30.4	24.0		[19]
	DZP		35.9	34.0		[27]
	DZ + 2D		30.1	24.3	16.7	[26]
	DZ + 2P	49.5	38.9			[22]
H_3P	DZP	15.5	13.7	6.5		[19]
	DZP		8.5	6.7		[25]
	DZ + 2P	17.3	9.0			[23]
H_3As	DZP	14.9	8.1	6.2		[19]
	DZP		5.2	4.0		[25]

Correlated

H_3Z	Method	HF	HCl	HBr	HI	Ref.
H_3N	MP2	63.1	46.1			[22]
	ACCD		38.5	31.4	23.8	[26]
H_3P	MP 2	25.2	18.2			[23]
	CIPSI		11.0	10.0		[25]
H_3As	CIPSI		6.7	6.0		[25]

correlation. Jasien and Stevens [26] have gone on to the fourth row halogen, i.e. $H_3Z \cdots HI$, using an approximate coupled clusters technique which includes double excitations. Although differing in some quantitative aspects from the earlier work, their results confirm the progressively larger role of correlation for H-bonded systems containing atoms lower in the periodic table. Hinchliffe [19] has also considered systems up to and including $H_3As \cdots HBr$ but only at the SCF level.

One feature upon which all calculations, SCF or correlated, agree is that the $H_3N \cdots HF$ complex is the most strongly bound of the $H_3Z \cdots HX$ class. This feature is summed up in Table 1 which reports the results of binding energies for the various $H_3Z \cdots HX$ complexes by various groups. So as to avoid prejudicing our analysis by the use of any one specific set of calculations, the results of a number of different theoretical approaches are included in the table. It may be noted that for any base, the H-bond energy diminishes rather smoothly as the acid changes from HF to HI. With regard to the bases, the substitution of NH_3 by PH_3 leads to a dramatic weakening of the bond for any given acid. On the other hand, replacing PH_3 by AsH_3 appears to lead to only a relatively small further decrement.

In summary, substitution of the N and F centers in the $H_3N \cdots HF$ complex by atoms below them in the periodic table leads to a progressive weakening of the H-bond. At the same time, correlation becomes more important, influencing the interaction not only directly but also through correlation-induced changes in the optimized geometry. Quantitative accuracy for these heavier systems undoubtedly requires a consistent treatment of electron correlation throughout, as well as the use of a fairly extensive basis set.

2.2 $H_2Y \cdots HX$

Calculations of comparable accuracy are somewhat less abundant for the complexes that pair a hydrogen halide with a base of the type YH_2 [34–40]. Overall the trends exhibited in Table 2 are much like the previous complexes involving H_3Z as a base. Beginning with $H_2O \cdots HF$, substitution of HF by HCl and then by HBr leads to a steady weakening of the bond. Similarly, replacement of H_2O by H_2S weakens the interaction to a large degree. Data for $H_2Se \cdots HX$ are limited to the SCF level [34] but indicate these complexes are only slightly weaker than the $H_2S \cdots HX$ analogues. Comparison of the data in Tables 1 and 2 suggests that H_3N is a stronger base than is H_2O. That is, $H_3N \cdots HF$ is more strongly bound than is $H_2O \cdots HF$ with similar comparisons noted for HCl and HBr. The comparisons between H_3Z and H_2Y for the other bases are less clear due to remaining uncertainties in their binding energies.

Also similar to the $H_3Z \cdots HX$ complexes is the relative importance of electron correlation to both the energetics and optimized geometries of these $H_2Y \cdots HX$ systems. While the $R(O \cdots F)$ distance in $H_2O \cdots HF$ is reduced by only 5 pm by correlation, this bond contraction increases to about 17 pm in both $H_2S \cdots HF$ and $H_2O \cdots HCl$, and to 34 pm in $H_2S \cdots HCl$ [24]. Along a similar line, correlation contributes 27% of the total binding energy in $H_2O \cdots HF$, over 40% in $H_2S \cdots HF$ and $H_2O \cdots HCl$, and 69% in $H_2S \cdots HCl$.

Table 2. Computed binding energies D_e of $H_2Y \cdots HX$ (kJ/mol)

SCF

H_2Y	Basis set	HF	HCl	HBr	Ref.
H_2O	DZP	37.8	22.7	18.2	[34]
	DZ + 2P	32.5	17.4		[35]
	6-31 + G(2d, 2p)	31.8	17.2		[36]
H_2S	DZP	14.7	10.0	6.6	[34]
	DZ + 2P	16.3	9.0		[37]
H_2Se	DZP	14.2	8.4	4.7	[34]

Correlated

H_2Y	Method	HF	HCl	Ref.
H_2O	MP 2	40.3	27.6	[35]
	MP 4	38.1	23.4	[36]
H_2S	MP 2	26.4	20.7	[37]

A very recent thorough set of calculations by Del Bene [36] has applied a large doubly polarized basis set, in tandem with full fourth-order MP theory, to binary complexes including HOH. Her results suggest that going beyond the second order has very little influence upon the calculated energetics of these systems, regardless of the basis set employed.

Whereas the geometry of the $H_3Z \cdots HX$ complexes (C_{3v}) is fairly clearcut and makes sense from the standpoint of the collinearity of the H—X bond and the single lone pair of ZH_3, the equilibrium structures of the $H_2Y \cdots HX$ systems are not so obvious on first principles alone. Were the H_2Y molecule to orient itself so as to have one of its two lone pairs point toward the HX molecule, one might expect the "bent" structure pictured in Fig. 1a in which α is on the order of 125°. On the other hand, placing the two hydrogens of H_2Y as far as possible from HX yields the "linear" C_{2v} geometry of Fig. 1b wherein $\alpha = 180°$.

Calculations reveal the equilibrium angle α in complexes containing H_2O, viz. $H_2O \cdots HF$ and $H_2O \cdots HCl$, is approximately 140°, while the complexes containing H_2S are more "perpendicular" in nature with angles in the range 90–110°. This distinction between the two types of complexes is borne out by experimental measurements in the gas phase [41–50]. Two simple explanations have been offered for this finding. The first makes use of hybridization arguments wherein H_2O is described by a set of tetrahedrally disposed sp^3 orbitals about the oxygen. The angle α expected for this arrangement is 125°, not far from the observed angle of 140°.

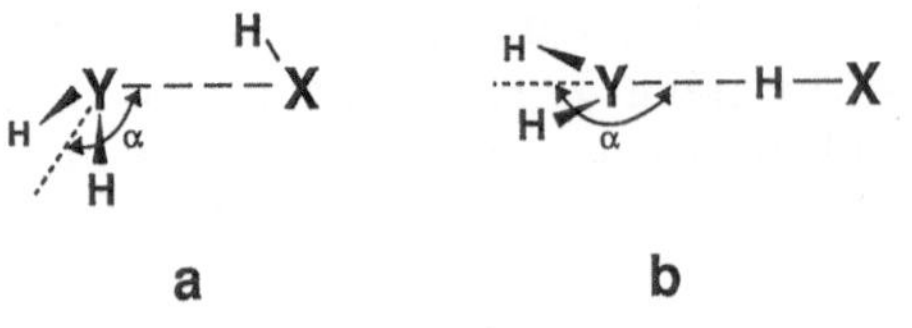

Fig. 1a, b. Bent (a) and linear (b) configurations of $H_2Y \cdots HX$. The α angle extends from the $Y \cdots X$ axis to the HYH bisector

On the contrary, the internal HSH angle in isolated H_2S is 92°, consistent with nearly pure p-orbitals participating in each S—H bond [41]. The third mutually orthogonal p-orbital, containing a lone pair of electrons, could then orient itself toward the HY, leading to the observed angle of α near 90°.

An alternative viewpoint arrives at a similar conclusion but by focusing on the electrostatic interaction between the two subunits. The HX molecule contains a strong dipole, with the hydrogen at its positive end. The dipole moment of H_2Y coincides with the HYH bisector. The optimal alignment of the dipole moments of the two molecules would thus favor the "linear" arrangement in Fig. 1b. If we consider next the quadrupole moment of the H_2Y molecule, it is found to have negative out-of-plane components. The interaction between the HX dipole and the H_2Y quadrupole would thus favor the perpendicular arrangement wherein $\alpha = 90°$. As pointed out by Singh and Kollman [39], the final result can be thought of as a compromise between these two opposing trends. HSH has a rather high ratio of quadrupole to dipole moment; hence, the former is more important and an angle α of nearly 90° is the result. The two moments are more nearly balanced in HOH so the equilibrium angle is a compromise between 90° and 180°. Indeed, a number of papers have appeared of late in the literature that argue that electrostatic considerations of the type mentioned here are responsible for the relative orientations of a wide array of H-bonded and van der Waals complexes [51–58]. The essential equivalence of the two types of descriptions, viz. electrostatic vs HOMO-LUMO, has been discussed recently by Legon and Millen [54].

Electron correlation does not appear to be of crucial importance to evaluation of the optimal angular characteristic of these complexes. The α angle changes by about 10° or less when second-order MP theory is employed. On the other hand, if one wishes to calculate the height of the energy barrier for inversion separating the two equivalent C_s geometries, correlation does become important. For example, the SCF value of this barrier is only 45 cm^{-1} for $H_2O \cdot\cdot$ HF, as compared to an MP2 value of 144 cm^{-1} [35].

2.3 $H_3Z \cdot\cdot HYH$

In complexes that pair a H_3Z hydride with one of H_2Y type, it is conceivable that either molecule could act as proton donor. However, in the only complex of this type to have its structure determined in the gas phase, HOH was observed to act as proton donor and NH_3 as acceptor [59]. This finding is logical from the standpoint that HOH is a stronger proton donor than is NH_3.

Little ab initio work has been performed on this complex since the mid 1970s. A set of calculations by Latajka and Scheiner [60] in 1987 carefully examined the sensitivity of the energetics of the $H_3N \cdot\cdot HOH$ complex to basis set. Using sets of the polarized double-ζ type, the authors observed that the SCF interaction energy varies between 19 and 26 kJ/mol, depending upon particulars of the basis set. For example, adding a second set of d-functions to the standard 6-31G** set reduced the H-bond energy by some 5 kJ/mol. Del Bene has recently obtained results in the same range [36]. Similar sensitivity to details of the basis set was observed at the correlated MP2 level. On the other hand, this degree of sensitivity is diminished dramatically if the energetics are corrected for basis set superposition error (see below).

The best estimate of the binding energy, after correction of this error [60], is approximately 23 kJ/mol, 5 of which are associated with electron correlation. Del Bene's recent calculation [36], which employed a $6-31+G(2d, 2p)$ basis set and full fourth-order Møller-Plesset treatment of correlation, found a value of 28 kJ/mol (although superposition error was not removed). Within the context of hydrides of first-row atoms, this complex is the most weakly bound, with the order as follows:

$$H_3N \cdots HF > H_2O \cdots HF > H_3N \cdots HOH$$

At the present time, there is no experimental information available concerning the structure or energetics of other complexes of the type $H_3Z \cdots HYH$. Ab initio calculations are limited to fairly small basis sets with the exception of Del Bene's high-level calculation of $H_3P \cdots HOH$ [36]. The binding energy was computed to be 6.7 kJ/mol at the SCF level, increasing to 9.2 kJ/mol at the full MP4 level.

2.3.1 Basis Set Superposition Error

This error is due to the fact that whenever a finite basis set is used, the electronic description of either subunit in a complex can be improved if it makes use of the orbitals associated with its partner. The resulting lowering of the energy is due not to a genuine physical interaction but rather to a mathematical artifact of the use of a basis set of limited flexibility. Boys and Bernardi [61] proposed in 1970 a functional counterpoise correction technique for extracting this error wherein the energy of each subunit is calculated within the context of the basis set of the entire complex, i.e. including the orbitals of the subunit itself as well as the "ghost" orbitals of its partner.

A problem soon manifested itself in that due to computational limitations of the early 1970s, it was difficult for ab initio calculations of H-bonded systems to make use of basis sets of moderate or larger size. Quite expectedly, small basis sets, e.g. minimal, are subject to large basis set superposition error (BSSE). When this error was subtracted, the remaining interaction was found to be rather weak, even repulsive in certain cases. The finding that the corrected potential was in poorer agreement with experiment than the raw uncorrected result led a number of investigators to suggest that the counterpoise technique provided an overestimate of the true BSSE and even to ideas about "damping" this correction term [62, 63]. However, it is now clearly recognized that small basis sets are notoriously ill-suited to study molecular interactions of H-bonding type. Since by their very nature they must underestimate a number of components of the interaction and ignore others, a severely underattractive potential is in fact the "correct" result of a calculation of this type. That is, it is the quality of the basis set which is at fault and not the counterpoise correction at all [64–66].

An impressive body of literature has built up in the ensuing years to demonstrate that from a formalistic as well as numerical point of view, the counterpoise technique represents a realistic means of estimating and removing the artifacts introduced into the potential by the basis set superposition error [67–73]. While this fact has come to be widely accepted with regard to the SCF level, there remains some residual controversy concerning correlated levels of calculation. This problem is particularly irksome since while it is possible to reduce the SCF BSSE to negligible proportions by use of large basis sets, the correlated counterpoise term seems to be quite tenacious, remaining disturbingly large, even with extremely large and flexible sets [74, 75]. Nevertheless, the weight of recent evidence points to the necessity of removing the correlated BSSE by the full counterpoise correction as well [76–81].

Further analysis into the superposition of the basis sets of two subunits leads to a second problem. The superposition error described above results from the lowering of the energy of each subunit by the variation principle as its basis set is artificially expanded by the approaching orbital set of its partner molecule. This type of error is denoted as "primary" BSSE to distinguish it from the "secondary" error, also known as basis set extension effects [82–88] which are less direct. The approaching orbitals of the partner cause a change in the molecular properties of the molecule, e.g. multipole moments and polarizability. These perturbations of the molecular properties lead in turn to modification of the various contributions to the total interaction energy. Whereas the primary superposition error always produces an artificial deepening of the interaction potential, the effects of secondary BSSE are considerably more complex. A recent article [86] has provided a three-dimensional visualization of both the primary and secondary BSSE that one can expect around a typical H-bonding molecule such as HF.

Latajka and Scheiner [87] have recently investigated this problem in the context of a simpler system composed of a single base such as NH_3 or OH_2 and the Li^+ ion. In a simple case such as this, it is possible to study the secondary BSSE largely in terms of the perturbations of only one of the partners (the base) since the compact electronic structure of Li^+ makes it subject to only very minor superposition error. The largest contribution to the secondary BSSE then arises from the perturbations in the permanent dipole moment of the base and the consequent change in the ion-dipole electrostatic term. The results emphasized that secondary BSSE can be quite large, comparable in magnitude to the primary error at both the SCF and MP2 levels. The data also argued against prior suppositions that secondary BSSE offers an improvement in the interaction energy and against presumed cancellation with the primary BSSE. The authors demonstrated that one should attempt to remove both 1° and 2° BSSE whenever possible, since their data did not show any "overcorrection" by such a procedure. They further suggested the beneficial effects of including a diffuse sp shell in the basis set since this leads to lowering of both types of BSSE and to a certain level of cancellation between them. Of course, it is always preferable to use basis sets which minimize BSSE of both primary and higher-order type; Szczesniak and Scheiner have discussed means of doing so [88].

2.4 HX··HX′

2.4.1 $(HF)_2$

There is a great deal of experimental and theoretical information concerning $(HF)_2$ [89–101]. The equilibrium geometry of the complex is sketched in Fig. 2a. The H-bond is not quite linear since the hydrogen atom of the proton donor molecule lies some 10° from the F ·· F axis. The proton acceptor molecule is nearly perpendicu-

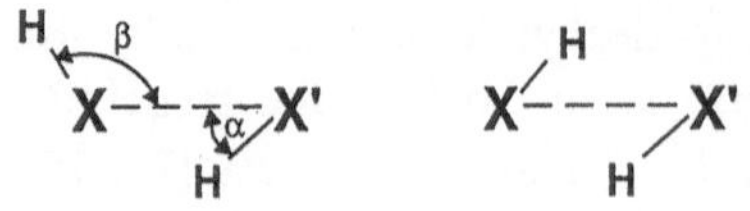

Fig. 2a, b. Equilibrium geometry of HX ·· HX′ (**a**) and transition state for conversion of one structure to its equivalent (**b**)

lar to this axis, making a β angle of about 117°. This geometry is reproduced rather well by ab initio theory [93–100] as may be seen in the upper portion of Table 3. Especially notable is the recent calculation by Kofranek et al. [96] who showed that a correlated wave function using a fairly large basis set is capable of quantitatively reproducing various facets of the molecular geometry.

Table 3. Calculated equilibrium geometry of $(HX)_2$

$(HF)_2$ Basis set	Level	R(FF), pm	α, Degs	β, Degs	Ref.
[531/31]	SCF	279	7.4	122.4	[93]
[642/41]	SCF	283	5.7	123.5	[93]
DZP	SCF	279	7.9	117	[94]
DZP	SCF	280	8.2	115.4	[95]
	CI	272	9.1	105.1	[95]
[862/41]	CPF	279	6.8	114.5	[96]
expt		279	10 ± 6	117 ± 6	[91]

$(HCl)_2$ Basis set	Level	R(ClCl), pm	α, Degs	β, Degs	Ref.
DZP	SCF	396	2.6	97.0	[101]
[442/31]	CEPA-SD	381	<2	97	[102]
$+ VP^S(2d)^S$	SCF	421	11	90	[103]
	MP 2	384			[103]

Del Bene [98] and Latajka and Scheiner [100] have examined the sensitivity of the calculated H-bond energy to the level of theory employed. Due to near cancellation between third and fourth-order terms in the Møller-Plesset expansion, calculations limited to second order are nearly identical to full fourth-order energies. Del Bene's computed values of ΔE exhibit a disturbing level of fluctuation with relatively minor changes in the basis set at both the SCF and MP2 levels. However, Latajka and Scheiner demonstrate that a good deal of this fluctuation is due to the basis set superposition error; when BSSE is removed by the counterpoise correction, the energies are much more stable and behave in a manner compatible with properties computed for each monomer with that particular basis set.

At this point, the most reliable estimate of the binding energy of $(HF)_2$ is 18.1 kJ/mol [96] which compares rather favorably with an experimental estimate [89] of 19.1 kJ/mol.

2.4.2 (HCl)$_2$

Ab initio studies of this homodimer [101–106] are less abundant, many of them being restricted to rather low levels of theory. As may be noted in the lower section of Table 3, a DZP SCF level calculation by Allavena et al. [101] led to an equilibrium geometry with a slightly more linear H-bond than in (HF)$_2$: $\alpha = 2.6°$. The proton-acceptor molecule is closer to perpendicular; $\beta = 97°$. Their raw interaction energy of 33.5 kJ/mol is reduced dramatically to 15.1 kJ/mol when the counterpoise correction is introduced. Votava et al. [102] have improved on this calculation by introducing electron correlation via a CEPA approach and a basis set with 2 sets of d-functions on Cl. They find a nearly linear H-bond with α less than 2° and the same value of β, viz. 97°. Their binding energy, however, is much smaller, only 7.9 kJ/mol. This value is in nice agreement with experimental estimates of 9.0 kJ/mol [107] and 9.5 kJ/mol [89]. More recent calculations by Latajka and Scheiner [103] agree with the perpendicular nature of the complex and place the binding energy at a slightly smaller value of 6.7 kJ/mol. The reader should note the rather strong influence of correlation in contracting the distance between the two HCl molecules.

2.4.3 Mixed Dimers

Experimental work by Janda et al. [108] has indicated that HCl acts as proton donor when paired with HF. With regard to ab initio studies, Hobza et al. [106, 109] found from calculations with fairly small basis sets that HF $\cdots$ HCl is likely to be somewhat more stable than HCl $\cdots$ HF, i.e. HCl acts as proton donor. More recent work using doubly polarized sets [103] indicates the difference in energy between these two geometries is quite small, perhaps 0.25 kJ/mol. The angular features of the complexes are reminiscent of the (HF)$_2$ or (HCl)$_2$ homodimers above. In either case, the hydrogen of the proton donor remains within 7° or 8° of the F $\cdots$ Cl axis. The orientation of the proton acceptor molecule seems to be fairly independent of the donor. That is, HCl is nearly perpendicular to the F $\cdots$ Cl axis ($\beta = 93°$) while this angle is 120° for HF.

The best ab initio estimate of the binding energy of the HF $\cdots$ HCl complex, incorporating both electron correlation (MP2) and correction of BSSE, is 10.3 kJ/mol [103]. The order of stability of these complexes is therefore calculated to be

$$HF \cdots HF > HF \cdots HCl > HCl \cdots HCl .$$

That is, replacement of each HF molecule by HCl lowers the binding energy somewhat. Comparison with the heterodimers discussed above suggests that HX acts as a considerably weaker base than does H$_2$Y. For example, the binding energy of HF $\cdots$ HF is roughly half that of H$_2$O $\cdots$ HF.

The structure which combines HF and HI has been determined by Bumgarner and Kukolich [110] by microwave spectroscopy. HF acts as proton donor with this hydrogen approximately along the I $\cdots$ F axis. Curiously, the angle β is only 70°, i.e. the hydrogen of HI is turned in towards the HF molecule. This interesting observation will certainly be the subject of theoretical scrutiny in the near future.

2.4.4 Inversion

A reversal in the roles of the two HX molecules as proton donor and acceptor may be accomplished primarily by motion of the two H atoms. Such an inversion will pass through a "cyclic" structure similar to that depicted in Fig. 2b as a transition state. Ohashi and Pine [111] have clear evidence for extremely rapid tunneling between the two equivalent forms of HCl $\cdot\cdot$ HCl, implying the existence of a low barrier separating them. Similar motion was noted in HF $\cdot\cdot$ HF [90, 92].

While all reliable calculations agree that the inversion barrier for $(HF)_2$ is rather small, there is a good deal of sensitivity to the precise theoretical approach. For example, using geometries optimized for the C_s minimum and the C_{2h} transition state at the SCF level and with various standard basis sets, Del Bene et al. [94, 98] found that the SCF barrier varies between 1.1 and 5.0 kJ/mol. However, when correlation is applied directly to these same geometries, the results become quite erratic; in some cases the barrier is negative, i.e. the C_{2h} geometry is more stable than C_s. Inclusion of correlation effects directly into the optimization leads to inversion barriers of 4.5 [97] and 4.3 [96] kJ/mol.

The analogous barrier for $(HCl)_2$ has been examined in less rigorous detail but appears to be a bit lower. Frisch et al. [94] computed values ranging between 0.4 and 1.3 kJ/mol at the MP4 level with a polarized double-ζ basis set, consistent with more recent calculations of Latajka and Scheiner [103] employing a doubly polarized set at the MP2 level. However, it must be noted that in neither case were the geometries optimized at the correlated level. Nonetheless, it is safe to conclude that the barrier is quite low in $(HCl)_2$, in nice agreement with experiment.

It is not entirely clear that both HF $\cdot\cdot$ HCl and HCl $\cdot\cdot$ HF represent minima on the potential energy surface of this mixed complex. An early calculation by Hobza et al. [109] indicated that the barrier separating these two putative minima will disappear when dispersion and zero-point vibrations are considered. Later calculations by Latajka and Scheiner [103] using a larger basis set and explicit inclusion of correlation arrived at a barrier of only 0.6 kJ/mol after zero-point correction. While the question of whether both of these minima indeed exist remains open to higher levels of theory, it seems clear that any barrier which might exist will be exceedingly small.

The structure of the transition state has been characterized as "cyclic" since the F $\cdot\cdot$ FH angles are approximately 54° [95–97]; the analogous angle for the C_{2h} geometry of $(HCl)_2$ has been calculated recently to lie between 50 and 55° [103]. The mixed (HF) (HCl) complex exhibits similar angles [103]. Along the inversion pathway, the distance between heavy atoms is subject to change. At the SCF level, Gaw et al. [95] calculated a contraction in $(HF)_2$ of 7 pm. In contrast, Michael et al. found a smaller reduction of only 4 pm at the correlated level [97] while Kofranek et al. find almost no contraction at all [96]. Turning to $(HCl)_2$, Hobza et al. [105] calculated a contraction of 21 pm with a 4-31G basis set at the SCF level. More recent work [103] suggests the contraction lies between 1 pm and 10 pm, depending on the level of theory. It therefore appears that while the angles of these transition states are pretty much resolved at 50–55°, the amount of distortion in the intermolecular distance remains open.

2.5 $H_2Y \cdots HY'H$

2.5.1 $(H_2O)_2$

In part due to the great importance of water in everyday chemistry and biology, the water dimer is perhaps the most intensively studied complex by ab initio methods. Early on, a number of different geometry types were considered for this system which included:

i) the "standard" linear H-bond wherein the hydrogen of the proton donor is located approximately along the O $\cdots$ O axis;

ii) a cyclic complex contains a pair of distorted H-bonds, each molecule acting simultaneously as both proton donor and acceptor;

iii) a bifurcated structure also contains two distorted H-bonds but both bridging protons are covalently bound to one oxygen atom.

Calculations using semiempirical methods or ab initio results with small basis sets produced conflicting results about the relative stabilities of these geometries [112–114]. As more reliable methods were employed, it soon became apparent that only the linear structure represents a true minimum on the potential energy surface; the other two are characterized as stationary points of order 1 or higher and are hence not minima [94]. The microwave spectrum interpreted by Dyke et al. also pointed unequivocally to the linear structure as the stable geometry of the dimer in the gas phase [115–119]. This geometry is depicted in Fig. 3.

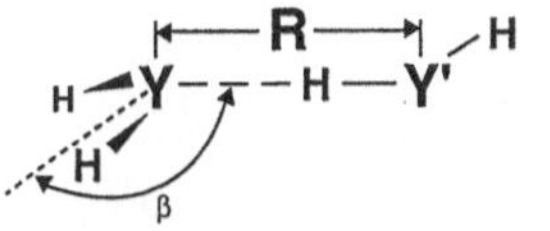

Fig. 3. Structure of $H_2Y \cdots HY'H$. R represents distance between Y and Y', while β measures angle to HYH bisector

Recent ab initio studies have refocused their attention upon resolving questions such as the binding energy of the complex, refinement of the details of the molecular geometry such as the equilibrium orientation of the proton acceptor molecule, and reasons why the O $\cdots$ O distance in the dimer is substantially longer than in condensed phases [120–123].

In 1979 Kołos demonstrated that even minimal basis sets could be of some use in studying the water dimer, provided that the counterpoise corrections are included [124, 125]. However, the same does not apply to minimal sets of the STO-nG type which produce erratic results even after correction. This conclusion has been verified in later work [66, 126]. Szczesniak and Scheiner used the water dimer as a testing ground to demonstrate that subtraction of the full counterpoise correction at both the SCF and MP2 levels leads to a much lessened sensitivity of the interaction energy to small changes in basis set [80]. Newton and Kestner [123] found that after proper correction is made for correlation and the superposition error, the intermolecular distance in the water dimer agrees very well with the microwave structure. Their results clearly pointed to many-body effects as causing the shorter O $\cdots$ O distances in the liquid. This finding has been confirmed by other calculations [120]. A number of schemes have been developed in attempts to incorporate the counterpoise corrections into the various contributions to the total interaction energy [70, 127, 128] and to eliminate the basis set superposition entirely [129].

Baum and Finney [130] estimated that correlation and the basis set superposition error each amount to roughly 10% of the total interaction energy and suggested that a nearly correct result could be obtained at the SCF level by a fortuitous cancellation of these two effects. However, they stressed that both terms must be properly treated on an individual basis when studying extended regions of the potential energy surface. On the other hand, they did note that the superposition error does not appear to change much with molecular orientation, leading to expectations that counterpoise correction may not be a crucial factor in determining the angular characteristics of the energy minimum. Their results verified that the only minimum on the surface is the trans linear structure illustrated in Fig. 3, a conclusion supported by other calculations [94, 131].

In the past few years, a number of papers have appeared which systematically investigate the dependence of the interaction energy of the water dimer upon the level of theory [80, 98, 100, 126, 132–135]. The most thorough study is that of Frisch et al. [133] in that they also studied the effects of each improvement in theory upon the equilibrium geometry. Some of their results are excerpted in Table 4.

Table 4. Calculated equilibrium geometries and binding energies of the water dimer (from [133])

Basis set	$R(O \cdots O)$, pm		β, Degs		$-\Delta E$, kJ/mol	
	SCF	MP 2	SCF	MP 2	SCF	MP 2
STO-3G	274.0	—	124.0	—	24.8	—
3-21G	279.7	280.2	124.6	107.9	45.9	52.9
6-31G*	297.1	291.3	117.5	102.7	23.5	31.0
6-31 + G*	296.4	290.1	130.3	128.9	22.5	29.7
6-311 + + G**	299.9	291.0	143.1	135.8	20.2	25.5
6-311 + + G(2d, 2p)	303.5	291.1	130.8	123.2	17.0	22.8
6-311 + + G(3df, 2pd)	302.6		133.2		16.7	
expt	296 ± 1	[123]	123 ± 10			

Focusing first upon the interoxygen distance, the first column demonstrates the overall lengthening effect caused by enlargement of the basis set. A distance of 302.6 pm is listed in the last row with their largest set; however, this value cannot be taken as the Hartree-Fock limit since it is still subject to some 1.5 kJ/mol SCF superposition error. One should take note of the strong underestimates of the distance with the small (STO-3G and 3-21G) basis sets. Second-order correlation effects produce a significant contraction of the H-bond. (The inappropriateness of including correlation with a small unpolarized basis set like 3-21G is underscored by the opposite result of a correlation-induced *lengthening*.) The correlated distance calculated with the doubly polarized 6-311 + +G(2d, 2p) set of 291.1 pm is probably a slight underestimate of a "best" theoretical value since a total superposition error of 4.1 kJ/mol contaminates this result. Higher orders of perturbation theory are likely to uphold this statement since third and fourth-order terms tend to cancel one another. The ab initio approach thus yields a H-bond length somewhat smaller than the experimental value of 296 pm [123]. It should be mentioned in the latter regard, however, that recent work by Odutola et al. [119] suggests that the experimental structure may warrant reexamination, with more attention paid to tunneling effects.

Enlargement of the basis set leads to fairly inconsistent changes in the intermolecular orientation angle β. Correlation seems to consistently lower the angle, perhaps due to the correlation-induced lowering of the dipole moment of water. Values in the MP2 column and with high-quality basis sets fall within the experimental range of this angle; the most accurate value listed is 123.2°.

Disregarding the STO-3G value as meaningless, the binding energies exhibit a decreasing trend as the basis set is enlarged. In all cases, correlation increases the calculated binding energy of the water dimer, due in large measure to the introduction of the attractive influence of dispersion. Since the MP2 result with the largest basis set is subject to over 4 kJ/mol BSSE and since additional enlargements seem certain to further depress this quantity, the value of 22.8 kJ/mol may be considered an upper bound to the true binding energy. Calculations at higher levels by Frisch et al. [94] suggest that higher orders of Møller-Plesset perturbation theory will not appreciably affect the binding energy. After removing their basis set superposition error, Harrison and Bartlett [134] arrive at a highly correlated complexation energy of 18.3 kJ/mol. Taking all this into consideration, one might guess that ab initio methods will eventually produce a binding energy of about 17 or 18 kJ/mol for the water dimer, quite close to the best calculated interaction energy in $(HF)_2$.

2.5.2 $(H_2S)_2$

Gas-phase spectroscopic measurements [115] had been unable to clearly determine the structure of the H_2S dimer, but had offered the suggestion of a linear or bifurcated geometry. Frisch et al. [94] compared these two theoretically and found indications that the bifurcated structure represents a transition state on the surface, about 1 kJ/mol higher in energy than the linear structure. The dimer is only very weakly bound, by less than 6 kJ/mol and by an even smaller amount (3.3 kJ/mol) when zero-point vibrational energies are considered. Accentuating the weakness of the interaction, the S ·· S distance was calculated to be quite long at 451.4 pm.

Fernandez et al. [136] later reexamined the linear and bifurcated structures, comparing them also with a "relaxed" bifurcated and cyclic type of geometry. In contrast to Frisch et al., they found that with their largest basis set (6-31G*), the bifurcated structures are more stable than linear by 0.5 kJ/mol. Although this energy difference is diminished at the MP2 level, the bifurcated remains most stable.

Probably the largest source of error in these ab initio studies is that, although rather high levels of theory are used to calculate the energetics, viz. polarized double-ζ basis sets and inclusion of correlation up through fourth-order Møller-Plesset perturbation theory, the geometries have been optimized only at the SCF level. Since the data clearly point to a very strong influence of correlation, failure to include correlation directly in the scanning of the potential energy surface could easily lead to mischaracterizations.

Woodbridge et al. [137] have carried out measurements of H_2S in solid O_2 matrix and found indications that the equilibrium geometry is linear. They have gone beyond the previous calculations and included correlation directly into their geometry optimization. Consistent with expectations, strong effects are observed: the S $\cdots$ S distance contracts by 44 pm. The proton-accepting molecule is nearly perpendicular to the S $\cdots$ S axis, similar to the $H_2S \cdots$ HF complex described above. $(H_2S)_2$ is calculated to be bound by 9.6 kJ/mol, in comparison to an experimental estimate of 7.1 $\pm$ 1.3 kJ/mol.

In summary, the $(H_2S)_2$ complex does appear to be linear in nature but the potential energy surface is extremely shallow and the complex is quite weakly bound. In fact, one might question whether a dimer of this type should be classified as H-bonding at all.

2.5.3 Mixed Dimers

Geometry optimizations of the two complexes in which either H_2O or H_2S acts as proton donor have been carried out using a 6-31G** basis set by Amos [138]. The two structures were found to lie within about 4 kJ/mol of one another, with a slight energetic preference for the case where H_2S acts as proton donor. Consistent with this difference is the shorter R(O $\cdots$ S) distance for $H_2O \cdots$ HSH (362.2 pm) as compared to $H_2S \cdots$ HOH (381.1 pm). The equilibrium angle β for this case was computed to be 131°. When H_2S acts as proton acceptor, on the other hand, a value of 100° was obtained for β. However, because of the recognized need for a higher level of calculation, Amos was unwilling to conclude that the energy difference between the two configurations is meaningful. Del Bene's recent work [36] confirms Amos's conclusions. Indeed, with a large, doubly polarized basis set, Del Bene finds that $H_2O \cdots$ HSH is lower in energy than $H_2S \cdots$ HOH by only 0.4 kJ/mol at the SCF level. This difference is increased to only 0.8 kJ/mol at the full MP4 level and is unchanged when zero-point vibrational energies and temperature corrections are included. Calculations with a smaller 4-31G basis set by Chin and Ford [139] considered only the case where H_2S acts as proton donor.

2.6 $(ZH_3)_2$

2.6.1 $(NH_3)_2$

Early molecular beam electric deflection behavior of $(NH_3)_2$ had led Odutola et al. [140] to postulate that its equilibrium structure contains a single, approximately linear H-bond. Cook and Taylor [141] estimated the binding enthalpy of this dimer to be 19 $\pm$ 2 kJ/mol. In 1985, Fraser et al. [142] published a microwave spectrum indicating that the dipole moment of $(NH_3)_2$ is 0.74 D which argued against the traditional linear dimer where a much larger value would be expected. Followup work [143–146] led the Klemperer group to suggest an equilibrium geometry which had not been given much consideration previously.

Using Fig. 4 to define the geometrical parameters, Klemperer proposed a structure wherein the angles θ_1 and θ_2 take the values 49° and 64°, respectively. This geometry

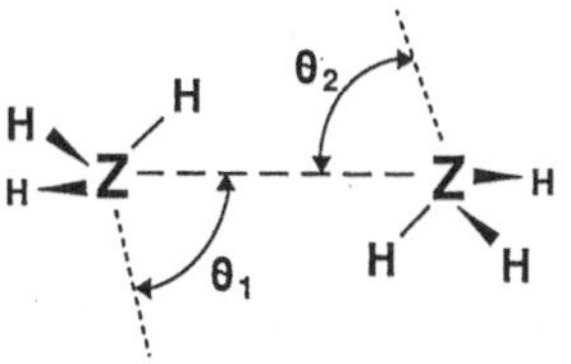

Fig. 4. Geometry of $(ZH_3)_2$. θ_1 and θ_2 refer to angles between $Z \cdots Z$ axis and C_3 symmetry axis of each subunit

is similar to a cyclic structure in that it contains a pair of distorted H-bonds in which each N atom acts simultaneously as proton donor and acceptor. The inequality of the two angles explains the small but nonzero dipole moment of the dimer. (Although they have no evidence concerning the internal rotation of the two NH_3 molecules about their respective C_3 symmetry axes, Klemperer et al. propose a structure in which the molecule for which $\theta = 49°$ is rotated by $180°$ relative to that illustrated in Fig. 4. Such a change would effectively break any "H-bond" in which this rotated molecule serves as proton donor.) The close similarity in structural detail observed for various isotopomers implies that vibrational averaging does not have a drastic effect upon the equilibrium geometry.

The experimental findings have motivated a good deal of theoretical inquiry. While there is some disagreement concerning the position of the absolute minimum, all ab initio calculations of $(NH_3)_2$ suggest the potential energy surface is extremely flat [94, 98, 133, 147–151]. Studies considering only the C_s (linear) and C_{2h} (cyclic) geometries have found very similar energies for the two; close enough indeed that the relative order of stability is found to reverse upon small changes in basis set, inclusion of correlation, or counterpoise correction [94, 98, 149–151]. A potential energy surface calculated at a correlated level with a doubly-polarized double-ζ basis set suggested a C_{2h} cyclic structure, with $\theta_1 = \theta_2 = 70°$, is the global minimum [147]. An exceedingly shallow valley connects this point on the surface directly with a linear-type structure ($\theta_1 = 112°$, $\theta_2 = 0°$) which is within 1 kJ/mol of the global minimum. Moreover, there is no energy barrier for passage from one point to the other through the valley. This finding led Latajka and Scheiner to propose that zero-point vibrational excursions away from the cyclic equilibrium structure might be the source of the nonzero dipole moment of the dimer [147], a notion later supported by the photoelectron spectroscopic work of Carnovale et al. [152].

While there remains some dispute among various ab initio calculations concerning the details of the lowest energy structure of $(NH_3)_2$, it is unanimous that the potential energy surface is quite flat. Theoretical estimates of the binding energy vary between 13 [149] and 18 [98, 133, 147] kJ/mol. This value is diminished by perhaps 7 kJ/mol when zero-point vibrations are included [133]. The resulting $\Delta H°$ is in satisfactory coincidence with revised experimental estimates of less than 12 kJ/mol [143–146].

The flatness of the potential energy surface and ensuing large-amplitude oscillations would be expected to cloud interpretation of the spectroscopic measurements. Questions have indeed been raised recently by Snels et al. [153a] and by Liu et al. [148]. On the other hand, data obtained by Süzer and Andrews [153b] suggest a structure like that proposed by the Klemperer group for the gas phase may be present in low-temperature matrices as well. It is clear that the details of the potential energy surface of the ammonia dimer warrant further experimental and ab initio

work which will hopefully shed additional insights into the nature of the interaction.

2.6.2 $(PH_3)_2$

Work on the PH_3 simer is much less abundant. Whereas Frisch et al. [94] had
suggested a linear equilibrium structure for $(NH_3)_2$, the same level of theory provided
a cyclic geometry of C_{2h} symmetry for $(PH_3)_2$. Both PH_3 molecules are rotated by
180° about their C_3 symmetry axes from those depicted in Fig. 4. This geometry
certainly casts doubt on any interactions that might be classified as a H-bond.
Further, the interaction energy is extremely small, only 3.3 kJ/mol at the MP4
level with a 6-31+G(d, p) basis set. This value is reduced to 1.3 kJ/mol when
zero-point vibrations are considered, making it unlikely that this complex will be
observed in the gas phase.

2.7 $H_2O \cdots HCH_3$

Although the extremely weak proton-donating ability of CH_4 precludes the development of a traditional H-bond with water, analysis of the very weak nature of this
interaction is interesting nonetheless. Latajka and Scheiner [60] applied a number of
different basis sets to $H_2O \cdots HCH_3$, all of polarized double-ζ quality or better.
They found that due to the very shallow interaction potential, the basis set superposition error can have a profound effect upon the predicted equilibrium intermolecular separation. At the SCF level, counterpoise correction of the BSSE increased
the $R(O \cdots C)$ distance by up to 30 pm; the corresponding correction of the MP2
equilibrium distance ranged up to 40 pm.

In terms of energy, the SCF BSSE varied between 0.04 and 3.5 kJ/mol, with a
similar range computed for the MP2 analogue [60]. While these quantities are not
very large in an absolute sense, they are comparable in magnitude and, in some
cases even larger than, the actual binding energy. For example, the uncorrected
interaction energy computed for $H_2O \cdots HCH_3$ with the 6-31G** basis set is
4.1 kJ/mol at the MP2 level but this value is reduced to 1.6 kJ/mol when the
counterpoise correction is applied. After suitable correction, the binding energy
seems relatively stable with respect to modification of basis set. The best calculated
value for this quantity lies in the range between 1.4 and 1.9 kJ/mol. Of this total,
somewhat more than 50% is due to correlation.

2.8 Summary

Ab initio methods appear capable of reproducing many features of H-bonded
complexes, given a high enough level of theory. Although early calculations using
fairly small basis sets and ignoring the effects of electron correlation or superposition
error were successful in certain regards, this circumstance was largely a matter of
fortuitous cancellation between opposing forces. For example, it is possible that the
exaggerated attraction imposed by basis set superposition can make up for certain
aspects of the true attractive character of dispersion which is neglected at the SCF
level. While such a combination can occasionally produce an equilibrium geometry

or complexation energy which is in reasonable coincidence with experiment, the different distance dependencies of each term will produce severe distortions in other regions of the potential energy surface.

Recent calculations have demonstrated that correlation plays an important role in H-bonding. Not only does correlation introduce dispersion energy into the picture, but it also leads to modifications of other terms which are present at the SCF level. The contribution of electron correlation has been shown to increase as one goes down in any column of the periodic table with respect to the nonhydrogen atoms involved in the interaction. While correlation is quite important in calculating both the complexation energy and the equilibrium separation of the pair of subunits, its effect upon the angular characteristics of the equilibrium geometry is much weaker. In fact, the relative orientations can usually be predicted to a fair level of accuracy on the basis of electrostatic (and exchange) forces alone.

Although small systematic changes in basis sets can frequently lead to erratic alterations of the quantitative nature of the H-bonding phenomenon, subtraction of the superposition error by the counterpoise procedure leads to much more uniform behavior. This argument applies to the correlated as well as SCF level of treatment. Perturbation theory of the Møller-Plesset type furnishes an efficient and accurate means to account for electron correlation. The literature indicates that MP2 calculations are quite reliable for H-bonds, due in large measure to the opposite effects generally observed for the third and fourth-order terms.

The preferred geometry of the H-bond is "linear". This is rigorously true of complexes of the type $H_3Z \cdots HX$ which belong to the C_{3v} point group: both the bridging proton and the Z lone pair lie directly along the $Z \cdots X$ axis. However, even in other types of complexes, the bridging proton usually lies within about 10° or so of the axis connecting the heavy atoms. The relative orientation of the proton-acceptor molecule can usually be predicted on the basis of the directions of its lone electron pairs. A more or less equivalent electrostatic picture describes the geometry in terms of the interactions between the various elements of the multipole expansion of each subunit. In a number of cases, the linear type of arrangement is not most stable but rather a cyclic or bifurcated type of structure better represents the equilibrium geometry. However, such arrangements are limited to very weakly bound complexes, e.g. $(NH_3)_2$ or $(H_2S)_2$, in which one might argue that a H-bond is not present at all.

The most strongly bound complex considered here is $H_3N \cdots HF$. Replacing either N or F with second or third-row analogues leads to a progressive weakening of the interaction energy. Similar patterns are noted in complexes other than the $H_3Z \cdots HX$ type. A lessening of the H-bonding interaction is associated with substitution of H_3Z by the weaker bases H_2Y and HX. With regard to the bond strengths of the homo-dimers, $(H_2O)_2$ and $(HF)_2$ are found to be comparable whereas the interaction in $(NH_3)_2$ is quite weak and the potential energy surface rather flat.

3 Vibrational Spectra

Introduction of gradient techniques has greatly facilitated calculation of the vibrational frequencies of a number of chemical systems. Over the years, it has come to be recognized that frequencies calculated at the SCF level often exaggerate the experimental values. A large part of this discrepancy is due to the SCF approximation itself which tends to overestimate the dependence of the energy upon a stretching or bending parameter which takes the system away from equilibrium. A second factor is the common assumption of harmonicity, i.e. that the energy is a simple quadratic function of the normal coordinate. Many investigators compensate for this systematic error by introducing a "scaling factor" of approximately 0.9 so as to maximize coincidence with experiment. However, such an approach can be misleading since the overestimate is not always uniform from one mode to the next.

In the case of molecular interactions such as H-bonding, complete normal mode analyses are not as abundant as for single molecules. The relatively weak nature of the interaction, difficult to capture quantitatively by ab initio methods, generally makes for poorer reproduction of experiment. The problem is compounded by the fairly strong anharmonic effects that can be expected for these weak bonds.

Calculations of vibrational intensities have begun to spring up in the literature more and more often of late. These intensities, when combined with the calculated frequencies, can often be of great assistance in assignment of various bands in the experimental spectra. Most calculations of vibrational intensities make use of what is termed the "double harmonic" approximation, viz. both mechanical and electrical [154]. The validity of this approximation, the accuracy expected from a given basis set or level of theory, and the interpretation of the various facets of the vibrational spectra are rapidly emerging as an exciting new area of research into H-bonding.

For many years, it has been known that formation of a H-bond produces a marked red shift and manyfold intensity enhancement in the $X-H$ stretching mode of the proton donor [154, 155]. As discussed below, ab initio calculations have attempted to provide some fundamental insights into the cause of this phenomenon. Perturbations of the other internal vibrations of the subunits are perhaps not as dramatic nor as universal but offer some valuable information nonetheless. Another window into the nature of the interaction resides in the intermolecular modes that are directly associated with the H-bond.

3.1 HX⋯HX′

3.1.1 (HF)₂

The first column of Table 5 lists the frequencies calculated for the HF monomer with a variety of different theoretical schemes. This information is followed by the stretching frequencies of the two HF molecules in the dimer. A scan of the SCF data in the upper portion of the table reveals the familiar overestimation of all frequencies in comparison to the experimental values in the last row of the table. Incorporation of electron correlation lowers the SCF frequencies by several hundred wave numbers, taking them into the vicinity of the experimental values. A primary source of disagreement is the use of the harmonic approximation in calculating these frequencies. Nevertheless, the correlated frequencies calculated by Michael et al. using a triple-ζ polarized basis set [97] fall quite close to the experimental range.

Table 5. Calculated intramolecular frequencies (cm^{-1}) in monomer and dimer of HF

Basis set	Monomer	Dimer		Ref.
		Acceptor	Donor	
SCF				
DZ	4103	4070	4021	[95]
6-31G**	4495	4454	4406	[156]
DZP	4440	4397	4349	[95]
+VP^S	4488	4448	4394	[103]
correlated				
DZ	3732	3773	3743	[95]
DZP	4150	4153	4095	[95]
TZP	3994	3976	3920	[97]
expt				
	3961[a]	3929[b]	3868[b]	

[a] [157]   [b] [158]

Whereas ab initio methods require a very high level of accuracy in conjunction with anharmonic corrections to reproduce experimental vibrational frequencies, perturbations in these frequencies caused by H-bond formation may be somewhat more amenable to accurate estimation at only modest levels of theory. Table 6 compiles the calculated shifts in frequency caused in the two HF molecules as a result of dimerization. It may first be noted that the SCF results show surprisingly little sensitivity to basis set. In fact, even the rather small 4-31G basis set appears capable of reproducing the experimental shifts surprisingly well despite the absence of correlation or consideration of anharmonicity effects. This situation is undoubtedly the result of fortuitous cancellation between terms. Indeed, the second portion of Table 6 provides evidence that correlation does produce a sizeable change in the frequency shifts. However, the effect is not consistent from one basis set to another. For example, the correlated shifts are 2 or 3 times smaller than the SCF values for the DZ basis set whereas very small increases are observed when polarization functions are added. All calculations agree that the shift in the frequency of the proton donor molecule is considerably larger than that of the acceptor. Moreover, Kurnig et al. [156] found a similar distinction for the calculated intensities: while the proton acceptor exhibits only a small increase, that of the donor increases nearly threefold.

The intermolecular vibrational frequencies are listed in Table 7 for $(HF)_2$. Due to the weakness of the H-bond interaction, one may expect these vibrations to exhibit substantial anharmonic character. It is therefore not surprising that the theoretical methods which make use of the harmonic approximation are somewhat in error. Nevertheless, the level of agreement exhibited in Table 7 provides an encouraging starting point for the future. For example, whereas most of the experimental estimates of v_1, the bending motion of the proton donor molecule, are rather high when compared to an experimental measurement, the correlated value obtained with a polarized triple-ζ basis set is rather good. The theoretical estimates of v_3 are likewise too high but are subject to improvement. It should be pointed out that correlation has very little effect on the frequencies calculated with the DZ basis set. In contrast, when polarization functions are included, the correlated frequencies are substantially larger than the SCF values.

Table 6. Frequency shifts (cm^{-1}) caused by dimerization for proton acceptor and donor molecules in $(HF)_2$

Basis set	Acceptor	Donor	Ref.
SCF			
4-31G	−36	− 79	[159]
DZ	−33	− 82	[95]
6-31G**	−41	− 86	[156]
DZP	−43	− 91	[95]
+VP^s	−40	− 86	[103]
correlated			
DZ	−10	− 40	[95]
DZP	−47	−105	[95]
TZP	−18	− 74	[97]
expt			
	−32	− 93	[158]

Table 7. Calculated frequencies (cm^{-1}) of intermolecular vibrational modes of $(HF)_2$

Basis set	v_1	v_2^a	v_3	v_4	Ref.
SCF					
DZ	519	475	189	165	[95]
6-31G**	601	455	230	127	[156]
DZP	529	442	193	143	[95]
+VP^s	522	433	206	142	[103]
correlated					
DZ	523	459	196	166	[95]
DZP	607	486	218	156	[95]
TZP	420	—	167	127	[97]
expt					
	401	—	127		[160]

^{a} Out-of-plane mode

It should finally be mentioned that some recent calculations of $(HF)_2$ [161] have indicated that basis set superposition error is likely to produce only minor problems when calculating IR intensities. This circumstance results from a cancellation between the artifactual enhancement of the dipole derivative of one molecule and an opposite trend in the other. On the other hand, this same study suggested that Raman intensities are likely to be subject to very large errors; further study of this problem is recommended.

3.1.2 $(HCl)_2$

Although the calculations for this system are somewhat more limited than for $(HF)_2$, the trends mentioned above reappear although not as pronounced. A comparison of the two systems is presented in Table 8, using a similar basis set for purposes of consistency. It may first be noted that the intermolecular frequencies are uniformly lower for $(HCl)_2$ than for $(HF)_2$, reflecting the weaker binding of the former. A similar pattern is noted in the perturbations of the monomer frequencies caused by dimerization. The frequency shifts in the proton donor and acceptor molecules are respectively 94 and 40 cm^{-1} for $(HF)_2$ whereas the analogous shifts are 24 and 9 cm^{-1} for $(HCl)_2$. The latter values agree rather well with experimental measurements of 69 and 6 cm^{-1} [89, 111].

With regard to the intensities, that of the proton donor stretch is enhanced by a factor of nearly 3 in $(HF)_2$ whereas this magnification is closer to 2 in $(HCl)_2$. Little change is noted in the intensity of the acceptor stretch in either dimer. It should also be noted that the calculated intensities of the intermolecular modes are generally smaller for $(HCl)_2$ than for $(HF)_2$.

Table 8. Calculated frequencies (cm^{-1}) and intensities (km/mol) of $(HF)_2$ and $(HCl)_2$, compared with monomers; from Ref. [103]

	$(HF)_2$		$(HCl)_2$	
	v	A	v	A
Monomer	4488	168	3174	57
intramolecular modes				
p-acceptor stretch	4448	167	3165	65
p-donor stretch	4394	459	3150	124
intermolecular modes				
p-donor, in-plane bend	522	219	314	1
p-donor, out-of-plane bend	433	230	179	43
p-acceptor, bend	206	160	156	40
H-bond stretch	142	6	48	31

3.1.3 Mixed Dimers

Vibrational data for the complex containing HF and HCl in which one or the other molecule acts as proton donor are presented in Table 9. The shift in the HF frequency when this molecule acts as proton donor in HCl ·· HF is less than the same quantity in HF ·· HF; the intensity of this band is also smaller in the former complex. Both trends are indicative of the weaker binding in HCl ·· HF in comparison to $(HF)_2$. Consistent with the stronger interaction in HF ·· HCl than in HCl ·· HCl, the intensity of the p-donating HCl stretch is larger in the former system (although no substantive difference is observed in the frequencies). The frequencies and intensities of the intermolecular modes in the heterodimers are generally intermediate between the large values observed in $(HF)_2$ and the smaller ones in $(HCl)_2$, again verifying the relationship between vibrational spectra and binding strength.

Table 9. Calculated frequencies (cm^{-1}) and intensities (km/mol) of mixed dimers of HF with HCl; from Ref. [103]

	HCl ·· HF		HF ·· HCl	
	v	A	v	A
intramolecular modes				
HF stretch	4426	398	4464	203
HCl stretch	3158	71	3152	161
intermolecular modes				
p-donor, in-plane bend	413	101	289	115
p-donor, out-of-plane bend	295	163	242	79
p-acceptor, bend	132	98	121	115
H-bond stretch	80	1	78	2

3.2 H_2Y·· $HY'H$

3.2.1 $(H_2O)_2$

In order to provide some insights into the intensity enhancement undergone by the H—Y stretch of the proton donor molecule in a H-bond [162, 163], Zilles and Person made use of atomic polar tensors in their study of the water dimer [164]. These quantities are defined in terms of the effect on the dipole moment of a small displacement of each atomic center of the system under examination. Further information arises from a partitioning into charge, charge flux, and overlap contributions (CCFO). Even though their work was limited to a rather small basis set (4-31G), they were able to extract some very useful information. Comparison of certain elements of their calculated spectra with experimental data led them to believe that the 4-31G basis set correctly reproduces the essential ingredients.

Zilles and Person note [164] that dimerization leads to a significant increase in the atomic charge of the bridging hydrogen, consistent with a good deal of prior work. Hence, stretching of this proton away from the oxygen atom causes a proportionately larger increase in the dipole moment of the dimer than it does in the monomer. More important, though, is the effect on the charge flux term. The same stretch of the bridging hydrogen away from the O atom causes its charge to become much more positive in the dimer than in the isolated monomer. The latter increasing positive charge may be attributed to a sort of "freezing" of the electron cloud which inhibits its following of the proton within the complex.

In the same year, Swanton et al. published a similar type of calculation [165] of the water dimer using a somewhat larger basis set including polarization functions on all atoms. Their results largely confirmed the prior 4-31G data of Zilles and Person. Calculations of various deuterated species led to the conclusion that the intensity of the OH stretch involving the bridging hydrogen drops significantly upon deuteration of this atom. Their approach to studying the effect of dimerization upon the spectrum of water was to compare their H_2O dimer with a "non-interacting dimer" in which the two molecules assume the correct dimer geometry and the "dimer" normal modes occur but each molecule is assumed to behave as it does as an isolated monomer. The only modes in which the H-bond interaction plays an active role are the H-bond stretch and the out-of-plane H-bond shear, i.e. the two modes which directly involve the H-bond. Swanton et al. also tested an electrostatic model and concluded it is not capable of accurately describing infrared intensities due to the important role played by exchange and charge transfer effects. These authors extended their studies of the water dimer the following year with an analysis of the polarizability tensor and Raman scattering activities [166].

The water dimer has been reexamined more recently by Amos [138] using the 6-31G** basis set as well as extending it by the addition of some functions with low exponents and a second set of polarization functions. They compared their computed frequencies with a previous calculation [94] as well as with experimental results in inert gas matrices [167, 168] and found fairly good agreement, especially when comparing frequency *shifts* caused by dimerization. (It should be mentioned at this juncture that the experimental assignment of the various bands in the water dimer is a point of some controversy at present [169, 170].)

Of perhaps particular interest to workers in the field is the sensitivity of the calculated intensities to basis set. The first portion of Table 10 lists the intensities calculated for the three normal modes of the water monomer with a variety of different basis sets. (The extensions to 6-31G** in the second column refer to diffuse s and p functions and a second set of polarization functions on all atoms as mentioned above.) Although v_2 is in the range between 90 and 100 km/mol with all three basis sets, a significant amount of variability exists in the other two modes. Note also that the calculated intensities are significantly higher than the experimental values in the last column of the table.

The second portion of the table reports the enhancement experienced by the intensity of each mode when the dimer is formed. There is qualitative agreement concerning the magnitude of this factor for all modes although disagreement is noted in certain quantitative particulars. For example, all basis sets agree the strongest magnification occurs in the OH stretch of the proton donor, but the range of this

Table 10. Calculated intensities of water monomer and intensity enhancements in dimer

Basis set		6-31G**	6-31G ext	[541/41]	expt
Intensities in monomer (km/mol)					
	v_1	17	14	3.7	2.2
	v_2	97	91	100	54
	v_3	58	81	59	45
Intensity enhancements (A_{dim}/A_{mon})					
p-donor	v_1	10.4	13.6	54.1	
	v_2	0.91	0.71	0.96	
	v_3	1.79	1.57	1.71	
p-acceptor	v_1	1.71	1.57	2.84	
	v_2	1.06	1.21	0.92	
	v_3	1.52	1.27	1.37	
Ref.		[138]	[138]	[165]	[164]

factor extends between 10 and 54. With regard to the other two modes of the donor, v_2 experiences a small diminution whereas a nearly twofold increase is noted in v_3. v_1 of the acceptor is predicted to intensify by a factor between 1.5 and 3, v_3 undergoes a smaller increase, and v_2 remains approximately unaffected by dimerization.

The vibrational data for the intermolecular modes of the water dimer are compiled in Table 11. The [541/31] basis set predicts lower frequencies than 6-31G** and its extended derivative for all modes. There is some disagreement among the various calculations concerning the ordering of the three modes of lowest frequency although all agree that they are rather close together. Another point of contention is the relative intensities of the intermolecular modes. The largest 6-31G ext set predicts the a' stretch is most intense while the two smaller sets indicate the a' bend is strongest. All agree that the a'' torsion is of lowest intensity.

Table 11. Intermolecular vibrational frequencies (cm^{-1}) and intensities (km/mol) calculated for $(H_2O)_2$

Basis set	6-31G**		6-31G ext		[541/31]	
	v	A	v	A	v	A
a'' shear	605	185	589	154	541	188
a' shear	375	78	356	88	322	103
a' stretch	175	127	174	257	169	102
a'' torsion	145	64	153	17	105	31
a' bend	142	237	142	64	96	231
a'' bend	121	142	129	180	113	148
Ref.	[138]		[138]		[165]	

In summary, ab initio methods using basis sets of moderate size appear capable of calculating internal vibrational frequencies rather well, particularly shifts due to dimerization. Intermolecular frequencies are somewhat more difficult, due in part no doubt to their high contribution from anharmonic effects. Whereas perturbations of intensities resulting from H-bonding are reasonably consistent from one basis set to another, there remains a fairly high degree of variability in the absolute intensities.

3.2.2 Dimers containing H_2S

Amos [138] has used similar theoretical methods to calculate the vibrational frequencies of $(H_2S)_2$ as well as mixed dimers containing both H_2O and H_2S. The intermolecular frequencies, all calculated with the same 6-31G** basis set, are reported in Table 12. There is a clear trend evident in the data in that frequencies decrease from left to right, similar to the decreasing binding energy as each H_2O molecule of the water dimer is substituted with H_2S. There is no such clear pattern in the case of the intensities.

The spectrum of the mixed dimer of H_2O and H_2S has been observed in rare gas matrices by Nelander [167] and Barnes et al. [171], prompting an ab initio calculation by Chin and Ford [139]. They considered only the $H_2O \cdots HSH$ case where H_2S acts as proton donor, making use of a 4-31G basis set. Their analysis concentrated on changes induced in each subunit by complexation and did not report any data for the intermolecular modes.

Table 12. Intermolecular vibrational frequencies (cm^{-1}) calculated for $H_2Y-HY'H$ with 6-31G** basis set (from [138])

	H_2O-HOH	H_2O-HSH	H_2S-HOH	H_2S-HSH
a" shear	605	371	345	217
a' shear	375	190	221	130
a' stretch	175	113	70	43
a" torsion	145	131	68	67
a' bend	142	98	88	62
a" bend	121	78	59	40

3.3 $(NH_3)_2$

The controversy about the nature of binding in the ammonia dimer has inspired a number of ab initio calculations of its vibrational spectrum. Unfortunately, both of these studies [172, 173] have been limited to the relatively small 4-31G basis set at the SCF level which cannot be expected to provide a good description of the binding. The hope is that the basis may nevertheless lead to information about the vibrational spectrum useful in discriminating one type of structure from another.

Sadlej and Lapinski [172] point out that the vibrational levels of each monomer are split in the dimers. Whereas all components are IR-active for the linear dimer, the symmetry of the cyclic structure makes one component of each pair IR-active and the other Raman-active. They note that the v_1 and v_3 stretching frequencies are lowered upon dimerization, particularly in the case of the proton donor. The v_4 bending frequency of the proton acceptor increases slightly and that of the donor more so.

Yeo and Ford [173] emphasize that whereas a number of vibrational modes in the linear dimer could easily be distinguished as involving only one molecule or the other, all normal modes are highly coupled in the cyclic structure. The intensities calculated for the intramolecular vibrations of the linear ammonia dimer vary between 3 and 60 km/mol. Comparison to the monomer shows magnifications of up to a factor of 4 as well as weakening of other modes by one order of magnitude. The intensities of the cyclic dimer are overall somewhat stronger, varying between 15 and 93 km/mol.

Calculated vibrational frequencies and intensities for the intermolecular modes are presented in Table 13 for both the linear and cyclic structures. The data are presented in pairwise fashion so that the results of two independent calculations using an identical basis set may be compared and the sensitivity to details of the theoretical procedure assessed. Due to the weakness of the interaction, the results are rather variable from one calculation to the next. For example, the intensity of the first a″ mode of the linear dimer differs by a factor of three. The two calculated values of the second a″ frequency show a variation by a factor greater than two. Another discrepancy is the elucidation of the mode corresponding to the imaginary frequency; one finds this to be of a′ symmetry and the other a″.

It thus appears that whereas reasonable results may be obtained for the perturbations of the vibrational spectrum of each molecule in a weakly bound complex, one cannot expect a similar level of accuracy for the intermolecular modes where minor differences in details of the calculation can produce qualitatively different conclusions.

One piece of information that these calculations can add to the question of relative stability of the cyclic and linear geometries is the zero-point vibrational energy contributions. These sets of calculations [172, 173] are consistent with one another in that they both indicate a larger amount of such vibrational energy is contained

Table 13. Calculated frequencies (cm^{-1}) and intensities (km/mol) of intermolecular modes in linear and cyclic dimers of NH_3

Linear					Cyclic			
	v		A			v		A	
a′	395	381	96	78	a_g	633	477	0	0
a″	393	314	17	51	a_u	388	255	104	87
a″	293	136	56	41	a_g	204	137	0	0
a′	i	134	—	49	a_u	199	96	9	26
a′	69	112	44	32	b_u	197	93	229	235
a″	170	i	27	—	b_g	164	159	0	0
Ref.	[173]	[172]	[173]	[172]		[173]	[172]	[173]	[172]

[i] Imaginary frequency

within the cyclic structure. The data of Sadlej and Lapinski lead to a difference of 0.7 kJ/mol whereas a larger value of 3.6 kJ/mol arises from Yeo and Ford's frequencies. In either case, this finding lends support to the earlier contention of Latajka and Scheiner [147] that the cyclic geometry is the only true minimum on the electronic potential energy surface and that vibrational energy considerations tend to shift this equilibrium structure a small amount along the pathway toward the linear geometry. Such a shift could account for the small dipole moment observed in the dimer.

On the other hand, earlier calculations by Frisch et al. [94] with somewhat larger basis sets had indicated that the relative amounts of zero-point vibrational energy in the cyclic and linear dimers can reverse upon small changes in basis set, e.g. addition of diffuse functions. Furthermore, the use of the harmonic approximation on a surface as flat as this one is dubious, so the question remains incompletely resolved.

3.4 $H_2O \cdots HX$

The vibrational spectrum of $H_2O \cdots HF$ has recently been calculated using a triple-ζ doubly-polarized basis set and another somewhat smaller one by Somasundram et al. [174]. Comparison of these data with a recent calculation of $H_2O \cdots HCl$ [175] can provide some insight into the effect of H-bonding upon the spectrum.

Table 14 reports the shifts in the frequencies of each molecule as a result of formation of the complex. Both v_1 and v_3 are red-shifted by several wave numbers whereas a small positive shift occurs in v_2. The near insensitivity of these frequencies to perturbations induced by formation of the H-bond is consistent with prior experimental measurements [176]. Certainly the largest change occurs in the stretch of the proton donor. Somasundram et al. calculate a red shift of some 260 cm^{-1}, somewhat smaller than the experimental measurement of 353 cm^{-1} [177]. Latajka and Scheiner's calculated shift for HCl of 105 cm^{-1} is similarly smaller than an experimental value of 216 cm^{-1} [178]. Whereas the experimental shift for $H_2O \cdots HF$ corresponds to a gas-phase system, the latter value was measured in N_2 matrix which might explain some of the discrepancy. A primary point of comparison is that all of the frequency shifts are smaller in the $H_2O \cdots HCl$ complex, consonant with the weaker binding.

Table 14. Frequency shifts (cm^{-1}) caused in each subunit of $H_2O \cdots HX$ by complexation[a]

		$H_2O \cdots HF$		$H_2O \cdots HCl$
		DZP	TZ + 2P	$+VP^s(2d)^s$
H_2O	v_1	− 7	− 9	− 4
	v_2	+ 5	0	+ 1
	v_3	− 10	− 10	− 3
HX		−259	−264	−105

[a] Data for $H_2O \cdots HF$ from [174]; $H_2O \cdots HCl$ from [175]

Table 15. Intensity enhancements caused in each subunit by complexation, reported as $(A_{complex}/A_{monomer})^a$

		$H_2O \cdots HF$	$H_2O \cdots HCl$
		TZ + 2P	$+VP^S(2d)^S$
H_2O	ν_1	5.5	1.9
	ν_2	1.0	0.9
	ν_3	1.6	1.3
HX		4.6	6.4

[a] Data for $H_2O \cdots HF$ from [174]; $H_2O \cdots HCl$ from [175]

A similar analysis of the intensity changes induced by complexation is contained in Table 15. The ν_1 band of H_2O undergoes a substantial intensification while the other vibrations of this subunit are relatively static. The expected strengthening of the stretching band of the proton donor is evident in Table 15. One curious feature of the data is the greater magnification of the HCl stretch as compared to HF despite the weaker binding of the former molecule. This unexpected result may be an artifact of the use of different basis sets for the two different systems. Another possibility has to do with the much larger mass of Cl compared to F which might change the character of the normal mode. An analysis of this finding using atomic polar tensors within the context of a single basis set would be most enlightening.

The calculated data for the intermolecular modes are presented in Table 16. The stretching frequency for $H_2O \cdots HF$ is considerably higher than in the HCl analogue, concurring again with the stronger binding in the former complex. The calculated frequency of 220 cm^{-1} is in nice agreement with a previous gas-phase measurement of 198 cm^{-1} [179]. Similarly for $H_2O \cdots HCl$, there is good agreement between the calculated value of 118 cm^{-1} and 100 cm^{-1} measured in a matrix [175, 178] as well as a gas-phase frequency of 119 cm^{-1} for the very similar $(CH_3)_2O \cdots HCl$

Table 16. Vibrational frequencies (cm^{-1}) and intensities (km/mol) of $H_2O \cdots HX^a$

	$H_2O \cdots HF$			$H_2O \cdots HCl$	
	ν		A	ν	A
	DZP	TZ + 2P	TZ + 2P	$+VP^S(2d)^S$	
stretch	224	220	87	118	3
	794	786	194	459	77
	654	644	226	351	38
	226	234	3	143	33
	150	182	155	94	28

[a] Data for $H_2O \cdots HF$ from [174]; $H_2O \cdots HCl$ from [175]

[180]. It is pertinent to mention here that previous calculations [35, 181] have found that inclusion of anharmonic coupling of the HF and F $\cdots$ O stretches, coupled with electron correlation, have the effect of slightly increasing this stretching frequency.

The assignment of the other modes is not a simple one and there may be significant differences between the two systems. Hence, no labels are used in Table 16; the modes are instead listed in order of decreasing frequency. There is a clear pattern of all frequencies being uniformly higher for $H_2O \cdots HF$. In contrast, there is no obvious pattern exhibited by the intensities.

A previous calculation [35] has attempted to quantitatively reproduce the experimentally observed frequencies of the low-frequency in-plane bending mode corresponding to water inversion in $H_2O \cdots HF$. This was successfully accomplished by first calculating the potential energy profile corresponding to this motion at SCF and higher levels. The vibrational levels were then calculated for this profile. It was found that whereas the frequency of the fundamental is poorly reproduced by the SCF potential, including electron correlation leads to very good agreement with experiment [182].

Another result to emerge from this study [35] was the resolution of the question as to the equilibrium structure of $H_2O \cdots HF$. The proximity of the ground vibrational state to the top of the barrier in the inversion potential leads to a wave function which is especially flat in the vicinity of its maximum, with considerable amplitude in the regions corresponding to both a pyramidal and "planar" C_{2v} geometry. It is thus easy to understand why vibrational averaging of this function does not unambiguously identify the pyramidal structure of this complex even though, strictly speaking, it represents the equilibrium geometry.

Ab initio calculations have more recently been applied to another question arising from spectroscopic measurements. The main feature of the gas-phase vibrational spectrum of $(CH_3)_2O \cdots HF$ is a very strong band assigned to the HF stretch. This band is accompanied by sidebands of weaker intensity whose origins have spurred speculation. Combination bands involving bending vibrations were suggested by Arnold and Millen [183]. Stepanov [184] invoked anharmonic coupling with the F $\cdots$ O stretch, later supported by Bouteiller and Guisani [185]. Bouteiller et al. [186] have readdressed this problem by first computing the energy and dipole moment of the system on a grid of points including variations of both $R(O \cdots F)$ and $r(HF)$. A fairly large polarized basis set was used at the SCF level. These properties were then fit to a polynomial expansion, yielding force constants up to fourth order in both geometrical parameters. A variational treatment, including mechanical anharmonicity, was then used to obtain all the necessary vibrational levels. The spectrum arising from this procedure, incorporating elements of both frequencies and intensities, reproduced very well all the essential features of the stretching fundamental of both $(CH_3)_2O \cdots HF$ and $(CH_3)_2O \cdots DF$. The comparison with experiment was not as favorable for the first overtone, however. The authors attribute the latter discrepancy to the use of an SCF wave function which cannot treat quantitatively the larger distortions from the equilibrium geometry involved in the overtones.

3.5 $H_3Z \cdot\cdot HX$

Bouteiller et al. [187] have recently carried out an investigation of complexes which couple HCl or HBr with an amine such as NH_3 or NH_2CH_3. This work included the effects of electron correlation explicitly for the $H_3N \cdot\cdot HCl$ system where CI induced a decrease in the HCl stretching frequency of some 250 cm^{-1} whereas an increase was observed in the N $\cdot\cdot$ Cl stretch. The authors noted as well that anharmonicity contributes a great deal (340 cm^{-1}) to the reduction in v_{HCl}. Despite this rather high level of theory, the anharmonic correlated value of the N $\cdot\cdot\cdot$ Cl stretching frequency of 201 cm^{-1} remains higher than the experimental values which fall in the range 130 to 166 cm^{-1} [188, 189]. The XH stretching frequencies in the two larger systems, $CH_3NH_2 \cdot\cdot HCl$ and $H_3N \cdot\cdot HBr$, are also lowered by several hundred wave numbers by anharmonicity effects; the calculated N $\cdot\cdot$ X frequencies are 144 cm^{-1} for both systems. The authors conclude that correlation and anharmonicity are both necessary for quantitative accuracy in calculating these frequencies.

A complete elucidation of all the normal vibrational modes, including intensities and frequencies, has been described by Kurnig et al. for $H_3N \cdot\cdot HF$ and $H_3N \cdot\cdot HCl$ at the SCF level using a 6-31G** basis set [156]. The frequencies and intensities calculated for the intermolecular modes are exhibited in Table 17. Consistent with patterns described earlier and the stronger binding energy of $H_3N \cdot\cdot HF$ as compared to $H_3P \cdot\cdot HF$, the frequencies are uniformly higher in the former complex. The calculated frequencies of 876 and 467 cm^{-1} in the first row, corresponding to a bend of the proton-donor molecule, agree rather well with experimental measurements of 916 and 477 cm^{-1}.

Table 17. Frequency (cm^{-1}) and intensity (km/mol) of intermolecular modes in $H_3Z \cdot\cdot HF^a$ (from [156])

	$H_3N \cdot\cdot HF$		$H_3P \cdot\cdot HF$	
	v	A	v	A
bend (2)	876	208	467	134
bend (2)	236	10	108	6
stretch	240	3	113	1

^{a} Degeneracy of mode indicated in parentheses

With regard to the intramolecular frequencies, Table 18 reports the changes in these properties relative to the uncomplexed monomers. While the stretching frequencies of NH_3 are essentially unaffected by complexation, the analogous quantities in PH_3 increase by some 30 cm^{-1} or so; the latter changes agree nicely with experimentally measured increases between 20 and 32 cm^{-1} [190]. One of the bending frequencies of NH_3 increases by 93 cm^{-1}, in comparison to the experimental shift of 120 cm^{-1}; the other remains nearly constant. PH_3 exhibits an opposite pattern in that the first bending frequency is slightly red-shifted while the second increases

Table 18. Changes in frequency[a] and intensity[b] in subunits caused by formation of H-bond[c] (from [156])

H₃Z	$H_3N \cdots HF$		$H_3P \cdots HF$	
	ν	A	ν	A
stretch	0	35	26	0.4
stretch (2)	−2	16	30	0.7
bend	93	1.0	−13	1.4
bend (2)	−7	1.2	94	1.0
HF				
stretch	−432	7.1	−134	4.0

[a] $\nu_{complex} - \nu_{monomer}$, cm^{-1}
[b] $A_{complex}/A_{monomer}$
[c] Degeneracy of mode indicated in parentheses

by a large amount. The stretching frequency of HF is very much red-shifted by complexation, more so with NH_3 than with PH_3, as expected due to the stronger bonding in the former case.

Turning next to the intensities, it may be noted in Table 17 that the $H_3N \cdots HF$ intermolecular bands are uniformly more intense than the comparable vibrations in $H_3P \cdots HF$. More curious, though, are the intensity changes occurring in the intramolecular modes in Table 18. Whereas the stretching modes of NH_3 become very much stronger when complexed to HF, the PH_3 vibrations *lose* intensity. Only small changes, primarily increases, are found in the intramolecular bends of H_3Z. The HF stretching mode undergoes a substantial strengthening, again more so in $H_3N \cdots HF$ than the PH_3 analogue.

Kurnig et al. [156] made use of atomic polar tensors to help understand these intensity changes, in much the same way as Zilles and Person had examined the water dimer previously [164]. It was possible to attribute the intensity enhancement observed in the HF stretch to an increase in the sensitivity of the molecular dipole moment to the HF bond length. Part of this increased sensitivity is due to the greater positive charge on the proton when it acts as a bridge in a H-bond. Since this charge is related to the strength of the bond, it is not surprising to see a more intense band in $H_3N \cdots HF$ than in $H_3N \cdots HCl$. The second component in the sensitivity arises from the ability of the electron cloud to follow the bridging proton as it moves away from F. In a result quite similar to the earlier finding of Zilles and Person, formation of the H-bond appears to "freeze" the charge cloud, disabling it from following the proton to the same extent as in the free HF monomer.

Atomic polar tensors were also used to analyze the opposite intensity patterns noted for the complexes involving NH_3 and PH_3 [156]. Beginning with the former monomer first, the N—H stretching bands of NH_3 are rather low in intensity. This appears to be due to a cancellation between two opposing effects. Since a partial positive charge resides on the H atoms, a stretch away from N will raise the bond moment. However, this increase is counteracted by the charge flux associated with this stretch which reduces the positive charge of the hydrogen, thereby acting to *lower* the bond moment. Upon formation of a H-bond to HF, each hydrogen of NH_3 becomes more positively charged. A second effect of the interaction is a reduced ability of the electron cloud to migrate toward these H atoms as they stretch away from the nitrogen nucleus. These two effects reinforce one another in that both lead to a rapid increase in the N—H bond moment as the bond is stretched, and consequently to the strong intensifications of the NH_3 stretching modes described in Table 18.

The situation is rather different in PH_3 in that the hydrogen atoms are slightly negatively charged in that monomer. As they are stretched away from P, the electron density shifts toward the P (as in NH_3), increasing their negative charge. The P—H stretches therefore produce a substantial change in the molecular dipole moment, explaining the relatively intense bands for the monomer. Complexation with HF pulls electron density from PH_3, just as it did in the case of NH_3. In contrast to the situation in NH_3 where the hydrogens are positively charged and become more so in the complex, the lost density results in a decreased (negative) charge on the PH_3 hydrogens. Their motion thus produces a *smaller* change in the molecular moment, resulting in the weakening of the PH stretches detailed in Table 18.

The very weak intensities for the intermolecular stretches of both systems in the last row of Table 17 motivated an inquiry and comparison with other systems. It was concluded [156] that such low intensities will generally be the case unless either i) the two subunits produce a substantial perturbation in the electronic structure of one another or ii) the stretching mode includes elements of other motions that tend to reorient the two subunits. Such a reorientation would be capable of altering the dipole moment vector of the entire complex even if the individual contributions of the subunits are unchanged. The C_{3v} symmetry of the $H_3Z \cdots HF$ complexes prohibits any such reorientation from contributing to the intermolecular stretch, resulting in the low intensities. On the other hand, it is just such a reorientation which was deemed responsible for the rather high intensities of the intermolecular bend of the proton donor in the first row of Table 17.

3.6 Summary

In contrast to the structures and energetics of these complexes which can be treated rather well at modest levels of theory, the vibrational spectra are much less tractable. It is well known that SCF frequencies are uniformly too large; this is true in H-bonded complexes as well as in single molecules. Inclusion of electron correlation, coupled with use of a sufficiently flexible basis set, can bring the internal frequencies of the individual subunits quite close to experimental values. On the other hand, if one is interested primarily in *shifts* in these frequencies induced by formation of a H-bond, the SCF treatment seems to provide quite reasonable results in many cases, even with relatively modest-sized basis sets.

The intermolecular modes are much more difficult to calculate as they involve a good deal of anharmonic coupling. The importance of the latter effect appears to grow more important as the interaction becomes weaker. Treatment of anharmonicity, in tandem with well-correlated wave functions, can provide quantitative accuracy, e.g. a good reproduction of the band shape in $(CH_3)_2O \cdot\cdot HF$. Accuracy with lower levels of theory is far from assured although qualitative trends seem reasonable.

Quantitative reproduction of vibrational intensities remains elusive, especially for the intermolecular modes. Nevertheless, modest levels of theory can frequently reflect many of the patterns correctly. Moreover, analysis of the calculated intensities using atomic polar tensors can shed valuable insights into the fundamental nature of the H-bond. For example, a large part of the well known intensification of the proton donor H—X stretch has been identified with the diminished ability of the electron cloud to follow the bridging proton within the context of a H-bond. Whereas some of the intensity changes occurring within the proton acceptor molecule seem at first sight rather anomalous, a similar type of analysis has offered a simple explanation in terms of electronic redistributions accompanying formation of the H-bond.

Taking a global view of the large quantity of vibrational data generated with a host of different basis sets and levels of theory, a number of clear trends do seem to emerge. The red shift of the proton donor HX stretching frequency is directly related to the strength of the interaction. The intensity enhancement of this mode varies between a factor of 2 and 7, with the greater intensities generally associated with stronger bonds. The intermolecular H-bond stretching frequency appears to obey a like relationship with respect to bond strength although the trend is obfuscated by variability of the data.

4 Comparison with Li-Bonding

4.1 Structure and Energetics

A fundamental question which frequently arises in connection with the H-bond is its uniqueness: Are there other atoms that can replace hydrogen as a bridge between two different molecules? The most obvious possibility for such a substitute is Li. A systematic comparison of H-bonds with Li-bonds was carried out for complexes composed of NH_3 on one hand and HF, LiF, HCl, and LiCl on the other [22]. The pertinent data are collected together in Table 19.

The entries in the first row of the table indicate that Li bonds tend to be considerably longer than H-bonds. That is the $R(N \cdot\cdot F)$ distance is nearly 100 pm longer in $H_3N \cdot\cdot LiF$ than in $H_3N \cdot\cdot HF$; the difference is roughly 80 pm for $R(N \cdot\cdot Cl)$. This distinction is not surprising since Li contains a fully occupied core of 1s electrons. One interesting feature concerns the influence of electron correlation upon the calculated properties of the two complexes. Note that the MP2 lengths of both H-bonds are shorter than their SCF counterparts whereas an opposite effect is noted in $H_3N \cdot\cdot LiF$ where the MP2 value of $R(N \cdot\cdot F)$ is slightly (10 pm) longer.

Inspection of the second row of Table 19 reveals that in all cases, correlation increases the AX bond length (A = H, Li). However, of greater fundamental interest is the comparison of this bond with its length in the isolated monomer. The third

S. Scheiner

Table 19. Calculated properties of H-bonded and Li-bonded complexes (from [22])

	$H_3N \cdots HF$		$H_3N \cdots LiF$		$H_3N \cdots HCl$		$H_3N \cdots LiCl$
	SCF	MP2	SCF	MP2	SCF	MP2	SCF
$R(N \cdots X)$, pm	272.8	269.3	365.2	366.5	329.7	314.4	411.8
$r(AX)$, pm	92.2	95.0	158.2	159.2	129.3	131.7	208.2
$\Delta r(AX)^a$, pm	2.2	2.8	1.8	1.4	2.3	4.0	2.0
$-\Delta E^b$	49.5	63.1	95.8	98.4	38.9	46.1	111.5
μ, D	4.39		8.74		3.90		10.16
$\Delta\mu^c$, D	0.97		0.91		1.13		1.14

[a] $A = H, Li; \Delta r = r_{complex} - r_{monomer}$
[b] in kJ/mol, using geometry optimized at indicated level of theory (SCF or MP2)
     vector sum of dipoles of isolated monomers

row of the table, labeled $\Delta r(AX)$, contains the stretching of this bond which results from formation of the complex. This stretch is about 2.2 pm for both H-bonded complexes at the SCF level but increases substantially when correlation is included. At the MP2 level, the HF bond elongates by 2.8 pm upon complexation with NH_3 while the stretch of the HCl analogue is 4 pm. Correlation produces an opposite effect on the Li bond, *diminishing* the LiF stretch by 0.4 pm.

The interaction energies reported in the fourth row of Table 19 are considerably larger for the Li-bonded complexes than for their H-bonded analogues. It is interesting also that whereas in the case of the H-bonds, $H_3N \cdots HCl$ is less tightly bound than is $H_3N \cdots HF$, the Li bonds follow an opposite pattern wherein the interaction is stronger in $H_3N \cdots LiCl$ than in $H_3N \cdots LiF$. Although correlation strengthens both types of interactions to some extent, it clearly plays a more important role in H-bonds.

The trends observed in the dipole moments of the various complexes are surprisingly similar to the interaction energies. The Li-bonded systems contain much higher moments than their H-bonded analogues, larger by a factor of perhaps 2 or more. In the case of Li bonds, substituting LiF by LiCl raises the moment while an equivalent replacement of HF by HCl leads to a reduction in the moment. The last row of Table 19 reports the enhancement of the dipole moment caused by formation of the complex. Specifically, it lists the difference between the moment calculated for the complex and that obtained by summing together in vector fashion the moments of the individual isolated monomers. The dipole moment enhancement does not seem to indicate any fundamental difference between H-bonds and Li-bonds; one sees only a larger value for complexes containing ACl as compared to AF.

The results concerning the structure and energetics of these two types of bonds may be summarized as follows. Li-bonds are substantially stronger than their H-bonded counterparts. This greater strength is due in large part to the much higher dipole moments of the LiX monomers, as compared to HX. This higher moment adds to the electrostatic energy which makes up the bulk of the total interaction energy in Li bonds. It is their largely electrostatic nature which seems to make Li bonds fairly insensitive to correlation effects. In contrast, H bonds contain substantial contributions from other components such as dispersion energy. The latter term is introduced by correlation effects and it is largely due to its influence that the H bond is significantly strengthened by electron correlation, i.e. increased interaction energy, shorter $R(N \cdots X)$, larger $r(HX)$ stretch.

4.2 Vibrational Spectra

A later set of calculations by Szczesniak et al. [191] extended the comparison of H and Li bonds to the vibrational frequencies and intensities of $H_3N \cdots HCl$ and $H_3N \cdots LiCl$. The changes induced in the internal vibrations of each monomer by formation of these complexes are reported in Table 20. This study computed the frequencies and intensities at both the SCF and correlated MP2 levels; the relevant data are supplied for each.

The trends in the SCF data for the $H_3N \cdots HCl$ system are quite similar to those described in an earlier section for the very similar $H_3N \cdots HF$ complex. In either case, the stretching frequencies of NH_3 are very little affected by complexation whereas the intensities increase markedly. In contrast, formation of a Li bond produces notable red shifts of these stretches as well as strong intensifications.

Table 20. Changes in frequency[a] and intensity[b] in subunits caused by complexation[c] (from [191])

		$H_3N \cdots HCl$		$H_3N \cdots LiCl$	
		H_3N			
		v	A	v	A
stretch	SCF	0	27	-13	173
	MP2	-3	20	-10	37
stretch (2)	SCF	0	15	-26	111
	MP2	-3	36	-26	60
bend	SCF	74	0.9	176	2.6
	MP2	71	0.7	131	1.1
bend (2)	SCF	-7	1.3	-12	3.7
	MP2	-14	1.6	-15	1.6
		ACl			
	SCF	-424	9	73	2.5
	MP2	-898	50	93	0.9

[a] $v_{complex} - v_{monomer}$, cm^{-1}
[b] $A_{complex}/A_{monomer}$
[c] Degeneracy of mode indicated in parentheses

The first bending frequency is substantially blue-shifted by H-bonding; this shift is approximately doubled in the Li-bonded complex. Whereas the intensity of this mode is weakened in $H_3N \cdots HCl$, the opposite effect of greater intensity is observed in $H_3N \cdots LiCl$. A small red shift occurs in the second bending mode of NH_3 in either case, coupled to a small intensity enhancement.

It is evident from comparison of successive rows of Table 20 that most of the frequency shifts are not significantly changed when electron correlation is included directly. A major exception to this rule is associated with the HCl stretch: the MP2 frequency shift is approx. twice that calculated at the SCF level. The red shift of the HCl stretch in $H_3N \cdots HCl$, measured in Ar matrix, is 1517 cm^{-1}, considerably higher than our correlated shift [192]. A major source of this discrepancy is no doubt the ability of the matrix to stabilize a charge-separated species such as $H_3NH^+ \cdots Cl^-$. The proton is thus free to move further away from the Cl atom in the matrix than in the gas phase, thereby lowering the stretching frequency. The effect of the matrix is emphasized by prior work on $(CH_3)_2O \cdots HCl$ where the frequency shift was found to be 1.8 times larger in an inert matrix as compared to the gas phase [180, 193–196].

It is probably in the frequency shift of the HCl or LiCl subunits that the most dramatic distinction between H and Li bonds appears. In contrast to the very strong red shift of the HCl stretch of at least several hundred wave numbers, the LiCl stretch is shifted in the opposite direction, i.e. toward higher frequencies, and by a much lesser amount. This contrast is not limited to SCF level calculations but is confirmed by inclusion of electron correlation and agrees with prior experimental observations [197]. A second major distinction concerns the intensification of this stretch upon formation of a complex. While the HCl stretch strengthens by a factor of perhaps 50 or so, there is little change calculated for the LiCl stretch, again conforming to experimental measurements [197].

The intensity enhancements of the NH_3 modes associated with formation of a H-bond are affected to only a small degree when correlation is included. In the case of the HCl stretch, however, the MP2 intensification is several times larger than SCF calculations would suggest. In contrast to the lack of sensitivity of the intensities of the NH_3 vibrations to correlation in $H_3N \cdots HCl$, the MP2 intensity enhancements are generally a factor of two or so smaller than the SCF values in $H_3N \cdots LiCl$.

The calculated frequencies and intensities for the intermolecular modes are exhibited in Table 21. The higher frequencies obtained for the Li-bond stretch as compared to the H-bond are consistent with the stronger nature of the former inter-action. Note also that correlation induces a significant increase in the H-bond stretching frequency, just as the MP2 interaction energy is greater than the SCF value; the Li-bond parameters are relatively unaffected by correlation. Another indication of the stronger character of the Li bond is the larger frequencies for the first bending mode in Table 21, corresponding to a bend of the NH_3 molecule. The pattern of SCF vs. MP2 frequencies for this bend is also reminiscent of the trends for the H-bond stretch. The final bending motion refers to the HCl or LiCl molecule. The frequencies in the two systems are vastly different, due in large part to the much larger mass of Li as compared to H.

The intensity of the stretching mode appears to be fairly low for both systems. Correlation has a strong damping effect on all intensities for $H_3N \cdots LiCl$ whereas SCF and MP2 intensities are not drastically different for the H-bonded system

Table 21. Frequency (cm^{-1}) and intensity (km/mol) of intermolecular modes in $H_3N \cdots ACl^a$, A = H, Li (from Ref. [191])

		$H_3N \cdots HCl$		$H_3N \cdots LiCl$	
		v	A	v	A
stretch	SCF	173	10	248	24
	MP2	200	43	238	10
bend (2)	SCF	239	46	525	550
	MP2	276	44	499	208
bend (2)	SCF	700	150	56	80
	MP2	912	114	70	22

^{a} Degeneracy of mode indicated in parentheses; intensities have been premultiplied by this factor

(with the exception of the stretch where the SCF value is magnified by a factor of four).

Szczesniak et al. [191] have made use of atomic polar tensors to provide insights into the reasons for the above observations concerning the intensities. As discussed earlier in the context of H-bonds, the strong intensification of the HCl stretch is due in large part to the fact that the H atom becomes more positively charged as it moves away from the Cl, due to the reduced ability of the electron density to move along with it once the H-bonded complex has been formed. This effect is magnified by correlation effects, explaining the greater intensification of the HCl stretch at the MP2 level.

The situation for the Li bond is quite different. Considering first the isolated LiCl monomer, this molecule is highly polar with Li bearing nearly a full positive charge. A stretch of the bond further increases this charge, which translates into a fairly sizeable charge flux term already in the monomer. Complexation with NH_3 increases the charge flux, but to only a fraction of the amount observed in HCl. Moreover, some of the electron density extracted from the NH_3 winds up on the Li atom, lowering its positive charge. The latter diminished charge produces a damping effect upon the intensity of the LiCl stretch, counteracting any increase arising from the charge flux; the net result is that complexation leads to no substantial change in intensity.

Another fundamental distinction between H and Li bonds is associated with the charge distributions occasioned by formation of the complex. Szczesniak et al. [191] explored these redistributions by way of "spectroscopic" atomic charges which act to mimic experimental quantities such as vibrational intensities. They noted that whereas most of the charge extracted from the ammonia in $H_3N \cdots HCl$ was picked up by the Cl atom, it is the Li atom in $H_3N \cdots LiCl$ that is the ultimate sink of electron density. The authors were able to discern a relationship also between the total charge transferred from NH_3 to the electron acceptor and the calculated intensity of the intermolecular stretch. This finding conforms to the requirement of a changing molecular dipole moment in order to lend intensity to this vibration.

5 Multiple H-Bonds

It had been well established early on that sequential trimers in which the central molecule acts as proton acceptor in one H-bond and as donor in the other are energetically favored over other types in which the central molecule acts as donor in both bonds or as acceptor in both. This finding had been simply rationalized on the basis of redistributions of electron density which accompany each interaction [8]. A second question concerns the relative energies of open chains as compared to cyclic systems in which the two ends of the chains interact to form an additional H-bond. Formation of the last bond occurs at the expense of angular distortions of the remaining bonds, so the answer to this question is not entirely clear a priori.

5.1 (HX)$_3$

The trimer of HF has served as a general model of the cooperativity engendered by two H-bonds. Early ab initio calculations had not been able to provide an unequivocal answer to the relative stabilities of cyclic vs. open-chain (HF)$_3$ due to certain shortcomings such as small basis sets or incomplete geometry optimizations [198–200]. Karpfen et al. [93] used gradient optimization techniques and relatively minor symmetry restrictions, together with polarized basis sets, to try to settle the question. They found the cyclic structure more stable than the open trimer by some 9 to 14 kJ/mol. The importance of nonadditive effects was clear in that the total complexation energy of the open trimer (containing two H-bonds) was larger by 5 or 6 kJ/mol than twice the H-bond energy of a dimer.

As illustrated by the $-\Delta E(HF)_3$ entries in the first two rows of Table 22, the computed complexation energy of the cyclic trimer (from three monomers) is in the neighborhood of 50 or 60 kJ/mol. This structure benefits from the same sort of head-to-tail cooperativity which is present in the open sequential trimer. On the other hand, each of the three bonds must be distorted so as to form the ring. If the former cooperativity is stronger than the deformation effect, one would expect the total complexation energy to exceed the total of three individual H-bond energies (within the context of an optimized dimer). The difference between these two quantities is reported as Δ in the fourth column of Table 22. With either basis set, this quantity is negative. Thus, Karpfen et al. found that the energy required to distort each H-bond so as to form a cyclic trimer is not quite compensated by the cooperativity resulting from the sequential nature of these bonds.

In the following year, Gaw et al. [95] reinvestigated the cyclic trimer of HF, focussing their attention upon the vibrational frequencies. They found that the internal stretches of the HF subunits are redshifted by between 200 and 300 cm^{-1}, in comparison to the much smaller shifts of less than 100 cm^{-1} in the dimer. In contrast to the earlier result, the cooperativity effect within the cyclic trimer seems to be more important than the strain energy imposed on each H-bond, as indicated by the positive entries for Δ in Table 22. These investigators also found that the energy barrier for a synchronous exchange of all three protons is surprisingly low, only 123 kJ/mol.

Table 22. Calculated binding energy of $(HF)_3$ and $(HCl)_3$ including comparison of cooperativity and strain energies (in kJ/mol)

Basis set	Level	$-\Delta E(HX)_3$[a]	Δ[b]	Ref.
		$(HF)_3$		
[531/31]	SCF	60.2	−1.3	[93]
[642/41]	SCF	48.5	−1.7	[93]
DZ	SCF	77.3	4.8	[95]
DZP	SCF	62.4	3.4	[95]
TZP	SCF	54.1	2.3	[201]
TZP	ACCD	58.6	1.3	[201]
+VP^S	SCF[c]	49.4	−1.0	[103]
+VP^S	MP2[c]	52.3	0.9	[103]
		$(HCl)_3$		
+VP^S	SCF[c]	12.2	4.8	[103]
+VP^S	MP2[c]	15.1	2.0	[103]

[a] $\Delta E(HX)_3 = E(HX)_3 - 3E(HX)$
[b] $\Delta = -\{\Delta E(HX)_3 - 3[E(HX)_2 - 2E(HX)]\}$; Δ is defined so as to be positive if the binding energy of the trimer is greater than 3 times that of an optimized dimer
[c] Corrected for BSSE

Liu et al. [201] improved upon the earlier work by including electron correlation into evaluation of the relative energetics (although not into the geometry optimizations). They found indications that an open chain type of structure would collapse into the cyclic geometry with little or no energy barrier. As may be seen by the appropriate two rows of Table 22, correlation does not appear to significantly affect the interaction energies nor the conclusion that the positive cooperativity of the three sequential H-bonds is only slightly more important than the distortion energy associated with the bond strain. Their calculations of the tetramer demonstrated a sharp increase in the average H-bond energy, indicative of the smaller bond strain required to form the tetramer.

Scuseria and Schaefer [202a] followed up some earlier work by attempting to determine whether the open trimer of HF represents a true minimum in the potential energy surface. Within the context of their double-ζ basis set, it was found that the *cis* open trimer is indeed a minimum (Hessian matrix has all positive eigenvalues), whereas the *trans* structure is a transition state. However, the situation is decidedly different when polarization functions are added to the basis set wherein the *trans* geometry becomes a transition state and the *cis* type does not correspond to a stationary point of any order. Nor does relaxing the fully planar restraint lead to a *cis* stationary point. Vibrational frequencies and intensities were reported for the stationary points.

More recently, Latajka and Scheiner [103] have examined the cyclic trimers of HCl as well as HF for purposes of comparison. The effects of electron correlation upon the geometries were considered using polarized basis sets designed to minimize basis set superposition errors. Correlation was found to reduce the F ·· F distances in $(HF)_3$ by 2 pm relative to the SCF bond lengths. In considering nonadditivity in the trimers, the authors were careful to remove BSSE which is capable of severely contaminating the results and leading to incorrect conclusions. As may be seen in Table 22, a total complexation energy of 52 kJ/mol was computed for $(HF)_3$, only 3 kJ/mol of which is associated with correlation energy. In comparison, $(HCl)_3$ is much more weakly bound (15 kJ/mol). However, correlation plays a considerably stronger role here, contributing 19% of the total binding energy. As in the earlier work, the total complexation energy of $(HF)_3$ is comparable to three times the dimerization energy. We note from Table 22 that in $(HCl)_3$, the cooperativity exceeds the strain energy by 2 to 5 kJ/mol.

Latajka and Scheiner analyzed their results also on the basis of a more rigorous definition of many-body effects in which relative geometries are frozen in the cyclic structure; moreover, their prescription deals explicitly with BSSE [103]. They found that the three-body term is not negligible in either trimer. This term is approximately half as large as the two-body interaction energy in $(HF)_3$ while about $^1/_4$ as large in $(HCl)_3$. The three-body term appears to be adequately described at the SCF level since correlation does not change it appreciably.

Vibrational frequencies and intensities were calculated for the cyclic trimers of both HF and HCl [103]. As noted earlier by Gaw et al. [95], the red shifts of the HF stretchers are much larger in the trimer than in the dimer. This trend extends to the HCl analogues although the numerical values of the shifts are uniformly smaller. The intensities of the HX stretches, too, exhibit a marked increase in the trimer as compared to the dimer. Similar comparisons extend to the intermolecular modes as well in that both the frequencies and intensities are considerably greater in the trimers. Kolenbrander et al. [202b] have provided evidence of separability of the in-plane and out-of-plane molecular vibrations in $(HF)_3$. They estimate a barrier of at least 30 kJ/mol separates two equivalent conformations of this trimer, some 10 times higher than the interconversion barrier in $(HF)_2$.

5.2 $(H_2O)_3$

Experimental work [203–205] has demonstrated that the trimer of water, like those of the hydrogen halides, is cyclic. In particular, Engdahl and Nelander [203] have made infrared spectral measurements in Ar and Kr matrices which indicate the three water molecules are equivalent. Honegger and Leutwyler [206] have recently supplemented these experimental measurements with ab initio data in which the geometry of $(H_2O)_3$ was optimized at the SCF level using both the 4-31G and 6-31G* basis sets. Rather than a pure equilateral triangle, they find a slight distortion in that the distance between one pair of molecules is about 1 pm longer than the other two pairwise distances. These three $O \cdots O$ distances are approximately 10 pm shorter than in the dimer. The three bridging protons are all within 3° of the oxygen plane but deviate from the corresponding $O \cdots O$ axis by about 20°. Upon going from the dimer to trimer, the r(OH) bond of the bridging protons stretches by an additional 0.4 pm. The total binding energy of the trimer is calculated with the larger 6-31G* basis set to be 48.1 kJ/mol, 4.9 kJ/mol greater than three times the binding energy computed for the optimized dimer. Thus, the effects of cooperativity in this trimer are greater than the distortion energies imposed upon each H-bond. As noted in the previous section for $(HX)_3$, the red shifts of the O—H stretching frequencies are considerably larger in $(H_2O)_3$ than in the dimer.

A number of workers have tried to understand the cooperativity in larger clusters, ice in particular, on the basis of ab initio calculations. Early calculations by Clementi et al. [207a] suggested that nonadditivity effects in water trimers might be small. Newton [121] later performed calculations on the types of dimer and trimer arrangements likely to occur in ice and then used a statistical averaging procedure. His derived potential function, including electron correlation, a dispersion term, and correction for BSSE, accounted for over half of the O ·· O compression observed on going from $(H_2O)_2$ to ice. Yoon et al. [120] considered a similar problem and incorporated three-body terms and variation in the OH distance as well as in O ·· O; a quite large basis set was used. They concluded that a good deal of the O ·· O shrinkage is a result of three-body interactions. Koehler et al. have more recently considered the tetramer of water [207b] and found the 4-body term to be much smaller than the 3-body term. They further confirmed the finding of Latajka and Scheiner [103] that cooperativity is not much influenced by correlation.

5.3 $H_3Z \cdot\cdot HX \cdot\cdot HX$

The cooperativity in heterotrimers combining NH_3 with a pair of HF or HCl molecules has been considered recently by Hinchliffe [20]. Unfortunately, these calculations were limited to the SCF level, did not allow for relaxation of the internal geometries of NH_3 or PH_3 within the complex, and the entire system was assumed to be linear, i.e. C_{3v} symmetry; the potential energy surface was explored no further. His analysis concentrated on Mulliken populations and revealed substantial reorganization of charge density.

Kurnig et al. [21] optimized the geometries of the complexes $H_3Z \cdot\cdot HF \cdot\cdot HF$, $Z = N, P$ using a 6-31G** basis set. Only one minimum was identified in either surface, corresponding to a nearly cyclic type of structure, illustrated in Fig. 5. An instructive means of analyzing this geometry is via comparison with the two dimeric complexes of which it is composed. That is, the geometry of the $H_3Z \cdot\cdot HF$ portion of the complex is compared with the same system in the absence of the terminal HF molecule. The HF ·· HF segment is similarly compared with the optimized geometry of $(HF)_2$.

The nonlinearity of the complex is not surprising if one considers that the HF ·· HF dimer is itself quite bent. That is, the optimized $H_iF_iF_0$ angle in $(HF)_2$ is 102° with the 6-31G** basis set. This same angle is only slightly smaller, 84° in

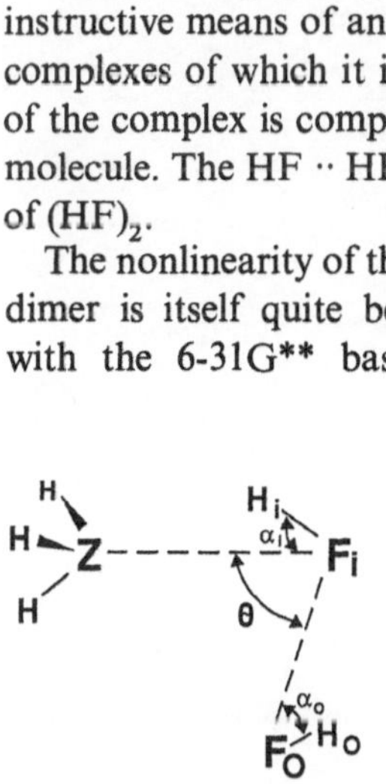

Fig. 5. Structure of $H_3Z \cdot\cdot HF \cdot\cdot HF$. α angles measure deviation of indicated hydrogens from axes connecting heavy atoms

$H_3N \cdots HF \cdots HF$ and $88°$ in $H_3P \cdots HF \cdots HF$. This reduction probably reflects a small amount of attractive interaction between the H_3Z molecule and the terminal HF unit. This bending force acts to pull H_i off of the $Z \cdots F_i$ axis, by $12°$ when $Z = N$ and by $21°$ when $Z = P$. The local C_3 symmetry axis of H_3Z likewise deviates from the $Z \cdots F_i$ axis as shown.

A number of features of the geometries provide clear indications of the cooperativity between the two H-bonds in this complex. First of all, $H_3Z \cdots HF$ acts as a stronger base than a single HF molecule in that the terminal HF bond is stretched much more in the full trimer than in $(HF)_2$. Similarly, the central HF bond in the trimer is stretched twice as far as in $H_3Z \cdots HF$, indicating that prior addition of a second HF molecule to HF enhances its proton-donating ability. The distances between heavy atoms, too, reflect the cooperativity within the complex. Combination of $(HF)_2$ with ZH_3 reduces the $F \cdots F$ distance by 12 pm ($Z = N$) or 6 pm ($Z = P$). A contraction of the $Z \cdots F$ distance in $H_3Z \cdots HF$ of between 13 and 17 pm is likewise observed upon addition of the terminal HF molecule.

Kurnig et al. [21] probed the energetic manifestations of cooperativity as well. The relevant data are summarized in Table 23 where the first several rows contain the calculated energies of each reaction. The total complexation energy of the $H_3N \cdots HF \cdots HF$ system is 98.2 kJ/mol, nearly twice as large as the P analogue. The cooperativity reported in the next row of the table refers to the greater exothermicity of adding an HF molecule to a preformed $H_3Z \cdots HF$ dimer, as compared to adding one HF to another. By Hess's law, this value is equivalent to the difference between adding H_3Z to $(HF)_2$ and adding H_3Z to HF. In either case, the cooperativity is equal to 23.8 kJ/mol for $H_3N \cdots HF \cdots HF$, amounting to 1/4 of its total complexation energy. The proportion is quite similar in $H_3P \cdots HF \cdots HF$ even though the absolute values are smaller.

Another measure of the cooperativity is the three-body term which does not allow the internal geometries of the dimers or monomers to differ from their structure in the trimer. Moreover, this term accounts explicitly for the mutual interaction

Table 23. Interaction energies and measures of cooperativity (in kJ/mol) of $H_3Z \cdots HF \cdots HF$ (from [21])

	Z = N	Z = P
route 1		
1a: $H_3Z + HF \rightarrow H_3Z \cdots HF$	−49.4	−17.3
1b: $H_3Z \cdots HF + HF \rightarrow H_3Z \cdots HF \cdots HF$	−48.8	−36.7
route 2		
2a: $HF + HF \rightarrow HF \cdots HF$	−25.0	−25.0
2b: $H_3Z + HF \cdots HF \rightarrow H_3Z \cdots HF \cdots HF$	−73.3	−29.0
total[a]	−98.2	−54.0
coop[b]	−23.8	−11.8
3-body[c]	−17.1	− 5.9

[a] 1a + 1b = 2a + 2b
[b] 1b − 2a = 2b − 1a
[c] Using internal geometries of the trimer throughout

between the H_3Z and terminal HF subunits. From the last row of Table 23, it is evident that whereas the 3-body term is somewhat smaller than the preceding quantity, it also supports the finding of a fair amount of cooperativity in both trimers.

5.3.1 Spectroscopic Features

One of the more common means of investigating complexes of the type discussed here is via vibrational spectroscopy. Previous work had shown that the frequency shifts of the HX subunits are larger in trimers than in dimers [208, 209], providing some evidence of cooperativity but unable to relate these frequencies to structural or energetic quantities. Kurnig et al. [21] were able to obtain good linear correlations between the experimentally observed frequencies and a) the calculated H—F bond stretches (complex vs. monomer) and b) the calculated harmonic stretching force constants for these bonds. Such correlations open the door to future estimates of geometrical details based upon spectroscopic measurements.

Workers from the same laboratory later extended the previous calculations by carrying out full normal vibrational mode analyses of these two complexes, including frequencies and intensities [156]. Table 24 exhibits the changes undergone by the HF stretching mode of the central HF molecule as a result of formation of various complexes. The first row of the table reports that the frequency of this stretch is red shifted by 432 cm^{-1} when HF forms a H-bonded complex with H_3N and by 134 when combined with H_3P. If a second HF molecule is then added to the binary complex, the frequency undergoes a further red shift approximately equal to the first. The magnitude of this second shift is particularly interesting as the pertinent molecule is acting as proton acceptor rather than donor to the new bond which is being formed.

Dimerization of HF has only a small effect on the frequency of the proton-accepting subunit, as shown in the third row of Table 24. On the other hand, this dimerization does make the frequency inordinately susceptible to a red shift when it forms a H-bond to H_3Z. Comparison of the last with the first row of Table 24 indicates that the frequency shift in the HF dimer is nearly twice that occurring in the monomer, a

Table 24. Frequency shifts and intensity enhancements engendered in HF stretch of central subunit of $H_3Z \cdots HF \cdots HF$ (from [156])

	$\Delta\nu$, cm^{-1}		A_p/A_r [a]	
	Z = N	Z = P	Z = N	Z = P
route 1				
1a: $H_3Z + HF \rightarrow H_3Z \cdots HF$	−432	−134	7.1	4.0
1b: $H_3Z \cdots HF + HF \rightarrow H_3Z \cdots HF \cdots HF$	−389	−143	1.0	1.1
route 2				
2a: $HF + HF \rightarrow HF \cdots HF$	−41	−41	1.4	1.4
2b: $H_3Z + HF \cdots HF \rightarrow H_3Z \cdots HF \cdots HF$	−780	−236	5.2	3.1

[a] Intensity in product divided by that in reactant

dramatic piece of evidence for cooperativity within the context of the vibrational frequencies. It is interesting also that the trends discussed above are consistent with the patterns noted earlier for the optimized H—F bond lengths.

The last two columns of Table 24 report analogous data for the intensities of these same modes. As expected, complexation with H_3Z produces a marked strengthening of the HF stretch, seven-fold for H_3N and four-fold for H_3P. Adding a second HF molecule to $H_3Z \cdots HF$ has no significant effect upon this intensity, as compared to the frequency which is changed by several hundred wave numbers. A second large intensification is observed in the last row which corresponds to addition of H_3Z to HF $\cdots$ HF. However, unlike the frequencies where the shift is greater if H_3Z is added to HF $\cdots$ HF than to the monomer, the opposite is found in the intensities. That is, the intensity of the HF stretch is strengthened less if the molecule is already acting as proton acceptor to a second HF molecule. Hence, whereas a positive cooperativity is associated with the frequency shifts, the intensities are subject to a negative cooperativity.

Another good indicator of the cooperativity is the set of intermolecular modes, the relevant calculated data for which are displayed in Table 25. Beginning with the frequencies, the first row of the table indicates that addition of a second HF molecule to $H_3N \cdots HF$ increases the N $\cdots$ F stretching frequency from 240 to 290 cm^{-1}, an increment of 21%. A similar increase of 24% is observed in the P analogue systems. The bending frequencies of the $H_3N \cdots HF$ H-bond also show a clear increase upon addition of the second HF. Some of these increments are quite sizable. For example, the $v_{\beta 1}$ frequency of $H_3P \cdots HF$, corresponding to a H_3P bend, more than doubles. The intensities, too, are substantially enhanced when the second HF molecule is added to $H_3Z \cdots HF$. The only exception is the $v_{\beta 2}$ mode of $H_3P \cdots HF \cdots HF$.

5.4 Oligomers of HCN

Experimental work of various sorts indicates that the dimer, trimer, and probably tetramer of HCN are all fully linear, consistent with what might be expected in terms of the collinearity of each HCN molecule and the N lone pair [210–216]. (There is

Table 25. Frequencies and intensities of intermolecular vibrational modes in $H_3Z \cdots HF \cdots HF$[a] (from [156])

	v, cm^{-1}				A, km/mol			
	$H_3Z \cdots HF$		$H_3Z \cdots HF \cdots HF$		$H_3Z \cdots HF$		$H_3Z \cdots HF \cdots HF$	
	Z = N	Z = P	Z = N	Z = P	Z = N	Z = P	Z = N	Z = P
v_σ	240	113	290	140	3	1	13	5
v_b	876	467	1162	800	208	134	264	247
v_t	877	464	1024	603	208	134	245	275
$v_{\beta 1}$	236	108	417	247	10	6	49	10
$v_{\beta 2}$	238	117	307	175	10	6	13	0.1

[a] v_σ refers to Z $\cdots$ F stretch, v_b and v_t to bends of proton donor, $v_{\beta 1}$ and $v_{\beta 2}$ to bends of the acceptor

intriguing evidence that $(HCN)_3$ coexists as both a linear and cyclic complex [14, 214, 216].) The linear nature of these structures permits the analysis to be free of complications involving subtle (or not-so-subtle) changes in angle as one progresses to larger clusters.

While the dimer has been studied extensively by ab initio calculations at a reasonable level [174, 217, 218], the higher oligomers had been essentially ignored until the recent work of Kofranek et al. [219, 220] which made use of basis sets ranging from double-ζ up to near Hartree-Fock quality for the valence shells and including two sets of polarization functions. Geometries were fully optimized at the SCF level. The following results are concerned with fully linear geometries, consistent with experimental observation. The data presented in Table 26 were calculated by Kofranek et al. with their double-ζ basis set I which was used for oligomers up to n = 5. Additional calculations with the larger basis sets were limited to smaller oligomers but the trends were very much like those observed with basis set I.

The table focuses on the changes induced in the properties of the cluster upon each addition of another HCN. For example, upon formation of the dimer, the C—H bond of the proton-donating HCN was stretched by 0.61 pm. A further stretch of only 0.16 pm accompanies addition of a third HCN molecule, followed by much smaller changes as the chain grows further. The intermolecular $r(H \cdots N)$ distances also exhibit a cooperativity in that each successive addition of another HCN molecule produces a further reduction in the H-bond length involving the terminal proton donor molecule. As reported in Table 26, these H-bond contractions are: 5, 1, 0.5 pm as n varies from 3 to 5. The 5 pm contraction on going from dimer to trimer has recently been confirmed by pulsed-nozzle Fourier transform measurements [215]. The energetics follow a very similar pattern of cooperativity. The values listed for ΔE correspond to the increase in total complexation energy arising from each new molecule. The dimerization energy is 23.4 kJ/mol; elongation of the chain produces progressively larger increases, up to 31.7 kJ/mol on going from n = 4 to n = 5.

Kofranek et al. [219] also examined cyclic geometries of the trimer and tetramer of HCN. In contrast to $(HF)_3$ where the cyclic geometry is most stable, the linear trimer of HCN is preferred by 4.6 kJ/mol with the larger basis set. Nevertheless, the authors noted that the cyclic structure represents a true minimum on the potential energy surface of $(HCN)_3$ with all vibrational frequencies real. (This finding is

Table 26. Effects of adding additional HCN molecule to growing linear cluster (from [219])

n	r(C—H), pm	r(H $\cdots$ N), pm	$-\Delta E$, kJ/mol	v(C—H), cm^{-1}	v(H $\cdots$ N), cm^{-1}
2	0.61[a]	—[b]	23.4	−80[c]	—[d]
3	0.16	−4.76	29.0	−30	43
4	0.05	−1.34	30.8	−28	23
5	0.03	−0.49	31.7	−16	14

[a] 114.03 pm in monomer
[b] 223.25 pm in dimer
[c] 3654 cm^{-1} in monomer
[d] This frequency is equal to 122 cm^{-1} in $(HCN)_2$

encouraging in light of recent experimental data suggesting that both linear and cyclic trimers are present in gas-phase mixtures [14].) The situation in the tetramer is different in that the cyclic and linear geometries are nearly equal in energy (only the double-ζ basis set was used).

Kofranek et al. [219] also performed a complete vibrational analysis so as to obtain the frequencies of all modes within the harmonic approximation. The stretching frequency of the HCN monomer was calculated to be 3654 cm^{-1} with the double-ζ basis set. Formation of the dimer lowers the frequency of the proton-donor molecule by 80 cm^{-1}, as indicated in the first row of Table 26. This frequency drops further by 30 cm^{-1} when a third molecule is added (to the other end of the chain); subsequent red shifts are 28 and 16 cm^{-1} as the chain continues to grow. The intermolecular H $\cdot\cdot$ N stretching frequency is fairly low, only 122 cm^{-1}, for $(HCN)_2$. Commensurate with the positive cooperativity exhibited and the increasing complexation energy, the frequency of this stretch increases as the chain elongates. The shift is 43 cm^{-1} in the transition from dimer to trimer, with smaller but significant increases observed for subsequent additions of HCN.

Kofranek et al. later enlarged the scope of their studies of $(HCN)_n$ by including electron correlation explicitly and obtaining information about the vibrational intensities as well [220]. Three different basis sets were used, all of at least polarized triple-ζ quality, to examine the linear monomer, dimer, and trimer. The authors found that electron correlation has a negligible effect on the C—H bond elongations that accompany dimerization and trimerization; the contraction of the H $\cdot\cdot$ N H-bond length on going from dimer to trimer is 5 pm at the SCF level, as compared to the slightly smaller 4 pm at the CPF level.

The shifts in the C—H stretching frequency noted above at the SCF level are not affected appreciably when correlation is included. Dimerization increases the intensity of the C—H stretch of the proton-donating molecule by a factor of 4 or 5. In the trimer, the C—H stretches of both proton-donating molecules combine together. The intensity of the lower frequency mode (the symmetric combination) is enhanced by a factor of 10 with respect to the monomer while that of the antisymmetric combination vanishes almost completely. The C—H stretch of the terminal molecule which does not act as proton donor has the highest frequency; its intensity is not much affected by complexation. The intensity of the H $\cdot\cdot$ N stretch in the dimer is calculated to be quite low, 1.3 km/mol at the SCF level and 2.8 km/mol when correlation is included. The two H $\cdot\cdot$ N stretches in the trimer form symmetric and antisymmetric combinations. The former has zero intensity and the other is quite weak as well with an intensity of 2 km/mol at either level.

Kofranek et al. conclude that while electron correlation must be included for quantitative agreement with experiment, the overall changes accompanying oligomerization are well described at the SCF level. Whereas the effects of cooperativity upon the geometries and energetics are not particularly strong, there is nonetheless evidence of dramatic changes in the vibrational spectra.

Anex et al. [216] have recently completed a combined theoretical and experimental probe of HCN oligomers, focusing their attention on the vibrational frequencies. Their ab initio calculations, carried out with a 6-31G** basis set, agree well with the earlier data of Kofranek et al. They too find that correlation has a fairly small effect on frequency shifts although an improvement is noted in the absolute frequencies. The authors conclude that while the vibration of the terminal "free" C—H bond remains relatively pure in the trimer, there is considerable mixing between the stretches of the two molecules acting as proton donors. In comparison with a number of other theoretical treatments of vibrational analysis, the ab initio approach was most successful.

6 Concluding Remarks

Our picture of the nature of the hydrogen bond has evolved significantly over the last several years. The wealth of gas-phase experimental information has added a great deal to our compendium of knowledge concerning the structure and dynamics of H-bonded systems. An important window into their internal electronic structure has been opened by spectral data, both in vacuo and in rare-gas matrices. This work has provided a stimulus and useful yardstick for ab initio calculations which can, in turn, supply complementary information and assist in interpretation of the experimental data.

A simple, yet powerful, framework for understanding the overall structure of each complex arises from the electrostatic interaction between the two subunits. Methods relying almost exclusively upon the latter term have been demonstrated to correctly predict the geometry of a wide range of molecular complexes. The success of this picture rests on the fact that the electrostatic component of the total interaction energy is highly anisotropic, in contrast to the much lesser dependence of the other components upon intermolecular orientation. While alternative pictures, in which other components play a more important role, can also be reconciled with many experimental observations, the simplicity and solid formal underpinning of the electrostatic viewpoint makes it the method of choice for "back-of-the-envelope" predictions of structures in simple as well as larger and more complicated systems.

Of course, the electrostatic interaction is only one of many components of the interaction energy. Accurate reproduction of each of these components is essential to a reliable calculation of the energetics as well as to the details of the geometry. High-quality ab initio calculations seem capable of reproducing these properties to high precision. The calculated results conform with the conventional wisdom that the binding energy is intimately related to the strength of both the proton donor and the base. In the case of weaker interactions, the geometry and other properties are such that the complex would perhaps more appropriately be designated as van der Waals type rather than as H-bonding.

The most exciting new arena for ab initio study of molecular complexes is perhaps that of vibrational frequencies and intensities. Accurate calculation of these quantities requires a much higher level of theory than does the structure and energetics. Frequencies are generally overestimated at the SCF level, even with large basis sets. Electron correlation provides much more accurate results, even if one retains the harmonic approximation. A common alternative is "scaling" the SCF data; however, this approach must be exercised with caution as errors are not necessarily uniform. Lower levels of theory are useful if one is interested in frequency shifts resulting from a perturbation such as formation of a H-bond rather than in absolute frequencies.

The errors encountered with a given level of theory in calculation of vibrational intensities are considerably more profound than those in the frequencies. Meaningless results are common unless the basis set is of double-ζ quality and contains polarization functions. The double-harmonic approximation is another source of error. Nonetheless, the effects of H-bonding upon the intensity of each mode can be profitably studied within this framework. Analysis of these intensities sheds light upon the electronic density perturbations caused by formation of the bond.

Much of the recent progress of ab initio methods is due to the current ability to study the effects of electron correlation on more or less a routine basis. When used in conjunction with large polarized basis sets and correction of basis set superposition error, a new level of accuracy is now possible in treating these systems. Perhaps of greater importance, calculations of this level provide a clear physical picture of each of the various components of the interaction. This situation contrasts with much of the earlier work where a reasonable total interaction energy was frequently obtained by cancellation of errors, e.g. exaggeration of one term at the expense of another which is underestimated, or contamination of certain components by superposition error.

It is likely that progress along these lines will continue in the future as computer power improves and as programming becomes more efficient. Some recommendations have recently been made by this group to guide future investigators in their choice of ab initio technique to study H-bonding [88]. In general, Møller-Plesset perturbation theory, even if restricted to second order, appears to capture most of the correlation energy important to such interactions. Basis sets should be chosen to correctly represent the molecular properties of each subsystem, provide an efficient framework for electron correlation, and minimize basis set superposition error (of both primary and secondary type). The "isotropic" (i.e. sp-set or non-polarization) part of the basis can be chosen as a well-tempered expansion of Gaussian functions with well-spaced exponents. Such a set covers a range which includes very compact functions to describe the core well and thereby minimize superposition error, as well as diffuse functions for optimal flexibility. Exponents of polarization functions can be selected from this series. If one is restricted to a small number of such functions, one set may be chosen by maximizing the perturbation correlation energy of each subunit. Maximization of the dipole polarizability of the subunit can be used as a criterion for a second set. When possible, a polarization function of higher value of l (e.g. f-function) should be added; a small exponent which maximizes the molecule's quadrupole polarizability is recommended.

The work described in this review has concentrated upon the equilibrium geometries of each complex. It is frequently the case that improperly balanced theoretical approaches can offer a reasonable description of this particular structure. However, more complete understanding of the nature of the H-bond or fitting of an empirical function for later use in simulations requires the calculation of extended regions of the potential energy surface. It is here that the above recommendations become especially important.

Acknowledgements: The results from this laboratory quoted here were made possible only through the efforts of my collaborators, in particular Drs. Szczesniak and Latajka from the University of Wrocław. I am also grateful to the National Institutes of Health for generous support of this work through projects GM29391 and GM36912.

7 References

1. Pulay P (1987) in: Lawley KP (ed) Ab initio methods in quantum chemistry, vol 2, Wiley, New York, p 241
2. Schlegel HB (1981) in: Csizmadia IG, Daudel R (eds) Computational theoretical organic chemistry, Reidel, Dordrecht, p 129
3. Schlegel HB (1987) in: Lawley KP (ed) Ab initio methods in quantum chemistry, vol 1, Wiley, New York, p 249
4. Wilson S (1984) Electron correlation in molecules. Clarendon, Oxford
5. Carsky P, Urban M (1980) Ab initio calculations. Springer, Berlin Heidelberg New York
6. Møller C, Plesset MS (1934) Phys. Rev. 46: 618
7. Binkley JS, Pople JA (1975) Int. J. Quantum Chem. 9: 229
8. Scheiner S (1983) in: Wyn-Jones E, Gormally J (eds) Aggregation processes in solution. Elsevier Scientific, Amsterdam, p 462
9. Legon AC (1983) Ann. Rev. Phys. Chem. 34: 275
10. Dyke TR (1984) Top. Curr. Chem. 120: 85
11. Peterson KI, Fraser GT, Nelson DD Jr, Klemperer W (1985) in: Bartlett RJ (ed) Comparison of ab initio quantum chemistry with experiment for small molecules. Reidel, New York, p 217
12. Millen DJ (1986) Int. J. Quantum Chem. 29: 191
13. Legon AC, Millen DJ (1987) Acc. Chem. Res. 20: 39
14. Miller RE (1988) Science 240: 447
15. a) Emsley J (1980) Chem. Soc. Rev. 9: 91
    b) Weber A (ed) (1986) Structure and dynamics of weakly bound molecular complexes. D. Reidel Publishing Co., Dordrecht
16. Beyer A, Karpfen A, Schuster P (1984) Top. Curr. Chem. 120: 1
17. Newton, MD (1986) Transactions ACA 22: 1
18. Pople JA (1982) Faraday Discuss. Chem. Soc. 73: 7
19. Hinchliffe A (1985) J. Mol. Struct. (Theochem) 121: 201
20. Hinchliffe A (1983) J. Mol. Struct. (Theochem) 105: 335
21. Kurnig IJ, Szczesniak MM, Scheiner S (1986) J. Phys. Chem. 90: 4253
22. Latajka Z, Scheiner S (1984) J. Chem. Phys. 81: 4014
23. Latajka Z, Scheiner S (1984) J. Chem. Phys. 81: 2713
24. Szczesniak MM, Latajka Z, Scheiner S (1986) J. Mol. Struct. (Theochem) 135: 179
25. Alabart JR, Caballol R (1987) Chem. Phys. Lett. 141: 334
26. Jasien PG, Stevens WJ (1986) Chem. Phys. Lett. 130: 127
27. Brciz A, Karpfen A, Lischka H, Schuster P (1984) Chem. Phys. 89: 337
28. Goodwin EJ, Howard NW, Legon AC (1986) Chem. Phys. Lett. 131: 319
29. Howard NW, Legon AC (1987) J. Chem. Phys. 86: 6722
30. Howard NW, Legon AC (1988) J. Chem. Phys. 88: 4694
31. Legon AC, Willoughby LC (1983) Chem. Phys. 74: 127
32. Legon AC, Willoughby LC (1982) J. Chem. Soc. Commun. 997
33. Willoughby LC, Legon AC (1983) J. Phys. Chem. 87: 2085
34. Hinchliffe A (1984) J. Mol. Struct. (Theochem) 107: 361
35. Szczesniak MM, Scheiner S, Bouteiller Y (1984) J. Chem. Phys. 81: 5024
36. Del Bene J (1988) J. Phys. Chem. 92: 2874
37. Szczesniak MM, Scheiner S (1985) J. Chem. Phys. 83: 1778
38. Amos RD, Gaw JF, Handy NC, Simandiras ED, Somasundram K (1987) Theor. Chim. Acta 71: 41

39. Singh UC, Kollman PA (1984) J. Chem. Phys. 80: 353
40. Scheiner S (1983) J. Chem. Phys. 78: 599
41. Viswanathan R, Dyke TR (1982) J. Chem. Phys. 77: 1166
42. Cazzoli G, Favero PG, Lister DG, Legon AC, Millen DJ, Kisiel Z (1985) Chem. Phys. Lett. 117: 543
43. Goodwin EJ, Legon AC (1984) J. Chem. Soc., Faraday Trans. 2, 80: 51
44. Legon AC, Willoughby LC (1983) Chem. Phys. Lett. 95: 449
45. Legon AC, Willoughby LC (1982) Chem. Phys. Lett. 92: 333
46. Legon AC, Millen DJ, North HM (1987) Chem. Phys. Lett. 135: 303
47. Willoughby LC, Fillery-Travis AJ, Legon AC (1984) J. Chem. Phys. 81: 20
48. Bevan JW, Kisiel Z, Legon AC, Millen DJ, Rogers SC (1980) Proc. R. Soc. Lond. A372: 441
49. Kisiel Z, Legon AC, Millen DJ (1982) Proc. R. Soc. Lond. A381: 419
50. Kisiel Z, Legon AC, Millen DJ, Rogers SC (1983) J. Chem. Phys. 78: 2910
51. Buckingham AD, Fowler PW (1985) Can. J. Chem. 63: 2018
52. Buckingham AD, Fowler PW (1983) J. Chem. Phys. 79: 6426
53. Hurst GJB, Fowler PW, Stone AJ, Buckingham AD (1986) Int. J. Quantum Chem. 29: 1223
54. Legon AC, Millen DJ (1987) Chem. Soc. Rev. 16: 467
55. Brobjer JT, Murrell JN (1983) J. Chem. Soc., Faraday Trans. 2, 79: 1455
56. Rendell APL, Bacskay GB, Hush NS (1985) Chem. Phys. Lett. 117: 400
57. Spackman MA (1986) J. Chem. Phys. 85: 6587
58. Czerminski R (1987) Int. J. Quantum Chem. 31: 649
59. Herbine P, Dyke TR (1985) J. Chem. Phys. 83: 3768
60. Latajka Z, Scheiner S (1987) J. Comput. Chem. 5: 674
61. Boys SF, Bernardi R (1970) Mol. Phys. 19: 553
62. Johansson A, Kollman P, Rothenberg S (1973) Theor. Chim. Acta 29: 169
63. Morokuma K, Kitaura K (1981) in: Politzer P, Truhlar DG (ed) Chemical applications of atomic and molecular electrostatic potentials, Plenum, New York, p 215
64. Kołos W (1979) Theor. Chim. Acta 51: 219
65. Kołos W (1980) Theor. Chim. Acta 54: 187
66. Alagona G, Ghio C, Cammi R, Tomasi J (1987) Int. J. Quantum Chem. 32: 207
67. Gutowski M, van Duijneveldt FB, Chałasinski G, Piela L (1987) Mol. Phys. 61: 233
68. Gutowski M, van Duijneveldt FB, Chałasinski G, Piela L (1986) Chem. Phys. Lett. 129: 325
69. Leclercq JM, Allavena M, Bouteiller Y (1983) J. Chem. Phys. 78: 4606
70. Cammi R, Bonaccorsi R, Tomasi J (1985) Theor. Chim. Acta 68: 271
71. Alagona G, Ghio C, Cammi R, Tomasi J (1987) Int. J. Quantum Chem. 32: 227
72. Ostlund NS, Merrifield DL (1976) Chem. Phys. Lett. 39: 612
73. Kurdi L, Kochanski E, Diercksen GHF (1985) Chem. Phys. 92: 287
74. Wells BH, Wilson S (1983) Mol. Phys. 50: 1295
75. Wells BH, Wilson S (1985) Mol. Phys. 54: 787
76. Hobza P, Schneider B, Sauer J, Carsky P, Zahradnik R (1987) Chem. Phys. Lett. 134: 418, 553
77. Chałasinski G, Gutowski M (1985) Mol. Phys. 54: 1173
78. Gutowski M, van Lenthe JH, Verbeek J, van Duijneveldt FB, Chałasinski G (1986) Chem. Phys. Lett. 124: 370
79. van Lenthe JH, van Duijneveldt-van de Rijdt JGCM, van Duijneveldt FB (1987) in: Lawley KP (ed) Ab initio methods in quantum chemistry, Vol II, Wiley, New York, p 521
80. Szczesniak MM, Scheiner S (1986) J. Chem. Phys. 84: 6328
81. Cole SJ, Szalewicz K, Purvis GD III, Bartlett RJ (1986) J. Chem. Phys. 84: 6833
82. Dacre D (1978) Mol. Phys. 36: 541
83. Karlström G, Sadlej AJ (1982) Theor. Chim. Acta 61: 1
84. Fowler PW, Buckingham AD (1983) Mol. Phys. 50: 1349
85. Tolosa Arroyo S, Garcia JE, Olivares del Valle FJ (1986) J. Mol. Struct. 136: 99
86. Latajka Z, Scheiner S (1987) Chem. Phys. Lett. 140: 338
87. Latajka Z, Scheiner S (1987) J. Chem. Phys. 87: 1194
88. Szczesniak MM, Scheiner S (1988) Coll. Czech. Chem. Commun. 53: 2214
89. Pine AS, Howard BJ (1986) J. Chem. Phys. 84: 590
90. Pine AS, Lafferty WJ, Howard BJ (1984) J. Chem. Phys. 81: 2939

91. Howard BJ, Dyke TR, Klemperer W (1984) J. Chem. Phys. 81: 5417
92. Dyke TR, Howard BJ, Klemperer W (1972) J. Chem. Phys. 56: 2442
93. Karpfen A, Beyer A, Schuster P (1983) Chem. Phys. Lett. 102: 289
94. Frisch MJ, Pople JA, Del Bene JE (1985) J. Phys. Chem. 89: 3664
95. Gaw JF, Yamaguchi Y, Vincent MA, Schaefer, HF III (1984) J. Am. Chem. Soc. 106: 3133
96. Kofranek M, Lischka H, Karpfen A (1988) Chem. Phys. 121: 137
97. Michael DW, Dykstra CE, Lisy JM (1984) J. Chem. Phys. 81: 5998
98. Del Bene JE (1987) J. Chem. Phys. 86: 2110
99. Redmon MJ, Binkley JS (1987) J. Chem. Phys. 87: 969
100. Latajka Z, Scheiner S (1987) J. Comput. Chem. 5: 663
101. Allavena M, Silvi B, Cipriani J (1982) J. Chem. Phys. 76: 4573
102. Votava Chr, Ahlrichs R, Geiger A (1983) J. Chem. Phys. 78: 6841
103. Latajka Z, Scheiner S (1988) Chem. Phys. 122: 413
104. Hobza P, Sauer J (1984) Theor. Chim. Acta 65: 279
105. Hobza P, Carsky P, Zahradnik R (1979) Coll. Czech. Chem. Commun. 44: 3458
106. Hobza P, Zahradnik R (1981) Chem. Phys. Lett. 82: 473
107. Rank DH, Sitaram P, Flickman WA, Wiggins TA (1963) J. Chem. Phys. 39: 2673
108. Janda KC, Steed JM, Novick SE, Klemperer WG (1977) J. Chem. Phys. 67: 5162
109. Hobza P, Szczesniak MM, Latajka Z (1980) Chem. Phys. Lett. 74: 248
110. Bumgarner RE, Kukolich SG (1987) J. Chem. Phys. 86: 1083
111. Ohashi N, Pine AS (1984) J. Chem. Phys. 81: 73
112. Scheiner S (1980) Theor. Chim. Acta 57: 71
113. Morokuma K, Pederson L (1968) J. Chem. Phys. 48: 3275
114. Thiel W (1978) Theor. Chim. Acta 48: 357
115. Dyke TR, Mack KM, Muenter JS (1977) J. Chem. Phys. 66: 498
116. Dyke TR, Muenter JS (1974) J. Chem. Phys. 60: 2929
117. Dyke TR (1977) J. Chem. Phys. 66: 492
118. Odutola JA, Dyke TR (1980) J. Chem. Phys. 72: 5062
119. Odutola JA, Hu TA, Prinslow D, O'dell SE, Dyke TR (1988) J. Chem. Phys. 88: 5352
120. Yoon BJ, Morokuma K, Davidson ER (1985) J. Chem. Phys. 83: 1223
121. Newton MD (1983) J. Phys. Chem. 87: 4288
122. Newton MD (1983) Acta Cryst. B39: 104
123. Newton MD, Kestner NR (1983) Chem. Phys. Lett. 94: 198
124. Kołos W (1979) Theor. Chim. Acta 51: 219
125. Kołos W (1980) Theor. Chim. Acta 54: 187
126. Zahradnik R, Hobza P (1986) Int. J. Quantum Chem. 29: 663
127. Roszak S, Sokalski WA, Hariharan PC, Kaufman JJ (1986) Theor. Chim. Acta 70: 81
128. Alagona G, Ghio C, Cammi R, Tomasi J (1987) Int. J. Quantum Chem. 32: 227
129. Mayer I, Vibok A (1987) Chem. Phys. Lett. 140: 558
130. Baum JO, Finney JL (1985) Mol. Phys. 55: 1097
131. Singh UC, Kollman PA (1985) J. Chem. Phys. 83: 4033
132. Del Bene JE, Mettee HD, Frisch MJ, Luke BT, Pople JA (1983) J. Phys. Chem. 87: 3279
133. Frisch MJ, Del Bene JE, Binkley JS, Schaefer HF III (1986) J. Chem. Phys. 84: 2279
134. Harrison RJ, Bartlett RJ (1986) Int. J. Quantum Chem., QCS 20: 437
135. Hobza P, Schneider B, Carsky P, Zahradnik R (1986) J. Mol. Struct. (Theochem) 138: 377
136. Fernandez PF, Ortiz JV, Walters EA (1986) J. Chem. Phys. 84: 1653
137. Woodbridge EL, Tso T-L, McGrath MP, Hehre WJ, Lee EKC (1986) J. Chem. Phys. 85: 6991
138. Amos RD (1986) Chem. Phys. 104: 145
139. Chin S, Ford TA (1985) J. Mol. Struct. (Theochem) 133: 193
140. Odutola JA, Dyke TR, Howard BJ, Muenter JS (1979) J. Chem. Phys. 70: 4884
141. Cook KD, Taylor JW (1979) Int. J. Mass Spectr. Ion Phys. 30: 345
142. Fraser GT, Nelson DD Jr, Charo A, Klemperer W (1985) J. Chem. Phys. 82: 2535
143. Nelson DD Jr, Fraser GT, Klemperer W (1985) J. Chem. Phys. 83: 6201
144. Nelson DD Jr, Fraser GT, Klemperer W (1987) Science 238: 1670
145. Nelson DD Jr, Klemperer W, Fraser GT, Lovas FJ, Suenram RD (1987) J. Chem. Phys. 87: 6364
146. Nelson DD Jr, Klemperer W (1987) J. Chem. Phys. 87: 139

147. Latajka Z, Scheiner S (1986) J. Chem. Phys. 84: 341
148. Liu S-Y, Dykstra CE, Kolenbrander K, Lisy JM (1986) J. Chem. Phys. 85: 2077
149. Sagarik KP, Ahlrichs R, Brode S (1986) Mol. Phys. 57: 1247
150. Sadlej J, Lapinski L (1986) J. Mol. Struct. (Theochem) 139: 233
151. Latajka Z, Scheiner S (1987) J. Comput. Chem. 5: 674
152. Carnovale F, Peel JB, Rothwell RG (1986) J. Chem. Phys. 85: 6261
153. a) Snels M, Fantoni R, Sanders R, Meerts WL (1987) Chem. Phys. 115: 79
     b) Süzer S, Andrews L (1987) J. Chem. Phys. 87: 5131
154. Sandorfy C (1984) Top. Curr. Chem. 120: 41
155. Hadzi D, Bratos S (1976) in: Schuster P, Zundel G, Sandorfy C (eds) The hydrogen bond —
     Recent developments in theory and experiments, North-Holland, Amsterdam, p 565
156. Kurnig IJ, Szczesniak MM, Scheiner S (1987) J. Chem. Phys. 87: 2214
157. Guelachvili G (1976) Opt. Commun. 19: 150
158. Pine AS, Lafferty WJ (1983) J. Chem. Phys. 78: 2154
159. Curtiss LA, Pople JA (1976) J. Mol. Spectrosc. 61: 1
160. Andrews L (personal communication)
161. Szczesniak MM, Scheiner S (1986) Chem. Phys. Lett. 131: 230
162. van Thiel M, Becker EP, Pimentel GC (1957) J. Chem. Phys. 27: 486
163. Tursi AJ, Nixon ER (1970) J. Chem. Phys. 52: 1521
164. Zilles BA, Person WB (1983) J. Chem. Phys. 79: 65
165. Swanton DJ, Bacskay GB, Hush NS (1983) Chem. Phys. 82: 303
166. Swanton DJ, Bacskay GB, Hush NS (1984) Chem. Phys. 83: 69
167. Nelander B (1978) J. Chem. Phys. 69: 3870
168. Bentwood RM, Barnes AJ, Orville-Thomas WJ (1980) J. Mol. Spectrosc. 84: 391
169. Nelander B (1988) J. Chem. Phys. 88: 5254
170. Wuelfert S, Herren D, Leutwyler S (1988) J. Chem. Phys. 88: 5256
171. Barnes AJ, Bentwood RM, Wright MP (1984) J. Mol. Struct. 118: 97
172. Sadlej J, Lapinski L (1986) J. Mol. Struct. (Theochem) 139: 223
173. Yeo GA, Ford TA (1986) S.-Afr. J. Chem. 39: 243
174. Somasundram K, Amos RD, Handy NC (1986) Theor. Chim. Acta 69: 491
175. Latajka Z, Scheiner S (1987) J. Chem. Phys. 87: 5928
176. Ayers GP, Pullin ADE (1976) Spectrochim. Acta Part A 32: 1641
177. Thomas RK (1975) Proc. Roy. Soc. A344: 579
178. Ault BS, Pimentel GC (1973) J. Phys. Chem. 77: 57
179. Bevan JW, Legon AC, Millen DJ, Rogers SC (1975) J. Chem. Soc. Chem. Commun. 341
180. Bertie JE, Falk MV (1973) Can. J. Chem. 51: 1713
181. Bouteiller Y, Allavena M, Leclercq JM (1981) Chem. Phys. Lett. 84: 361
182. Legon AC, Millen DJ (1982) Faraday Discuss. Chem. Soc. 73: 71
183. Arnold J, Millen DJ (1965) J. Chem. Soc. 503: 510
184. Stepanov BI (1945) Zh. Fiz. Khim. 19: 507; (1946) 20: 907
185. Bouteiller Y, Guissani Y (1980) Chem. Phys. Lett. 69: 280
186. Bouteiller Y, Mijoule C, Szczesniak MM, Scheiner S (1988) J. Chem. Phys. 88: 4861
187. Bouteiller Y, Mijoule C, Karpfen A, Lischka H, Schuster P (1987) J. Phys. Chem. 91: 4464
188. Barnes AJ, Kuzniarski JNS, Mielke Z (1984) J. Chem. Soc., Faraday Trans 2 80: 455
189. Barnes AJ, Wright MP (1986) J. Chem. Soc., Faraday Trans 2 82: 165
190. Arlinghaus RT, Andrews L (1984) J. Chem. Phys. 81: 4341
191. Szczesniak MM, Kurnig IJ, Scheiner S (1988) J. Chem. Phys. 89: 3131
192. Barnes AJ, Beech TR, Mielke Z (1984) J. Chem. Soc., Faraday Trans 2 80: 455
193. Millen DJ, Schrems O (1983) Chem. Phys. Lett. 101: 320
194. Bertie JE, Millen DJ (1965) J. Chem. Soc. 497
195. Lassegues JC, Huong PV (1972) Chem. Phys. Lett. 17: 444
196. Barnes AJ, Wright MP (1986) J. Mol. Struct. (Theochem) 135: 21
197. Ault BS, Pimentel GC (1975) J. Phys. Chem. 79: 621
198. Del Bene JE, Pople JA (1971) J. Chem. Phys. 55: 2296
199. Swepston PN, Colby S, Sellers HL, Schäfer L (1980) Chem. Phys. Lett. 72: 364
200. Clark JH, Emsley J, Jones DJ, Overill RE (1981) J. Chem. Soc. Dalton Trans. 1219
201. Liu S-Y, Michael DW, Dykstra CE, Lisy JM (1986) J. Chem. Phys. 84: 5032

202. a) Scuseria GE, Schaefer HF III (1986) Chem. Phys. 107: 33
     b) Kolenbrander KD, Dykstra CE, Lisy JM (1988) J. Chem. Phys. 88: 5995
203. Engdahl A, Nelander B (1987) J. Chem. Phys. 86: 4831
204. Dyke TR, Muenter JS (1972) J. Chem. Phys. 57: 5011
205. Wülfert S, Herren D, Leutwyler S (1987) J. Chem. Phys. 86: 3751
206. Honegger E, Leutwyler S (1988) J. Chem. Phys. 88: 2582
207. a) Clementi E, Kołos W, Lie GC, Ranghino G (1980) Int. J. Quantum Chem. 17: 377
     b) Koehler JEH, Saenger W, Lesyng B (1987) J. Comput. Chem. 8: 1090
208. Andrews L (1984) J. Phys. Chem. 88: 2940
209. Johnson GL, Andrews L (1982) J. Am. Chem. Soc. 104: 3043
210. Hopkins GA, Maroncelli M, Nibler JW, Dyke TR (1983) Chem. Phys. Lett. 114: 97
211. Maroncelli M, Hopkins GA, Nibler JW, Dyke TR (1985) J. Chem. Phys. 83: 2129
212. Buxton LW, Campbell EJ, Flygare WH (1981) Chem. Phys. 56: 399
213. Georgiou K, Legon AC, Millen DJ, Mjoberg PH (1985) Proc. Roy. Soc. London A399: 377
214. Jucks KW, Miller RE (1988) J. Chem. Phys. 88: 2196
215. Ruoff RS, Emilsson T, Klots TD, Chuang C, Gutowsky HS (1988) J. Chem. Phys. 89: 138
216. Anex DS, Davidson ER, Douketis C, Ewing GE (1988) J. Phys. Chem. 92: 2913
217. Pettit BA, Boyd RJ, Edgecombe KE (1982) Chem. Phys. Lett. 89: 478
218. Tse JS (1982) Chem. Phys. Lett. 92: 144
219. Kofranek M, Karpfen A, Lischka H (1987) Chem. Phys. 113: 53
220. Kofranek M, Lischka H, Karpfen A (1987) Mol. Phys. 61: 1519

The Extramolecular Electrostatic Potential.
An Indicator of the Chemical Reactivity

Jacopo Tomasi[1], Rosanna Bonaccorsi[2] and Roberto Cammi[3]

[1] Dipartimento di Chimica e Chimica Industriale, Via Risorgimento 35, I-56126 Pisa, Italy.
[2] Istituto di Chimica Quantistica ed Energetica Molecolare CNR, Via Risorgimento 35, I-56126 Pisa, Italy.
[3] Istituto di Chimica Fisica, Viale delle Scienze 1, I-43100 Parma, Italy.

1 Introduction

The theoretical study of chemical reactivity has two distinct aims which in the current daily work are often intermingled and not clearly expressed. One goal is the interpretation of the behaviour of matter at the molecular level, the comprehension of the relative influence of the different factors which rule chemical interactions. The second goal is the obtainment of numerical quantities which describe the essential aspects of these interactions. Clearly the two goals are connected. It is not possible to get numerical evaluations of any sort if there is not an interpretative model, a theory, at the basis of the computational procedure The elaboration of interpretative models requires, on the other hand, continuous controls which are normally done by the computation of selected quantities.

The worry is, however, that there is a shift of emphasis on the various aspects of the theoretical problem when one is primarily involved in obtaining numerical data or in the elaboration of models. The misgiving elegantly expressed by Coulson in a famous talk in 1959 about the future of quantum chemistry [1] seems now to have been to a good extent dispelled. Coulson noticed a separation of chemists working with theoretical tools into three groups: the first group interested in computations "in depth", and "prepared to abandon all conventional chemical concepts" to "achieve their desire for complete accuracy" (according to the Coulson's words), the second interested only in understanding the nature of some basic and simple chemical concepts, without any desire to make these concepts more rigorous. The third group, hinted at in Coulson's vision of the future of theoretical chemistry, is even rigorous than group II, being projected towards biology.

This was perhaps a partisan view, but 30 years after that talk we may state that group I has not lost its connections with real chemistry, and that groups II and III have been able to reformulate chemical concepts using theoretical tools derived from the activity of group I.

The molecular electostatic potential (MEP) represent just one of the attempts of connecting in "depth" studies on molecular systems with basic and elementary concepts of chemistry. It is no mere coincidence that the first proposals of using detailed MEP values in the study of inter- and intra-molecular processes [2] has immediately followed the obtaining of effective computer codes for ab initio calculations of the electronic stucture of polyatomic molecules.

The link between group I and group II provided by the suggestion of using MEP values has been followed by many others and now we can state that the gap between group I and II lamented by Coulson has been almost completely filled. Also group III benefits from the cooperative effort of the two preceding groups, even if, being "biological systems much more perverse than any laboratory chemical system" (according, again, to the Coulson's words) there are levels of the quantum theory which cannot be directly employed in biological studies.

Several theoretical tools developed by group II are also of current use in biological studies: the use of MEP is among them. One could ask if there are stimuli coming from group III which may prompt elaboration and innovations in traditional quantum chemistry which will be able to fill some of the still existing gaps between methods, and people concerned with "laboratory chemical systems" and those methods and people concerned with substructures of living bodies.

The answer is positive, and there is many people working in this direction. The institution of new links, the elimination of gaps, mean that also the theory must be almost in part renewed, and enlarged. This is, in our opinion, the perspective for the next decade. We shall examine in the following sections topics belonging to the past, more or less recent, but part of these subjects will be, with great probability, still employed in the future advances hinted above.

2 The Electrostatic Observables

The charge distribution, the electrostatic potential, the electrostatic field, and the electric field gradient belong to the category of physical observables related to a static description of material systems.

As physical observables they have a special status in the quantum theory. They are not related to specific levels of approximation of the theory, nor to special or specific numerical manipulations of quantities defined only for a given approximation.

There are, however, differences among the expectation values of these observables which deserve a few comments. The charge distribution has the aspect of a local quantity, expressed as a scalar. We shall make use in the following of the total (electrons and nuclei) charge distribution:

$$\Gamma_M(r) = \varrho_M(r) + \sum_{\alpha \in M}^{nuc} Z_\alpha \delta(r - r_a) \qquad \alpha \in M \ . \tag{1}$$

The local aspect of the property, and its scalar characteristics, are both important in the process of reduction of the information, which is a basic point in the elaboration of interpretations of quantum systems and properties. The electronic components of Γ_M, i.e. ϱ_M, is the diagonal term of the first order electronic density matrix, corresponding thus to a severe reduction of the information present in the original wavefunction.

The molecular electrostatic potential is still a spatial quantity, of scalar nature, missing in part the local character of Γ_M. In fact, by definition, the MEP $V(r)$ is given by an integral over the entire charge distribution

$$V_M(r) = \int \Gamma_M(r')/(r - r') \, dr' \ . \tag{2}$$

This means that V_M at point r feels all the components of the charge distribution, giving more emphasis to the nearest elements of Γ_M, but with considerable contributions even by distant elements on the molecular distance scale, the coulombic attenuation factor being related to the first inverse power of r only. This feature has a noticeable influence on the use of V_M in the interpretation of chemical phenomena, as will be discussed later on.

The molecular electric fields $F_M(r)$ is again a spatial quantity, but it is by defintion a vector:

$$F_M = -\nabla \cdot V_M \,,$$

$$F_{M\alpha}(r) = -\frac{\partial}{\partial\alpha} V_M(r) \,. \tag{3}$$

This fact makes the information carried by $F_M(r)$ less easy to manage for a visual inspection.

The field gradient $Q_M(r) = -\nabla \cdot F$ is still a local quantity, with a more complex mathematical structure.

It may be remarked that in passing from V to F and to Q the influence of the distant components on the local value decreases: V is the electrostatic property most sensitive to distant contributions.

The expectation values of all the observables mentioned above play a role in the prediction and interpretation of the properties of molecules, in a variety of elaborations and combinations depending on the formulation of the theory for the specific theme under examination (theory of the chemical bond, theory of the geometrical equilibrium structure, theories for the various spectroscopic properties, for molecular dynamics, for reactivity, for the chemical reaction mechanisms, etc.). We shall limit our attention to the theories for chemical reactivity; a very important subject, but limited with respect to the variety of problems of interest in the molecular sciences shortly summarized above. This limitation of the theme also implies a limitation in the use of electrostatic quantities. The emphasis is placed here on the energy of molecular interactions and on the forces acting on the molecular interacting units.

Molecular charges, potential and field obviously play an important role in assessing the intermolecular interaction energy and forces; but they are not the unique ingredients of a quantum theory of these phenomena. There is however a coherent attempt by many researchers to stress the importance of the classical elements in quantum description of the molecular interactions. This strategy is not limited to weak molecular interactions but covers other aspects of the molecular theory, from the chemical bond to the reaction mechanisms. In another chapter of this series of volumes we attempted a summary of the results obtained with this strategy for intramolecular interactions. Here, we simply remark that the use of MEP for the study of intermolecular interactions is justified not only by intuitive and empirical physical considerations, but also by a more general strategy.

3 Theories for Intermolecular Interactions

Molecular charges, electrostatic potential and field belong to the class of "chemical concepts" in use in the early stages of the molecular theory. The adoption of the methods of a rigorous quantum approach for the study of chemical problems has not consigned these concepts to oblivion. The quantum counterparts of these quantities, which we could name the "electrostatic observables", play an important role in the theorie and methods currently used for the study of weak molecular interactions.

The history of the development of this area of theoretical research in chemistry is too well known to be fully examined here. For our purposes it is sufficient to recall that the two main approaches, the supermolecule approach which considers the whole interacting system (in the simplest case a couple of molecules A + B) as a unique entity, and the perturbation theory approach, which focus its attention on the interaction itself, taking as known the properties of the isolate systems A and B, ultimately converge toward a substantially homogeneous view of the interaction act.

There are however differences in the emphasis given to the various components of the interaction energy ΔE and on the practical procedures to obtain the separate contributions to ΔE.

The reasons of these differences derive from intrinsic difficulties in the perturbation theory treatment of the molecular interaction problem and on the incomplete development of the theories for decomposing the supermolecule variational results.

In the perturbation theory, the models of wider use consider the Hamiltonian of the interacting system, H_{AB} as composed of an unperturbed Hamiltonian H_{AB}^0, corresponding to the Hamiltonian of the two separate partners ($H_{AB}^0 = H_A + H_B$) and by a perturbation operator, V_{AB}, responsible for the mutual interaction energy. H_{AB}^0 does not reflect however the full permutational symmetry of H_{AB}, and this fact makes rather involved the solution of the perturbation problem when the overlap between the terms constituting the zeroth order function $\Psi_{AB}^0 = \Psi_A^0 \cdot \Psi_B^0$ is not negligible.

This problem has been treated by many authors, and it is the subject of many detailed reviews (see e.g. Refs. [3–8]). It is not necessary to repeat here an analysis of the various methods thus far proposed; a shortened discussion will be sufficient. Let us look firstly at the conventional Rayleigh Schrödinger perturbation theory neglecting the effects of the higher permutational symmetry of H_{AB} (i.e. neglecting the exchange of electrons between A und B). This approximation is called "long-range approximation" (LR), because intuitively it should be applicable at long distances between A and B, where the electronic overlap, or the electron exchange, is negligible (see, however, Chalasinski et al. [9]); it is also called "polarization approximation", a name coined by Hirschfelder [10], which we shall not use in the following. The wavefunction may be expressed in terms of the products $\Psi_A^r \Psi_B^s$ of the wavefunction related to the two Hamiltonians H_A and H_B (product state approximation).

The expression for the interaction energy $\Delta E = E_{AB} - (E_A^0 + E_B^0)$ is easily obtained

$$\Delta E = \sum_{n=1}^{\infty} E_{LR}^{(n)} = \langle \Psi_A^0 \Psi_B^0 | V_{AB} | \Psi_A^0 \Psi_B^0 \rangle \qquad \text{I order}$$

$$- \left[\sum_k \frac{\langle \Psi_A^0 \Psi_B^0 | V_{AB} | \Psi_A^k \Psi_B^0 \rangle}{E_A^k - E_A^0} + \sum_l \frac{\langle \Psi_A^0 \Psi_B^0 | V_{AB} | \Psi_A^0 \Psi_B^l \rangle}{E_B^l - E_B^0} \right. \qquad \text{II order}$$

$$\left. - \sum_k \sum_l \frac{\langle \Psi_A^0 \Psi_B^0 | V_{AB} | \Psi_A^k \Psi_B^l \rangle}{(E_A^k - E_A^0)(E_B^l - E_B^0)} \right]$$

$$+ \text{ higher order terms}. \qquad (4)$$

The first order contribution may be rewritten in the following form:

$$E_{LR}^{(1)} = \langle \Psi_A^0 \Psi_B^0 | \, V_{AB} \, | \Psi_A^0 \Psi_B^0 \rangle$$

$$= \int \int \frac{\Gamma_A^{(1)} \Gamma_B^{(2)}}{|\vec{r}_1 - \vec{r}_2|} \, d\vec{r}_1 \, d\vec{r}_2 = \int V_A(2) \, \Gamma_B^0(2) \, d\vec{r}_2 = \int \Gamma_A^0(1) V_B(2) \, d\vec{r}_2 \,, \quad (5)$$

in which we have introduced the total charges distributions Γ_A^0 and Γ_B^0 of the unperturbed partners, and the corresponding electrostatic potentials V_A and V_B. This term, the classical interaction energy between two rigid charge distributions, will be called in the following coulombic, or electrostatic interaction energy, E_{ES}.

The next two terms give a first order approximation to classical polarization energy. Actually the real polarization energy should take into accounts all the effects due to the mutual polarization of the two charge distributions Γ_A^0 and Γ_B^0: other components related to this physical effect are to be found in the following terms of the expansion (4).

The last second order term describes energy contributions associated with instantaneous fluctuations in the charge distributions of A and B: it is called dispersion energy (to the first order). In conclusion:

$$E_{LR}^{(2)} = E_{pl}^{(2)} + E_{dis}^{(2)}. \quad (6)$$

This simple scheme is not valid however, for the reasons said above. The introduction of the correct permutational properties of the system may be formally described by associating with each term of the expansion (4) a further term, accounting for the effects of the electron exchange (Symmetry Adapted Perturbation Theories, SAPT):

$$\Delta E = \sum_{n=1}^{\infty} E_{LR}^{(n)} + E_{exc}^{(n)} \,. \quad (7)$$

There is no unique generally accepted definition of $E_{exc}^{(n)}$ except for $n = 1$. Moreover, the higher order contributions to E_{exc} are intermingled with polarization, or dispersion, contributions giving origin to terms currently named $E_{pl-exc}^{(n)}$, $E_{dis-exc}^{(n)}$. The contribution of these terms to ΔE may be not negligible for non-covalent interactions of mean strength (as the hydrogen-bond interactions, for example).

For $n = 1$ the term in question may be written

$$E_{exc}^{(1)} = \frac{\langle \Psi_A^0 \Psi_B^0 | \, V \, | \mathscr{P} \Psi_A^0 \Psi_B^0 \rangle - \langle \Psi_A^0 \Psi_B^0 | \, V \, | \Psi_A^0 \Psi_B^0 \rangle \langle \Psi_A^0 \Psi_B^0 | \, \mathscr{P} \Psi_A^0 \Psi_B^0 \rangle}{1 + \langle \Psi_A^0 \Psi_B^0 | \, \mathscr{P} \, | \Psi_A^0 \Psi_B^0 \rangle} \,,$$

$$(8)$$

where $\mathscr{P}$ is the sum of all the permutations interchanging at last one pair of electrons between A and B. This expression is often approximated by the first terms of an expansion of $\langle \mathscr{P} \rangle$ in the powers of the overlaps.

The first order contribution to ΔE may be recast in the following form, which will be used later on

$$E^{(1)} = E^{(1)}_{ES} + E^{(1)}_{exc} = \frac{\langle \Psi^0_A \Psi^0_B| \, V \,| \mathscr{A} \Psi^0_A \Psi^0_B\rangle}{\langle \Psi^0_A \Psi^0_B| \, \mathscr{A} \Psi^0_A \Psi^0_B\rangle} \, . \tag{9}$$

All that was said above supposes a previous knowledge of the exact wavefunction of A and B. Generally speaking this is not true. What we know is an approximate expression of the corresponding Hartree-Fock equations. Neglecting the approximate nature of these Hartree-Fock wavefunctions (the accuracy of the approximation may be nowadays pushed, for small systems, near the Hartree-Fock limit) the perturbation theory may be recast in a more complex form, using the double perturbation theory, in which the perturbation operators correspond to the electronic correlation and to the intramolecular interaction [6]. The analysis of the resulting expressions gives some interesting results, bringing in evidence terms easily computed and terms difficult to obtain.

The extraction from the expansion of ΔE of all the terms belonging to the H-F theory:

$$\Delta E = \Delta E^{HF} + \sum_n E^{(n)} \text{ (non HF)}, \tag{10}$$

gives rise to terms of non definite physical meaning, the various dispersion-polarization, exchange-polarization, etc. contributions being partitioned between the H-F and non H-F portions of ΔE.

In conclusion, the perturbation theory provides a physical meaning to various contributions to ΔE (the meaning being less evident for $n \geq 2$) and a precise recipe for the electrostatic term alone. The exchange term is clearly defined for its first order contribution and the polarization term is clearly defined at the first order approximation in the LR formulation of the theory.

The supermolecular approach which gives ΔE as a difference between two large quantities, computed separately:

$$\Delta E = E_{AB} - (E^0_A + E^0_B), \tag{11}$$

it is in principle more subject to errors than the perturbation theory approach. The relevant energies may be however computed at any level of the quantum molecular theory (H-F or post H-F methods), and, at least for small systems the objection with regard to the precision is now settled.

The partition of the variational ΔE is however fully elaborated only at the H-F level.

There are several variants of the same basic approach which is based on the calculation of some intermediate energies. We shall summarize here the Kitaura-Morokuma [11] version of the method which is the most popular.

The most concise and elegant formulation of the K-M method given in terms of the Fock matrix F of the system $A \cdot B$, written in terms of the MOs of the subsystems, with a partition into blocks referring to occupied and vacant orbitals (A_0, A_V, B_0,

B_V). From F and F^0 (the corresponding matrix with the subsystems at infinite separation) an interation matrix Σ is defined:

$$\Sigma = (F - \varepsilon S) - (F^0 - \varepsilon S^0). \tag{12}$$

This matrix has the following block form:

	A_0	A_V	B_0	B_V
A_0	ESX	PLX	EX'	CT
A_V		ESX	CT	EX'
B_0			ESX	PLX
B_V				ESX

By deleting some blocks of this matrix and by solving the corresponding pseudo H-F equation one obtains an energy E^x (the upper index x stays for the blocks retained in the calculation). Making the difference between suitable E^x and E^y values and using also $E^0 = E^0_A + E^0_B$, the reference energy of the separate subsystems, one defines the E_k terms of the decomposition of ΔE.

The M-K decomposition assumes the form

$$\Delta E = E_{ES} + E_{PL} + E_{EX} + E_{MIX}. \tag{13}$$

A further decomposition of the residue E_{MIX} has been elaborated by Nagase et al. [12]

$$\Delta E = E_{ES} + E_{PL} + E_{EX} + E^I + E^{II} + E^{III} + E^{IV} + E_{RES}. \tag{14}$$

The four additional terms are interpreted, respectively, as the coupling between the polarization of the monomer M with the charge transfer $M \to N$ (E^I and E^{IV}) and the coupling between the polarization of a monomer M with the charge transfer $N \to M$ (E^{II} and E^{III}). The details can be found in the reference papers or in one of the numerous reviews on this subject [13].

It is convenient to reconsider some of the terms of (13) in a different formulation. We define a set of normalized wavefunctions and the corresponding energies calculated as expectation value of the full Hamiltonian H_{AB}.

1) A simple Hartree product built up from the H-F wavefunctions of the separate partners, both internally antisymmetrized

$$\Psi_1 = \Psi^0_A \Psi^0_B, \tag{15}$$

$$E_1 = \langle \Psi_1 | H_{AB} | \Psi_1 \rangle. \tag{16}$$

From this energy we derive E_{ES} as a difference:

$$E_{ES} = E_1 - E_0 , \tag{17}$$

E_{ES} has the same identical expression of $E_{LS}^{(1)} \equiv E_{ES}$ defined in Eq. (5).

2) A simple product between two internally antisymmetrized wavefuctions each optimized (at the H-F level) in the field of the other

$$\Psi_2 = \Psi'_A \Psi'_B , \tag{18}$$

$$E_2 = \langle \Psi_2 | H_{AB} | \Psi_2 \rangle , \tag{19}$$

$$E_{PL} = E_2 - E_1 . \tag{20}$$

3) A fully antisymmetrized product of the unperturbed wavefunctions of A and B (put at the desidered distance and orientation):

$$\Psi_3 = \mathscr{P} \Psi_A^0 \Psi_B^0 , \tag{21}$$

$$E_3 = \langle \Psi_3 | H_{AB} | \Psi_3 \rangle , \tag{22}$$

$$E_{EX} = E_3 - E_1 , \tag{23}$$

$E_3 - E^0$ may be recast in a form completely equivalent to that we have reported for $E^{(1)}$ (Eq. 9) establishing another link between perturbation theory and K-M decomposition of the variational ΔE.

It is not convenient, for our purposes, to recast in this more expanded form the algorithms giving origin to E_{CT} and to E^I, E^{II}, E^{III} and E^{IV}.

Instead, it may be of some help to our discussion to report a scheme of the orbital interactions in the K-M method [13].

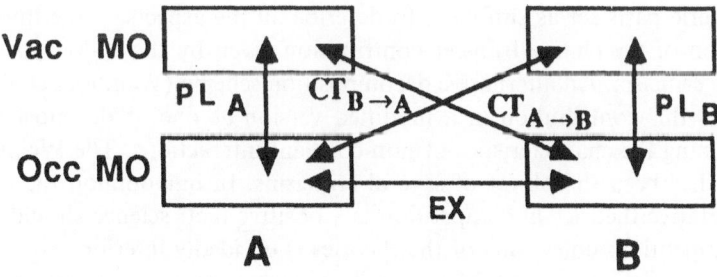

The charge transfer contribution $E_{CT} = E_{CT\,B \to A} + E_{CT\,A \to B}$ is derived from the mixing of occupied orbitals of one partner with the virtual orbitals of the second. Such a term is not present in the rigorous formulations of the perturbation theory which assume that the expansion basis of each partner is complete. It is present however in theories dealing with uncomplete expansions. If one looks at the orbital diagrams describing the contributions to $E^{(2)}$ (see e.g. Daudey et al. [14] or Tomasi [15] it becomes evident that some of the contributions interpreted as belonging to the dispersion-exchange or dispersion-polarization terms have a charge-transfer character.

Similar considerations may be put forward for the four additional terms introduced by Nagase et al. [12].

It is time to draw some conclusions from this examination of the K-M decomposition scheme.

The first term in the decomposition of the supermolecular interaction energy E_{ES} has the same formal expression as the corresponding term in the perturbation theory (Eq. 5). The basic expression for the use of the electrostatic potential as an index of chemical reactivity does not depend upon the theory adopted in describing molecular interactions.

The second term E_{PL} differs from the corresponding LR perturbation theory, $E_{pl}^{(2)}$ because it is not limited to first order effects in the polarization. The SCF procedure in fact allows a complete mutual relaxation of the charge distributions of A and B under the electrostatic effect of the partner (at the H-F level).

The third term, E_{EX}, is identical to $E_{exc}^{(1)}$, as we have remarked above.

In a different version of the decomposition procedure [16] the term corresponding to the exchange contributions (called E_{OV} in Ref. [16]) collects terms belonging to higher orders in the perturbation theory. E_{EX} in fact is defined in terms of the unperturbed wavefunction Ψ_A^0 and Ψ_B^0, while E_{OV} is defined over the mutually polarized wavefunctions Ψ_A' and Ψ_B'; in this way some physical contributions intermingled in all the perturbation theories with polarization or dispersion contributions are here separated and collected with the contributions of analogous physical origin described by $E_{exc}^{(1)}$.

Charge transfer (CT) contributions are somewhat neglected by the formal elaborations of the perturbation theory. We know that they have an important role in the interpretation of many reactions (see. e.g. Fukui [17] and Klopman [18]) and represent an important factor in specific, albeit limited, classes of noncovalent interactions. The reason of this neglection of CT terms in the development of the symmetry-adapted version of the perturbation theory is due to the fact that attention has been focussed on small systems for which it was possible to consider the monomeric basis set as sufficient to describe all the aspects of the interaction. The definition of the charge-transfer contribution given by the K-M procedure is also open to criticism. An alternative decomposition scheme (Weinhold et al. [19]) in fact indicates this contribution in a modified version as one of the most important in determining the characteristics of non-covalent interactions. The Weinhold analysis in turn has been the object of several criticisms. In our opinion the occurrence of interpretative theories in competition is a positive fact: science should benefit from this competition, unless one of the theories is decidedly inferior.

The main defect of the K-M (or similar) decomposition of the supermolecular ΔE is, as anticipated at the beginning of this section, that the theory has been developed only within the Hartree-Fock approximation.

It should be clear, from the remarks made above, that this is not a real problem for the term which constitutes the main object of this discussion, E_{ES}. Equation (5) may be directly applied, whatever the level of accuracy is in the description of electrostatic potential V_A and of the charge distribution Γ_B.

Also the second classical term E_{PL} could be easily recast in a similar form. More complex is the decoupling of terms involving contemporary excitations of the partners in a CI description of the supermolecule. There are in the literature several suggestions in this vein (a brief discussion stating our views can be found in [20]), but a completely worked out and satisfying solution of this problem is not available at present.

The main consequence is that the dispersion terms are missing in the decomposition of the variational interaction energy.

Dispersion terms may be computed separately and added to the H-F value of ΔE. There are several methods for the calculation of E_{DIS}, but it is not necessary to go into details here [21, 22]. The admixture of contributions computed with different levels of the theory does not contradict the usual practice of quantum chemistry. It remains however that a coherent treatment, based on a well-defined expansion basis set is preferable.

All the considerations made above refer to the formal aspects of the theory. The theory is not completely refined, nor in the perturbational nor in the variational version, but sufficient as guide for studies on systems of actual chemical interest. Systems of interest in chemistry are normally larger than the model systems employed to formulate and to test a theory. This means that the researcher is often obliged to use, in the calculations, basis sets which are very far from being complete.

This gives rise to another problem, which must be briefly discussed because it is relevant to the use of electrostatic potentials in the study of molecular reactivity.

In normal ab initio calculation the functions of the basis set are centered on the atoms of the system, and in the study of an interaction act in function of the distance (and orientation) R between the partners, their position is changed accordingly.

Thus, at large R, there are two separate basis sets describing each a monomer, while at short distances r the union of the two basis sets

$$\chi_{AB} = \chi_A \oplus \chi_B, \tag{24}$$

describes the whole system AB. The dimer is thus better described than the separate monomers (there is more flexibility in describing $n_A + n_B$ electrons with χ_{AB} than in describing n_A electrons with χ_A and n_B electrons with χ_B). When the basis sets are small, the different level of accuracy in the results is numerically important.

In consequence the potential energy surface describing $\Delta E(R)$ is somewhat distorted in favor of the more compact geometries. This is called the Basis Set Superposition Error (BSSE), present either in perturbation and variational approaches. It is nowadays common practice to resort to the counterpoise procedure (CP) for a correction of this error. The CP procedure, in its original definition [23], consists of using the dimeric basis set χ_{AB} also for the calculation of the reference energy E^0 which depends now on R:

$$E^{0CP}(R) = E_A^{0CP}(\chi_{AB}; R) + E_B^{0CP}(\chi_{AB}; R). \tag{25}$$

E^{0CP} is more negative than E^0, and of consequence the CP corrected interaction energy more positive

$$\Delta E^{CP}(R) = E_{AB}(R) - E^{0CP}(R) \tag{26}$$

$$= E_{AB}(R) - (E^{0CP}(R) + \delta^{TOT}(R)), \tag{27}$$

where

$$\delta^{TOT}(R) = E^0 - E^{0CP}(R), \tag{28}$$

is the "full" CP correction. The Potential Energy Surface (PES) of AB is not changed if referred to the statutary zero energy reference of quantum chemistry: separated electrons and nuclei. What changes is the intermediate reference energy, the isolated molecules; this change of reference energy is no longer a simple energy shift, but a function of R (intermolecular coordinates).

Clearly the CP correction to the BSSE is an empirical device, not related to the physics of the problem, which must be accepted or refused on practical grounds, and in particular on the basis of its effectiveness (better results with acceptable computer costs). The study of an interaction act is not limited however to obtaining numerical values (e.g. the stabilization energy ΔE_{eq} and the corresponding geometry R_{eq}) but also involves the interpretation of the act. For this reason an empirical correction recipe, as the procedure CP is, must not introduce elements which contradict the accepted picture of the event and should preserve (or improve) the tools employed in the analysis before the introduction of the correction.

An objection to CP based on the mathematical description of the interaction has been raised by several authors: the method employs some portions of the functional space twice. For example a portion of the molecular space χ_A is employed to describe the electrons belonging to A (wavefunction related to E_A^{CP}) and then employed again to describe electrons belonging to B (wavefunction related to E_B^{CP}). Several variants of the CP procedure excluding this disturbing feature have been proposed (see e.g. [24–25]). The corresponding "uncomplete" CP procedure is, however, less efficient in reducing the BSSE. It seems, at present, that the latest formulation proposed by Olivares del Valle and coworkers [26] is capable of giving good BSSE corrections without using the occupied subspaces twice. This formulation exploits the CP corrections to the K-M decomposition scheme which we shall consider below. It must be said, however, that this objection has been contradicted by many authors; a sharp rebuttal is expressed in the wide and detailed review by van Lenthe at al. [27].

The second condition for the acceptance of the CP corrections regards the decomposition of the interaction energy.

In the K-M and related decomposition schemes the CP correction may be introduced by using for each E_x term, which derives, as said above, from a difference between separately evaluated E^x terms and the reference energy E^0, a specific reference energy obtained by a *partial* enlargement of the monomeric basis set, dictated by the physical nature of the component E_x. The algorithm may be put in a form giving the CP correction as a sum of separate contributions

$$\delta^{CP} = \delta^{TOT} = \sum_x \delta^x , \tag{29}$$

with

$$E_x^{CP} = E_x + \delta^x , \tag{30}$$

and

$$\Delta E^{CP} = \Delta E + \delta^{CP} , \tag{31}$$

with the CP correction applied to the desired level of "completeness" (details in refs. [28–29]). In our opinion, there are no physical reasons suggesting a CP correction to the classical interaction terms E_{ES} and E_{PL} (there is no mixing with the orbitals of the partner in the corresponding calculations, see Eqs. (17) and (20)). Accordingly, the correction is limited to the other terms; i.e. in the more detailed NYFK decomposition scheme:

$$\delta^{TOT} = \delta^{EX} + \delta^{CT} + \delta^I + \delta^{II} + \delta^{III} + \delta^{IV} + \delta^{RES} . \tag{32}$$

As said above, we support a partitioning of δ^{CP} which leaves E_{ES} unchanged. There is another proposal, advocating the calculation of E_{ES}, i.e. of $V_A(r)$ and $\Gamma_B(r)$, Eq. (5) with the full dimeric space [30], and then introducing a new correction term, related to E_{ES}.

This proposal is based on an analysis of $E_{ES} + E_{EX}$ with the first order correction $E^{(1)}$ in the SA perturbation theories (see Eqs. (5) and (9)). The double use of the occupied subspaces is accepted here, of course. Using the terminology employed here for this proposal, our δ^{EX} correction is split into two components which we could call $\tilde{\delta}^{ES}$ and $\tilde{\delta}^{EX}$

$$\delta^{EX} = \tilde{\delta}^{ES} + \tilde{\delta}^{EX} . \tag{33}$$

All the CP corrections of Eq. (32) are positive, while $\tilde{\delta}^{ES}$ and $\tilde{\delta}^{EX}$ may have the opposite sign.

We stress again that decompositions and corrections not based on the use of physical observables have a degree of arbitrariness, and that different definitions may be accepted, when employed in a coherent way. With this point in view, a CP correction to E_{ES} could be acceptable. This proposal introduces, however, some features in the interpretation of the reaction act and potentially destroys the use of molecular electrostatic potentials as indicators of chemical reactivity.

$V_A(r)$ and $\Gamma_B(r)$, will depend, in fact, on the position of the second partners, which fixes the position of the "gosth" basis set functions. Let us consider the case of $V_A(r)$. As we shall discuss in the next section V_A may be described in terms of maps drawn on selected planes, or in terms of a multipolar expansion. Let us consider a molecule provided of some elements of symmetry (says, H_2CO); the symmetry will be present also in Γ_A and V_A. When this molecule interacts with a partner B (says HCl) with an orientation corresponding to a lower symmetry, the V map will be deformed and the multipole expansion too.

This deformation has nothing to do with the polarization effects due to the second partner. Clearly this interpretation of a partial CP correction to E_{ES} (a δ^{ES} contribution in our terminology) makes impossible the use of V_A as a molecular index of reactivity. It is not compulsory to use V_A in an interpretation of the interaction act, but the introduction of CP correction to E_{ES} deprives the researcher of a tool, without doing anything in exchange, the $\delta^{EX} = \tilde{\delta}^{ES} + \tilde{\delta}^{EX}$ being intuitively and physically more directly related to E_{EX}. The CP correction to E_{ES} also makes a concept often invoked by experimentalists less clear, the "enhancement" of the monomer's dipole (or other terms of the multipolar expansion) due to the interaction with a partner.

4 Electrostatic Interaction Energy and Molecular Electrostatic Potential

The theory of non covalent molecular interactions, as we have already said, is a classic topic, extensively studied and the subject of several specialized textbooks. From the outline reported in the preceding section we have tried to convey to the reader the notion that the theory is not firmly established in all the details. Some corrections to the basic components of the two-body interaction between rigid molecules (i.e. higher order terms in the perturbation theory approach, or coupling terms of high order in the variational supermolecule approach) are still ill-defined and not yet fully analyzed in molecular systems of current chemical interest.

In addition, we have completely neglected in our concise exposition the analysis of the effects of the interaction on the internal geometry of the partners; these effects may be of some importance also for non-covalent interactions of moderate strength, and often play an important role when the non-covalent interaction is a preliminary step in a true chemical interaction which affects the chemical bonds.

Another aspect not considered in our exposition deals with the many body effects, i.e. the effects specific of the simultaneous non-covalent interaction of many molecules. The limitations of the present state of the theory are even more remarkable in the analysis of non-additive many-body contributions to the interaction.

In conclusion, there are still remarkable possibilities of improving the theory and of enlarging our knowledge of non-covalent interactions. In spite of these limitations, the theoretical description available at present may be used for many investigations.

For a relatively large range of interactions (practically all the non-covalent interactions, except the very weak ones, with stabilization energies lower than 2 kcal/mol) the leading term is the electrostatic one, E_{ES}. The definition of this term is the same in all the theoretical approaches, and depends upon the definition of two observables of the separate partners, $V_A(r)$ and $\Gamma_B(r)$, having both a well-defined physical status (we have examined, and discarded, in the preceding section, a recent alternative proposal which makes E_{ES} no longer independent of the properties of the separate partners).

E_{ES} is the contribution with the highest long-range character: so it will be the most important term at greater distances.

At intermediate distances, the leading character of E_{ES} cannot be deduced by formal relationships; the empirical evidence deriving from numerous decomposition of ΔE strongly suggests that E_{ES} still is in this region — the most interesting from the chemical point of view — the most important term in the description of the interaction as well as in the description of the changes in the interaction due to chemical substitutions in the partners A and B.

So, it has some sense to look at E_{ES} alone, discarding the other therms, for a first analysis of the interaction. The experience will learn, or the prudence will suggest, when the examination of other contributions to ΔE is advisable.

The expression of E_{ES} in terms of molecular electrostatic potential (MEP) and charge distribution of the two partners (see Eq. 5):

$$E_{ES}(r) = \int V_A(r) \, \Gamma_B(r) \, dr \,, \tag{34}$$

assumes a simpler form when the charge distribution $\Gamma_B(r)$ is reduced to a point charge q_B:

$$E_{ES}(R) = q_B V_A(R) \,. \tag{35}$$

We have changed the notation for the space variable r, to emphasize the analogy betweeen the position of q_B, where the electrostatic potential of A is sampled, and the set R of coordinates which in our notation indicates the position and orientation of B in a reference frame solidal with A.

Is an accepted convention to call "maps of V_A" the isovalue contours plotting of $E_{ES}(R)$ (Eq. 35), drawn in a given plane, with $q_B = +1$ (in atomic units) [2]. This convention, which assimilates electrostatic potential to energy, has been maintained in other graphical and numerical representations of $V_A(R)$.

We may envisage different uses of the MEP in the study of intermolecular interactions. We indicate here three different lines of exploitation, with short comments. First, an inspection of the $V_A(R)$ values, with regard to the recognition of the general shape of the MEP and/or to the identification of some special features, like the location and depth of the minima, the numerical value of $V_A(R)$ at some preselected points, etc. Second, the use of $V_A(R)$ in connection with an appropriate description of $\Gamma_B(R)$ to get an appreciation of $E_{ES}(R)$ and then of $\Delta E(R)$. Third, a decomposition of $V_A(R)$, at selected values of R, to gain information on the role played by the submolecular components of A in defining the value of this observable.

1) The local inspection of $V_A(R)$ is made easier by the scalar character of the MEP, a property which we have already indicated as being extremely useful in practical applications.

   The "global" inspection leads to the comparison of the MEP of different molecules. The search for similities, and of differences, may be greatly improved by the use of pattern research and artificial intelligence strategies, based on topological and topographical techniques. The characterization and classification of molecules via the MEP is at present largely exploited, with success, in commercial applications. Practical motivations stimulate, of consequence, a faster growth of this kind of application.

2) The direct calculation of $E_{ES}(R)$ via Eq. (34) is a straightforward step in supermolecular calculations, but it is not convenient when the full calculation of ΔE is not requested. Equation (34) requires in fact the evaluation of a noticeable portion of the two-electron repulsion integrals over the dimeric basis set, for each position $\boldsymbol{R}$. There are other ways of getting approximate values of $E_{ES}(R)$. The simplest expression formally makes use of a point charge approximation of $\Gamma_B(R)$. If Γ_B is reduced to a finite set of point charges:

$$\Gamma_B(r) \simeq \sum_k q_k \delta(r - r_k), \tag{36}$$

the coulombic energy is reduced to

$$E_{ES}(R) = \sum_k q_k V_a(r_k). \tag{37}$$

Other expansions of Γ_B, as well as of V_A, may be employed. This topic will be discussed in the appropriate section.

A practical recipe, based on the use of MEP values only, is sometimes employed, although formally incorrect. The recipe consists of replacing the interaction energy with an auxiliary function $F(R)$ defined as the algebraic sum of potential products:

$$F(R) = \sum_i V_A(r_i) V_B(r_i). \tag{38}$$

The applications are generally restricted to the search for the optimal orientation of a small molecule B (for example, water) at opportune sites of a larger molecule A. The method was shown to be justified in the proportionality found empirically between the hydrogen bond strength and the product of the two MEPs at special reference points [31–32]. For some applications of the F function method see Nágy [33–34], and Alagona et al. [35].

When the charge distribution Γ_B may be described by a point dipole, it is convenient to explicitly use the molecular electric field (MEF) $\vec{F}_A$:

$$E_{ES}(R) = -\vec{F}_A \cdot \vec{\mu}_B(R). \tag{39}$$

The definition of $\vec{F}_A$ is shown in Eq. (3).

Maps of $E_{ES}(R)$ computed according Eq. (39), with a dipole moment $\vec{\mu} = 1$ (in atomic units) colinear to the direction of $\vec{F}_A$ are commonly called "maps of the MEF", in analogy with the "MEP maps". The display of the isoenergy curves is often accompanied by a graphical representation of the direction of $\vec{F}$, representation made nowadays easier by the computer graphic facilities.

3) The information carried by $V_A(r)$ may be profitably analyzed in terms of submolecular contributions (chemical groups, or other subunits). According to the definition of $V_A(r)$ (Eq. 2), also relatively distant groups give significant contributions to the MEP at a given point r.

Such an analysis has proved to be quite useful for the interpretation and prediction of chemical reactivity. The partition of the charge distribution $\Gamma_A(r)$

is arbitrary, but when a partition has been selected, the decomposition of $V_A(r)$ follows immediately. In fact, from

$$\Gamma_A(r) = \sum_g \gamma_g^A(r),\tag{40}$$

one has (Eq. 3):

$$V_A(r) = \sum_g \int [\gamma_g^A(r')/(r - r')]\, dr'.\tag{41}$$

A coherent and comprehensive description of molecular entities in terms of molecular subunits is reported in another volume of this series [20]: some points will be considered again in a later section.

We have thus identified and shortly commented on three uses of the MEP in the field of molecular reactivity. This selection of topics has, however, neglected other themes, still related to molecular interactions and molecular reactivity, giving rise to other specialized chapters in which an exhaustive review of the use of MEP could be organized.

Crystal packing forces are to some extent ruled by electrostatic interactions. The calculation of the Madelung constant is just one of the possible applications of the semiclassical concepts based on the consideration of the leading classical terms of ΔE, and then of the MEP. The presence of a regular and infinite lattice of charge subunits introduces new elements not present in the simple case of a dimeric interaction considered in Sect. 3 (see, e.g. Taylor [36]) and Ghio et al. [37]. We shall not consider in the following sections problems concerning the electrostatic potential and energy in perfect crystal lattices, nor problems related to reactions in crystals or to the presence and influence of defect, occlusions, etc.

The same basic approach may be adopted also to study the interactions of molecules with the surface of massive bodies. According to the nature of the interaction and the characteristics of the body, the elaboration of the methodology highlights other features of the problem, but the MEP, and similar concepts, continue to play an important role. Interactions of solutes with charged metallic surfaces, interactions of small molecule with zeolites, interactions of molecular substrates with enzymes or other biological receptors, phenomena involving membranes, are some scattered examples of a very large field of phenomena whose characteristics depend on the interaction of molecules with larger bodies.

Of even wider occurence and larger importance in chemistry is the interaction of a molecule with a solvent. Solvent polarization effects may be described in terms of the values of $V_A(r)$ (or $F_A(r)$) at some specific positions around the molecule. We summarized in an earlier contribution to this series [20] some points of our methodological proposals in this field, but much work has been done by many other researchers with other variants of the approach.

Every theme considered here and many others of similar nature deserve a detailed review. The number of publications is so large, that we felt ourselves justified for not giving here bibliographical references, which would be lamentably incomplete.

A computer search of the literature concerning the use of the MEP and similar semiclassical elctrostatic concepts has given as output more than 20,000 items over the last four years. A comprehensive review is clearly impossible.

5 Calculation of the Electrostatic Potential and the Search for Simpler Formulations

The MEP, being the expectation value of a one-electron observable, is related to the evaluation of the penetration integrals over the functions of the expansion basis sets:

$$V_A(r) = \int \Gamma_A(r') |r' - r|^{-1} dr'$$

$$= \sum_\alpha^{nuc} Z_\alpha |R_\alpha - r|^{-1} - \int \varrho_A(r') |r' - r|^{-1} dr'$$

$$= \sum_\alpha^{nuc} Z_\alpha |R_\alpha - r|^{-1} - \sum_{r,s} D_r \langle \chi_r | r' - r|^{-1} | \chi_s \rangle , \qquad (42)$$

where the D_r coefficients are the elements of the first order electronic density matrix describing A.

The calculation of V_A at a given point r may be thus done by using the same program employed to get the molecular wavefunction of A, at a marginal cost with respect to that necessary to obtain the wavefunction itself. The one electron penetration integrals are collected in a square matrix $V(r)$ with the same dimensions m of the basis set, and the electronic component of $V_A(r)$ may be reduced to a more compact expression

$$V_A(r)^{el} = tr D V(r) . \qquad (43)$$

The quality of the $V_A(r)$ values depends, of course, on the quality of the D matrix. Much attention has been paid to the influence of the basis set in SCF calculations: an analysis of this topic may be found in early reviews on the MEP [32, 38, 39]. The overall conclusion is that the general shape of $V_A(r)$ is not heavily affected by the basis set: the numerical values (e.g. the values of the minima) are more sensitive to the basis set but the number of known examples in which these changes affect the picture of chemical reactivity is quite limited. $V_A(r)$, like all the one-electron observables is correct to the first order in the electron correlation effects. This property is not sufficient to ensure insensivity to correlation effects because second order corrections could be large. The information on this point was scarce until recently. Daudel et al. [40] in a study of the inclusion of CI corrections to the wavefunction of the ground and first excited states of H_2CO showed that the MEP values in the region of chemical interest are little affected by the number of configurations. The minimum for the ground state passes from -50.3 kcal/mol to -50.2 when the number of configurations increases from 634 to 2474. These findings have been confirmed by a recent and detailed study by Luque et al. [41] in which abundant numerical evidence on the correlation and basis set effects on MEP values is analyzed: electron correlation has a relevant effect on $V_A(r)$ near the nuclei but produces only small changes in the regions of chemical interest.

In practically all the chemical applications it is however necessary to compute $V_A(r)$ in a large number of points, which especially in the most recent algorithms of automatic search of the relevant features of the MEP (see for example Ref. [42]) may be numbered by thousands. When geometry changes in A are also taken into account, or when the MEP is employed in molecular simulations, the number of points may easily surpass a million.

Things are even worse when A is very large. Ab initio calculations for large molecules are limited by the computer possibilities, and even if it were possible to get a D_A density matrix, the repeated evaluation of $V_A(r)$ via Eq. (42) becomes a too exacting task.

It is necessary to find elaborations of the basic formulation able to reduce in a sizeable way the computation times. Efficient strategies to obtain $V_A(r)$ for very large molecules without previous calculation of D_A must also be sought. Much effort has been spent to reach these objectives, and now we shall examine some proposals.

The years that have elapsed from the beginning of an extensive use of MEP values from molecular ab initio calculations (1969) have show great changes in the computer hardware and software, and several proposals have become outdated. We shall contain within reasonable limits the quotations, especially of the older literature.

The methods of calculation of the one electron integrals on the atomic basis set, appearing in Eq. (2) have been greatly improved since 1969. The most recent method which we know has applications to the calculation of MEP values is by Obara and Saika [43]: this paper also reports a concise discussion of preceding methods.

The integration of the Poisson equation:

$$\nabla^2 V_A = -4\pi \Gamma_A,\tag{44}$$

may be exploited to simplify the analytical form of the one-electron penetration integrals. This possibility was pointed out by Srebenik et al. [44], and adopted later on in several programs developed specially for MEP calculations (for further comments see Ref. [45]). The computational times are reduced by a factor of 3–5 and this reduction increases for atomic functions with a higher angular moment.

Approximate expression of the one-electron integrals are sometimes employed, especially in connection with semiempirical wavefunctions. The first proposal was made by Carbó and Martin [46], and several other variants have been suggested over the years.

The reduction of computational times given by the use of atomic pseudopotentials has been also exploited for MEP calculations. The first attempts were made by the Weinstein's [47] and the Pullman's [48] groups. More recent pseudopotential basis sets apparently give good representation of $V_A(r)$ in molecules containing heavy atoms (unpublished results).

The use of semiempirical methods may be considered a different way of reducing computational time. Practically all the semiempirical methods have been tested. The questions related to the reliability of the MEP values obtained with semiepirical methods are intermingled with technical problems related to approximations introduced in the semiempirical methods. Giessner-Prettre and A. Pullman in a series of well-known papers [49–51] have examined a detailed strategy to get MEP values from wavefunctions computed within the ZDO approximation, defining several levels of approximation. Similar analyses on the CNDO wavefunctions have been performed by other authors [52–54]. The performance of the INDO method has been examined by Weinstein et al. [55], Petrongolo [56], and Culberson and Zerner [57], amongs others. More recent studies by Luque et al. [58–60] and by Ferenczy et al. [61] deals with the performances of the MNDO and AM1 methods. PCILO and EH results have been reviewed by Petrongolo [56]. The EH method has been more recently considered by Weber et al. [62, 63], with an improved parametrization. The work of the Weber's group deserves mention also for the nice graphical representation of MEP values on the molecular surface.

This concise set of references contains papers with methodological interest. The number of papers concerned with applications is much larger.

6 Simplified Analytical Expressions of the Molecular Electrostatic Potential

6.1 The Multipole Expansions

All the methods considered in the preceding section relies, strictu sensu, on analytical expressions of the MEP. This section is dedicated to a specific strategy of elaboration of V_A into simpler analytical expressions via multipole expansions. The use of a multipole expansion may reduce the computation time by several order of magnitude. Multipole expansions are however approximate, and may led to considerable errors. There is no unique way of getting a multipole expansion once $V_A(r)$ (or $\Gamma_A(r)$) is known, and the different methods vary greatly in reliability, computer time, and other characteristics. The variety of the approaches, and the importance of the subject suggest that we treat this topic in a specific section. The expression of the MEP in terms of the so-called multipole expansion is a classic topic [64–65], described in many textbooks [7, 66–68].

The one-centre expansion of the MEP is not adequate, however, for the large number of chemical applications we mentioned in Sect. 4. In fact the expansion theorem holds for points r lying outside a sphere containing all the elements of the charge distribution. In molecules, this condition is never formally satisfied, because $\varrho_A(r)$ has an exponential decay. The difference between $V_A(r)$ and the *exact* multipole expansion of $V_A(r)$ is generally called the *penetration term*:

$$V_A(r) = V_A^{mult}(r) + V_A^{pen}(r), \tag{45}$$

with

$$V_A^{mult}(r) = \sum_{l=1}^{\infty} \sum_{m=-l}^{l} Q_l^m r^{l+1} Y_l^m(\theta, \varphi), \tag{46}$$

where Q_l^m are the elements of the various multipolar terms, and $Y_l^m(\theta, \varphi)$ are spherical harmonic functions (see Berrondo et al. [69] for a detailed study of a simple case).

$V^{mult}(r)$ has the correct asymptotic behavior: when the number of terms in a truncated expression of $V^{mult}(r)$ is kept fixed, the description improves at larger distances from the expansion origin. $V^{mult}(r)$ is also convergent at large values of r: this remark is not a pleonasm, because convergency is not ensured for analogous multipole expansions concerning higher terms of the interaction energy (e.g. polarization and dispersion contributions [70–71]).

Convergency and asymptoticity of $V^{mult}(r)$ are not sufficient for chemical purposes. In studies of chemical reactivity the most important regions of the outer molecular space are near the van der Waals surface of the molecule, or just inside it. The reactive regions are often surrounded by peripheral groups, and of consequence the point of larger interest are deeply inside the sphere containing the largest portion of the $\varrho(r)$ charge distribution. In these cases the portion of $V_A(r)$ not included in the expansion (i.e. V^{pen}) is excedingly large, and the addition of higher terms to the multipolar expansion may produce worse results.

The use of one centre multipolar expansions should of consequence be limited to very restricted classes of systems, in particular to material systems of small size, interacting at large distances, in other words, to problems of physical and not chemical interest.

Better results for chemical reactivity studies are obtained be using many-centre expansions. The adoption of many-centre expansions means that the total charge distribution $\Gamma_A(r)$ is segmented into smaller portions. The convergence radius of these fragments of charge distributions is smaller than the molecular one, and so it is easier to get a sensible description of $V_A(r)$ for the crevices of a corrugated molecule. It may also be assumed that a small number of terms in each expansion series will be sufficient to reach a reasonable description of $V_A(r)$.

This said, the number of the local expansions, their location, the number of expansion terms for each centre, and the method to get the numerical coefficients for each expansion term must be defined. There are no formal constraints, and the strategy should be selected on the basis of its efficacy: computer time and precision. A detailed and clear exposition of the problems involved in, and of the options open to the definitions of local expansions has been recently done by F. Vigné-Maeder and the late P. Claverie [72]. We shall follow in part this exposition, giving more emphasis, at the end, to the use of atomic monopole expansions (i.e. atomic charges) and to mixed representations, which represent, in our opinion, the most versatile method for chemical reactivity problems.

It is convenient to start from the analytical expression of $\varrho_A(r)$, as done by quantum calculations over a finite basis set of simple analytical functions. The basis set functions are in general of Gaussian type, and they have special mathematical properties which may be exploited here [73, 74]. The extension of this analysis to other basis functions, such as the Slater type orbitals, is however possible.

The electronic density distribution may be dissected into elementary charge distributions:

$$\varrho(r) = \sum_{s,t} D_{st} \chi_s^*(r) \chi_t(r), \tag{47}$$

where the D_{st} coefficients are elements of the first order density matrix (see Eq. 42). A decoupling of the contraction factors, if necessary, is here introduced. The elementary distribution functions

$$\chi_s^*(r)\,\chi_t(r)\,, \tag{48}$$

are the product of two Gaussian functions centred at r_s and r_t respectively. This basic charge distribution may be expressed, exploiting the mathematical properties of the Gaussian functions, as a sum of Gaussian functions centred at a well defined overlap centre r_g. For example, when the two Gaussian functions are of s type with exponent coefficients α_s and α_t, the distribution (48) is reduced to a single Gaussian function with exponent coefficient $\alpha_g = \alpha_s + \alpha_t$ centred at a point determined by the position vector

$$r_g = (\alpha_s r_s + \alpha_t r_t)/(\alpha_s + \alpha_t) \tag{49}$$

In general, the elementary charge distribution (49) is reduced to the sum of a finite number of Gaussian functions, from which it is immediate to derive an exact multipole expansion composed by a *finite* number of terms: the upper term is the sum of the angular momentum quantum numbers of the two functions. There is still, of course, a penetration term, related to the exponential decay of the functions. Hall has shown, with numerical examples, that this penetration term is reasonably small [75].

A description of $\varrho_A(r)$ in terms of a finite number of Gaussian functions may be thus replaced by an exact expansion composed by a finite number of terms centred at a finite number of places. Generally, the expansion basis sets are of atomic nature, with a limited number of functions, in some cases, not centred on the nuclei (the so-called bond-functions). There will be, of consequence, many elementary distributions centred on unique centre (the nuclei, for example) which may be reduced to single functions centred at the same place. In conclusion, we shall have expansion centers on the lines connecting the basis set centres, and on the basis set centres themselves. So, if a molecule A has N basis set centres (N atoms or M atoms and M' bond orbital centres, with $M + M' = N$), we shall have $N(N - 1)/N$ joining segments on which a variable number of expansion centres is placed, and N vertexes, also acting as expansion centres. To these expansion sets one has to add N point monopoles, corresponding to the nuclear charges.

This decomposition, satisfactory from the formal point of view, cannot be used in practice, because the number of expansion centres is excessively large. A few numerical values may help the reader to realize how large the number of centres is: for the H_2O molecule the expansion of the charge distribution in terms of the STO-3G basis set requires 46 expansion centres; the use of an analogous basis set without the $s = p$ restriction requires 64 centres; the Clementi's minimal basis set [76] requires 58 centres, the 6–31G** basis set 134 centres. The number of expansion centres for benzene and adenine with the Clementi's basis set are 1325 ad 2125 respectively [72].

Fortunelli and Salvetti, in a recent paper [77] which had more ambitious aims, the simplification of ab initio computation methods with Gaussian basis sets, present a different method of reducing the number of the centres of expansion of the elementary charge distributions, based on the "asymptotic completeness" theorem valid for appropriately selected sets of Gaussian functions [78]. This approach has not yet been completely developed and tested for the evaluation of MEP values; we shall continue our discussion following the simplifications of the elemtary distribution expansions proposed and tested in recent years.

Some of the elementary distribution expansion centres may be simply delected (each distribution is weighted by the corresponding D_{st} coefficient, and many of these coefficients, especially for distant χ_s and χ_t functions, are small). Other expansion centres may be eliminated by applying a further expansion procedure. A elementary multipole expansion, centred at r_g, may be thus eliminated by expansion at another position, r_s, already employed for a different elementary expansion. In some case it is convenient to "share" the multipole expansion at r_g among two or more centres, with different r_s values, using appropriate weights. These further expansions of the elementary multipole expansions are no longer bounded by the angular momentum l and are then subjected to truncation errors. There is of course a complete freedom in the selection of the new r_s expansion centres, but this freedom is limited by some practical considerations. The distance between the r_g and the r_s centre must be not too large, otherwise the expansion will contain important contributions with high l values. The convergence properties of a r_s expansion is ensured only at the exterior of a sphere containing all the pertinent r_g centres. A reduction in the number of the r_s centres requires, to maintain a given approximation in the results, an increase in the maximum rank of the expansion functions, and inversely the decision of keeping low the maximum rank implies a less drastic reduction of the r_s centres.

Vigné-Maeder and Claverie [72] examined 17 contraction schemes, but the number of proposals available in the literature is higher.

This approach was employed for the first time, to the best of our knowledge, by Hall and coworkers. In Refs. [75, 79] the approach is applied to molecular wavefunctions expressed in terms of floating spherical Gaussian orbitals (FSGO [80]), with a shrinking of the resulting spherical distributions (in number equal to $n(n + 1)/2$, where n is the number of centres for the FOSGO's) to point charges. The number of centres was reduced to n by Shipman [81], who still used FSGO wavefunctions. Other point charge representations of FSGO charge distributions and potentials were proposed by Amos and Yoffe [82] and by Friedmann et al. [83]. The extension of the approach to molecular distributions containing Gaussian functions of higher angular momentum was considered by Martin and Hall [84], still maintaining the shrinkage to point charges.

Further progress of the Hall analysis will be reviewed later on.

The distributed multipole analysis (DMA) proposed by Stone [85] follows an analogous line of approach. The nuclei and the bond midpoints are selected in the DMA as unique r_s expansion centres.

Stone and Alderton [86] made some considerations which deserve a mention here. These authors remarked that minor changes in the basis set produce large changes in the value of the multipole components at the various DMA sites, even when the strategy of selection of the expansion sites remain unchanged. This fact is a consequence of having selected a "basis-set-oriented" method. An analogous dependence on the BS is present also in the Mulliken atomic charges, which are an extreme simplification of a distribute multipole expansion. One-electron observables, as for example the MEP, are less dependent on the BS, because the available basis functions are combined in a different way to maintain the electronic density almost constant. Stone and Alderton tried to exploit this effect by using a division of the physical space to get multipole expansions less depending on the basis set. The results confirm this analysis. Expansions based on the partition of the physical space, however, produce local multipole moments larger by orders of magnitude than those obtained with the usual DMA. The convergence properties in the description of the MEP are of consequence worse. The same analysis applies to the use of the Bader surfaces [87] exploited by Cooper and Stutchbury [88]. In this last case, in addition, the computational time is exceedingly large (several orders of magnitude larger than normal DMA calculations) due to the difficulty of performing volume integrations with complicated boundaries.

We have reported this discussion in some detail because it combines several points treated separately by many authors. These remarks shed some light on the difficulty of getting local multipole expansions which are reliable, easy to handle, and at the same time transferable from molecule to molecule. We have in fact considered, until now, the problem of getting multipole expansions from an already known $\varrho_A(r)$ function, without touching the more important problem of formulating local expansions transferable from molecule to molecule. We shall see later how these problems may be partially solved.

Stone's DMA method has been applied in several other papers. Our review cannot be exhaustive but we would like to quote two additional papers using this approach because they give additional information on the basic problems of the electrostatic approach. Price, Harrison and Guest [89] examined the DMA description of the MEP of a large molecule, with formula $C_{63}H_{113}N_{11}O_{12}$, obtained from a 3–21G SCF wavefunction. The description of the electrostatic potential obtained in such a way is comparable to that obtained with potential derived atomic charges (PD-AC) to which we shall refer later on in more detail. The superiority of a distribute multipole description, in describing the anisotropic contributions to the MEP on the van der Waals surface is shown clearly.

In another paper Faerman and Price [90] present a very detailed set of DMA results (especially in the supplementary material) concerning peptides and amides (from 3–21G wavefunctions). The main interest of this contribution is the attention paid by the authors to the transferability of the multipole descriptions. Both papers demonstrate the inadequacy of Mulliken charges for the description of the MEP as well as in representing the transferability properties of the groups.

Coming back to the 17 partition schemes numerically tested by Vigné-Maeder and Claverie [72], we found among them several schemes which have had previous elaboration and many applications. We quote among them the OMTP procedure (overlap multipole truncated potential) [91–94] in which the expansion centres r_g are limited to one per segment, with expansion terms until the quadrupole term ($l \leq 2$), and to the nuclei. For a molecule with N atoms, the expansion centres are thus $N(N + 1)/2$. Langlet et al. [94] introduced a further simplification: the expansion terms are limited to the nuclei, to the true chemical bonds, and to a point on the segments connecting two atoms chemically linked via a third atom. The OMPT procedure has been applied in several studies on systems of biological interest, by the Pullman group and by independent researchers.

This kind of expansion is considered by Vigné-Maeder and Claverie as the best compromise between simplicity and accuracy. A simpler expression, considered by these authors still acceptable for many applications, limits the expansion centres to the nuclei alone, always with $l_{max} = 2$.

The technique of starting from the elementary charge distributions has been adopted by many other authors. We quote, as a significant example of the evolution of the methods over the years, the works of Rabinowitz et al. [95–99]. In an early paper, a multipole expansion centred on the atoms, and not based on a progressive reduction of elementary distributions, was adopted [100]). This method has been applied in many studies on molecular system (see e.g. Rein [101]. The elementary charge distribution expansions were introduced in Ref. [96], reduced in number via a selection on the resulting charges in Ref. [97], and refined by allowing the Gaussian distributions with lower orbital exponents to be left unchanged [98]. The results are fairly good [99].

The elimination of multipole expansion for the elementary distributions regarding Gaussian functions with a low orbital exponent, α, introduced in Ref. [98], pays attention to a problem we have already mentioned. The exact potential V may be divided into two components (Eq. 46) in which one component, V^{pen}, which in the expansion method is neglected, may be considered to be the error intrinsic in the multipole expansion. A large contribution to V_A^{pen} is given by the exponential tails of elementary distributions with low α's, which are quite spread. An exact representation of these components of $V_A(r)$ reduces the extent of the neglected contributions to the MEP.

The idea of keeping uncontracted diffuse Gaussian contributions derives from G. G. Hall. In a set of papers [102–106] this author analyzed some aspects of the approximation of molecular electronic densities (not all the papers are concerned with the MEP). Hall devoted much attention to the problems of fitting the electron densities: the quoted set of papers presents several points of interest. As far as the MEP is concerned, Hall elaborated a method based on a fitting of the approximate MEP via the Dirichlet functional of the error field [103] and suggested a method which maintains the shrinkage to point charges of most of the elementary charge distributions, accompanied by the conservation of diffuse elementary distributions.

The reduction of the multipolar expansion terms to the atomic centres has been adopted by many authors. This selection of centres is reasonable under many aspects; the electronic charge is essentially condensed around the nuclei, with an important spherical contribution: higher multipole terms should be relatively smaller than with other partitions. Moreover, there is a greater chance that these atomic multipolar distributions present some degree of transferability. Several of the already quoted papers present and discuss methods and results obtained with atomic multipole expansions. We add here a paper by Mezei and Cambell [107] in which the expansion is pushed to very high orders, and several methods of reduction to atomic expansions are compared, including also a "very extreme split" in which all the overlap from an atomic centre is eliminated, and a "sharp split", based on the partition of a bond with its median plane: each method shows advantages and defects. Atomic multipole expansions are able, in general, to give a fairly good description of the MEP in the region outside 4.5 a.u. [108], at the expense, however of the inner region [106]. The systematic work of Sokalski et al. [108–110] deserves a particular mention: the numerous expansions elaborated by this group are also available upon request to the authors. Other expansions of this type may be found in Refs. [111–113].

6.2 Atomic Charges

The extreme case of segmental multipole expansions is given by the expansion into atomic charges. The number of expansion centres is small, and each expansion is limited to the first term, the monopole, or charge.

The use of atomic charges in the interpretation of chemical phenomena has accompanied the progresses of chemistry over the past two centuries. An almost satisfactory status to atomic charges in modern theoretical chemistry was provided several decades ago by R. Mulliken [114]. The reviews on atomic charges have been quite numerous, and the steady progress requires periodical updating; some recent reviews are reported in Refs. [115–117].

In recent times more attention has been paid to the ability of atomic charges to reproduce $V_A(r)$. The atomic charges derived from the Mulliken population analysis, called in short Mulliken charges, are in general disappointing, giving a representation of the MEP of relatively poor quality. Several shortcomings of the Mulliken charges were also known in the past, and there is a large amount of literature suggesting alternative partitions into atomic contributions of the computed molecular charge distribution, some of these definitions, (see Refs. [115–116] for more details) which preserve the molecular dipole moment, or other similar properties, give better representations of the MEP. Several of the papers quoted in Sect. 6.1 give definitions of atomic charges better than the Mulliken ones for the description of $V_A(r)$. One of the reasons for the poor performances of the Mulliken charges is that this quantity is a base oriented property of the molecular wavefunction (see Sect. 6.1) for the definition of which no particular attention has been paid with regard to their possible role as numerical sources of the MEP.

Atomic charges may be, however, defined in other ways. A score of experimental methods can be used as a source of "experimental" charges (actually they derive from the application of theoretical models to experimental data). ESCA, NMR, IR, Mossbauer, X-ray diffraction experimental results, among others, are sometimes employed to obtain atomic charges in molecule [118–125]. The values obtained are often compared with charges obtained from other sources (in general Mulliken charges) or combined to get an extimate of the molecular dipole moment and compared then to other values of μ; a systematic comparison with reliable values $V_A(r)$, which would be the most direct check, have not yet been attempted, with the exception of X-ray results [39].

Atomic charges may be also derived, with simple computational techniques, by applying electronegativity equalization procedures. There are several variants of the method [126–131]. A discussion of its theoretical justification may be found in a review by Bergmann and Hinze [132]. The approach is flexible enough to allow further improvements in the parameters: in our opinion, instead of aiming at a better correlation with Mulliken charges, a redefinition of the parameters based on a comparison with $V_A(r)$ would be advisable. The results obtained with the most recent versions seem to indicate that the approach is promising.

Another unexpensive approach to obtain atomic charges derives from the application of the Hückel method [133] for π electrons, and its extension to σ electrons proposed by Del Re several years ago [134–135] (see also Berthod and Pullman [136]) which requires a determination of atom electronegativity and atom-atom exchange integrals. The Huckel-Del Re parameters have been reparametrized by the Pullman group in Paris [137–138] to get monopole expansions (i.e. atomic charges) which reproduce at the best the MEP obtained with the ab initio OMPT expansion quoted above [91–93]. The method has been applied to a considerable number of biological problems, in which the size of the involved molecules makes impossible ab initio molecular calculations. The comparisons displayed in the source papers, and in some of the application papers, indicate that the optimization procedure, essentially performed by optimizing MEP and MEF on a large number of points on the van der Waals surface, has been successfull.

Independently, the Hückel-Del Re parameters have been reoptimized by Houser and Klopman [139] to mimic the CNDO charges. These authors do not use MEP values as a check, but the CNDO charges give in general a reasonable qualitative description of the main features of $V_A(r)$.

Another simple procedure to get atomic charges was proposed several decades ago by Smith and Eyring [140]. The method uses bond polarizabilities, covalent atomic radii, and parametrizes the charges on the molecular dipole moments. Improvements were introduced by Allinger and Wuesthoff [141], by Micovič and Allinger [142] and more recently by R. J. Abraham et al. [143–145].

The optimization of the parameters in semiempirical formulas (all the approaches considered in the last pages belong to this category) proceeds slowly, because parameters optimized for one class of compounds cannot, in general, be applied to another class without some modifications. All the methods based on the parametrization of simple expressions (electronegativities, polarizabilities, etc.) are addressed to specific chemical classes of compounds: the extension to other classes requires a revision of the parametrization, or, often, a novel distinct parametrization.

As the last approach to obtaining atomic charges we shall consider the so-called potential derived (PD) methods. Actually such methods constitute a reversal of the problem discussed until now: the definition of atomic charges to get representations of the MEP. The PD methods start from MEP values, obtained by calculation, and derive atomic charges fitting best the molecular potential. This formulation of the problem – a fitting of MEP with point charges limited to the atoms – was first explored by Momany in 1978 [146], and gained in a short time a considerable popularity: ten years later the material was considered sufficient to write a detailed review on this specific topic [117]. The main reason of the interest in such a specific area of a by far wider domain consists of the fact that PD charge distributions have been adopted in a considerable number of computer packages for the evaluation of molecular interactions and molecular conformations. It is beyond the limits of this chapter to review the impressive progress in the elaboration and use of integrated computational packages for molecular modelling. The analysis of the simple case of bimolecular interactions, sketched in the preceding sections of this chapter, has been applied, with the opportune changes, additions and simplifications to many programs concerned with inexpensive calculations of molecular conformations of large molecules and to the evaluation of the geometric and energetic properties of molecular complexes. The experience has tought us that it is convenient to consider intramolecular interactions as a specific case of intragroup interactions, for which the experience gained in the study of intramolecular interactions may be profitably exploited. In recent years, the semiclassic programs (molecular mechanics), have been implemented as a consequence by electrostatic interaction terms, often neglected in the earlier versions.

The atomic charges, which as we have said are often derived from the MEP, are thus often employed for two distinct, even if connected, purposes, namely the evaluation of the energy (or better, the electrostatic component of the energy) and obtaining the MEP. In a very accurate representation, it is correct to use a physically well-defined quantity for two different aims, but approximate representations could be good for one type of application and completely inadequate for the second. Apparently this is not the case for the PD atomic charges, which are often employed for both purposes.

In the large amount of literature concerning PD charges, we shall select a limited number of quotations, focussing mainly on the technical details. Momany, in the already quoted paper [146] makes a least square fitting, with the additional constraint of preserving electronegativity. The addition of a further constraint concerning the preservation of the molecular dipole components makes the fitting worse; a partial recovering of the quality of the preceding fitting has been obtained by scaling the ab initio $V_A(r)$ values by a fixed value. The fitting makes attention mainly to the distant regions of space, and introduces a penalty function for points lying slightly within the van der Waals surface.

The selection of the set of points where the MEP was to be evaluated and then fitted affects the quality of the fitting. A compromise between accuracy and the number of points must be reached. This selection of point locations, the consideration of additional constraints, the mathematical method adopted to perform the fitting, represent the main methodological differences in a sizable set of papers regarding the obtaining of potential derived atomic charges (PD-AC, we are using here the terminology introduced by Williams and Yan [117]). The evolution of the approach had led to the necessity of introducing other acronyms, such as PD-LP (Potential Derived atomic charges with inclusion of Lone Pair charge sites), PD-BC (as above, with additional Bond Charge sites), etc. All these methods may be grouped in the PD/LSF (Potential Derived/Least Square Fitting) methods.

Some earlier papers (Bonaccorsi et al. [147], Alagona et al. [148], Kollman [149], Agresti et al. [150], Smit et al. [151], Sauer and Morgeneyer [152], Thole and van Duijnen [153] do not exactly correspond to this scheme because other considerations and approximations were also taken into account, but the above mentioned papers present all results which may be considered as belonging to the PD approach.

Cox and Williams [154] pay attention to the effect of the basis set and to the comparison of PD-AC values with Mulliken charges. In fact the PD/LSF procedure is still time-consuming, and the possibility of coming back to Mulliken charges computed on smaller basis sets, with an appropriate scaling with respect to more precise results must not be ruled out as irrelevant for numerical applications. In their paper, Cox and Williams examine the performances of three basis sets, STO-3G,

6–31G and 6–31G** for 14 molecules of small size. The analysis shows the larger sensitivity of the Mulliken charges to the basis set. The scaling procedure allows us to get, for the set of molecules here examined, a fairly good agreement between STO-3G or 6–31G PD charges and the 6–31G** PD charges taken as reference. The fitting procedure is based on a least-square analysis over points lying outside the van der Waals surface in a layer with thickness $t = 1$ Å. No constraints over the value of the dipole moment, or other observables, have been introduced.

Singh and Kollman [155] adopt a Levenberg-Marquardt non-linear optimization procedure to perform the fitting and examine the use of successive shells of various multiples of the van der Waals radii in selecting the points on which the fit is performed. The analysis is performed over 18 molecules also of medium size (nucleic acid bases, sugars and phosphates) using basis sets of different quality, from STO-3G and Clementi's MB to 6–31G*. Not all the molecules are treated at the same level of accuracy. For simpler molecules, such as H_2O, CH_3OH, CH_2O, also models with a large number of charges are analyzed, with or without additional constraints on μ and Θ. It is worth remarking that models including charges for the heteroatom lone pairs have optimal positions often at variance with chemical intuition, and strongly dependent upon the basis set. Some results of this analysis have been employed to implement the AMBER molecular mechanics program for conformational studies on nucleic acids and proteins [156]; more recent versions of this program adopt modifications in the definition of the charges, the evaluation of the electrostatic term being the key point assessing the quality of a molecular mechanics program.

Chirlian and Francl [157] follow the Cox and Williams approach [154], with the introduction of a more efficient fitting procedure, based on the search of the stationary points of a Lagrangian function associated to the square deviations. The use of Lagrangian multipliers permits us to avoid a preliminary guess at the charges. The program, named CHELP (CHarges from ELectrostatic Potentials) is available through QCPE [158]. The results discussed in the paper have been obtained without the use of additional constraints: the values of the dipole moment are employed as additional check. On the basis of results concerning 16 molecules of small dimensions and 5 basis sets, it turns out that the 3–21G BS gives consistently good results in a reasonable amout of time. The fitting must be extended to cover the region up to 3 Å from the van der Waals surface.

Woods et al. [159] again use the Lagrangian multipliers method, adding the constraint on the dipole moment. Attention is focussed on the numerical aspects of the method, in particular on the effect of number and location of the sampling points, and on the variance of the results with respect to a rotation of the position of these points. The paper extends the analysis to several molecules of medium size, opening the way to the problem of transferability of the PD-AC values obtained for small molecules.

Breueman and Wiberg [160] present a variant of the CHELP program, named CHELPG, which is shown to be less dependent upon molecular orientation than the original program. Both versions were employed for a study of changes in atomic charges due to the internal rotation in formamide about the C-N bond. This paper opens new perspectives for the use of PD-AC values.

The papers reviewed thus far refer to PD-AC values obtained with ab initio methods. The approach may be used also starting from MEP values obtained from semiempirical wavefunctions. We have already considered the application of semi-empirical methods to compute $V_A(r)$ values (Sect. 5): here we shall restrict the subject to obtaining PD-AC values. Ferenczy, Reynolds and Richards [61] apply the already mentioned CHELP program [158] to AM1 MEPs, and compare the results with those obtained using ab initio STO-3G and 6–31G** basis sets. The quality of the correlation is fairly good in both cases (and better than that among the corresponding Mulliken charges), even if the authors put the emphasis on the qualitative agreement between AM1 and STO-3G values. It is worth remarking that these authors use, in the evaluation of AM1 MEP values, the same approximation used in the evaluation of the AM1 wavefunction and energy, at variance with respect to other papers on similar subjects, in which the interpretation of the wavefunction given in Ref. [49] and the related deorthogonalization recipes are adopted. (The deorthogonalization procedure has been recently reviewed by Chuny-Phillips [161] for CNDO wavefunctions with positive conclusions). The calculation of $V_A(r)$ with the semiempirical approximations greatly reduces the computational time. We have numerical evidence (not yet published) supporting the view that this computationally simpler approach gives acceptable results also in the calculation of the solvation energy with our polarizable continuum model: the solvation energy is actually a different probe of the electrostatic potential (and field) at a region near the van der Waals surface [162].

Besler, Merz and Kollman [163] examine PD-AC values obtained with the MNDO and AM1 methods (using the procedure of deorthogonalization of the semiempirical wavefunctions) and compare them with STO-3G and 6–31G* values. The authors find that MNDO charges are superior to AM1 charges in correlating with PD-AC values obtained from 6–31G* wavefunctions. The analysis is with 20 compounds, and the fitting exploits the Lagrangian multipliers method.

Orozco and Luque, in a set of recent papers [164, 165], compare again AM1 and MNDO PD-AC values with STO-3G and 6–31G* PD charges over 21 molecules. Both semiempirical methods give fairly good results, but the correlation between MNDO and 6–31G* values is considered "excellent" by the authors [164]. The analysis is extended in Ref. [165] to the comparison of E_{ES} values obtained using MNDO, AM1 and 6–31G** charges with SCF values obtained using the Morokuma's decomposition scheme in a set of hydrogen bonded dimers (including the two nucleic acid base pairs). PD charges give very good values (the AM1 ones are slightly better than the MNDO ones) while the Mulliken charges are poor.

It should be clear from the short exposition given above, that the derivation of PD-AC values has been a very popular subject in the last years. The computational techniques seem at present consolidated and there is a general agreement among the conclusions of the various researchers. The representation of electrostatic potentials and electrostatic interactions via atomic charges is very fast and some people are willing to accept some limitations in the quality of the results if the method is fast enough. The same people for other applications may require better approximations. We have already quoted a paper by Price et al. [89] touching on this point: atomic charges are unable to describe anisotropies of the charge distribution around the nuclei. This limitation may partially be overcomed by resorting to PD-LP or to PD-BC formulations of the method (for the meaning of the acronyms, vide supra). The addition of point charges with the aim of giving a better description of the charge distribution – and of consequence of the MEP – near atoms with lone pairs represents a small increment of computational times in the derivation and subsequent use of PD point charge distributions. Bonaccorsi et al. [166], Agresti et al. [150], Williams and Weller [167] among others, adopt this description with satisfactory results. The first two papers adopt a further constraint, of keeping the charge value equal to $-2e$ and optimizing the position only. Williams and Weller, consider as variable parameters the charge value and the distance from the nucleus, with the constraint of maintaining the local geometry. Least square fittings including lone pair charges may lead to unphysical results: we have already mentioned the examples given by Singh and Kollman [155] who found "inverted" lone pair sites, i.e. location of the point charge in the direction opposite to that of the main component of the lone pair. Several other models including lone pair charges, but not related to a fitting of MEP values, are available in the literature: for a review see Williams and Yan [117]. Point charges centred near the middle of the bond have been considered by Bonaccorsi et al. [166], Agresti et al. [150]. Williams and Houpt [168] introduces additional sites along the axis of the C-F bonds in fluorocarbons (PD-BC method) and along the extension of the axis (PD-EC method); the models have been tested with success in the study of crystal structures.

A further step in the elaboration of PD descriptions is exemplified by a paper by D. E. Williams [169]. The last square fitting is performed over models consisting of charges, dipoles and quadrupoles on the nuclei, with different selection of the multipole terms (models $C + D, C + Q, C + D + Q$). In parallel, a model composed of bond dipoles only is also considered. This last model seems to be a convenient alternative to the PD-AC model. The Williams' programs are available through QCPE [170].

We have thus closed a loop: the least square fitting was adopted to improve simple descriptions of the molecular charge distribution, and subsequently refined to reach levels of complexity in the models very similar to those employed in the segmented multipole expansions of the charge distribution. The presence of a loop does not means that there has not been progress in the description of the MEP: on the contrary, the PD/LSF approaches represent one of the most promising ways of getting less expensive MEP descriptions of acceptable quality.

6.3 Molecular Fragments

We shall consider here, in a separate subsection, still another approach to get MEP values at a lower computational cost. The basis idea is to start from a description of molecular subunits. Theoretician are obliged to divide molecules into smaller components. The main goal of these partitions is to provide the basic elements for the interpretation of chemical phenomena, but they will be considered here under a different viewpoint, namely their possible utilization to get computationally simpler methods of obtaining MEP values. We shall consider briefly Bader's zero flux surface and the localized orbitals partitionings.

The zero flux surface gives an unambiguous partition of the physical space, based on clear physical concepts, very sensitive to the chemical environment and able to provide detailed information on the molecular structure [171]. We have already remarked (Sect. 6.1) that this partition is not appropriate for a multipolar expansion in the conventional approach. The integration of the electronic charge belonging to a close portion of space defined as the intersection of zero flux surfaces with a constant density envelope gives sensible and interesting results for atomic populations [172]. We are not aware of applications to the MEP, but further extensions of this approach surely are worth examining.

A different definition of molecular fragments, which has been explicitly formulated for the approximate description of charge densities was proposed several years ago by Hirshfeld [173]. This method has given satisfactory results when coupled to experimental X-ray diffraction measurements of molecular charge densities and potentials.

The partition of the electronic charge distribution into subunits defined in terms of localized moleculare orbitals (LO) may be use more directly to get approximate expressions of the MEP. The expressions are particularly simple when the molecular wavefunction is decribed at the Hartree-Fock level. In such an approximation, the density function is the sum of the orbital density functions, and correspondingly the D matrix (Eq. 47) may be written as a sum of orbital matrices D_i. The total matrix is invariant with respect to non-singular transformation of the orbitals, while each orbital matrix depends on the transformation. This properties introduces a certain degree of arbitrariness in the description one selects (remark the contrast with the univocally defined partition following the zero-flux surfaces), but permits us to search for the most convenient definition of the molecular orbital. Localization procedures — here are several methods giving, in general, results not too different from each other, see e.g. [174] — give orbital charge densities mainly localized on molecular fragments of chemical significance. Smaller contributions to the LO density deriving from distant atoms can be treated with appropriate techniques, as detailed in Ref. [20] and summarized here below.

When the molecular wavefunction is decribed at a post Hartree-Fock level, the treatment is more complex, but there are several methods which can recover to a large extent the simple description summarized above.

One of the first attempts of use molecular orbitals for a multipole expansion of the MEP was by Cohan [175]. The MO charge centroids were selected as expansion centres and the formal expression of this segmental expansion were compared with those of the conventional one-centre multipole expansion. As a test case, Cohan selected H_2^+, and so the paper is simply indicative of a possible alternative to the multipole expansion strategy, being in the specific case the canonical MO coincident with the LO, and its charge centre coincident with the molecular centre of masses.

The use of LOs in the evaluation of MEP and of E_{ES} is bimolecular interaction was examined by Amos and Crispin [176] who suggested the use of localized electron distributions also for induction and dispersion contributions to ΔE. We have already quoted a paper by Mezei and Campbell [107] where the use of LOs is examined and compared to other partitions of the molecular charge distribution.

A method concerned with MEP calculations on large molecules, and based on a LO description of the charge distribution, was proposed by Pullman's group [177–178]. This method, called LMTP (Localized MulTiPole expansion) uses the centroids of each LO as an expansion centre. The number of expansion centres is lower than the OMPT method elaborated by the same group (see Sect. 6.1). Expansions limited to the quadrupole, or also including the octupole term are examined in the two source papers. This approach presents several advantages: it makes it easier to describe changes in the MEP due to conformational changes, and makes the description of the MEP of large molecules by superposition of molecular fragment contributions more reliable and effective. The LMTP method has in fact been exploited by Pullman's group in very important studies of biological problems.

In Pisa, we employed LO descriptions of the electronic charge distribution in connection with MEP representations at the same time as our first proposal of using MEP values in the quantum mechanical study of molecular properties [2].

The strategy which we examined more in detail in the following years may be summarized thus. The orthogonal transformation of canonical SCF MO orbitals $\varphi_i(r)$ into LO's $\lambda_i(r)$ is followed by a deletions of the "tails", i.e. of the components not describing the main portion of λ_i. This manipulation may be described as a projection of λ_i in the basis set subspace spanned by functions belonging to the chemically significant portion of λ_i (it may be an atom, in the case of inner shell or lone pair orbitals, i_x and l_x respectively, a couple of adjacent atoms, in the case of σ or "banana" orbitals, b_{AB} and β_{AB} respectively, or larger portions of the molecule in the case of LO of more complex or delocalized nature). From the main portions $\bar{\lambda}_i$, normalized to unity, we derive electronic charge distributions

$$\bar{\varrho}_i^{(M)} = n_i \bar{\lambda}_i^* \bar{\lambda}_i , \tag{50}$$

where n_i is the appropriate occupation number. One or more $\bar{\varrho}_i^{(M)}$ may be reassembled to get a description of the group g:

$$\bar{\varrho}_g^{(M)} = \sum_{i \in g} \bar{\varrho}_i^{(M)} , \tag{51}$$

and an approximate expression of the total charge distribution of the molecule M:

$$\varrho_M^0 = \sum_{g \in M} \bar{\varrho}_g^{(M)} , \tag{52}$$

which differs from the original description, ϱ_M, by the deletion of tails and the lack of orthogonality between orbitals.

By a suitable partition of the nuclear charges we define approximate expressions of the total density:

$$\Gamma_M^0 = \sum_g \gamma_g^{(M)} , \tag{53}$$

in which the single components

$$\bar{\gamma}_g^{(M)} = \bar{\varrho}_g^{(M)} + \text{nuc. charges} , \tag{54}$$

are in general electrically neutral (see Ref. [20] in the present series for a more detailed presentation).

The MEP may be described at several levels of decreasing precision and requiring decreasing computational times

1) the use of the LCAO (Linear Combination of Atomic Orbitals) expressions of
   $\bar{\varrho}_g^{(M)}$ (or $\bar{\gamma}_g^{(M)}$), in a form equivalent to the general expression (42);
2) the replacement of the $\bar{\varrho}_g^{(M)}$ distribution by a set of point charges. The number
   of point charges is limited to one or two for each $\bar{\varrho}_i^{(M)}$. The values of the point
   charges is in general restricted to integer values ($-2e$ or $-1e$). The location of
   the point charges is done by standard rules (position of the LO charge centroid,
   LO dipole and quadrupole moment): in the definition of these rules we have take
   profit of the minimization of the square deviation between MEP values deriving
   from the model and from $\bar{\varrho}_i^{(M)}$ (and also from $\varrho_i^{(M)}$, with tails). The electronic
   point charges are supplemented by the necessary positive point charges on the
   nuclei;
3) the partition of the electronic charge distribution makes it easy to derive atomic
   point charge descriptions for just a portion of the molecule. The definition of a
   portion of the molecule for which a rough description is sufficient may be done
   in terms of the group g defined above or with a partition which breaks one (or
   more) bonds. In the first case one (or more) atoms will be partly described at a
   better level (level 2 or level 1), in the second case we may exploit the fact that a
   localized orbital for a bond has the following general expression:

$$\bar{b}_{AB} = c_A h_A + c_B h_B,\tag{55}$$

where h_A and h_B are generalized hybrids centred on atoms A and B, and c_A and
c_B are numerical coefficients. By using the projection techniques described in Sect.
6.1 it is easy to derive from $c_A h_A$ a monopole (and if necessary a dipole) contribution
centred on the atom A. The utility of the LO description in making this kind of
partition of the molecular charge distribution has been already put in evidence in
commenting the LMPT method.

The second step in the elaboration of the model consists in replacing the $\bar{\varrho}_i^{(M)}$ and
$\bar{\varrho}_g^{(M)}$ description derived from the molecular wavefunction of M by *prototypes*, ϱ_g^0,
collected in a library and derived from calculations on some parent molecules.

So, the three levels of approximations described above, may be replaced by
analogous levels in which the numerical quantities involved in the definitions are no
more related to a calculation on the specific molecule M, but to a set of directly
transferable quantities. We put the emphasis on the direct transfer, because they may
be used directly, without further manipulation. The interested reader is referred to
Ref. [19] for some examples and for the bibliography.

The direct transfer of group contributions from a library does not permit the
appreciation of mutual interaction effects of groups inside a specific molecular
framework. A fairly large portion of these effects may be recovered, when necessary,
to resorting again to the expression of the localized orbitals $\bar{\lambda}_i^0$, without tails. For a
bond we have

$$\bar{\lambda}_i^0 = c_A^0 h_A^0 + c_B^0 h_B^0,\tag{56}$$

the difference with respect to Eq. (55) consists of the use made here of prototypes,
put in evidence by the superscript o. We have found that a considerable portion of

the intergroup interaction effects may be represented by changes in the coefficients c_A^0 and c_B^0, taking the hybrids unchanged. The changes in the coefficients are based upon empirical rules which relate the value of the coefficients to the electric field produced by the other groups on ϱ_i. We shall refer here below to an alternative device proposed by other authors to take into account intergroup effects. This treatment may be extended to the levels 2 (approximation of ϱ_g in terms of a set of point charges) and 3 (approximation in terms of atomic charges).

G. Náray-Szabó and coworkers have developed an analogous technique (decomposition into LOs, group description, use of transferable prototypes, etc.) with the main difference of evaluating the intragroup effects on the prototypes via bond polarities derived from zeroth-order PCILO calculations. We quote here a limited number of papers coming from the Náray-Szabó group [179–183] selecting those more addressed to methodological problems, but we cannot neglect recalling that this method has been largely employed by the Hungarian people, and by others, in many applications concerning a variety of chemical problems. One of the first versions of this program, made for a small computer of an old generation, was able to handle systems composed of 600 atoms.

We have not systematically exploited our version of this approach. Several results are scattered in different papers (a partial bibliography may be found in Ref. [20]) and we limit ourselves to recall that we have often found in our applications concerned with the evaluation of hydrogen-bond interactions, a reduction of the computational times larger than a factor 10,000 with respect to ab initio calculations, with errors in energy less than 0.3 kcal/mol (see e.g. Ref. [184]).

6.4 General Comments on the Simplified Expressions

We may make now some general comments, starting from the last approach developed in Pisa and in Budapest, and extending then to remarks on the other methods.

The approach presents a hierarchy of approximations, which may be profitably grouped together. Problems of chemical reactivity in large molecules often concern a restricted zone of the molecule which a detailed description of the interactions is needed (in this paper we have considered the E_{ES} term alone, this often being the leading term, but the remarks may be also extended to the other contributions to ΔE, which are of shorter range than E_{ES}). Approximate expressions are valid, in the best cases, at distances not shorter than the van der Waals radii. The MEP minima, which constitutes a remarkable feature of this observable, often used as reactivity index, lies well within the van der Waals surface. An LCAO expression has no problems for representation and represents the unique way of getting accurate descriptions at short distances. The introduction of directly transferable molecular subunits in the LCAO form makes it possible to obtain short-range descriptions also for very large molecules, for which a direct calculation of the whole wavefunction is not practical. The modification in situ of the "prototypes" $\gamma_g^0(r)$ via their polarization may give an even finer description. The juxtaposition of MEP values from molecules corresponding "grosso modo" to the desired molecular fragments is a crude substitute of this approach. A more promising and alternative approach was shown by N. L. Ostlund at the recent WATOC congress [185] for which we have not found bibliographical references. In short, the method consists of selecting in a large molecule the fragments of interest, and then in computing for the fragments a CNDO calulation with special values of the parameters for the atoms related to the bonds broken in the selection of the fragment. The MEP description derives directly from the CNDO wavefunction. We have not examined the method in detail, but surely it represents a different line of approach of interest for the new generation of computers. Coming back to the descriptions in terms of fragments, it is evident that a general use of the LCAO description is too costly. It is our impression that when one looks at a specific molecular site, the LCAO description of the site must be followed by a description at an intermediate level of the surrounding groups, leaving the simplest description (e.g. point charges) to distant groups. The intermediate level of description may be obtained with point charges or multipole distributions, centred on the atoms, on the bonds, or on other locations. The quality of the results in the site targeted for the observation is quite sufficient for all the most important methods we have reviewed above, and the difference in computation times among them is less than an order magnitude. The most distant portions of the large molecular systems under examination may be represented by atomic charges. The quality of these distant charges is not so important, and in very large molecular systems they may also be simply reduced to the relatively few groups bearing a net charge or a very large dipole moment. These drastic reductions are however possible for special problems (see e.g. Bonaccorsi et al. [186]) or for very large systems: for example we have found this approximation acceptable for histone octamers (unpublished results).

It seems to us important that all the steps of this sequence of approximations, in which the quality of the representation is inversely related to the distance from the target group, are methodologically connected, without abrupt changes of quality. This objective may be reached by different combinations of the method considered above, and the decisive point for the winning formulation will be made by other factors, such as the versatility of the program and the transferability of the components.

One could ask what is the collocation, in this framework, of the PD atomic charges: according to the opinion expressed above they are too simple for the accurate description of the target group. The main field of application, in our opinion, is in the domain of molecular mechanics or related fields. Here the description of the interaction must be fast, and accurate at the appropriate level. A quantitative analysis of the MEP values near the atoms of a given chemical group is not necessary, and PD-AC (or in general PD/LSF) values are sufficient for a first-order scanning of the molecular conformation, or of the geometry of non-covalent complexes, or of crystal packing, the level of description of the electrostatic component of the interaction obtained in such a way being sufficient. The use of PD-AC values for solvation problems is now under examination.

We indicated in Sect. 4 three main themes of methodological interest in the examination of the molecular electrostatic potential, with the remark that there is a larger number of subjects of chemical and physical interest for which the use of MEP values may be quite important. In Sects. 5 and 6 we have developed the analysis of only one of these themes. The space is not sufficient to treat, with the necessary detail, the other two themes nor, of course, the more specialistic subjects indicated in Sect. 4. The use of the molecular electrostatic potential, and of other related semiclassic quantities, of statical as well as of dynamical nature (electric field, polarization, etc.), is still in a period of expansion in theoretical chemistry, and surely other reviews, on more specialized topics, will be appear in the next future.

7 References

1. Coulson CA (1960) Rev Mod Phys 32: 170
2. Bonaccorsi R, Scrocco E, Tomasi J (1970) J Chem Phys 52: 5270
3. Pullman B (ed) (1978) Intermolecular Interactions: from Diatomics to Biopolymers, Wiley, Chichester
4. Van der Avoird A, Wormer PES, Mulder F, Berns RM (1980) Topics Curr Chem 93: 3
5. Arrighini GP (1982) Intermolecular forces and their evaluation by perturbation theory, Lect Notes in Chemistry 25, Springer, Berlin Heidelberg New York
6. Jeziorski B, Kołos W (1982) In: Ratajczak H, Orville-Thomas WJ (eds) Molecular Interactions, vol III, Wiley, New York, p 1
7. Kaplan IP (1986) Theory of molecular interactions, Elsevier, Amsterdam
8. Jeziorski B, Moszynski R, Rybak S, Szalewicz K (1989) In: Kaldor U (ed) Many body methods in Quantum Chemistry, Lect Notes in Chemistry 52, Springer, Berlin Heidelberg New York
9. Chałasinski G, Jeziorski B, Andzelm J, Szalewicz K (1977) Int J Quantum Chem 11: 247
10. Hirschfelder JO (1967) Chem Phys Lett 1: 325
11. Kitaura K, Morokuma K (1976) Int J Quantum Chem 10: 325
12. Nagase S, Fueno T, Yamabe S, Kitaura K (1978) Theoret Chim Acta 49: 309
13. Kitaura K, Morokuma K (1981) In: Politzer P, Truhlar D (eds) Chemical applications of atomic and molecular electrostatic potentials, Plenum, New York
14. Daudey JP, Claverie P, Malrieu JP (1974) Int J Quantum Chem 8: 1
15. Tomasi J (1982) In: Ratajczak H, Orville-Thomas WJ (eds) Molecular Interactions, vol III, Wiley, New York, p 119
16. Bonaccorsi R, Cimiraglia R, Palla P, Tomasi J (1983) Int J Quantum Chem 24: 307
17. Fukui K (1975) Theory of orientation and stereoselection, Springer, Berlin Heidelberg New York
18. Klopman G (1974) Chemical reactity and reaction path, Wiley, New York
19. Reed AE, Curtiss LA, Weinhold F (1988) Chem Rev 88: 899
20. Tomasi J, Alagona G, Bonaccorsi R, Ghio C, Cammi R (1990) In: Maksić Z (ed) Theoretical models of chemical bonding, vol 3, Springer, Berlin Heidelberg New York
21. Latham WA, Pack GR, Morokuma K (1975) J Am Chem Soc 97: 6624
22. Szcześniak MM, Scheiner S (1984) J Chem Phys 80: 1535
23. Boys SF, Bernardi F (1970) Mol Phys 19: 553
24. Daudey JP, Malrieu JP, Rojas O (1974) Int J Quantum Chem 8: 17
25. Magnasco V, Musso GF, Costa C, Figari C (1985) Mol Phys 56: 1249
26. Tolosa S, Esperilla JJ, Olivares del Valle FJ (1990) J Comp Chem 11: 576
27. Van Lenthe JH, van Duijneveldt-van de Rijdt JCM, van Duijneveldt FB (1987) Adv Chem Phys 69: 521
28. Cammi R. Bonaccorsi R, Tomasi J (1985) Theoret Chim Acta 68: 271
29. Cammi R, Tomasi J (1986) Theoret Chim Acta 69: 11
30. Tolosa S, Esperilla JJ, Espinosa J, Olivares del Valle FJ (1988) Chem Phys 127: 65
31. Kollman P (1976) J Am Chem Soc 99: 4875
32. Scrocco E, Tomasi J (1978) Adv Quant Chem 11: 116
33. Nagy P (1988) J Mol Struct (Theochem) 181: 36
34. Nagy P (1988) J Comp Aided Molec Design 2: 65
35. Alagona G, Ghio C, Nagy P (1989) J Mol Struct (Theochem) 187: 219
36. Taylor KF (1987) J Comp Chem 8: 291

37. Ghio C, Scrocco E, Tomasi J (1976) In: Pullman B (ed) Environmental effects on molecular structure and properties, Reidel, Dordrecht, p 329
38. Politzer P, Daiker KC (1981) In: Deb BM (ed) The force concept in chemistry, Van Nostrand Reinhold, New York, p 294
39. Politzer P, Truhlar D (eds) (1981) Chemical applications of atomic and molecular electrostatic potentials, Plenum, New York
40. Daudel R, Le Rouzo H, Cimiraglia R, Tomasi J (1978) Int J Quantum Chem 13: 537
41. Luque FJ, Orozco M, Illas F, Rubio J (private communication, to be published)
42. Sanz F, Manant F, José J, Segura J, Carbó M, De la Torre R (1987) J Mol Struct (Theochem) 170: 171
43. Obara S, Saika A (1986) J Chem Phys 84: 3963
44. Srebenik S, Weinstein H, Pauncz R (1973) Chem Phys Lett 20: 419
45. Baskarov AA, Varnek AA, 'Tsirel'son VG, Ozerov RP (1984) Zh Strukt Khim 25: 135
46. Carbó R, Martin M (1975) Int J Quantum Chem 9: 193
47. Weinstein H, Osman R (1977) Int J Quantum Chem QBS 4: 253
48. Pullman A, Gresh N (1978) Theoret Chim Acta 49: 283
49. Giessner-Prettre C, Pullman A (1972) Theoret Chim Acta 25: 85
50. Giessner-Prettre C, Pullman A (1974) Theoret Chim Acta 33: 91
51. Giessner-Prettre C, Pullman A (1975) Theoret Chim Acta 37: 335
52. Petrongolo C, Tomasi J (1973) Chem Phys Lett 20: 201
53. Caballol R, Gallifa RC, Martin M, Carbó R (1974) Chem Phys Lett 25: 89
54. Duben AJ (1981) Theoret Chim Acta 59: 81
55. Weinstein H, Chou DC, Kang S, Johnson CL, Green JP (1976) Int J Quantum Chem QBS 3: 135
56. Petrongolo C (1978) Gazz Chim Ital 108: 411
57. Culberson JC, Zerner MC (1985) Chem Phys Lett 122: 436
58. Luque FJ, Illas F, Orozco M (1990) J Comp Chem 11: 416
59. Luque FJ, Orozco M (1990) Chem Phys Lett 168: 269
60. Luque FJ, Orozco M (1990) J Comp Chem (in press)
61. Ferenczy G, Reynolds CA, Richards WG (1990) J Comp Chem 11: 159
62. Weber J, Fluckinger PF, Morgantini PY, Schaad O, Goussot A, Daul C (1988) J Comp Aided Mol Des 2: 235
63. Daul C, Goussot A, Morgantini PY, Weber J (1990) Int J Quantum Chem (in press)
64. Buckingham AD (1959) Quart Rev 13: 189
65. Buckingham AD (1967) Adv Chem Phys 12: 107
66. Hirschfelder JE, Curtiss CF, Bird RB (1954) Molecular theory of gases and liquids, Wiley, New York
67. Margenau H, Kestner NR (1969) Theory of intermolecular forces, Wiley, New York
68. Böttcher CFJ, Bordewijk P (1978) Theory of electric polarization, Elsevier, Amsterdam
69. Berrondo M, Eggleston SW, Carson EG (1989) Int J Quantum Chem 36: 749
70. Young RH (1975) Int Quantum Chem 9: 47
71. Ahlrichs R (1976) Theoret Chim Acta 41: 7
72. Vigné-Maeder F, Claverie P (1988) J Chem Phys 88: 4934
73. Boys SF (1950) Proc Roy Soc London A200: 542
74. Shavitt I (1963) In: Methods of computational physics, vol 2, Academic, New York, p 1
75. Hall GG (1973) Chem Phys Lett 20: 501
76. Clementi E, André JM, André MC, Klink D, Hahn D (1969) Acta Phys Hungar 27: 493
77. Fortunelli A, Salvetti O (1990) J Comp Chem (in press)
78. Klahn B, (1985) J Chem Phys 83: 5749
79. Tait AD, Hall GG (1973) Theoret Chim Acta 31: 311
80. Frost AA (1976) In: Schaefer HF (ed) Methods of Electronic Structure Theory, Plenum Press, New York, p 29
81. Shipman LL (1975) Chem Phys Lett 31: 361
82. Amos AT, Yoffe J (1975) Theoret Chim Acta 40: 221
83. Friedmann R, Brand W, Walther P (1986) Z Phys Chem (Leipzig) 267: 882
84. Martin D, Hall GG (1981) Theoret Chim Acta 59: 281
85. Stone AJ (1982) Chem Phys Lett 83: 233

86. Stone AJ, Alderton M (1985) Mol Phys 56: 1047
87. Bader RF, Anderson SE, Duke AJ (1979) J Am Chem Soc 101: 1389
88. Cooper DL, Stutchbury NCJ (1985) Mol Phys. 51: 569
89. Price SL, Harrison RJ, Guest MF (1989) J Comp Chem 10: 552
90. Faerman CH, Price SL (1990) J Am Chem Soc 112: 4915
91. Port GNJ, Pullman A (1973) FEBS Letters 31: 70
92. Pullman A, Perahia D (1978) Theoret Chim Acta 48: 29
93. Goldblum A, Perahia D, Pullman A (1979) Int J Quantum Chem 15: 121
94. Langlet J, Claverie P, Caran F, Boeuve JC (1981) Int J Quantum Chem 20: 299
95. Rabinowitz JR, Swissler TJ, Rein R (1972) Int J Quantum Chem S 6: 353
96. Rabinowitz JR, Namboodiri K, Weinstein H (1986) Int J Quantum Chem 29: 1697
97. Rabinowitz JR, Little SB (1986) Int J Quantum Chem QBS 13: 9
98. Rabinowitz JR, Little SB (1988) Int J Quantum Chem S 22: 721
99. Murray JS, Grice ME, Politzer P, Rabinowitz JR (1990) J Comp Chem 11: 112
100. Rein R (1973) Adv Quant Chem 7: 335
101. Rein R (1978) In: Pullman B (ed) Intermolecular Interactions: from Diatomics to Biopolymers, Wiley, Chichester, p 307
102. Hall GG, Martin D (1980) Israel J Chem 19: 225
103. Hall GG (1983) Theoret Chim Acta 63: 357
104. Hall GG, Smith CM (1984) Int J Quantum Chem 25: 881
105. Hall GG (1986) Int Rev Phys Chem 5: 115
106. Hall GG, Tsujinaga (1986) Theoret Chim Acta 69: 425
107. Mezei M, Campbell ES (1976) Theoret Chim Acta 43: 227
108. Sokalski WA, Poirier RA (1983) Chem Phys Lett 98: 86
109. Sokalski WA, Hariharan PC, Kaufman JJ (1987) Int J Quantum Chem QBS 14: 111
110. Sokalski WA, Savaryn A (1987) J Chem Phys 56: 526
111. Jug K (1975) Theoret Chim Acta 39: 301
112. Fernandez-Rico J, Alavrez-Colledo JR, Paniagua H (1986) Mol Phys 56: 1145
113. Einstein M (1988) Int J Quantum Chem 33: 127
114. Mulliken R (1955) J Chem Phys 23: 1833
115. Fliszar S (1983) Charge distribution and chemical effects, Springer, Berlin Heidelberg New York
116. Hall GG (1985) Adv At Mol Phys 20: 41
117. Williams DE, Yan JM (1988) Adv At Mol Phys 23: 87
118. Huheey JE (1970) Inorganic chemistry: principles of structure and reactivity, Harper and Row, New York
119. Jolly W, Perry WB (1974) Inorg Chem 13: 2686
120. Wiberg KB, Wendolowsky JJ (1978) J Am Chem Soc 100: 723
121. Gussoni M (1984) J Mol Struct 113: 324
122. Farnum DG (1975) Adv Phys Org Chem 11: 123
123. Nelson GL, Williams EA (1976) Progr Phys Org Chem 12: 229
124. Fliszqar S, Cardinal G, Baykara NA (1986) Can J Chem 64: 404
125. Coppens P, Hall MB (eds) (1982) Electron distribution and the chemical bond, Plenum, New York
126. Mortier WJ, Gosh SK, Shanikar S (1986) J Am Chem Soc 108: 4315
127. Mullay J (1986) J Am Chem Soc 108: 1770
128. Mullay J (1988) J Comp Chem 9: 399; 9: 764
129. Gastgeiger J, Marsili M (1980) Tetrahedron 36: 3219
130. Mortier WJ, Van Genechten K, Gastgeiger J (1985) J Am Chem Soc 107: 829
131. Hammarström LG, Liljefors T, Gastgeiger J (1988) J Comp Chem 9: 424
132. Bergmann D, Hinze J (1987) Struct and Bonding 66: 145
133. Hückel EP (1932) Z Phys 76: 628
134. Del Re G (1958) J Chem Soc 4031
135. Del Re G (1963) Theoret Chim Acta 1: 188
136. Berthold H, Pullman A (1965) J Chim Phys 62: 942
137. Lavery R, Zakrzewska K, Pullman A (1984) J Comp Chem 5: 363
138. Zakrzewska K, Pullman A (1985) J Comp Chem 6: 265

139. Houser JJ, Klopman G (1988) J Comp Chem 9: 893
140. Smith RP, Ree T, Magee JL, Eyring H (1951) J Am Chem Soc 73: 2263
141. Allinger NL, Wuesthoff T (1977) Tetrahedron 34: 3
142. Micovič LD, Allinger NL (1978) Tetrahedron 34: 3385
143. Abraham RJ, Griffiths L, Loftus P (1982) J Comp Chem 3. 407
144. Abraham RJ, Hudson B (1984) J Comp Chem 5: 562
145. Abraham RJ, Hudson B (1985) J Comp Chem 6: 173
146. Momany FA (1978) J Phys Chem 82: 592
147. Bonaccorsi R, Petrongolo C, Scrocco E, Tomasi J (1971) Theoret Chim Acta 20: 331
148. Alagona G, Cimiraglia R, Scrocco E, Tomasi J (1972) Theoret Chim Acta 25: 103
149. Kollman PA (1978) J Am Chem Soc 100: 2974
150. Agresti A, Bonaccorsi R, Tomasi J (1979) Theoret Chim Acta 53: 215
151. Smit PH, Derissen JL, van Dujneveldt FB (1979) Mol Phys 37: 521
152. Sauer J, Morgeneyer C (1983) Studia Biophisica 93: 253
153. Thole BT, Van Duijnen PT (1983) Theoret Chim Acta 63: 209
154. Cox SR, Williams DE (1981) J Comp Chem 2: 304
155. Singh UC, Kollman PA (1984) J Comp Chem 5: 129
156. Weiner SJ, Kollman PA, Case DA, Singh UC, Ghio C, Alagona G, Profeta S, Weiner P
     (1984) J Am Chem Soc 106: 765
157. Chirlian LE, Francl MM (1987) J Comp Chem 8: 894
158. Chirlian LE, Francl MM (1987) QCPE Bull 7: 324; QCPE program n 524
159. Woods RJ, Khalil M, Pell W, Moffatt SM, Smit Jr WH (1990) J Comp Chem 11: 297
160. Breueman CM, Wiberg KB (1990) J Comp Chem 11: 361
161. Chung-Phillips A (1989) J Comp Chem 10: 17
162. Miertuš S, Scrocco E, Tomasi J (1981) Chem Phys 55: 117
163. Besler BH, Merz KM, Kollman PA (1990) J Comp Chem 11: 431
164. Orozco M, Luque FJ (1991) J Comp Chem (in press)
165. Orozco M, Luque FJ (1991) J Mol Stuct (Theochem) (accepted for publ)
166. Bonaccorsi R, Scrocco E, Tomasi J (1977) J Am Chem Soc 99: 4546
167. Williams DE, Weller RR (1983) J Am Chem Soc 105: 4143
168. Williams DE, Houpt DJ (1986) Acta Crystall B42: 286
169. Williams DE (1988) J Comp Chem 9: 745
170. Williams DE (1989) QCPE Bull 9: 29. QCPE program n 568: PDM 88
171. Bader RFW, Nguyen-Dang TT, Tal Y (1981) Rep Prog Phys 46: 893
172. Bader RFW, Carroll MT, Cheseman JR, Chang C (1987) J Am Chem Soc 109: 7968
173. Hirshfeld FL (1977) Theoret Chim Acta 44: 129
174. Chalvet O, Daudel R, Diner S, Malrieu JP (eds) (1975) Localization and delocalization
     in quantum chemistry, vol I, Reidel, Dordrecht
175. Cohan NV (1969) Mol Phys 17: 307
176. Amos AT, Crispin RJ (1976) Mol Phys 31: 159
177. Lavery R, Etchebest C, Pullman A (1982) Chem Phys Lett 85: 266
178. Etchebest C, Lavery R, Pullman A (1982) Theoret Chim Acta 62: 17
179. Náray-Szabó G (1976) Acta Phys Hungar 40: 261
180. Náray-Szabó G (1979) Int J Quantum Chem 16: 265
181. Náray-Szabó G, Grofsick A, Kósak K, Kubinyi M, Martin A (1981) J Comp Chem 2: 58
182. Nágy P, Angyán JG, Náray-Szabó G, Peirel G (1987) Int J Quantum Chem 31: 927
183. Angyán JG, Ferenczy G, Nágy P, Náray-Szabó G (1988) Coll Czech Chem Comm 53: 2308
184. Bonaccorsi R, Scrocco E, Tomasi J (1985) J Biosc 8: 627
185. Ostlund NL (1990) Plenary Lecture at the WATOC Congress, Toronto
186. Bonaccorsi R, Scrocco E, Tomasi J (1986) Int J Quantum Chem 29: 717

Curve Crossing Diagrams as General Models for Chemical Reactivity and Structure

Sason S. Shaik

Department of Chemistry, Ben Gurion University[#], Beer-Sheva 84105, Israel

Philippe C. Hiberty

Laboratoire de Chimie Théorique[†], Bât. 490, Université de Paris-Sud, 91405 Orsay Cedex, France

This chapter reviews qualitative as well as quantitative aspects of the curve crossing model. The chapter starts with an introduction of key concepts in the Valence Bond (VB) language, such as the Heitler-London (HL) spin-pairing, the two-center bond, overlap repulsion in VB structures, and VB-type structures based on group orbitals. These terms are used to outline a detailed construction of curve crossing and avoided crossing diagrams, from VB and VB-type configurations for prototypical processes, involving three and four electrons delocalized over three centers. The diagrams are then generalized to any process which involves electronic and geometric reorganizations and applied to reactivity and structural problems. The applications involve the S_N2 reaction, radical addition to olefins, the trimerization of acetylene, the stability patterns of three-electron/three-center X_3 clusters, the stability patterns of hypercoordinated radicals with formally nine valence electrons, and a comparison of SiH_5^- with CH_5^-. Throughout the application part, and whenever available, we present ab initio computed diagrams using a multistructure VB method.

[#] The research at BGU was supported by the Basic Research Foundation administered by the Israel Academy of Sciences and Humanities.

[†] The Laboratoire de Chimie Théorique is associated with the CNRS (UA 506).

1 Introduction

Reactivity and structure are two important branches of chemistry which are conceptualized in terms of potential energy profiles on hypersurfaces. Potential energy profiles and surfaces are wonderful concepts but they replace the chemical complexity by a mathematical complexity. We are left, in fact, with the goal of understanding potential energy profiles and surfaces as the means to conceptualize chemistry. This has been an important conceptual goal in quantum chemistry [1] eversince the pioneering studies in the first half of this century [2].

In 1981 the Ben-Gurion University group has shown [3] that potential energy profiles can be constructed in a methodical manner by using Valence Bond (VB) type configurations which combine the useful concept of the bond from VB theory with the lucid concepts of orbital symmetry and orbital-coefficients from Molecular Orbital (MO) theory. The models of reaction profiles which have evolved since then are called curve crossing- and avoided crossing-diagrams, although other names are probably more known (SCD and VBCM diagrams [4]).

The curve crossing/avoided crossing diagrams have been applied by the Ben Gurion and the Orsay groups in qualitative and semi-quantitative manners to a variety of chemical problems, ranging from organic [4, 5] and organometallic reactivity [6] to the stability of hypercoordinated clusters [7]. These applications have been reviewed several times in the past [4, 5, 7, 8] and some of the material has recently been incorporated into the organic mechanism textbook by Lowry and Richardson [9] and into the electron transfer monograph by Eberson [10].

Recently, there has been a key development in this area when the Orsay and Polytechnique groups [11] developed a multistructure VB program which is capable of computing curve crossing- and avoided crossing-diagrams. An opportunity has thus emerged to implement the qualitative ideas into a rigorous computational scheme. Indeed the few applications that have so far been tried [12] project the potential of the model, and at the same time serve to unravel shades to which the qualitative model has been naturally blind. The melange of qualitative concepts and quantitative procedures which are at par with each other is a promising combination, and some of the resulting insight is presented in this chapter.

The construction of potential energy profiles from curve crossing and avoided crossing diagrams is based on the recognition that for any process, R(reactants) $\rightarrow$ P(products), which involves geometric- and electronic-reorganization, the Lewis wavefunctions Ψ_I which describe R and P intersect as shown in **1** along the reaction coordinate (RC), which is defined by the geometric transformation inherent in the structures of R and P. This curve crossing (**1**) along with the occasional crossing (**2**) of the Lewis curves by a structure which has a different character, than a Lewis bond, define the archetypal potential energy profiles which are necessary to describe common chemical transformations.

In this chapter we construct avoided crossing diagrams for prototypical cases and we then show how to take advantage of them to gain insight into problems of reactivity and structure. The discussion in the theme of chemical reactivity covers examples from S_N2 reactions, radical additions, and cycloadditions. The discussion in the structural theme includes analysis of the stability patterns of hypercoordinated

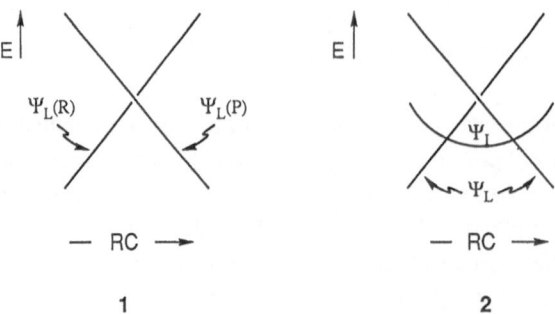

clusters of the type X_3, hypercoordinated radicals of the general formulae MH_{n+1} (M = C, Si, O, S, N, P), and the hypercoordinated anion SiH_5^-. The general aim of this chaper is to provide an overview of the ways to apply the diagrams to chemical problems, rather than an exhaustive review of the field or of the various chemical topics.

2 The Valence Bond Language

To construct an avoided crossing diagram it is essential to review some preliminary concepts which constitute part of the VB language [13] in the sense used in this chapter [8].

2.1 The Two-Center Bond

The essential event in a chemical reaction is the breaking or making of a bond between two centers, be they atoms or complex fragments. In the most common situation these two-center bonds are also two-electron bonds, known as Lewis electron-pair bonds [14]. Following Pauling [15], the Lewis bond A–B is described by a blend of the spin-paired Heitler-London (HL) form and the possible zwitterionic situations, as expressed in Eq. (1):

$$\Psi_L = (A-B) = C_1(A\!\!-\!\!\cdot B) + C_2(A\!:^- B^+) + C_3(A^+ :B^-). \quad (1)$$

Thus, depending on the relative magnitude of the mixing coefficients, C_1–C_3, a bond can be described to lie anywhere in the range spanned by pure covalency to pure ionicity – in accord with chemical wisdom [15, 16] and with theoretical analyses of bonding [17, 18].

In most common bonds in organic species the spin-paired HL configuration in Eq. (2) is bonded relative to the free atoms.

$$\Psi_{HL} = A\!\!-\!\!\cdot B = N_{HL} \left[|a\bar{b}| - |\bar{a}b|\right]. \quad (2)$$

The dominant part of the HL stabilization, relative to the free atoms, is the monoelectronic contribution which is defined in Eq. (3), where the quantity hs is the product of the resonance and overlap integrals between the two orbitals a and b which form the bond, as defined by Eq. (4). Further stabilization of the electron pair bond is provided by the mixing between the HL and the zwitterionic structures [8, 13].

$$\Delta E_{HL} \sim 2hs < 0, \tag{3}$$

$$h = \langle a|\, \mathbf{H}(1e)\, |b\rangle ; \qquad s = \langle a\,|\,b\rangle. \tag{4}$$

2.1.1 Representations of VB Configuration

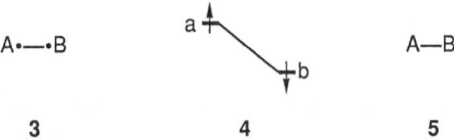

A•—•B	a, b	A—B
3	**4**	**5**

Drawings **3–5** summarize some representations of the HL configuration and the Lewis bond [8]. In **3** we show the most common representation of the HL configurations in terms of two electrons, symbolized by dots, with singlet-spin-pairing indicated by a line connecting the two dots. Another representation of Ψ_{HL} is **4** which shows the relative energy levels of the orbitals a and b and the spin-pairing of their resident electrons by a line which connects the orbitals, while the electrons are symbolized by the vertical arrows. This latter representation is called a bond diagram [3, 8, 19] which can be used to convey in the most general case a HL pairing of electrons which reside in orbitals which may be either localized atomic orbitals or delocalized group orbitals. Finally, **5** is the common illustration of a Lewis bond, whose explicit VB wavefunction is Eq. (1).

2.2 VB Configurations Involving Group Orbitals

As discussed above the key VB notion of spin-pairing can be carried over to the interaction of electrons which reside in delocalized group orbitals [8]. Consider, for example, the interaction between the three-electron species $(A \doteq B)^-$ and the radical, $C^{\cdot}$. Since the two fragments contain two odd electrons, we can pair them up in the HL fashion to give the bond diagram shown in Fig. 1a. The HL bond is seen to exist between the singly occupied hybrid of $C^{\cdot}$ and the singly occupied σ^* type orbital of $(A \doteq B)^-$.

A VB wavefunction which is based on delocalized fragment orbitals is a compact way of assembling a few VB configurations which are based on more localized orbitals [3, 8]. This involves simple transformations which have been discussed in the literature [3, 16, 17, 20]. Thus, Fig. 1b shows that the delocalized version of the VB wavefunction in part (a) is equivalent to a resonating linear combination of two localized VB structures, one involving a C·—·A bond, the other a long C· ·B bond. Compactization of VB wavefunctions by use of delocalized fragment orbitals has been discussed amply

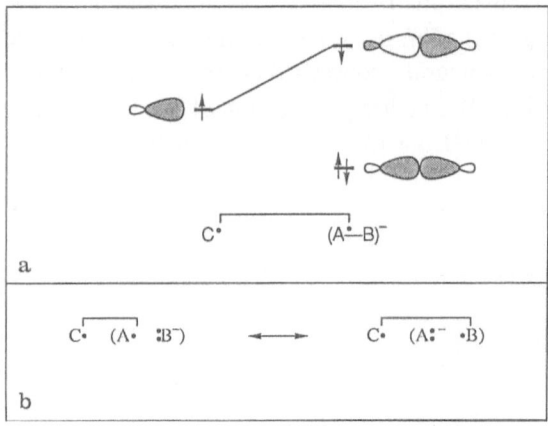

a

C⋅   (A⋅ :B⁻)   ⟷   C⋅   (A:⁻ ⋅B)

b

Fig. 1 a Bond diagrammatic representation of Heitler-London spin pairing for electrons in group orbitals. (**b**) The localized resonating VB structures which contribute to the wave function represented in (**a**)

in the literature in qualitative terms by Epiotis [19], Harcour [21], Linnett [22], and Shaik [3, 8] and in quantitative terms by Coulson-Fischer [23], Goddard [24], and Cooper-Gerratt-Raimondi [25].

2.3 Overlap Repulsion in Valence Bond Theory

In VB theory overlap repulsion is clearly associated with the Pauli exclusion principle and occurs in three distinct situations which are shown in drawings **6–8**, along with the corresponding monoelectronic expression of the overlap repulsion [8]. As may be seen, the overlap repulsion is given in $-2hs$ units for each pair of electrons which possess identical spins. In turn, the basic „unit" of overlap repulsion is identical in absolute magnitude to the HL stabilization in Eq. (3) (disregarding the effect of the normalization constants).

$$a\!\!\uparrow \quad \uparrow\!\!b \qquad\qquad a\!\!\uparrow \quad \uparrow\!\!\downarrow\!\!b \qquad\qquad a\!\!\uparrow\!\!\downarrow \quad \uparrow\!\!\downarrow\!\!b$$
$$\quad -2hs \qquad\qquad\qquad -2hs \qquad\qquad\qquad -4hs$$

$$\quad\quad 6 \qquad\qquad\qquad\qquad 7 \qquad\qquad\qquad\qquad 8$$

3 Construction of Avoided Crossing Diagrams

Having reviewed some preliminaries of the VB language we are ready to discuss the construction of avoided crossing diagrams and their application to chemical reactivity and structure. There are two types of avoided crossing diagrams which are based on drawings **1** and **2** and are useful for analyzing chemical problems. The first type is the two-curve diagram which considers primarily the Lewis structures of the bonds which undergo reorganization during the transformation. The second type is a many-curve model where additional structures become important along the reaction coordinate and are responsible for intermediate formation between the two Lewis structures [4, 5]. The following sections analyze the two model types for prototypical cases and apply them then to chemical problems.

3.1 Two-Curve Avoided Crossing Diagrams for Prototypical Processes

The prototypical processes which we consider are group transfer cases in which four or three electrons undergo reorganization over three centers in the chemical step. These models cover many of the classical reactions of organic chemistry as well as many of the clusters and structural problems which are discussed later in the text.

3.1.1 The Four-Electron/Three-Center Case. A Prototypical Ionic Exchange Process

Consider the process in Eq. (5) in which formally a group A^+ is transferred between two $X{:}^-$ groups, and constituting therefore a 4-electron/3-center process which embraces S_N2 reactions (e.g., $A = CH_3$), proton transfer ($A = H$), and so on [3–5, 8].

$$X_l{:}^- + A–X_r \rightarrow X_l–A + {:}X_r^- . \tag{5}$$

The identical X groups are labeled as l and r which stand for left and right, respectively. The choice of such an identity process is of course a matter of convenience and not of necessity.

Drawings **9–14** show the six effective configurations [8] which are necessary to describe the process in Eq. (5). The ground states' Lewis bonds can be made from the mixing of **9–13**. Thus, mixing of **9** with **11** and **13** generates the corresponding Lewis structure, made of the species, $X_l{:}^-$ and $(A–X_r)$. Similarly, mixing of **10** with **11** and **12** generates the second Lewis structure, that is, $(X_l–A)/{:}X_r^-$.

$$
\begin{array}{ccc}
X_l{:}^- \ A\text{---}\bullet X_r & X_l\text{---}\bullet A \ {:}X_r^- & X_l{:}^- \ A^+ \ {:}X_r^- \\
\mathbf{9} & \mathbf{10} & \mathbf{11}
\end{array}
$$

$$
\begin{array}{ccc}
X_l^+ \ {:}A^- \ {:}X_r^- & X_l{:}^- \ A{:}^- \ X_r^+ & X_l\bullet \ A{:}^- \ \bullet X_r \\
\mathbf{12} & \mathbf{13} & \mathbf{14}
\end{array}
$$

Figure 2a shows the behavior of the two Lewis structures along the reaction coordinate that stretches between R (reactants) and P (products) extremes and passes through the midpoint, M, where the distances $d(X_l...A)$ and $d(A...X_r)$ are equal. It is seen that the two structures generate two intersecting curves, each starting as the Lewis structure at the ground extreme and becoming a HL structure at the excited extreme. In fact, each curve is a usual homolytic bond dissociation curve, of the respective Lewis bond, augmented by the overlap repulsion generated by the approach of the anion $X:^-$ to $A^.$. These two effects, of bond dissociation and overlap repulsion, are the roots for the energy rise of each curve along its respective reaction coordinate $(R \rightarrow P$ or $P \rightarrow R)$.

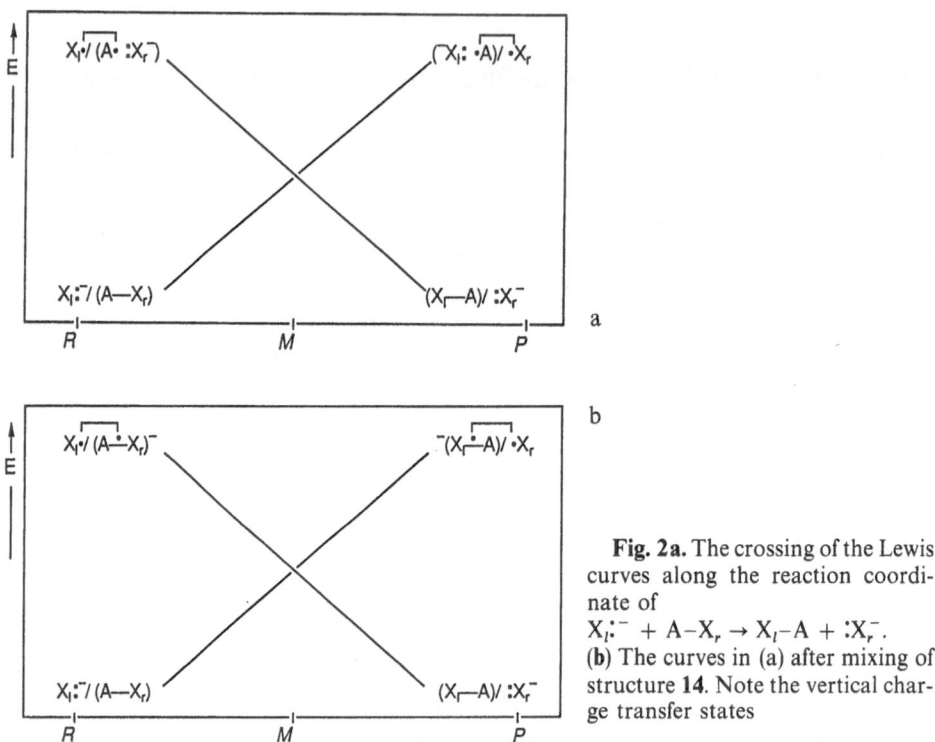

a

b

Fig. 2a. The crossing of the Lewis curves along the reaction coordinate of
$$X_l:^- + A-X_r \rightarrow X_l-A + :X_r^-.$$
(**b**) The curves in (a) after mixing of structure **14**. Note the vertical charge transfer states

The intersection of the Lewis curves forms the spine of the avoided crossing diagram. But the final form of the diagram requires the mixing of the remaining structure **14**, into the two curves. Following the VB mixing rules [8], it can be shown that **14** mixes into each curve at any point, along the reaction coordinate, except for the two ground state extremes. Thus, the two ground state extremes remain, as they should, Lewis structures whereas the rest of the curve points acquire some of the long bond character, **14**, leading to the final form of the curve crossing diagram in Fig. 2b. As can be seen, the two excited extremes are now the vertical charge transfer states relative to their corresponding ground states. Thus, each charge transfer state is generated from the state below it by transferring a single electron from the $X:^-$ anion to the $A-X$ bond. A charge transfer state of this kind is described above in Fig. 1.

In the final step we convert the curve crossing diagram of Fig. 2b to an avoided crossing diagram by allowing the two curves to mutually mix. The avoided crossing diagram is shown in Fig. 3 for two extreme situations which depend in turn on the size of the diagram gap G. This gap is the vertical charge transfer energy given in Eq. (6) in terms of vertical ionization potential (I^*) and the vertical electron affinity A^*.

$$G = I^*_{X:} - A^*_{A-X} .\tag{6}$$

The intrinsic gap, without solvation terms, is related to the bond strength, D_{A-X}, of the $A-X$ molecule [4b; 5a, b]. As a rough estimate, the vertical charge transfer energy gap can be taken in the range $D_{A-X} \leqq G \leqq 1.5 D_{A-X}$, with stronger binders having usually the proportionally higher gaps. The diagram in Fig. 3a shows the common expectation for cases where the gap is large. In these cases the symmetric $(X_l A X_r)^-$ species is expected to sit on a top of a barrier. Figure 3b shows an opposite extreme situation where the gap is very small. In this case the avoided crossing is expected to generate $(X_l A X_r)^-$ that is more stable than both reactants and products [7, 26].

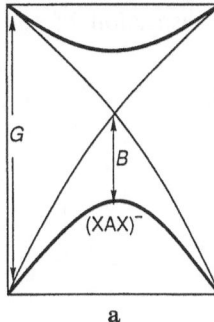
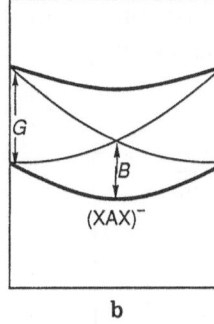

a   b

Fig 3 a, b. Avoided crossing diagrams for the process
$X_l^{.-} + A-X_r \rightarrow X_l-A + {:}X_r^-$.
The B is the avoided crossing interaction. (**a**) A case with large gap. (**b**) A case of small gap

We see therefore that the avoided crossing diagram can describe processes which proceed by surmouting a barrier, like Fig. 3a, as well as those which proceed via a stable intermediate, like Fig. 3b. Suffices it to mention a few examples, such as the S_N2 reactions which belong to case (a) with $(XCH_3X)^-$ transition states, and the halide exchange reactions which proceed through X_3^- intermediates (e.g., I_3^-, F_3^-, etc.) [7, 26]. Other Features of the model are discussed later in the application sections.

3.1.2 The Three-Electron/Three-Center Case. A Prototypical Spin Transfer Reaction

Consider the process in Eq. (7) which is a general process of atom abstraction and belongs to a broader class of radical reactions which include atom abstraction by radicals and radical additions. In all these reactions a net electron spin disappears on one atom and reappears on another, with a net reorganization of three electrons over the three centers. Therefore, in keeping with the classification of reactions in terms of electron- or group-transfers, these radical processes are referred to as spin transfer reactions.

$$X_l^{.} + A-X_r \rightarrow X_l-A + {}^{.}X_r .\tag{7}$$

The corresponding effective configurations which are needed for the description of the reaction complex are the Lewis structures **15** and **16**, and the two charge transfer configurations **17** and **18**. The Lewis structures are constructed as usual from optimized combinations of HL and zwitterionic configurations:

$$X_l\cdot \quad A\!\!-\!\!X_r \qquad X_l\!\!-\!\!A \;\cdot X_r \qquad X_l\!:^- A\cdot \; X_r^+ \qquad X_l^+ \; A\cdot \; :X_r^-$$

$$\quad\; \textbf{15} \qquad\qquad \textbf{16} \qquad\qquad \textbf{17} \qquad\qquad \textbf{18}$$

Figure 4 shows the behavior of the two Lewis structures along the reaction coordinate of the corresponding process in Eq. (7). Once again each Lewis curve starts, at the ground extreme, as a Lewis structure and ends up as a HL structure, at the excited extreme. The energy rise along the reaction coordinate is contributed as before by the homolytic dissociation of the A–X bond augmented by the overlap repulsion which is generated by the approach of one $X\cdot$ to $A\cdot$ which, in turn, is spin-paired to the other $X\cdot$. This overlap repulsion originates in the dominant triplet character of the uncoupled $X\cdot/A\cdot$ interaction and this can be revealed by inspection of the corresponding wavefunction at the excited extremes of the curves.

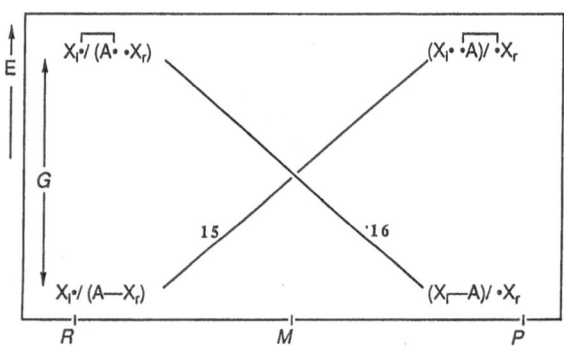

Fig. 4. The crossing of the two Lewis structures for
$$X_l^\cdot + A\!\!-\!\!X_r \rightarrow X_l\!\!-\!\!A + X_r^\cdot$$

Consider for example the excited extreme of the curve corresponding to **15** in Fig. 4. The corresponding wavefunctions is:

$$\Psi(15) = N[|x_l a \bar{x}_r| - |x_l \bar{a} x_r|].\tag{8}$$

It is seen that the $X_l\ldots A$ interaction is a pure triplet in the first determinant while in the second determinant it is 50% triplet and 50% singlet [7, 8, 12d]. Therefore, the $X_l\ldots A$ interaction is dominated by a triplet overlap repulsion which is added to the $A\!\!-\!\!X_r$ dissociation along the reaction coordinate, and contributes thereby to the energy rise of **15** along the $R \rightarrow P$ direction in Fig. 4. Exactly symmetric arguments exist for the curve corresponding to **16** along its respective reaction coordinate $(P \rightarrow R)$.

To complete the curve crossing diagram we need to mix the remaining two configurations **17** and **18** into each of the curves in Fig. 4. Following the VB mixing rules [8], the two configurations can mix into each curve along the reaction coordinate except for the ground and excited extremes where the mixing matrix elements vanish due to the infinite distances, either $X_l \ldots$ A at the R extreme or A $\ldots X_r$ at the P extreme.

Before concluding we need to relate the excited extremes of the curves to some spectroscopic states. Consider, for example, the excited extreme in the R end of the reaction coordinate in Fig. 4. What is described is a radical $X_i^{\cdot}$ which is singlet-paired to the A end of the A $\ldots X_r$ moiety which is 75% in the triplet state. The corresponding energy gap G can accordingly be expressed in terms of this relationship, as 3/4 of the singlet-triplet excitation of the A–X bond [7, 8, 12d, 27]:

$$G = (3/4) \, \Delta E_{st}(\text{A–X}) . \tag{9}$$

Returning to the same excited extreme in the R end of Fig. 4, we may realize that the spin-pairing of $X_i^{\cdot}$ to the A end of the AX_r moiety is arbitrary somewhat in view of the infinite $X_l \ldots AX_r$ distance. If we simply require a 50% chance of singlet coupling between $X_i^{\cdot}$ and each of the two ends of the A$^{\cdot\cdot}X_r$ moiety, the excited extreme will contain the A$^{\cdot\cdot}X_r$ moiety in a pure triplet state coupled to $X_i^{\cdot}$ to yield a net doublet state [8, 12d]. In this case the gap of the diagram becomes simply:

$$G = \Delta E_{st}(\text{A–X}) . \tag{10}$$

Thus, Eq. (9) emphasizes the use of a single bonding scheme for each individual curve. This definition of G is well adapted to the quantitative computations, as the crossing curves can be computed variationally. On the other hand, the definition in Eq. (10) has the conceptual advantage of connecting the ground states to spectroscopic excited states, but the computed crossing curves must be generated by rotations among the space of the adiabatic states, as has been done by Malrieu and collaborators [28] for H_3^-. We shall be using both definitions in this chapter, remembering that in any event the diagram gap for a three-electron/three center transformation is proportional to the singlet-triplet excitation of the bond which is broken during the reaction.

As in the 4-electron/3-center case, here too two extreme situations are expected for the avoided crossing diagram, by analogy to Fig. 3. Common cases with large singlet triplet gaps such as most of the atom abstraction reactions, are expected to display unstable species, $(X_lAX_r)^{\cdot}$, while stable $(X_lAX_r)^{\cdot}$ species are expected when the gap gets extremely small. Some cases of this latter situation such as Li_3 are discussed in the application sections [7].

3.1.3 Generalizations of Two-Curve Avoided Crossing Diagrams

The above procedure of construction can be repeated for any reaction or transformation which involves electronic and geometric reorganization. Thus in any single step process one can construct the diagram by defining the Lewis wave function of all the bonds which are broken and formed during the reaction. There will be one Lewis structure for the reactants' bonds and one for the products' bonds. The energy of each structure is then followed along the respective reaction coordinate and this defines a Lewis curve. The Lewis curves are then complemented by secondary structures to define the final curves of the curve crossing diagram. Each of the so resulting curves is generally typified by a ground and an excited states of well-defined characters.

A generalized curve crossing diagram is Fig. 5a. Here, the ground states are Ψ_R and Ψ_P with R and P referring to reactants and products, respectively. A qualitative understanding of the excited states in the diagram allows to assign them qualitatively and generate the diagram without having to compute each entire curve [8]. As shown later, much chemical knowledge can be generated and patterned from the knowledge of the diagram's excited states. It is essential therefore to go over them in some detail.

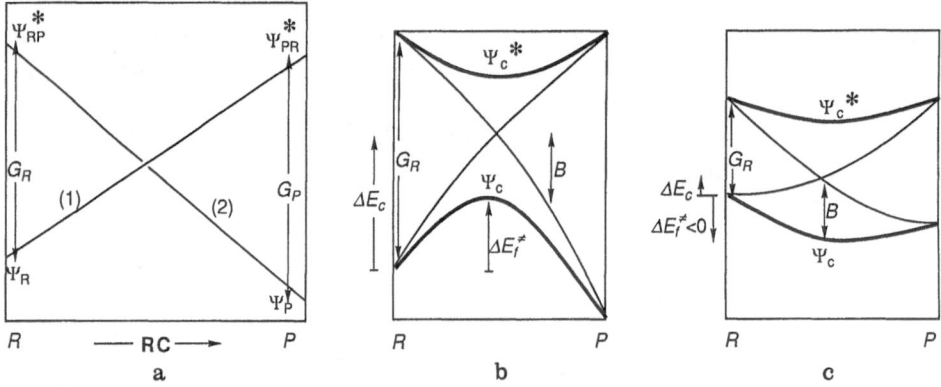

Fig. 5a. A general curve crossing diagram with two ground states and two "image" excited states which are defined in *STATEMENTS 1* and *2*. (**b**) and (**c**) are avoided crossing diagrams for cases of a large gap, (**b**), and a small gap, (**c**)

The excited states in Fig. 5 are designated as Ψ^*_{RP} and Ψ^*_{PR}. The double subscript denotes, in order, the geometry and spin-pairing pattern of the state. Thus, Ψ^*_{RP} describes a species with the same geometric features as R — the reactants — but with the spin-pairing pattern of the products $-P$. Similarly, Ψ^*_{PR} has the geometry of the products with the spin-pairing pattern of the reactants. This is the reason why each excited state correlates with the ground state end of the respective correlation line. We call these excited states the "image states" [8].

The foregoing description means that the image states of the diagram can be generated from the ground states, directly below them, by particular electronic excitations which prepare the requisite spin-pairing pattern, while conserving the total spin of the ground state [3, 8]. The simple principle is; that for each bond that is broken during the reaction we require one excitation, so that the total number of excitations in the image state is equal to the total number of bonds which are broken during the reaction. The two elementary excitations which are employed to this end were described above for the prototypical cases, that is the triplet unpairing of a Lewis bond, and the charge transfer excitation from a donor center to an acceptor moiety. The choice of the elementary excitation depends on whether or not the reaction centers in question undergo also formal changes in their "oxidation state".

The formal change in the oxidation states can be ascertained most clearly by comparing the HL structures of reactants and products.

Consider for example the HL structures for the S_N2 reaction in Eq. (11).

$$Y\!:^- + R\cdot\!-\!\cdot X \to Y\cdot\!-\!\cdot R + :X^- . \tag{11}$$

It is seen that $Y\!:^-$ undergoes a formal oxidation — a loss of one electron — while $X^\cdot$ experiences a formal reduction — a gain of a single electron. Thus, the reaction is typified by a formal "redox" of two reaction centers (X, Y). This in turns means that a vertical charge transfer excitation, from $Y\!:^-$ to R–X, is required to generate the requisite image state in the diagram. Since only one bond is broken in the reaction, then only one excitation is required, and this in turns means that the excited image states in the S_N2 diagram will be vertical and singlet-paired charge transfer states of reactants and products.

The following statements summarize the nature of the excited image states in the generalized curve crossing diagrams of Fig. 5a.

STATEMENT 1: For an isovalent reaction which is attended by bond exchanges without changes in the formal oxidation states of the reaction centers, the image states involve only triplet excitations, one for each bond which has to be broken in the reaction.

STATEMENT 2: For a reaction which is attended by both bond exchanges as well as changes in the formal oxidation states of the reaction centers, the image states involve a charge transfer excitation for each pair of centers that undergo a formal redox. The rest of the excitations will be triplets. The total number of excitations must equal the number of bonds which are broken during the reaction.

Knowledge of the excited states in the avoided crossing diagram provides us with the diagram gaps (G_R and G_P in Fig. 5a). These in turns help us to decide about the height of the barrier or the stability of the intermediate which is formed after avoided crossing.

3.1.4 The Activation Process and Expressions for Barrier Heights

Turning back to the curve crossing diagram in Fig. 5a, the corresponding avoided crossing diagrams are shown in Fig. b and c. Once again we show the expected extremes which derive from the size of the excitation gaps of the diagram. Thus, large gaps lead to Ψ_c species which are transition states for the $R \to P$ transformation. On the other hand, when both diagram gaps are small, as in part (c) of the figure, the avoided crossing leads to a stable intermediate, Ψ_c. Of course, parts (b) and (c) describe the extremes, and some intermediate situations can be envisioned where the $R \to P$ transformation is barrierless and downhill. A situation of the latter type is shown in **19** and is seen to be typified by two different gaps, one very small and the other one large. A case of this latter type is, for example, the barrierless process of gas-phase nucleophilic addition $HO^- + H_2C = O$ [29]. Thus an ensemble of cases, transition states, intermediates, and barrierless processes can be conceptualized in a unified manner using the gaps in the avoided crossing diagram.

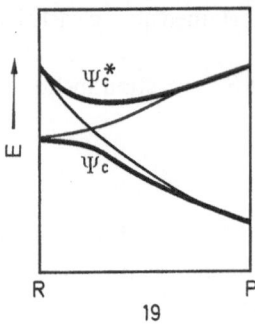

19

The diagram in Fig. 5 includes some additional features. An important quantity in the Fig. 5 is B which represents the avoided crossing interaction, i.e., the stabilization of the state Ψ_c relative to the crossing point. The quantity B is the stabilization energy due to the mixing of the two localized bonding situations, one reactant-like and the other product-like, which are degenerate at the crossing point. As such B is the Quantum Mechanical Resonance Energy (QMRE) of the state Ψ_c, be this a transition state as in Fig. 5a, an intermediate as in Fig. 5c, or a delocalized state like in 19 [7].

Having this physical meaning for B we can define a general equation for the energy of the delocalized state Ψ_c relative to the reactant state, and the symbol $\Delta E^{\neq}$ will be used for all the spectrum in Fig. 5 including both $\Delta E^{\neq} > 0$ and $\Delta E^{\neq} < 0$.

For any given direction of the process, the $\Delta E^{\neq}$ may be written in terms of the corresponding height of the crossing point and the QMRE. For the forward direction $\Delta E_f^{\neq}$ is given in Eq. (12) as a balance between the height of the crossing point (ΔE_c) and the QMRE (B).

$$\Delta E_f^{\neq} = \Delta E_c - B \tag{12}$$

Furthermore, the height of the crossing point can be most simply related to the gap, G_R, as a fraction (f) of this gap, that is:

$$\Delta E_c = fG_R \tag{13}$$

The expression for the barrier becomes then:

$$\Delta E_f^{\neq} = fG_R - B \tag{14}$$

Here, $\Delta E_f^{\neq}$ is described as a net result of bond distortion and overlap repulsion required to destabilize the ground state Ψ_R and bring it into resonance with the excited state, Ψ_{RP}^*. The activation process is viewed therefore as a total deformation effort (bond distortion and overlap repulsion) which is required to overcome the vertical gap G_R, and bring the two states Ψ_R and Ψ_{RP}^* into resonance [8]. The resonance causes delocalization of the electrons in the bonds that are broken and those that are formed, and allows thereby the transformation of reactants to products. In addition, the equation is seen to cover the range of cases in Fig. 5. Thus, whenever the relationship $fG_R > B$ exists, a transition state with a positive barrier is predicted.

On the other hand, whenever $fG_R < B$, the delocalized state Ψ_c will be below the reactants' complex (Ψ_R) and either an intermediate or a barrierless process are expected (e.g., **19**) [7, 29].

Equation (14) is seen therefore to offer us a useful way to reason about reactivity provided we understand how the parameters G, f and B vary. The parameter G is the simplest one, as it is related to a spectroscopic excitation energy of reactants or products. Indeed, in reactions involving electrophile-nucleophile combinations, Kochi et al. [30], Shaik [31], and Buncel et al. [29] have reported correlations between experimental barriers and the vertical charge transfer transitions of the reactants. Similarly, in spin-transfer processes and many-bond exchange reactions there exists a general correlation between the barrier ($\Delta E^{\neq} > 0$ and $\Delta E^{\neq} < 0$) and the singlet-triplet excitation gaps [7]. Thus, in many instances it is possible to rely only on the excitation gap for making reasonable predictions about barrier height and stability of delocalized clusters. Indeed, a good deal of reactivity and stability patterns in chemistry will be determined by the "energy gap rule", that is, the larger the gap the larger the barrier and vice versa.

The second quantity in Eq. (14) is f. Using the relation in Eq. (13), f is simply the ratio of the height of the crossing point to the gap, as follows:

$$f = \Delta E_c/G_R . \qquad \cdot(15)$$

Namely, f is a measure of how much of the gap enters under the crossing point by the deformations that cause resonance. This in turns depends on the shape and the curvature of the curves [4, 5] as illustrated in Fig. 6.

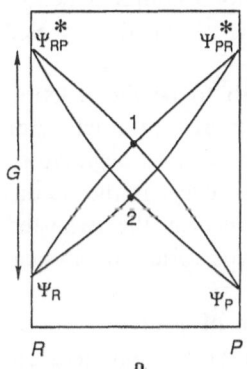 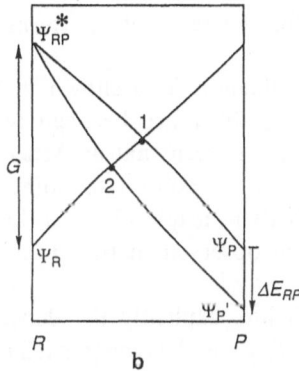

Fig. 6 a, b. The effect on the f quantity of: (a) the steepness of excited state's descent, and (b) the variation in the reaction energy (making the reaction more exothermic)

Figure 6a shows a case of two identity processes having the same gap (G) but differing in the way the curves descend from the excited states to the ground states. It is seen that the height of the crossing point is modulated by the curvature of descent of Ψ_{RP}^{*} toward Ψ_P (and Ψ_{PR}^{*} toward Ψ_R) along the reaction coordinate. We recall that the main factor which causes these excited states to descend in energy is the coupling of their infinitely separated spin-pairs into equilibrium Lewis bonds. Thus, if the odd electrons which are spin-paired are delocalized away from the atomic centers which should end up being bonded, then the bond coupling interaction of the spin-pairs will be weak and the descent will be moderate leading to a high-energy crossing point; point 1 in Fig. 6a. If, on the other hand, the odd electrons are localized on the union centers the descent of the excited states will be steep and lead thereby to a low-energy crossing point, 2 in Fig. 6a. In summation, large f is associated with delocalized spin-pairs in the excited image states of the curve crossing diagram, and the opposite is true for small f.

$$
\underset{\textstyle 20}{
\begin{array}{ccc}
\overset{\displaystyle 2e^-}{\overbrace{}} & & \overset{\displaystyle 3e^-}{\overbrace{}} \\[-2pt]
X_l\text{·}\quad (A\text{·}\;\; \text{:}X_r^-) & \longleftrightarrow & X_l\text{·}\quad (A\text{:}^- \;\;\text{·}X_r)
\end{array}
}
$$

To illustrate the effect of delocalization on the descent of the image states let us consider the ionic exchange process in Fig. 2b. Here the odd electron is delocalized in the $(A\dot{-}X)^-$ species over its two centers. Taking the explicit VB expression of this state we see in **20** that the bond coupling with the left-hand side $X_l^.$ encounters a repulsive three-electron interaction which counteracts the stabilizing effect of the two-electron interaction. Thus, the effect of delocalization in the image states is to place on the two union centers more than two electrons and to reduce thereby the two-electron stabilization. This in turns is expressed as a shallow descent of the image state, and hence as a high-energy crossing point and a large f [4, 5]. This aspect of f is called the "bond coupling delay", and variations in the bond coupling effect are among the key factors for the deviations of reactivity trends from the "energy gap rule". This blend, of "energy gap rule" with "bond coupling delay" effects, is the primary cause for the reactivity patterns in the set of identity S_N2 reactions [4, 5, 32].

The other key factor that influences f is shown in Fig. 6b to be the reaction energy, ΔE_{RP}, that is the energy difference between reactants and products. For a case where only ΔE_{RP} varies, it is seen that as ΔE_{RP} becomes more negative so will the crossing point decrease, from point 1 to point 2. This effect is the fundamental Bell-Evans-Polanyi (BEP) principle [2b, 33], known also as the Brønsted relationship [34], which is often observed in the Hammett-type series in physical organic chemistry.

There exist equations which link explicitly the dependence of f on the "bond coupling delay" effects and on ΔE_{RP} and the reader is referred to the original literature [4b, 5b, 8]. Of course, in the most general situation f will depend also on the curvature of ascent of the ground states toward the crossing point. Such an effect is related to overlap repulsion and is discussed later in the chapter by reference to the X_3 clusters. Other effects like electrostatic and steric interactions should enter into the final f value, but are not going to be discussed in this chaper.

The third quantity in Eq. (14) is B, the QMRE of the delocalized state. By appeal to Fig. 5b, c and to drawing **21** the general expression for B is:

$$
B = [(1 - S_{12})/2]\, \Delta E(\Psi_c, \Psi_c^*). \tag{16}
$$

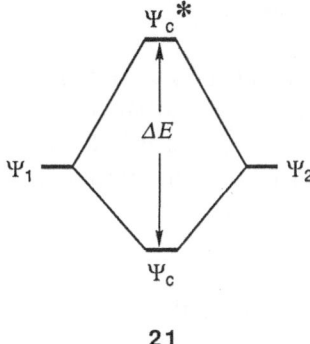

21

Here S_{12} refers to the overlap of the two bonding schemes of the curves at the crossing point, while $\Delta E(\Psi_c, \Psi_c^*)$ is the energy difference of the two states Ψ_c and Ψ_c^* resulting from the avoided crossing [8, 35]. Equation (16) leads in turn to two important relationship [7, 8, 12d, 35, 36] between B and fundamental properties of the delocalized state, Ψ_c, be it a transition state or a stable species.

The first relationship refers to the individual bonds which are reorganized during the transformation. By appeal to the three-center cases $X_l \ldots A \ldots X_r$, it is possible to show [35] that B is proportional to the $A \ldots X$ bond energy or alternatively to the singlet-triplet excitation energy of this bond, i.e.:

$$B \propto D(A \ldots X); \qquad B \propto \Delta E_{st}(A \ldots X). \tag{17}$$

The second relationship can be generated by expressing the $\Delta E(\Psi_c, \Psi_c^*)$ quantity in Eq. (16) in MO terms. Then the state energy difference becomes simply an energy difference between the MO's of the delocalized state [35] notably the HOMO and LUMO of Ψ_c as follows:

$$B = [(1 - S_{12})/2] (E_{LUMO} - E_{HOMO}). \tag{18}$$

This latter expression illuminates the connection between the avoided crossing diagrams and the orbital symmetry and related MO rules [1, 37]. It is apparent thus from Eq. (18) that B will get smaller the smaller becomes the HOMO-LUMO gap of the transition state (Ψ_c in the general case). Since "antiaromatic" transition states of forbidden reactions [37] possess small or vanishing HOMO-LUMO gaps, then according to Eq. (18) these transition states will possess much smaller B values than the "aromatic" transition states of allowed reactions [7, 8].

To summarize, the two-curve avoided crossing diagram is particularly suited to discuss the barrier problem, both positive and negative barriers. Thus, reactivity and stability patterns of hypercoordinated and delocalized species can be discussed in a unified manner by appeal to the total destabilization energy, ΔE_c, required to achieve resonance in the characteristic curve crossing diagram, and the magnitude of the resonance energy, B. The destabilization energy depends in turn on the diagram excitation gap and on the combination of the "bond coupling delay" effects and the reaction energy (Fig. 6).

3.2 Many-Curve Avoided Crossing Diagrams

Occasionally a chemical transformation is mediated by an intermediate which is electronically distinct from its Lewis-type precursor and successor reactants and products. Stepwise mechanisms in organic chemistry, such as S_N1, E1, $E1_{CB}$, and S_NV are common cases of this type. The intermediacies of pentacoordinated SiX_5^- species and of various MH_{n+1} radicals (e.g., M = P) are also of the same electronic reorganization type [4; 12b, c]. In all of these cases, the use of a many-curve model is both necessary and illuminating.

The most common category of intermediate occurs in the covalent-ionic transition, when one of the ionic VB configurations that belongs to the interchanging bonds becomes dominant in the central part of the reaction coordinate. This is typical to the classical stepwise mechanisms which involve bond heterolysis in organic chemistry, e.g., S_N1, and the stabilization of the ionic VB configuration can usually be predicted from qualitative chemical knowledge [4a, c].

Figure 7a shows the case of $(FHF)^-$ which has been computed recently [38] by Kabbaj, Sini, and Hiberty. In this case, the tri-ionic configuration $F^-H^+F^-$ crosses much below the HL structures and becomes especially stable at the mid-point of the reaction coordinate, due to the very small distance, $d(H^+F^-)$, which is made possible be the small "size" of H^+. As a result, the curve mixing and avoided crossings lead to the intermediate structure $(FHF)^-$ which is more stable than its precursor and successor Lewis structures, and possesses a dominant ionic character. The S_N1 and E1 mechanisms are analogs of $F^-H^+F^-$, involving instead a carbenium ion VB configuration which crosses below the two HL structures of reactants and products. In these cases, however, the intermediate configuration is made stable by the action of solvent [4a, 9] and is consequently not so stable leading thereby to a metastable intermediate, as shown in Fig. 7b in a schematic manner.

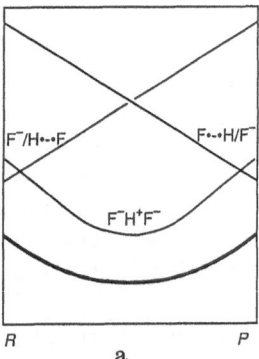 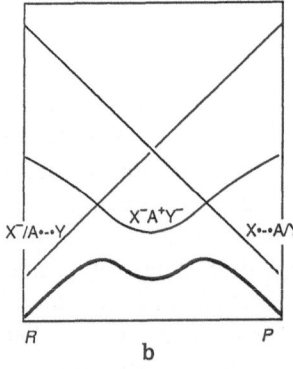

Fig 7 a, b. Curve crossings of the HL (covalent) and ionic structures and the formation of ionic type reaction intermediates in:
(a) $F^- + HF \rightarrow [FHF]^- \rightarrow FH + F^-$, according to ab initio VB computations, and in (b) a general situation (e.g., S_N1) where the ionic configuration is not so stable

Separate treatments of the HL and ionic configurations as distinct curves in a many-curve model is useful also in instances where there is no reaction intermediate. Thus, simple chemical knowledge of the substituent effect can help predicting the variation of the energy of the carbenium ion configuration in S_N2 or E2 mechanisms

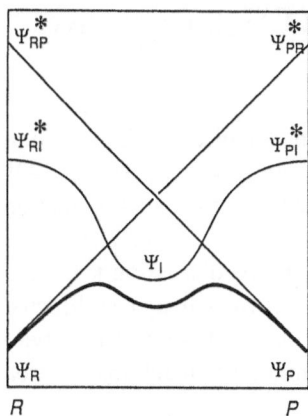

Fig. 8. A many-curve diagram where the Lewis curves are horizontally crossed by an intermediate curve for which the anchor excited states are not "image" states

and to analyze thereby the variation of the charge character of the transition state, and the S_N2-S_N1 or E2-E1 mechanistic transitions. These ideas have been treated in detail in several reviews and in a recent textbook [4, 9].

The second category of many-curve diagrams is used when a third curve, made of VB configurations of bonds other than those which interchange between reactants and products, crosses horizontally below the intersection point of the principal Lewis curves of the interchanging bonds. The general situation is illustrated in Fig. 8 which has one additional curve as compared to Fig. 5a. An example of this type is the S_NV reaction, **22** [39], where the principal curves involve the Lewis σ_{CX}, σ_{CY} bonds and look like Fig. 2b, having vertical σ charge transfer states as the image states. However, the reaction mechanism involves normally a π-attack via intermediary of a carbanion, **23**. This intermediate originates, in turn, in a π-charge transfer state, **24**, which is distinct, at the two extremes of the reaction coordinate, from the states of the principal curves. Thus, when such intermediate excited states – which are not image states of reactants and products – cross the principal curves they serve to mediate the $R \rightarrow P$ transformation with a lower barrier than the concerted transformation [4b, 39]. In addition to reaction intermediates, also stable hypercoordinated species, which involve formally nine and ten valence electrons [12 b, c; 40] and which we have examined, belong to this latter type discussed above.

22

23   **24**

4 Application to Nucleophilic Substitution of CH₃X Derivatives

The S$_N$2 reaction is one of the fundamental aspects in chemical education, and a most vivid research area in experimental and theoretical gas phase [41, 42], solution [43], and reaction dynamics studies [44]. Our main purpose in this section is to demonstrate the use of the two-curve diagram as a means to conceptualize the barrier and the structure of the transition state. As such, the discussion is limited to the class of identity reactions in the gas phase. The interested reader is directed to recent reviews and a monograph which give detailed accounts of S$_N$2 reactivity in the gas phase and in solvents for identity and nonidentity reactions as well [4b; 5 a, b].

4.1 The Identity S$_N$2 Reaction

The barrier of the identity S$_N$2 exchange, Eq. (19), has a fundamental significance as it represents the net outcome of electronic reorganization, unmasked by thermo-dynamic effects.

$$X\!:^- \; + \; R\!-\!X \; \rightarrow \; X\!-\!R \; + \; :X^- \, . \tag{19}$$

The observed reactivity patterns [4b; 5 a, b; 32; 41–43] display a few notable trends. The relative reactivity in a set of identity reactions follows the "leaving group ability" of X but is unrelated to its "nucleophilicity". In accord, small barriers are observed for X = I, Br, Cl, while large ones have been evaluated and computed for X = OH, CH₃O, HS, CH₃S, CN, CCH, PhCH₂. The reactivity trends appear to be medium independent and persist in the gas phase [41] in "vacuum" [42], and in a variety of solvents [43]. In the following sections we show how some of these trends can be reconstructed and predicted by use of the curve crossing diagram, and in what way are they related to charge transfer properties of the reactants.

4.1.1 The Activation Process

Figure 9 is the two-curve avoided crossing diagram for the identity S$_N$2 process [4b; 5 a, b; 32]. The simplest expression for the barrier is Eq. (20) which follows from the general discussion associated with Eq. (14) above.

$$\Delta E^{\neq} = \Delta E_c - B \, ; \qquad \Delta E_c = f(I^*_{X:} - A^*_{R-X}) \, . \tag{20}$$

The height of the crossing point ΔE_c in Eq. (20) corresponds to the total destabilization energy (bond distortion and overlap repulsion) which is necessary in order to raise the ground state so it achieves resonance with the descending charge transfer state [4b; 5 a, b; 32].

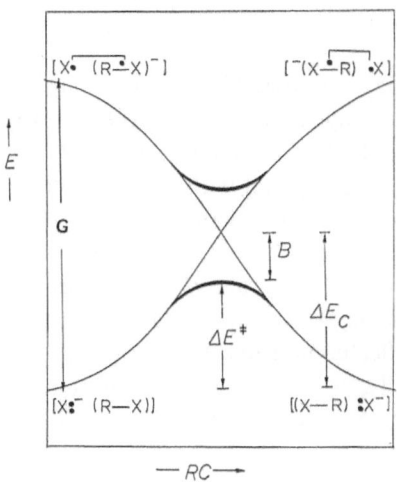

Fig. 9. A avoided crossing diagram for identity S_N2 reactions. The extremes of the *RC* refer to the ion-dipole geometries. Adapted from reference 5a with permission of Acta Chemica Scandinavica

The connection between the central barrier of the identity reaction and the distortion energy has been projected in an ab initio computational study [42c]. It has been shown that the central barrier of the gas phase reaction, $X:^- + CH_3X \rightarrow XCH_3 + :X^-$, correlates linearly with the corresponding distortion energy, ΔE_{dis} which is the computed difference between the energies of the CH_3X molecule, in its transition state geometry, relative to its ground state geometry within the $X:^-/CH_3X$ ion-dipole complex. The correlation which covers X groups like H, F, Cl, OH, HS, CH_3O, CN, NC, and HCC is nevertheless remarkably linear (r = 0.98) and reads as:

$$\Delta E^{\neq}(X:^-/(CH_3X) = \Delta E_{dis}(CH_3X) - (25.1 \pm 3.3 \text{ kcal/mol}). \tag{21}$$

It follows from Eq. (21) that the barrier derives from the distortion that is required to carry the CH_3X molecule to its transition state geometry, and that the interaction of $X:^-$ with the distorted molecule at the transition state is stabilizing and does not vary greatly. This in turns means that the QMRE of the transition state, namely *B* should be a narrow range variable, and its variation may be disregarded in qualitative considerations. Some justification for that conclusion comes from our discussion of *B* in Eq. (17) which shows that *B* is proportional to the C–X bond strength in the transition state. Since the bonds in the transition states are weakened relative to the ground state, and since the ground state C–X bond energies span not a very wide range (~ 70–108 kcal/mol; 1 kcal/mol = 4.184 kJ/mol), then *B* itself will vary in an even narrower range as is indeed reflected from Eq. (21). Thus, while we expect variations in the value of *B*, as a function of the X group, we also expect that this variation should be smaller than the corresponding variation in the height of the crossing point. This greatly simplifies the use of the curve crossing model because we need only understand the factors that govern the height of the crossing point in order to make predictions about the variations in the central barrier and all the associated reactivity trends.

4.1.2 Reactivity Factors in the Gas Phase

Using thermochemical considerations [4b, 5b, 32] it is possible to show that the variation of the electron transfer energy gap quantity is dominated by D_{C-X}, the bond energy of the CH_3X molecule. The stronger the bond, the larger normally the vertical charge transfer energy gap. The gap can be computed with a semiempirical VB scheme [4b, 5b, 32], or taken from a combination of experimental ionization potentials [45 a, b] of $X:^-$ and vertical electron affinity values of CH_3X molecules, derived from electron transmission resonance [45 c–i]. Alternative procedures are also available, such as a simple SCF estimation of the gap with a double zeta quality basis set [12a; 45 g, h]. Some of the values are collected in Table 1, and the different sets exhibit similar trends. The first set of values in Table 1 is the one used traditionally in the literature [4b; 5 a, b; 32] and will continue to serve us here.

Table 1. Vertical Charge Transfer Energies for the Gas Phase Reaction,
$X^- + CH_3X \rightarrow XCH_3 + X^-$

entry	X	$(I_{X:}^* - A_{R-X}^*)^a$		
		Calculated (I)[b]	Calculated (II)[c]	Experimental[d]
1	F	135	186	184–221
2	Cl	113	172	164
3	Br	99	147	141
4	I	81	126	99
5	CCH	146	241	–
6	OH	109	137	182
7	SH	95	162	129
8	CN	159	244	244

[a] In kcal/mol. 1 kcal/mol = 4.184 kJ/mol
[b] From Ref. [4b]
[c] Using the same semiempirical VB scheme as in (I), but with inclusion of overlap in all wave functions and matrix elements.
[d] Ionization potentials $(I_{X:}^*)$ from Ref. [45 a, b]. Electron affinities (A_{R-X}^*) from Ref. [45 c–i]

The f factor is the second quantity which determines the height of the crossing point in Eq. (20). Following the discussion in Sect. 3.1.4, f will be larger the more delocalized are the odd electrons in the charge transfer state (see drawing **20** and Fig. 6a).

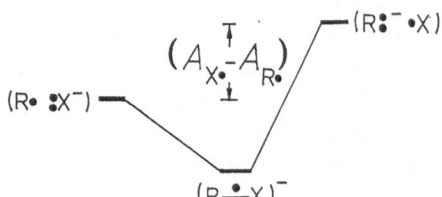

Fig. 10. A mixing diagram for the VB configurations which make up the radical anion, $(R \cdot X)^-$. Reprinted from reference 5a with permission of Acta Chemica Scandinavica

The radical anion $(H_3C \dot{\_} X)^-$ moiety of the charge transfer state can be described in VB terms as a hybrid of the corresponding three-electron structures with weights $W_{X:}$ and $W_{R:}$ ($R = CH_3$), as expressed by:

$$(R \dot{\_} X)^- = (W_{X:})^{1/2} [R^\cdot :X^-] \leftrightarrow (W_{R:})^{1/2} [R:^- \ ^\cdot X] ;$$

$$(W_{X:}) + (W_{R:}) = 1 . \tag{22}$$

An important property that governs $W_{R:}$ and $W_{X:}$ is the energy difference between the two VB structures, as schematized in their mixing diagram in Fig. 10. The energy gap is approximately the difference between the electron affinities of the $X^\cdot$ and $R^\cdot$ radicals ($A_{X^\cdot} - A_{R^\cdot}$). For a constant R in a series of reactions, the variation in this energy difference is determined by the electron affinity of $X^\cdot$ alone; a small $A_{X^\cdot}$ leads to large $W_{R:}$ and vice versa. When $W_{R:}$ is large, the odd electron has a smaller probability to be on the reaction center R, and as a result the bond coupling interaction, between $(R \dot{\_} X)^-$ and $X^\cdot$, becomes weak followed by a large f factor. These relationships can be expressed as in Eq. (23) where the relation to the electron affinity difference derives from a perturbational expression of the configuration mixing in Fig. 10:

$$f \propto W_{R:} ; \qquad W_{R:} \propto 1/(A_{X^\cdot} - A_{R^\cdot})^2 ; \qquad A_{R^\cdot} = \text{constant} . \tag{23}$$

For example, $X^\cdot$ radicals which have low electron affinities are H, NH_2, OH, CH_3O, SH and CH_3S. These groups will lead to radical anions with relatively large $W_{R:}$ and hence also to a large f factor. On the other hand, halogen atoms have large electron affinities and will accordingly generate radical anions with small $W_{R:}$ and lead to a small f [4 a, b; 5 a, b; 32].

According to Fig. 10, the $W_{R:}$ property should be determined also by the interaction matrix element between the two VB structures. The matrix element is related to the binding ability of X, and for a given electron affinity $A_{X^\cdot}$ a strong binder X will possess a larger $W_{R:}$ than weaker binder X (e.g., CCH, and CN vs I and Cl).

Other delocalization properties which lead to a high f factor are delocalization of the radical anion over a few identical C–X linkages as in CH_4^-, $CH_2Cl_2^-$, and so on [4b, 32b, 46]. Delocalization of the odd electron in $X^\cdot$, as for example in $PhCH_2^-$ leads to a large f factor. The relationship between the size of f and the extent of radical anion delocalization has been retrieved recently with the ab initio VB computations of the curve crossing diagram [12a] for $H^- + CH_4$, and the results are discussed later in this chapter.

4.1.3 Discussion of the Barriers in the Gas Phase

Table 2 shows a few data which illustrate the interplay of the vertical electron transfer energy gap and the bond coupling delay index, $W_{R:}$. Thus, for example, entries 3–5 form a group possessing approximately constant $W_{R:}$ and hence also constant f. As may be predicted from the above discussion, both the barrier and the distortion energy increase as the vertical electron transfer energy gap increases. On the other hand, in the comparison of Cl^- to OH^-, the gap is almost constant and both the

Table 2. Curve Crossing Factors, Distortion Energies and Central Barriers for the Gas Phase Reaction, $X^- + CH_3X \rightarrow XCH_3 + X^-$

Entry	X^-	$(I_{X:}^* - A_{R-X}^*)^{a,b}$	$(W_{R:})^b$	$(\Delta E_{dis})^{a,c}$	$(\Delta E^{\neq})^{a,c}$	$(\Delta E^{\neq})^{a,d}$
1	F^-	135	0.242	40.8	11.7	26.2; ~19[e]
2	Cl^-	113	0.251	28.9	5.5	~10; 13.2(2)[f]
3	HCC^-	145	0.362	72.5	50.4	~41
4	HO^-	109	0.357	52.1	21.2	~27
5	HS^-	95	0.340	38.8	15.6	~24

[a] In kcal/mol. 1 kcal/mol = 4.148 kJ/mol
[b] From Refs. [4b and 42c]
[c] From Ref. [42c]. These values are computed with the 4-31G basis set.
[d] From Ref. [41a]. These barriers are derived with the aid of the Marcus equation and the RRKM procedure.
[e] From Ref. [32b]
[f] From Ref. [41c]

barrier and the distortion energy increase in relation to the $W_{R:}$ quantity. The two types of reactivity trends provide the aforementioned blend of the energy gap and curvature factors, in the avoided crossing diagram. From a mechanistic point of view, the reactivity patterns project the two aspects of the S_N2 reaction [4a, 5, 32]; a transformation that involves simultaneously a single electron "movement" and bond coupling, or in short a single electron shift reaction as opposed to a single electron transfer reaction where the bond coupling is a separate event [4, 5].

Figure 11 shows a quantitative application of the model using the expression:

$$\Delta E^{\neq} = f(I_{X:}^* - A_{R-X}^*) - B ;$$

$$f = W_{R:} ; \qquad B = 14 \text{ kcal/mol (1 kcal/mol} = 4.184 \text{ kJ/mol)} . \qquad (24)$$

Since f is proportional to $W_{R:}$, and the latter value is in the range of algebraically reasonable f values [4b], it was decided to simply equate f with $W_{R:}$. The value of B was then estimated by using in Eq. (24) a barrier value, for Cl^-/CH_3Cl of 14.5 kcal/mol which appears to be the average value for the most recent sophisticated computational levels [5b, 47]. This B value is close to the ab initio VB computed [12a] value for $H^- + CH_4$, as discussed later, and was carried over to all the S_N2 reactions on carbon compounds. It is seen from Fig. 11 that the model expression gives a reasonable agreement with the barriers which are estimated from experimental rates by the RRKM procedure [41]. An equally good correlation is obtained if the model barriers are plotted against the ab initio computed barriers [5b]. Finally, reasonable barriers can be obtained also from the experimental vertical $(I_{X:}^* - A_{R-X}^*)$ gaps and the second set of calculated gaps in Table 1 [5b]. Thus, a naive semiquantitative analog of the curve crossing diagram based on the interplay of donor-acceptor capabilities and bond coupling effects captures the essence of the barrier problem.

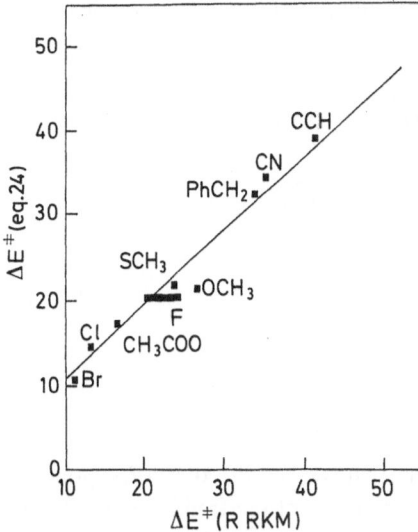

Fig. 11. A plot of the calculated central barriers [Eq. (24)] against the RRKM barriers derived from experimental data (1 kcal/mol = 4.184 kJ/mol). Adapted from reference 5a with permission of Acta Chemica Scandinavica

4.1.4 Transition State Geometries

As discussed above in Eq. (21), the molecular distortion is one of the major factors which promote the resonance between the ground and charge transfer states. Among the two possible distortions of the CH_3X molecule in **25**, the C–X stretching component dominates the total distortion energy [42c], and in such a situation it should be possible to discuss the geometry of the transition state in the same manner as we discuss barriers. To use a standard geometry scale we define in Eq. (25a) the percentage of bond stretching in the transition state relative to the ground state, in the ion-molecule complex. Equation (25b) defines the "looseness" of the transition state as the sum of percentages of bond stretching for the left (l)-and right (r)-hand side C–X bonds.

25

$$\%CX^{\neq} = 100 \, (d_{CX}^{\neq} - d_{CX}^0)/d_{CX}^0 , \tag{25a}$$

$$\%L^{\neq} = (\%CX^{\neq})_l + (\%CX^{\neq})_r . \tag{25b}$$

Table 3 shows curve crossing factors side by side with ab initio and semiempirical [48] $\%L^{\neq}$ indexes for a few transition states. For example, in the group of entries 1–3 or 4–6 $W_{R.}$ is constant and $\%L^{\neq}$ increases as the gap increases. If, however, we compare Cl^- to HO^-, now it is the variation of $W_{R:}$ that dominates and causes the transition state $(HOCH_3OH)^-$ to be looser than $(ClCH_3Cl)^-$.

Table 3. Curve Crossing Factors and Looseness Indexes for $(XCH_3X)^-$ Transition States

Entry	X^-	$(I^*_{X:} - A^*_{R-X})^a$	$W_{R:}$	$(\%L^{\neq})^b$	$(\%L^{\neq})^c$
1	F^-	135	0.242	50.0	—
2	Cl^-	113	0.251	42.4	38
3	Br^-	99	0.246	—	35
4	HCC^-	145	0.362	90.0	—
5	HO^-	109	0.357	60.8	—
6	HS^-	95	0.340	56.8	—

[a] In kcal/mol. 1 kcal/mol = 4.184 kJ/mol.
[b] Ab initio results from Ref. [42c]
[c] MNDO results from Ref. [48]

It is apparent that transition state looseness behaves as do the corresponding barriers, and this may be witnessed from the comparison of the trends in $\%L^{\neq}$ with those in the barriers (Table 2). In fact, the correlation between the barrier and the geometry of the transition state is a general feature which carries over to nonidentity reactions, and extends over $\sim 80\%$ of bond cleavage and more than 100 kcal/mol in reaction barrier [49]. This may be seen in Eq. (26) and (27) which correlate the ab initio data base [42b] (Hartree-Fock, 4-31G basis set) of barriers and percentages of bond cleavages for the forward and reverse direction of the nonidentity S_N2 reaction, $Y:^- + R-X \rightarrow Y-R + :X^-$. These relationships originate in the interplay of the curve crossing factors, so that both barriers and transition state geometries reflect the deformation energy that is required in order to destabilize the ground state and bring it into resonance with the charge transfer state.

$$\%CX^{\neq} = a\,\Delta E_f^{\neq} + b\,, \tag{26a}$$

$$\%CY^{\neq} = a\,\Delta E_r^{\neq} + b\,, \tag{26b}$$

$$\%L^{\neq} = \%CX^{\neq} + \%CY^{\neq} = a\,(\Delta E_f^{\neq} + \Delta E_r^{\neq}) + 2b\,;$$

$$a \sim 0.76\,(\text{kcal/mol})^{-1}\,; \qquad b \sim 14\%\,. \tag{27}$$

4.2 A Computational Test.
The Avoided Crossing Diagram for the S_N2 Reaction of CH_4 and H^-

The first direct computational test of the model has been carried out recently [12a] on the hydride exchange reaction, Eq. (28), using a multistructure VB method [11]. This method which is based on a code written by the Lefour and Flament [11c] is dissociation consistent [11b], and employs non-orthogonal configuration interaction among VB structures. The structures involve atomic orbitals purely localized on a single center or fragment, a feature which ensures correspondence between the mathematical functions and the concept of bonding scheme. As such, the method is well suited for the computations of curve crossing diagrams and for critically testing the qualitative ideas.

$$H^- + CH_4^- \rightarrow [CH_5^-] \rightarrow CH_4 + H^- . \tag{28}$$

The general question is whether the avoided crossing diagram is an oversimplified model or does it faithfully account for the barrier problem? Specifically: (i) Can the adiabatic curve be reproduced rigorously and shown to follow the energetic behavior that has been postulated by the qualitative theory? (ii) Do the f, G, and B parameters follow the predicted trends or estimations? (iii) Can a mixture of only two valence bond curves reproduce an accurate adiabatic potential surface?

4.2.1 Avoided Crossing Diagram and Valence Bond Structures

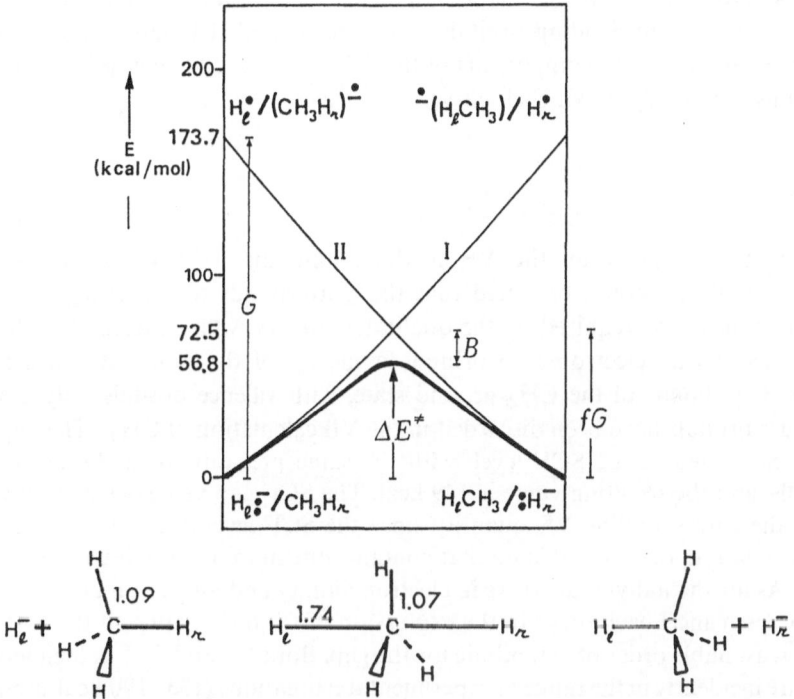

Fig. 12. Computed avoided crossing diagram for the H^- S_N2 exchange (1 kcal/mol = 4.184 kJ/mol). Reprint with permission from J. Phys. Chem. (1989), 93 5661. Copyright (1989) American Chemical Society

Figure 12 displays the calculated diagram for Eq. (28), using the 6-31G* basis set augmented with diffuse orbitals on the axial hydrogens [12a]. The diagram consists of two curves, I and II, which are anchored at two ground and two charge transfer states of reactants and products. The ground states of the two curves I and II are the Lewis structures **26** and **27**, where electrons are represented by dots and Lewis bonds by lines. Each Lewis structure represents an optimized mixture of a Heitler-London and two ionic configurations. The upper anchor points of the curves I and

26   **27**

II are valence-type charge-transfer states, which are composed of an H· and a radical anion, CH_4^-. In the latter species the negative charge and odd electron are equally distributed over the identical four C-H linkages. To generate the charge transfer state, we mix the Heitler-London component of the Lewis structure **26**, with two additional structures. One structure type possesses the charge in the axial orbital of the carbon atom, in the direction of the bond which is cleaved during the reaction. The second structural type is one in which the charge is located in the totally symmetric (a_1) combination of the σ^* antibonding orbitals of the inactive C-H bonds. These two latter structures are important components of the delocalized excited charge-transfer states and vanish gradually down each curve.

4.2.2 Results

The G factor, as calculated by the VB method, amounts to 174 kcal (1 kcal/mol = 4.184 kJ/mol). It should be noted that this corresponds to the energy of a vertical charge transfer, as required by the qualitative theory, which means that the CH_4^- species accepts an electron into orbitals made up of the same AOs which constitute the C-H bonds of the CH_4 ground state, with valence orbitals only. As such, the diffuse orbitals have been discarded in the VB calculation of CH_4^-. The gap may also be calculated at the SCF level, with the same precautions in the use of diffuse orbitals, and the resulting value is 149 kcal. The SCF gap value is expectedly smaller than the corresponding VB quantity since the SCF level disfavors, by lack of electron correlation, the ground state that contains one more bond relative to the excited state. As for the individual errors in electron affinity and ionization potential, these more or less cancel each other in the expression of G, so that after all the SCF value gives a reasonable order of magnitude for the gap. Both VB and SCF calculated values for G are incidently in the range of experimental estimations (133–190 kcal/mol) [45 a, e].

A large f factor was qualitatively predicted for the $H\!:^-/CH_4$ reaction [4b, 32b] because of the very delocalized nature of the odd electron in CH_4^-. The VB calculation substantiate the delocalize nature showing that the odd electron density on carbon, in the axial bond, is ≤ 0.29. Thus, as argued in the qualitative section, the bond coupling in the charge transfer state, $CH_4^- + H\cdot$, should be rather weak and its descent shallow. Indeed, the VB results show that the actual f value for the curves I and II is 0.42, while if pure Lewis curves are used, with no delocalization present in the three C–H linkages which remain intact in the reaction, the resulting f factor decreases to 0.27.

The third parameter of the avoided crossing diagram, the B factor, is the quantum-mechanical resonance energy (QMRE) of the CH_5^- transition state. While one cannot discuss the variance range of B on the basis of a single calculation, still we may note that the calculated value of 15.9 kcal/mol is in the order of magnitude of the value assigned above based on qualitative arguments. The fact that B is small in comparison with the value of the gap is an important feature which renders the avoided crossing diagrams qualitatively useful. This means that the final adiabatic curve stays close to the two curves which participate in the avoided crossing, and thereby the energetics of the adiabatic state curve can be easily deduced from knowledge of the behavior of the crossing curves.

Finally, the adiabatic state displays a barrier of 56.8 kcal, in close agreement with other estimations arising from ab initio calculations in the same basis set [12a], the electron correlation effects being fully taken into account. This means that no important factor is left out in the description of a chemical reaction in terms of VB structures, and that the barrier can be both conceptualized and computed in the same VB sense, namely as an outcome of avoided crossing of two curves which are anchored at the two ground and two vertical charge transfer states of reactants and products. Similar conclusions are being currently obtained for other $S_N 2$ reactions and their analogs, by the Orsay and the Toulouse groups [50].

4.3 $S_N 2$ Reactivity: A Summary

This section shows how the trends in the barriers and transition-state geometries of identity $S_N 2$ reactions can be conceptualized by reliance on two quantities of the avoided crossing diagram; the vertical electron transfer energy gap, and the delocalization properties of the charge transfer species which determine the "bond coupling delay" along the reaction coordinate [4 a, b; 5 a, b; 32]. Indeed, much of the "collage of $S_N 2$ reactivity patterns" [4b] can be conceptualized as the interplay of the trends which reflect the "energy gap rule", and follow the donor-acceptor capabilities of the reactants, with those trends which reflect the bond coupling and bond interchange effects, depending on the delocalization properties of the charge transfer species and in nonidentity reactions also on the reaction thermodynamics. The QMRE of the transition state is a significant quantity ($\geqq 14$ kcal/mol) which reflects the strong bonding in the transition state, as opposed to an actual single electron transfer reaction [5c, 10].

5 Application to Isovalent Bond Exchange Reactions

Having understood the interplay between the excitation gap and the bond coupling delay effect, we can now present some other features of the model. This is done by reviewing topics in spin transfer reactions and by an application to a cycloaddition reaction (both reactions are defined as isovalent types).

5.1 Regioselectivity of Radical Attacks

Radical attacks on substituted olefins [51], Eq. (29), proceed in two regiochemical pathways.

$$R^{\bullet} + \,>C=C<\, \rightarrow R-\overset{|}{\underset{|}{C}}-\overset{|}{\underset{|}{C^{\bullet}}} \tag{29}$$

The more common result is the attack on the less substituted carbon, and this in many cases is also the thermodynamically less favored pathway. Occasionally, the common trend is reversed and the attack proceeds on the more substituted carbon, e.g., in the reaction of CH_3 with $CF_2=CHF$. However, this result which is thermodynamically favored, is subject to switchovers to the "normal" regiochemistry. Thus, changing the attacking radical to CF_3, CCl_3, or H restores the superiority of the "normal" pathway, resulting in an attack on the less substituted site of $CF_2=CHF$ [51a]. Also, the CH_3 radical, which reverses the common order with $CF_2=CHF$, behaves normally with $CH_2=CHF$ and adds to the less substituted carbon [51a]. The following text explains how this jigsaw puzzle is shaped by the opposition between the "bond coupling delay" and the reaction thermodynamics (reaction energy) effects.

The avoided crossing diagram involves the triplet state of the olefin in the excited state, and the general situation in a regiochemical problem is schematized in Fig. 13. Thus, the barrier for the two regiochemical pathways is given as usual by the

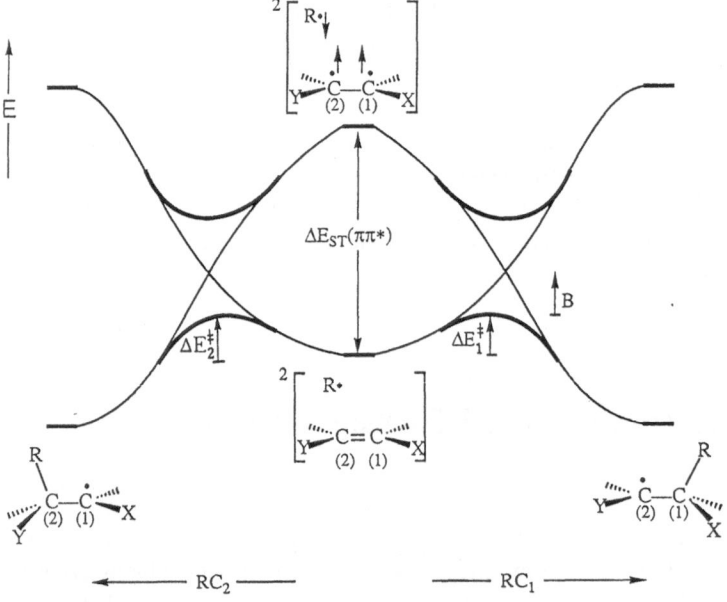

Fig. 13. Avoided crossing diagrams for the two regiochemical pathways of radical addition to olefins. The total spin is doublet.

difference between the height of the crossing point and the QMRE quantity, B, as follows:

$$\Delta E_{1,2}^{\neq} = \Delta E_c - B = f\Delta E_{st} - B. \tag{30}$$

Since the initial gap ΔE_{st} is common to the two regiochemical pathways, the factors which affect regioselectivity must arise from variations in f and/or B. As discussed above the B quantity is proportional to the sum of the bond energies in the transition state, that is Eq. (31), where $D^{\neq}$ represents the bond energy in the geometry of the transition state [52].

$$B \propto (D_{RC}^{\neq} + D_{\pi}^{\neq}). \tag{31}$$

In a structurally related series $D_{RC}^{\neq}$ and certainly $D_{\pi}^{\neq}$ do not vary greatly and their changes are assumed to counteract each other due to a trade off of bond strengths (this is equivalent to saying that there is a conservation of the total bond order in such series). Indeed, as we shall see later on, the variation of B is not dominant and can be disregarded in qualitative reasonings using Eq. (30).

It follows therefore that in order to predict regioselectivity in a given radical-olefin combination we can rely on f to make the predictions. We recall that f depends on the delocalization properties (Fig 6a) of the excited state and on the reaction energy (Fig. 6b). Thus, differences in these properties for the two regiochemical pathways should determine the regioselectivity of the addition. According to Eq. (30) the preferred regiochemical pathway will be the one which possesses a smaller f factor and hence a lower energy crossing point, ΔE_c.

Let us exemplify the argument with a simple case of the addition of CH_3 to monofluoroethylene. The computed [52] ΔE_c and B quantities, derived from a Morokuma analysis [53], are displayed for the two regiochemical pathways in **28a** and **28b**, in kcal/mol units (1 kcal/mol = 4.184 kJ/mol). The ΔE_c term gives the total energy of the localized reactant at the crossing point and is the sum of the Morokuma terms DEF, EX, ES, and PL. The B term is the difference between the SCF energy of the transition state and the energy of the localized reactant configuration at the same geometry. Thus, B is simply the sum of the Morokuma terms, MIX and CT. It is seen from **28a** vs **28b** that the pathway with the largest barrier corresponds to the pathway with the highest ΔE_c. It is apparent also that B is not a constant, but that its variation is smaller than that of ΔE_c. Since the reaction energies (ΔE) are seen to be almost the same for the two pathways (if at all, the ΔE is slightly more exoergic for the kinetically slower pathway, **28b**), what remains is to ascertain the role of electronic delocalization in the triplet state of fluoroethylene as the dominant effect on f and hence on the height of ΔE_c.

The spin density in the triplet state of a substituted olefin can be predicted by use of VB arguments, based on simple resonance theory. Consider a substituent X which possesses a lone pair (e.g., X = F, Cl, OR, etc.) or an X which can donate an electron pair in conjugation (e.g., CH_3, etc.). The $^3\pi\pi^*$ state can be described as a resonance hybrid of two forms in **29**. The principal form (left) is purely covalent and the secondary

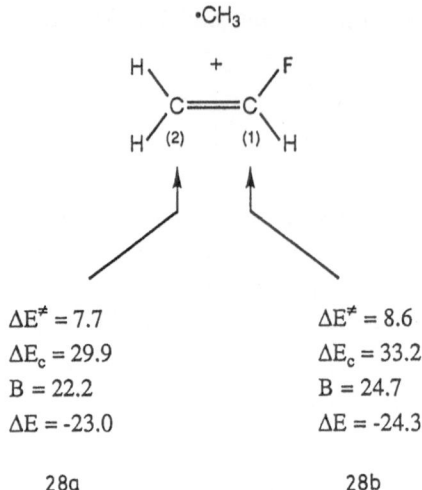

$$\Delta E^{\neq} = 7.7 \qquad\qquad\qquad \Delta E^{\neq} = 8.6$$
$$\Delta E_c = 29.9 \qquad\qquad\qquad \Delta E_c = 33.2$$
$$B = 22.2 \qquad\qquad\qquad\quad B = 24.7$$
$$\Delta E = -23.0 \qquad\qquad\qquad \Delta E = -24.3$$

$$\textbf{28a} \qquad\qquad\qquad\qquad\quad \textbf{28b}$$

form (right, $\lambda < 1$) is zwitterionic. There is another possible zwitterionic structure which keeps the repulsive triplet electrons adjacent to each other and weakens the stabilizing electrostatic interaction by further separating the charges, and is therefore less important than the zwitterionic structure shown in **29**. Similar arguments, for

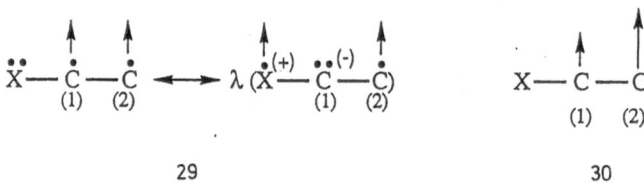

$$29 \qquad\qquad\qquad\qquad\qquad\qquad\qquad 30$$

other cases, can show that regardless of the nature of X, the net substituent effect is to take up spin density from C_1 and to increase thereby the C_2 spin density [52]. The qualitative spin density distribution in the $^3\pi\pi^*$ state is summarized in **30** with the sizes of the arrows indicating the relative spin densities. The qualitative conclusion summarized in **30** has been tested [52] for a variety of olefins by computing the triplet spin density and the results verify the qualitative picture that the more substitued carbon is the spin-poor site in the triplet $^3\pi\pi^*$ state.

For the case of fluoroethylene the spin densities are 1.083 on C_1 and 1.231 on C_2. The effect of the spin density is illustrated schematically in **31**. Thus the attack on the spin-poor site will cause a shallow descent of the excited state and a high crossing point will result. On the other hand, an attack on the spin rich site will result in a low energy crossing point. These are the roots for the difference in the ΔE_c's of the two pathways in **28a** vs. **28b**, and hence also for the relative barriers of the regiochemical attacks on monofluoroethylene.

An interesting consequence of the relative height of ΔE_c is the transition state geometries for the two regiochemical pathways. Thus, a large ΔE_c means a combination of high overlap repulsion between the radical and the olefin as well as a large degree

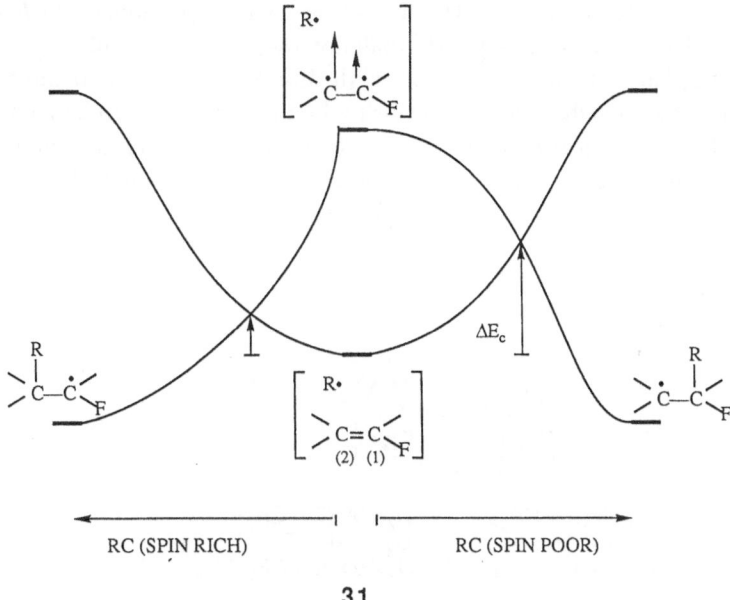

31

of bond distortion in both reactants. The corresponding transition states are shown in **32** vs **33**, and it is seen that the transition state for the attack on the spin-poor site, **33**, has a shorter CH_3−−C distance and a more distorted olefinic fragment, having a longer C−−C distance, in comparison with the transition state for the attack on the spin-rich site, **32**. The corresponding Morokuma quantities show that indeed the overlap repulsion, ΔE_{rep}(EX + ES + PL), and bond distortion, ΔE_{dis}(DEF) are larger for the transition state in **33**. Many of the regioselectivity results in radical

$$\Delta E_{rep} = 23.2$$
$$\Delta E_{dis} = 6.7$$

32

$$\Delta E_{rep} = 24.4$$
$$\Delta E_{dis} = 8.8$$

33

attacks on olefins follow the same pattern as in the fluoro-olefin case. The interesting cases of regioselectivity arise, however, when the spin density factor is counteracted by the opposite tendency of the reaction energies, ΔE, namely the reaction thermodynamic effect. Lets us take, for example, $CF_2 = CHF$. The ab initio computations show [52] that the spin density difference has a small preference (1.075 vs. 0.990) for the less substituted terminus, CHF. Therefore, a strong preference of ΔE for the more substituted terminus can easily tip the balance toward an "ab normal" regioselectivity.

The ΔE depends on the bond strength differences in the two radical products. Drawing **34** shows these differences using bond energies of model compounds [54]. Thus, when the radical R is $CH_3^\cdot$, ΔE is expected to be ~ 10 kcal/mol more exoergic for attack on C_2 (CF_2 terminus) than on C_1. However, for radicals like $H^\cdot$ and $CF_3^\cdot$ the ΔE should possess a very small directive preference, if all.

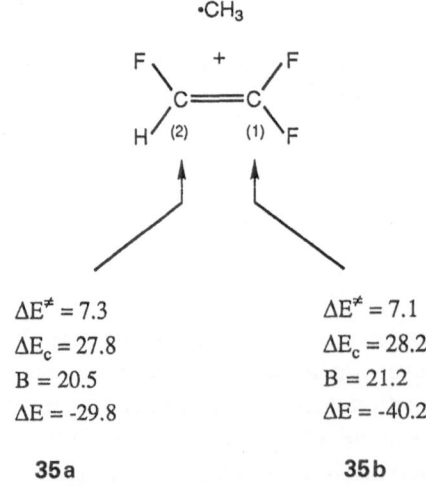

$D_1 = 87\ (R = CH_3)$ $D_2 = 99\ (R = CH_3)$
$D_1 = 100\ (R = H)$ $D_2 = 101\text{-}3\ (R = H)$
$D_1 = 99\ (R = CF_3)$ $D_2 = 99\ (R = CF_3)$

34

The reaction of $F_2C{=}CHF$ with $CH_3^\cdot$ has been computed with the 3-21G basis set [55] and the results, in terms of the usual quantities are displayed in **35**. It is seen that ΔE is computed to be ~ 10 kcal/mol more exoergic for a C_1 attack. This large difference now opposes a small directive power of the higher spin density on C_2, and the result is a tiny preference for a C_1 attack. Clearly, as the radical is varied from $CH_3^\cdot$ to $H^\cdot$ or $CF_3^\cdot$, the ΔE directive effect will diminish and the spin-density effect will take over to restore the normal regioselectivity of a C_2 attack (at the CHF terminus).

$\Delta E^{\neq} = 7.3$ $\Delta E^{\neq} = 7.1$
$\Delta E_c = 27.8$ $\Delta E_c = 28.2$
$B = 20.5$ $B = 21.2$
$\Delta E = -29.8$ $\Delta E = -40.2$

35a   **35b**

The regioselectivity of radical attack may be summarized in the rule of thumb in *STATEMENT 3*.

STATEMENT 3: Under normal circumstances, when the thermodynamic effect does not have a very large directive effect, the regioselectivity of radical attack will prefer the olefinitic site with highest triplet spin density (normally the less substituted site). An attack on the spin-poor site of the olefin will occur whenever the thermodynamic effect gains a strong preference for this site. Regioselectivity zigzags are expected whenever the triplet spin density and thermodynamic effects oppose one another.

5.2 The Trimerization of Acetylene. A Prototypical Multimolecular Process

The trimerization of acetylene to benzene, drawing **36**, is an interesting riddle [56]. Firstly this reaction is allowed by the "orbital symmetry" criterion [37], and secondly the reaction possesses an enormous thermodynamic driving force of $\Delta E \approx -140$ kcal/mol(1 kcal/mol = 4.184 kJ/mol) [56a]. With these favorable features it is somewhat puzzling that the barrier to this reaction is computed to be extremely high, ~ 62 kcal/mol at the MP3/6-31G* level [56b]. The curve crossing diagram provides a possible clue to this high barrier. As can be seen in Fig. 14, the excited state in the diagram involves unpairing to a triplet of the three acetylene molecules and pairing up the spins across the long linkages. The diagram gap which is approximately $3 \Delta E_{st}(\pi\pi^*)$ is enormous, ca. 375 kcal/mol. In addition, the triplet electrons are delocalized in the two π bonds of each acetylene. This leads to a high-energy crossing point and hence to a high barrier, despite the favorable thermodynamics. Since very large excitation gaps force a large destabilization energy upon the reactants, this usually means a combination of large distortion energies of the reactants and high overlap repulsion between them in the transition state. Both these effects are born out by the computational analysis of Houk et al. [56a] and of Bach et al. [56b].

36

The trimerization of acetylene is a prototypical multimolecular process where orbital symmetry allowedness is insufficient to render the reaction kinetically facile. In general, because the gap is a simple sum of the individual excitations for each bond that is broken during the reaction (*STATEMENTS 1* and *2*), multimolecular processes are expected to possess large excitation gaps. This means that such processes can become facile only if the reactants are linked in proximity to each other, at a distance

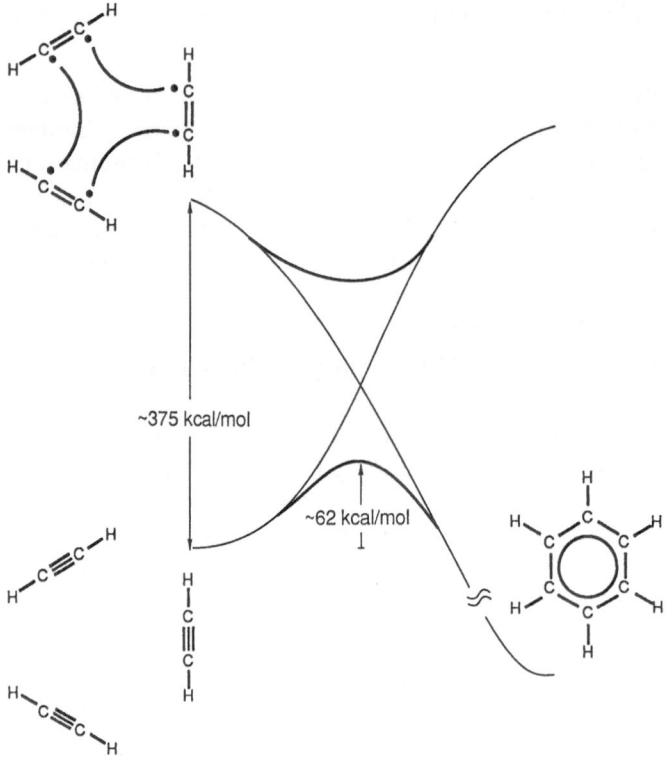

Fig. 14. A schematic avoided crossing diagram for the trimerization reaction of acetylenes (1 kcal/mol = 4.184 kJ/mol)

where large intermolecular repulsion would operate in a multimolecular process. In such proximity the excitation gap is reduced by the bonding effect of the spin-pairs in the excited states and by the built in overlap repulsion of the ground states. This is like starting the reaction in Fig. 14 somewhere in the middle of the reaction coordinate, where the gap is small, the barrier is thereby reduced and the reaction is catalyzed.

5.3 The Identity Exchange, $X^{\cdot} + X{-}X \rightarrow X_3^{\cdot} \rightarrow X{-}X + X^{\cdot}$, and Stability Patterns of $X_3^{\cdot}$ Clusters

The preceding applications assumed that when the bond energies of a related series do not vary greatly, then the variation of the QMRE quantity, B, ought to be smaller than the variations of the other curve crossing features. Let us turn now to analyze a seemingly more complicated problem, where the "energy gap rule" applies despite the great variation in the bond energies in the series and hence also in B. Such a series is the simplest spin-transfer reaction in Eq. (32) involving only monovalent atoms, e.g., H, Cl, I, Br, Li, and so on [7, 12 d, 26, 27, 36].

$$X^{\cdot} + X_2 \rightarrow [X_3^{\cdot}] \rightarrow X_2 + X^{\cdot}. \tag{32}$$

This section examines the stability patterns of the X_3^* clusters, in the light of the qualitative analyses in Sects. 3.1.2 and 3.1.4, and discusses the reasons for the applicability of the "energy gap rule" to predict the patterns in such series [7, 17].

5.3.1 From Stable Intermedites all the Way to Transition States

The scope of the problem is illustrated in Fig. 15. The radical trimers of very weak binders, alkali atoms and noble metals (X = Li, Na, K, Cu) are seen to be stable [7] relative to the reactants $X_2 + X^*$, with stabilization energies ($\Delta E < 0$) ranging from 6 to 20 kcal. On the other hand, a strong binder like H forms a H_3^* trimer which is a transition state in the identity exchange reaction, lying nearly 10 kcal/mol above reactants ($\Delta E > 0$). Halogens are intermediate-strength binders and their trimers exhibit a range of properties: a clearly unstable F_3^*, a slightly unstable [57 a, b] Cl_3^*, and stable or marginally stable I_3^* and Br_3^*.

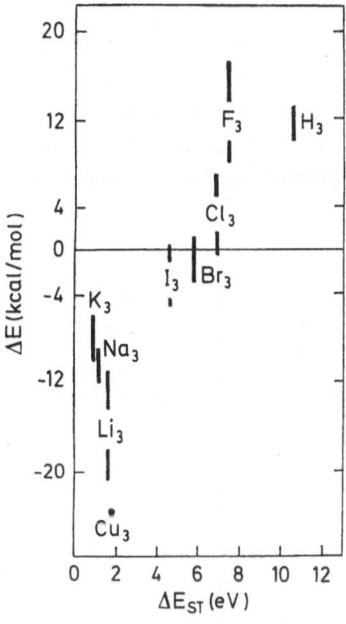

Fig. 15. A schematic plot of the stability (ΔE) of X_3 clusters (relative to $X_2 + X$) against the singlet to triplet excitation in the X_2 dimer. A stable cluster has $\Delta E < 0$, while an unstable cluster $\Delta E > 0$ (1 kcal/mol = 4.184 kJ/mol). Reprinted with permission from J. Phys. Chem. (1988), 92, 5086. Copyright (1988) American Chemical Society

Figure 15 shows also that ΔE_{st}, the singlet-triplet excitation of the X–X dimer, can be used as an organizing quantity for the series. Thus, the most stable trimers (X = Li, Na, K, Cu) in Fig. 15 correspond to very small singlet-triplet gaps of the dimers, below 2 eV. On the other hand, the unstable trimers (X = F, H) possess large ΔE_{st} values, and in the same line, the trimers exhibiting marginal or zero stability possess ΔE_{st} values in-between the two extremes. We recall that the gap G in the corresponding avoided crossing diagram (Fig. 4) is proportional to ΔE_{st} [Eq. (9)]. Figure 15 shows therefore the "energy gap rule" in action, and the G quantity appears to dominate the global behavior of the X_3^* series, in accord with the qualitative predictions in Fig. 5.

5.3.2 Computed Avoided Crossing Diagrams for the Limiting Cases of X = H and Li

The spectrum of energy barriers and stable intermediates displayed in the preceding section for the $X_3^{\cdot}$ systems raises a number of questions. First, it may seem a priori surprising that an avoided crossing of two curves should generate a stable intermediate as $Li_3^{\cdot}$, instead of a transition state as one is trained to expect from avoided crossings. Second, how is it possible that G dominates by itself the global behavior despite our expectation that other curve crossing parameters should change simultaneously, and in fact what are the laws of variation of these parameters, f and B? VB computations of the curve crossing diagrams for the two limiting cases, $H_3^{\cdot}$ and $Li_3^{\cdot}$, carried recently with the multistructure VB program [11], are presented here to provide some of the requisite insight [12d].

The computed diagrams for $H_3^{\cdot}$ and $Li_3^{\cdot}$ are shown in Fig. 16 a and b, assuming for simplicity a colinear approach for the two cases. The two curves in each diagram trace the energetic behavior of reactants' and products' Lewis structures, respectively. Since the orders of magnitude of the $\Delta E^{\neq}$, G and B parameters are so different in $H_3^{\cdot}$ and $Li_3^{\cdot}$, the diagrams are displayed with a logarithmically scaled energy axis to facilitate visualization of the mechanistic modalities of barrier vs energy well.

Let us first consider the energetics of the adiabatic curves. The $H_3^{\cdot}$ system displays a barrier of 15.3 kcal/mol, vs the 13.0 kcal/mol calculated in a full CI treatment with the same basis set [12d]. Similarly, the $Li_3^{\cdot}$ system correctly displays a stable intermediate, 3.8 kcal/mol lower than the reactants, vs 4.2 kcal/mol in a full CI treatment [12d]. It is apparent therefore, that the seemingly counterintuitive idea,

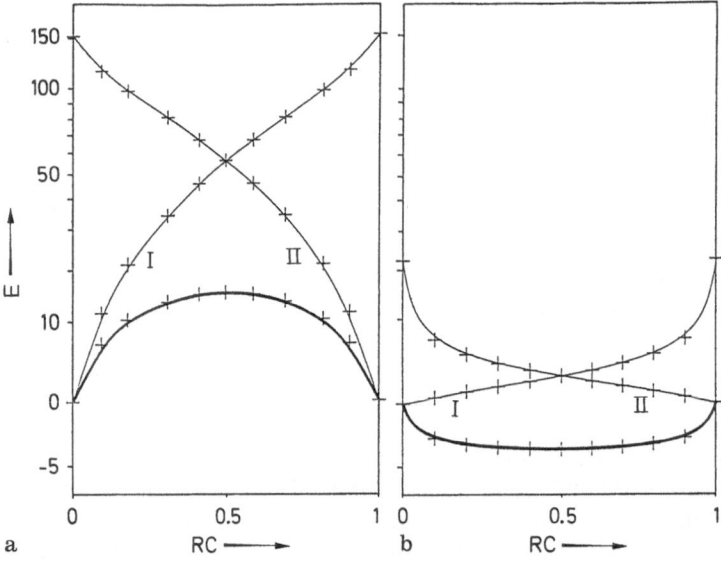

Fig. 16 a, b. Computed avoided crossing diagrams for, $X^{\cdot} + X_2 \rightarrow [X_3^{\cdot}] \rightarrow X_2 + X^{\cdot}$. In the cases of: X = H in (**a**), and X = Li in (**b**). The RC is defined as $(n_1 - n_2 + 1)/2$ where the n's are bond orders satisfying $n_1 = n_2 = 1/2$ at the geometry of $X_3^{\cdot}$. Reprinted with permission from J. Phys. Chem. (1990), 94, 4089. Copyright (1990) American Chemical Society

arising from the qualitative theory in Fig. 5, that an avoided crossing of two curves may generate a stable intermediate, is computationally affirmed.

The essential quantities of each diagram can be read on the corresponding energy scales of Fig. 16. Thus, the diagram gap for H_3^+ is 156.8 kcal which is 64% of the singlet-triplet excitation for H_2 (245 kcal/mol with our 6-13G basis set). On the other hand, the gap for Li_3^+ is 22.4 kcal and is 68% of the computed singlet-triplet excitation of Li_2 (32.9 kcal). It is seen that in both cases the gaps are related to the singlet-triplet gap of the dimer by a factor which is close to the one, 0.75, predicted by the qualitative picture in Eq. (9), and the small deviations are also accountable [7, 58].

The other diagram parameters, B and f, are also found to vary significantly between H_3^+ and Li_3^+. Thus for H_3^+, with the large gap, we obtain a large B value of 42.4 kcal/mol, while for Li_3^+, with the small gap, we obtain a small B of 6.6 kcal/mol. Similarly, the f values vary also in relation to the gap, being 0.37 for H_3^+ and 0.13 for Li_3^+. It is seen therefore that all the curve crossing quantities vary in proportion to the size of the gap G, and it is essential to understand how these variations arise and how do they coincide to render G the organizing quantity of the tendencies observed in the X_3^+ series.

5.3.3 Analytical Expressions for the B and f Quantities Using a Model VB Theory

Using Effective Heisenberg Hamiltonian theory [59], Malrieu [36] derived a very simple expression for B, as a function of $\Delta E_{st}'(X-X)$, the singlet-triplet energy gap of the $X-X$ dimer with an $X-X$ distance equal to the distance at the trimer geometry:

$$B = [\Delta E_{st}'(X-X)]/4 . \tag{33}$$

Our VB computed $\Delta E_{st}'(X-X)$ values are 177 kcal/mol for H_2 and 23 kcal for Li_2, respectively. If Eq. (33) is used with these values, the resulting B values are respectively 44 kcal/mol and 6 kcal/mol for the H_3^+ and Li_3^+ species, in close agreement with the rigorous valence bond calculations.

Interestingly, the ratios of $\Delta E_{st}'$ to the same quantity at the equilibrium distance of the dimer, are very similar for H_3^+ and Li_3^+ and close to 2/3. Thus, B the QMRE (quantum mechanical resonance energy) of the X_3 trimer is predicted and found to be proportional to the singlet-triplet excitation of the $X-X$ dimer at its equilibrium geometry. It follows that B and G are mutually proportional and correlate with ΔE_{st}. This trend is in line with the qualitative notions that both the excitation gap, in the avoided crossing diagram, and the B (QMRE) of the hypercoordinated species, be this a transition state or a stable cluster, should be related to and increase with the binding strength of the constituent atoms [see e.g., Eq. (17)].

The Effective Heisenberg Hamiltonian theory can be used also to derive an expression for f by appeal to the parameters ΔE_{st} and E_t of the dimer $X-X$, as illustrated in 37, where E_t is the triplet overlap repulsion. Using these terms, we obtain Eq. (34) where the primed quantities refer to the corresponding dimer values, at the trimer geometry, and the unprimed ones to the corresponding dimer values

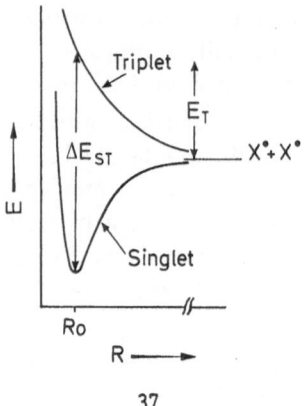

37

at the equilibrium bond length of the dimer itself.

$$f = 4/3 - \{[5\Delta E'_{st} - 4(2E'_t - E_t)]/3\Delta E_{st}\} .$$ (34)

E_t or E'_t can, in turn, be expressed as fractions, α, of the corresponding singlet-triplet gaps, ΔE_{st} and $\Delta E'_{st}$, as, for example:

$$E_t = \alpha \, \Delta E_{st} .$$ (35)

For strong binders like H, the overlap repulsion is as large or larger than the singlet-pairing stabilization and the α parameter is accordingly close to or larger than 0.5. On the other hand, for weak binders like Li, the overlap repulsion is significantly smaller than the singlet-pairing stabilization and the α parameter is accordingly smaller than 0.5 [36]. Using Eqs. (34) and (35) the new expression for f becomes Eq. (36), while the derivative of f with respect to α is Eq. (37):

$$f = (4/3) (1 - \alpha) - [(5 - 8\alpha) \Delta E'_{st}]/3 \, \Delta E_{st} ,$$ (36)

$$\partial f/\partial \alpha = (4/3) [2(\Delta E'_{st}/\Delta E_{st}) - 1] .$$ (37)

The derivative of f with respect to α is positive [12d] provided the ratio $\Delta E'_{st}/\Delta E_{st}$ is larger than 0.5. According to our computations this ratio is roughly a constant 2/3, so that the condition is expected to be met for most realistic intermediates or transition states. Moreover, since the ratio $\Delta E'_{st}/\Delta E_{st}$ is not expected to vary much from one system to another, the f parameter becomes roughly a linear function of α. Since weaker binders possess smaller α values than stronger binders it follows that the weaker binders will also possess smaller f parameters in their curve crossing diagram.

This behavior of f has a significance in terms of the mechanism of crossing. The crossing is promoted by overlap repulsion and bond distortion which destabilize the ground state, and simultaneously by singlet-pairing interaction and release of a triplet energy which stabilize the pseudo excited state. As mentioned above, by appeal to drawing 37 and Eq. (35), a small α as for weak binders means that for these atoms the triplet overlap repulsion is small in comparison with the bonding singlet-pairing interaction. This allows weak binders like Li to achieve a low energy crossing point, because a weak binder like Li can approach Li_2 almost to a bonding distance without effecting significant overlap repulsive destabilization of the ground state. At the same time, the excited state which is stabilized by the comparatively strong singlet-pairing interaction, descends relatively steeply to effect the crossing at a low energy, above the ground state. Thus, a small f value also means that for weak binders the X–X distances of the X_3 species are close (percentage-wise) to the equilibrium X_2 distances. Indeed the Li–Li-distances in the trimer are longer than those in the equilibrium dimer by only 7%.

The exact opposite reasoning is true for strong binders, since for them the triplet repulsion is larger than the singlet-pairing interaction. This prevents a close approach of H$^{\cdot}$ to H_2, to avoid excessive destabilization of the ground state, and in return much stretching of the H_2 bond is required to stabilize the excited state (by relaxing the largely triplet moiety of the pseudo excited state). The resulting large f value means, in turn, that for strong binders the crossing will involve $X_3^{\cdot}$ with X–X distances which, percentage-wise are highly stretched relative to the equilibrium distance at the dimer. Indeed, the H–H distances of $H_3^{\cdot}$ are 29% longer than the bond lengths in the corresponding dimers, as compared to only 7% in $Li_3^{\cdot}$ [12d]. It is thus the ratio of the triplet repulsion to singlet binding in the dimer that eventually determines the magnitude of f and the geometric features of the $X_3^{\cdot}$ cluster.

5.3.4 How Can a Stable Intermediate Arise from an Avoided Crossing of Two Curves?

According to Eq. (14) the relative energy of the trimer $\Delta E^{\neq}$ is given by the difference between the height of the crossing point, fG, and the QMRE quantity B. The variation of f appears to be a crucial factor for the mechanistic modalities of $H_3^{\cdot}$ and $Li_3^{\cdot}$. Had it been the case that f was a constant quantity then, all $\Delta E^{\neq}$ values in the $X_3^{\cdot}$ series would have been positive and proportional to G because B itself is proportional to G. It is the variation of f that causes the height of the crossing point, fG, to have a double dependency on G; a dependency which in turns, leads to two main results. First, since fG decreases faster than B with a decrease of G, a point is reached where B becomes larger than fG, that is, the energy of the $X_3^{\cdot}$ trimer will be lower than that of the reactants ($X_2 + X^{\cdot}$), and the cluster will thus constitute an intermediate in the transformation in Eq. (32). Second, the changes of f and B compensate each other in Eq. (14) and the net result is that G becomes the dominant factor and the organizing quantity of the $X_3^{\cdot}$ stability as predicted in general in Fig. 5.

The root cause for the stability patterns in X_3 clusters is summarized in *STATEMENT 4* which is in line with similar conclusions by Epiotis [60] and by Malrieu [36].

STATEMENT 4: Two fundamental atomic properties lead to the stability of clusters made of weak binders: the weak overlap repulsion between these atoms, and the high ratio between the singlet bonding interaction to the triplet overlap repulsion. In contrast, strong binders cannot form stable clusters because these atoms possess large overlap repulsion and a relatively low ratio between the singlet bonding interaction to the triplet overlap repulsion.

5.3.5 The Stability of X_n Clusters: A Summary and Generalization

Much of what is discussed above carries over to other clusters [7, 26, 58], such as X_6 and X_4, and the essence is summarized in *STATEMENT 5*.

STATEMENT 5: In each class of clusters, the gap G which is proportional to the singlet-triplet excitation of the dimer X_2 is the organizing quantity of the cluster stability. A large singlet-triplet excitation will lead to an unstable cluster which is a transition state for the corresponding isovalent bond exchange reaction [e.g., Eq. (32)]. These transition states will possess highly (percentage-wise) stretched X–X bonds, in comparison with the dimer, and large QMRE's (B's). In contrast, a very small singlet-triplet excitation will result in a stable cluster which constitutes an intermediate in the corresponding bond exchange reaction. These stable intermediates will possess only slightly stretched bonds (percentage-wise) and small QMRE's.

6 Application to Hypercoordination in $(MH_{n+1})^{\cdot}$ Radicals and $(MH_{n+1})^{-}$ Anions

Hypercoordination [61] refers generally to the phenomenon where certain main elements exceed the number of bonded ligands which are predicted by the "octet rule", as e.g., in PCl_5. If each bond is counted as a two electron bond then also the number of electrons in bonding exceeds eight, and the phenomenon is referred to as "hypervalency" which implies also the expansion of the valence shell to include d-orbitals [19b, 21, 61, 62].

The preceding section shows that hypercoordination occurs in clusters of weak binders and can be understood without invoking d-orbital participation. In this section we continue, in the same line, to explore the causes of hypercoordination in cases where the atoms are fairly strong binders. It is shown that hypercoordination, in these cases, is caused by non-Lewis-type configurations that become stable at the hypercoordinated geometry (see Figs. 7, 8). We are moving on then in this section from a two-curve to a many-curve model.

6.1 Hypercoordinated $(MH_{n+1})^{\cdot}$ Radicals

The hypercoordinated (9-M-(n + 1)) radicals of the general formula $(MH_{n+1})^{\cdot}$ may or may not be stable, with respect to the dissociation to a hydrogen atom and the normal-coordinated MH_n molecule, depending on the nature of the M atom. Three cases may be observed, as shown in **38–40**. In **38** the hypercoordinated species is more stable than the normal-coordinated constituents, while in **39** the species is metastable, and unstable in **40**.

While all the possible $(MH_{n+1})^{\cdot}$ valence species are unstable when M is a first row atom, the situation is different with the second row. For example, the $PH_4^{\cdot}$ valence radical has been observed [63], although it is not clear wether it is stable as in **38** or

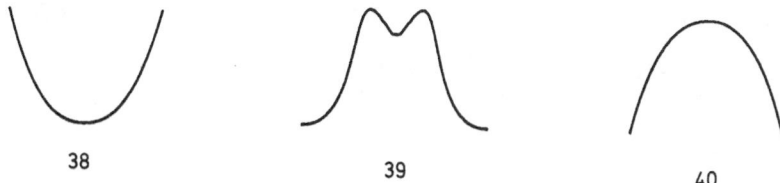

38   39
                                                                    40

metastable as in **39**. It is not a rule, however, that a second row M will form a stable $(MH_{n+1})^{\cdot}$. Thus, the isoelectronic species $SH_3^{\cdot}$ has never been observed, while the findings on $SiH_5^{\cdot}$ are contradictory [64]. On the computational side, the energies ΔE of $(MH_{n+1})^{\cdot}$ species relative to $H^{\cdot} + MH_n$, as calculated at a simple and coherent level of theory (UHF in 6-31G* basis set) [40], are respectively 31.8, 6.0, and 22.7 kcal/mol for the radicals $SiH_5^{\cdot}$, $PH_4^{\cdot}$, and $SH_3^{\cdot}$. In additions, the computations show that $PH_4^{\cdot}$ resides in a true minimum with substantial barriers to decomposition to the normal coordinated constituents. These tendencies have been confirmed at higher levels of the theory [65].

6.1.1 The Qualitative Model

In order to generate avoided crossing diagrams for the $(MH_{n+1})^{\cdot}$ species, let us consider them as intermediate structures in the course of the isovalent $H^{\cdot}$ exchange reaction of their normal valent constituents, as depicted in Eqs. (38–40):

$$H^{\cdot} + Si{-}H \longrightarrow \left[H{-}Si{-}H \right]^{\cdot} \longrightarrow H{-}Si + H^{\cdot} \qquad (38)$$

$$H^{\cdot} + P{-}H \longrightarrow \left[H{-}P{-}H \right]^{\cdot} \longrightarrow H{-}P + H^{\cdot} \qquad (39)$$

$$H^{\cdot} + S{-}H \longrightarrow \left[H{-}S{-}H \right]^{\cdot} \longrightarrow H{-}S + H^{\cdot} \qquad (40)$$

If we assume that the equatorial Si–H bonds, in Eq. (38), are not active and do not participate in the reaction, we shall be left then with three electrons and three orbitals: two atomic 1s orbitals of the axial hydrogens and a single orbital of the Si atom. Therefore, the same diagram as in the $H_3^{\cdot}$ case (see Fig. 16a above) can be drawn for this reaction and, in view of the strong Si–H bond, a high-energy transition state is expected from the avoided crossing of two Lewis curves.

The situation is different in the $PH_4^{\cdot}$ case [Eq. (39)], since one can think now of a third and low-energy curve, able to mix efficiently with the two Lewis curves, as in Fig. 8 and 7b. Indeed, the P atom bears a lone pair in an equatorial hybrid orbital, and it is not very costly in energy to promote one electron from this lone pair to one of the three singly occupied orbitals, either the $1s$ orbitals of the axial hydrogens as in **41**, or the axial orbital of P, as in **42**. The crucial intermediate configuration (Ψ_I) which is involved in the many-curve-crossing is a mixture of **41** and **42**, in variable proportions according to the geometry of the super system. At the reactant and product geometries **41** and **42** are strongly mixed, and form the anchor points of Ψ_I which correspond to the excited (n → σ^*) configurations **43** and **44**, respectively. At the hypercoordinated geometry, on the other hand, structure **41** is the one that constitutes the major character of Ψ_I, because this VB structure is stabilized both by electrostatic interactions as well as by axial 4-electron/3-center bonding, while no such stabilizing effects exist for **42**. The analysis of $SH_3^{\cdot}$ follows the exact same lines. Here too, one of the lone pairs on sulfur may play its role and an intermediate Ψ_I curve, made up **41**-like and **42**-like configurations, may mix in to stabilize the hypercoordinated state.

6.1.2 Analysis of the Stability Patterns for $SiH_5^{\cdot}$, $PH_4^{\cdot}$, and $SH_3^{\cdot}$

Following our discussion of many-curve avoided crossings (Sect. 3.2 and Figs. 7, 8), the impact of the intermediate curve will depend on the height of Ψ_I relative to the crossing point of the Lewis curves. Thus, a very stable Ψ_I will lead to a stable intermediate $(MH_{n+1})^{\cdot}$ like in **38**, while a higher energy Ψ_I will generate a metastable $(MH_{n+1})^{\cdot}$ as in **39**. The higher the intermediate configuration the closer becomes the situation to **40**, in which $(MH_{n+1})^{\cdot}$ is a transition state for the exchange process. One of the factors which determines energy of the intermediate (n → σ^*) curve relative to the Lewis curves is the lone pair ionization potential $I_{M:}$ of the central atom. The smaller the $I_{M:}$ the more stable is expected to be the hypercoordinated structure. We note that $I_{M:}$ is larger for SH_2 than for PH_3, (306 vs 244 kcal/mol) [66], and we accordingly expect the (n → σ^*) curve to be lower in energy and hence to mix more into the Lewis structures, for $PH_4^{\cdot}$ in comparison with $SH_3^{\cdot}$. A confirmation of this expectation is provided by a VB analysis of the MO wavefunctions [20, 67] of $PH_4^{\cdot}$ and $SH_3^{\cdot}$ which shows [40] that the relative weight of the (n → σ^*) configuration to the Lewis structures is 52:48 in the case of $PH_4^{\cdot}$, and 21:79 in the case of $SH_3^{\cdot}$.

Let us turn to discuss the stability patterns of the three title species, and initially reason without the third configuration Ψ_1 of PH_4^* and SH_3^*. In the framework of Lewis curves only, all the structures should be transition states because of large singlet-triplet $\sigma \rightarrow \sigma^*$ excitation gaps which originate in the fairly strong M-H bonds [16]. The ΔE values in a Lewis-only scenario should then increase with an increase of the bond strength. The strongest bonds are Si-H (90.3 kcal/mol) and S-H (87.7 kcal/mol), which should lead to equivalent ΔE's for the corresponding hypercoordinated radicals, SiH_5^* and SH_3^*. Since the P-H bond is weaker (76 kcal/mol), a smaller ΔE is expected for PH_4^*. Adding now the stabilizing effect of the third curve Ψ_1 on both PH_4^* and SH_3^*, we can reorder SiH_5^* and SH_3^* and predict the following trend of increasing ΔE values: $SiH_5^* > SH_3^* > PH_4^*$ which is in accord with computational results [40].

The curve crossing models predict some additional qualitative differences between the species. Thus, owing to the involvement of the intermediate curve below the crossing point of the Lewis structures (consult Fig. 8), PH_4^* is predicted to be a metastable intermediate of the general type in **39**, but with a rather low energy. The presence of the same intermediate configuration grants SH_3^* a chance of existing as a metastable intermediate like in **39**. However, because of the expected high energy of the intermediate configuration, then if SH_3^* does exists it will be a high energy intermediate with small barriers to decomposition to the normal-coordinated constituents. For the case of SiH_5^*, as discussed above, the avoided crossing is between the two Lewis curves only, which possess in addition a large excitation gap, and the model predict this species to be a transition state for the isovalent H* transfer reaction [Eq. (38)].

In agreement with the foregoing predictions, SH_3^* has been shown, by recent post-SCF computations, to display a tiny barrier for decomposition to $SH_2 + H$ [65 b]. As for SiH_5^*, the prediction of the curve crossing model is at variance with the interpretation of experimental observations by Nakamura et al. [64]. Some confirmation of our theoretical prediction is provided by a high-level ab-initio study [65 a] which has shown that SiH_5^* is a true transition state, above $SiH_4 + H$. Further experimental confirmation appears to be underway [68].

6.1.3 First-Row Hypercoordinated $(MH_{n+1})^*$ Radicals

A similar analysis can be used for the hypercoordinated radicals of first-row-atoms [40]. However, two major factors change and make these valence species less likely in comparison with their second row analogs. Firstly, the M-H bond energies are stronger for the first row elements, which means larger singlet-triplet excitation gaps between the Lewis curves. Secondly and more important, the lone pairs of a first row M are deeper in energy than in a second row M, which means high energy for the corresponding $(n \rightarrow \sigma^*)$ configurations. The combination of these two factors accounts for the fact that the three radicals CH_5^*, NH_4^* and OH_3^* do not possess stable valence states, although NH_4^* and OH_3^* have metastable Rydberg states [69].

6.2 Hypercoordinated Anions: The Different Bonding Features in SiH_5^- and CH_5^-

While CH_5^- is a high-energy transition state in the nucleophilic attack of CH_4 by H^- (vide supra), SiH_5^- is a stable intermediate which has been characterized in the gas phase [70]. These stability differences do not seem to originate from the absence or availability of d orbitals. Indeed a recent study by Reed and Schleyer [71] demonstrates that SiH_5^- remains stable even if calculated in a basis set devoid of d orbitals. Quantitative curve crossing diagrams provide a vivid account of the difference betwen the SiH_5^- and CH_5^- species in terms of an intermediate configuration, not based on d-AO's participation, which is stable for Si and very unstable in the carbon analog [12 b, c].

The diagrams are generated as usual by considering CH_5^- and SiH_5^- as intermediates in the identity exchanges in the below equations:

$$H^- + CH_4^- \rightarrow [CH_5^-] \rightarrow CH_4 + H^-, \tag{41}$$

$$H^- + SiH_4^- \rightarrow [SiH_5^-] \rightarrow SiH_4 + H^-. \tag{42}$$

The relevant VB structures which are required to generate the detailed diagrams are shown in **45–48** for the trigonal bipyramidal geometry. The usual Lewis structures are **45** and **46**, while **47** and **48** represent structures with a negative charge on the central MH_3 fragment. In **47**, there are two electrons in the p-AO of M, while in **48** one electron resides in the p-AO while the other in the $\sigma^*(MH_3)$ orbital of a_1' symmetry. It is seen that **48** is drawn with two resonating axial M–H bonds, p-type and σ^*-type, which can be formed potentially if efficient coupling can exist between the two odd electrons on the MH_3 fragment with those of the two axial H's.

The connection between the detailed four-curve diagram and the usual two-curve one is simple. In the two-curve diagram, the ground state reactant $H^- + MH_4$ is connected to an excited charge transfer state of the product, $MH_4^- + H^{\cdot}$, the latter being a mixture of **45**, **47**, and **48**. A two-curve diagram is thus simply obtained by allowing **47** and **48** to mix gradually with **45** starting from the ground state of the reactants all the way to the excited state of the products, and a similar procedure is being applied for the curve connecting the excited reactants to the ground stte products.

45   46

47

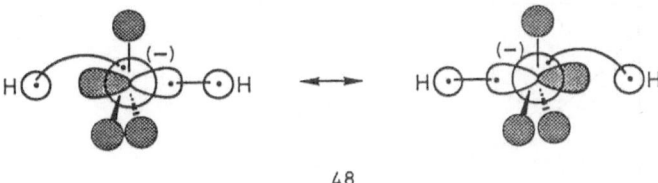

48

The ab initio-computed [12 b, c] four-curve diagram for Eq. (41) is shown in Fig. 17. It is apparent that structures **47** and **48** do not contribute cardinally to the ground- and transition-states, and mix efficiently only with the excited states. Therefore, CH_5^- arises mainly from the avoided crossing of curves **45** and **46**, and remains therefore a high-energy transition state. In this transition state the charge is delocalized mainly in the axial bonds and is virtually prohibited to delocalize into the equatorial bonds.

The diagram for SiH_5^-, in Fig. 18 is similar to that for CH_5^- as far as structures **45–47** are considered [12 b, c]. What makes the difference is structure **48** which collapses at the trigonal bipyramidal geometry to energies below those of the crossing point of **45** and **46**. As a result, the SiH_5^- intermediate is more stable than $SiH_4 + H^-$

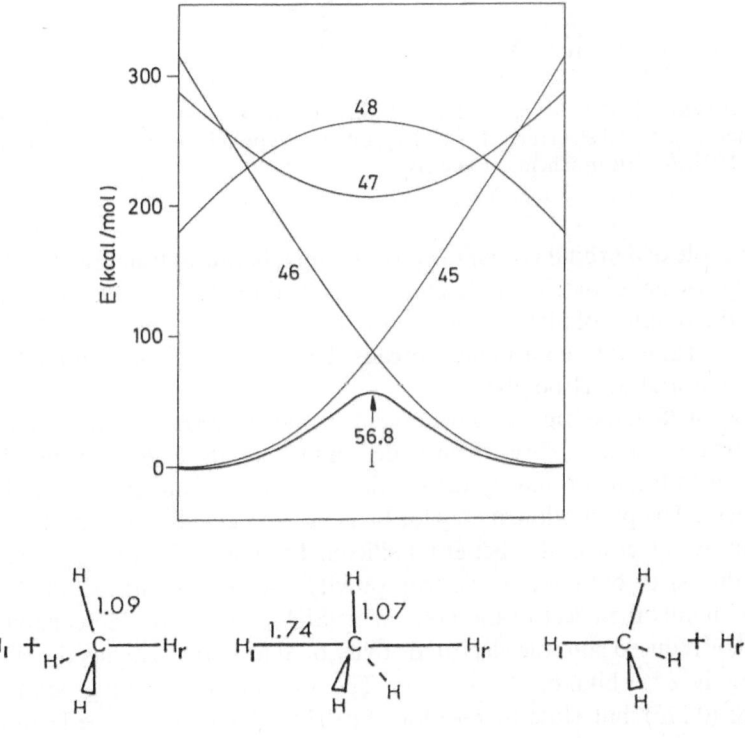

Fig. 17. A curve crossing diagram for $H^- + CH_4 \rightarrow [CH_5^-] \rightarrow CH_4 + H^-$ showing separately the Lewis and intermediate VB structures **45–48**. Reprinted with permission from J. Am. Chem. Soc. (1990), 112, 1407. American Chemical Society

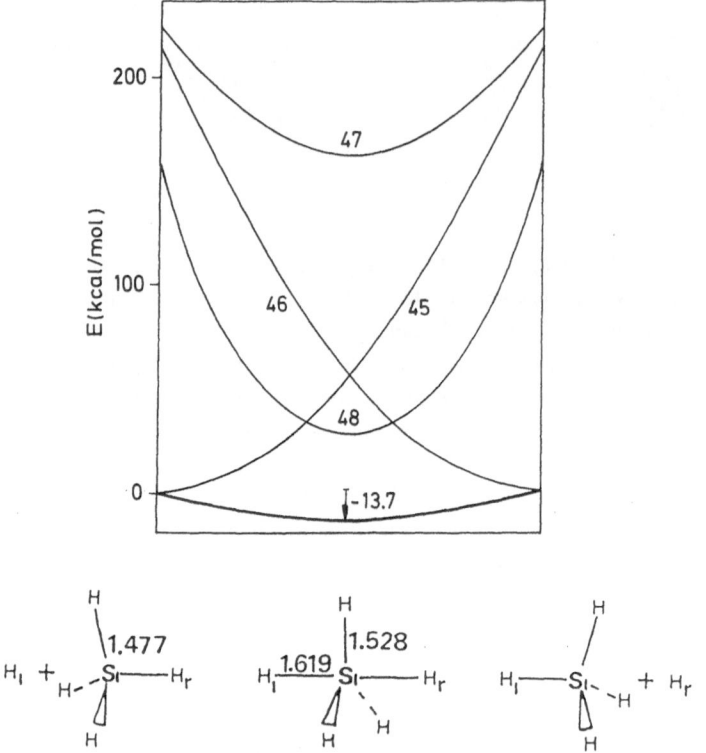

Fig. 18. A curve crossing diagram for $H^- + SiH_4 \rightarrow [SiH_5^-] \rightarrow SiH_4 + H^-$ showing separately the Lewis and intermediate VB structures **45–48**. Reprinted with permission from J. Am. Chem. Soc. (1990), 112, 1407. American Chemical Society

by 13.7 kcal. The role of d orbital is qualitatively marginal because structure **48** is very stable even in the absence of the d orbitals. Quantitatively though, d orbitals contribute 6.1 kcal/mol to the stability of SiH_5^-. As a result of the involvement of **48**, SiH_5^- finds an efficient delocalization mechanism and spreads the negative charge into the both the axial and equatorial Si–H bonds.

The low energy of **48** in the trigonal bipyramidal geometry depends on the bonding capability of $\sigma^*(MH_3)$, a capability which is determined by the $p \rightarrow \sigma^*$ promotion energy of the central fragment and by the overlap between σ^* and the s orbital of an axial hydrogens. The promotion energy for SiH_3^- is ~ 55 kcal lower than that for CH_3^-, and the overlap factor is also better for silicon. The reason for the latter effect has to do with the "size" of Si and its electropositivity. Due to the intrinsically long equatorial Si–H bonds the effect of the node in $\sigma^*(SiH_3)$, on its overlap capability with the axial H, is reduced, and the electropositivity of Si further augments the effect by concentrating the σ^* orbital on the Si center. The resulting σ^*-s overlaps are then significant for Si (0.175), but close to zero for C (0.037) which lacks these features. Because of the superior bonding capability of $\sigma^*(SiH_3)$ the electronic structure **48**, can maintain two axial bonds and become so stable in SiH_5^-. On the other hand, the $\sigma^*(CH_3)$ orbital is almost devoid of bonding capability and remains high in energy. The negligible overlap capability of the $\sigma^*(CH_3)$ orbital prevents also significant mixing of **48** into the two Lewis structures, thus probiting significant charge delocalization into the equatorial bonds.

The importance of the intermediate configuration can be further appreciated by comparing SiH_5^- and its corresponding radical. Thus, removal of one electron from SiH_5^-, to generate $SiH_5^\cdot$, changes drastically the bonding features. The hypercoordinated structure, **48**, now disappears and the bonding mechanism must be sustained by resonance of the two Lewis structures only. The consequence is that the $SiH_5^\cdot$ radical, as already analyzed, is a high-energy transition state. In the carbon analog the intermediate structure is unimportant and, in full accord, no dramatic effects follow when an electron is removed from CH_5^- to generate $CH_5^\cdot$, both species being now transition states in their respective reactions.

6.3 Hypercoordination: A Summary

We are just beginning to explore the fascinating area of hypercoordination, but a global picture with three classes of hypercoordination already emerges from the curve crossing model. The species of the different classes share one common feature, they all possess delocalized electrons and "multi-center bonding". The three classes differ though in the manners by which this electronic delocalization is stabilized or not with respect to the normal-coordinated constituents. *STATEMENT 6* summarizes the main features of these classes.

STATEMENT 6: The distinction between the classes of hypercoordination depends on two fundamental properties of the constituent atoms; their binding strength and their ability to form in bonding low-lying excited states, by virtue of having low ionization potentials, high electron affinities, and so on. One class of stable hypercoordination is formed by weak binders for which avoided crossing of the Lewis structures is sufficient to form a stable hypercoordinated species owing to very small excitation gaps (Fig. 5c). Typical members of this class are metallic- and heavy main element clusters, such as Li_n and Br_3. A second class of stable hypercoordination is formed by strong or weak binders which possess at least one low lying excited state which generates an intermediate curve that falls below the intersection point of the Lewis or Heitler-London structures (Figs. 7, 8). Bonding in this class can be described as a resonance hybrid between the intermediate structure and the Lewis forms of the normal-coordinated constituents. Typical members of this class are the radicals $PH_4^\cdot$, $AsCl_4^\cdot$, $SF_3^\cdot$, the anions SiH_5^-, SiF_5^-, $(FHF)^-$, and many of the intermediates in the classical mechanisms of organic chemistry, e.g., S_N1, and so on. In the third class we find the unstable hypercoordination which is formed by strong binders which have large excitation gaps between their Lewis-type curves and no low lying excited states. Bonding in this class can be described mainly as a resonance hybrid of the Lewis structures, but unlike the first class, here the resonance energy is smaller than the total destabilization energy which is required to promote the resonance. Belonging to this class are all the transition states for the single step interconversions of the normal coordinated constituents.

7 Concluding Remarks

The curve crossing model is a general way of looking at problems which are typified by geometric and electronic reorganization along a "reaction coordinate". In all of these problems — which cover single-step reactions, stepwise reactions and structural problems of hypercoordination vs normal-coordination — there are two extreme structures which undergo interconversion through a transition cluster which may be a transition state or a real intermediate. The model is conceptually compact, and two diagram types which are based on drawings **1** and **2** are sufficient to describe this entire ensemble of chemical problems.

For single step reactions one constructs the two Lewis curves and anchors them at specific excited states by mixing of additional secondary configurations (the excited states can be identified with the aid of *STATEMENTS 1* and *2*), as done in Figs. 2b and 5a. The height of the barrier can then be predicted and analyzed in terms of the balance between the total destabilization energy which is required to promote resonance (crossing) between the ground and excited states in the diagram and the quantum-mechanical resonance energy (QMRE) due to the avoided crossing interaction, B.

The total destabilization energy of the ground state at the crossing point is, in turn, a fraction, f, of the excitation gap, G. The excitation gap refers to a specific vertical excitation of the ground state reactants, and the quantity can be estimated from experimental data, from simple SCF computations, or from sophisticated VB computations of the diagram.

The quantity f depends on the curvature of the intersection curves, and though this is a complex entity which is more easily computed than predicted for each specific case, still we are able to identify three factors which affect the size of f. The first factor refers to the steepness of descent of the excited state and is called "the bond coupling delay effect"; f gets larger the more delocalized become the excited state's spin-pairs which need to be coupled into bonds, as the excited state descends toward the ground state. The second factor of f refers to the steepness of ascent of the ground states toward the excited states. This factor depends on the ratio of the singlet bonding interaction to the overlap repulsion, and as this ratio increases, as is the case in weak binders, the f factor decreases. Finally, the third factor is the reaction exothermicity, where a large exothermicity reduces the value of f.

Using the above basic ideas it is possible to discuss reactivity patterns and transition state structure in a variety of organic reactions, and to understand the variation of isoelectronic species from transition states to stable hypercoordinated clusters of metallic and nonmetallic elements.

In stepwise transformations the generalized Lewis curves or the Heitler-London-only curves are horizontally intersected by an intermediate or a third curve (Figs. 7 and 8). The intermediate curve can originate either in a low lying excited state that is not an electronic "image" of the Lewis states of reactants and products, or in a low energy ionic VB configuration. Using these diagram types it is possible to discuss many of the mechanisms of organic chemistry such as S_N1, S_NV, E1, $E1_{CB}$, etc., as well as many of the hypercoordinated structures such as pentacoordinated Si, bicoordinated H, and so on.

In addition to the wide chemical coverage of the curve crossing model, it brings also a measure of conceptual unity in the area of physical organic chemistry. The two-curve and many-curve models include many of the fundamental concepts of physical organic chemistry and show the root causes for the occasional breakdowns of these concepts [4 a, b], e.g., the Bell-Evans-Polanyi principle, the Leffler-Hammond postulate, and the potential energy surface contour diagrams. The model makes also a connection to other curve crossing models, the Evans, Polanyi, and Bell model [2b, 33], the Marcus model for electron transfer reactions [72], the Woodward-Hoffmann and Longuet-Higgins-Abrahamson diagrams [37, 73] for forbidden reactions, the Salem diagrams [74], the Devaquet-Sevin-Bigot diagrams [75], and to the Bernardi-Robb [76] and Malrieu's [38] diabatization models. Indeed, if the usefulness of a model would be judged by its ability to create unity, to pattern existing data and to predict new trends, then the curve crossing model should pass this test of usefulness.

The developments of the multistructure VB method [11, 12, 50 a] and the MO/CI diabatization method by Malrieu and collaborators [28, 50b] have allowed to test the model by rigorous quantitative means. As the applications accumulate they show that the qualitative diagrams can indeed be computed rigorously and isomorphically, and that therefore the model which is qualitatively useful has also a rigorous quantitative counterpart. On the quantitative side, it is still too early to conclude whether the qualitative estimates − and the assumptions associated with − the various curve crossing entities are generally correct or should be replaced by more refined features. A bonus of the quantitative applications is that they unravel features that the qualitative reasoning can handle only a posteriori, and we have thus much to look forward to in the future applications of the model.

Acknowledgements. S.S.S. thanks A. Pross for the exciting collaboration in the area of organic reactivity in the critical years 1980–1983. P.C.H. is indebted to his graduate students, O. K. Kabbaj, P. Maitre, G. Ohanessian, and G. Sini and to his Polytechnique associates, J. M. Lefour and J. P. Flament. The quantitative aspects of this chapter owe much to their ingenuity and perseverence. The authors are indebted also for the participation and contribution in various published works and in discussions to R. Bar, E. Buncel, E. Canadell, D. Cohen, A. Demolliens, O. Eisenstein, J. P. Malrieu, D. J. Mitchell, H. B. Schlegel, I. H. Um, F. Volatron, and S. Wolfe. Structures **3−8** were reprinted from reference 8 by permission of Kluwer Publishers, and structures **29, 30, 32−35, 37** and **41−48** were reprinted from references 52, 12d, 40, and 12c by permission of the American Chemical Society.

8 References

1. Salem L (1982) Electrons in chemical reactions, Wiley, New York, Chaps. 2, 4−6, and 8
2. (a) Wigner E, Witmer EE (1928) Z Physik 51: 851 (b) Evans MG, Polanyi M (1938) Trans Far Soc 34: 11 (c) Eyring H (1935) J Chem Phys 3: 107 (d) Laidler KJ, Shuler KE (1951) Chem Rev 48: 153

3. Shaik SS (1981) J Am Chem Soc 103: 3692
4. (a) Pross A, Shaik SS (1983) Acc Chem Res 16: 363 (b) Shaik SS (1985) Prog Phys Org Chem 15: 197 (c) Pross A (1985) Adv Phys Org Chem 21: 99
5. (a) Shaik SS (1990) Acta Chem Scand 44: 205 (b) Shaik SS, Schlegel HB, Wolfe S (1991) Theoretical aspecs of physical organic chemistry: Application to the S_N2 transition state, Wiley, New York (in press) (c) Pross A (1985) Acc Chem Res 18: 212 (d) Pross A, Chipman DM (1987) Free Radical Biology & Medicine 3: 55
6. Shusterman AJ, Tamir I, Pross A (1988) J Organomet Chem 340: 203
7. Shaik SS, Hiberty PC, Ohanessian G, Lefour JM (1988) J Phys Chem 92: 5086
8. Shaik SS (1989) In: Bertran J, Csizmadia IG (Ed) New concepts for understanding organic reactions, Kluwer, Dordrecht: NATO ASI series Vol. C267
9. Lowry TH, Richardson KS (1987) Mechanism and theory in organic chemistry, Harper and Row, New York, p 604–608; 359–360
10. Eberson L (1987) Electron transfer in organic chemistry, Springer-Verlag, Berlin
11. (a) Hiberty PC, Lefour JM (1987) J Chim Phys 84: 607 (b) Maitre P, Lefour JM, Ohanessian G, Hiberty PC (1990) J Phys Chem 94: 4082 (c) Lefour JM, Flament JP (1984–9) A multi structure VB program, D.C.M.R., Ecole Polytechnique, 91128, Paleseau Cedex, France
12. (a) Sini G, Shaik SS, Lefour JM, Ohanessian G, Hiberty PC (1989) J Phys Chem 93: 5661 (b) Sini G, Hiberty PC, Shaik SS (1989) J Chem Soc Chem Commun 772 (c) Sini G, Ohanessian G, Hiberty PC, Shaik SS (1990) J Am Chem Soc 112: 1407 (d) Maitre P, Hiberty PC, Ohanessian G, Shaik SS (1990) J Phys Chem 94: 4089
13. McWeeny R, Sutcliffe BT (1969) Methods of molecular quantum mechanics, Academic Press, London, 1969, Chap 6
14. Lewis GN (1916) J Am Chem Soc 38: 762
15. Pauling L (1939) The nature of the chemical bond, Cornell Univ Press, Ithaca, 1939
16. Sanderson RT (1983) Polar covalence, Academic Press, New York
17. Hiberty PC, Cooper DL (1988) J Mol Struct (THEOCHEM) 1969: 437
18. (a) Karafiloglou P, Malrieu JP (1986) Chem Phys 104: 383 (b) Lepetit MB, Oujia, B, Malrieu JP, Maynau D (1989) Phys Rev A 39: 3274, 3289
19. (a) Epiotis ND (1982) Lecture Notes Chem 30: 1-260 (b) Epiotis ND (1983) Lecture Notes Chem 34: 1-585
20. (a) Hiberty PC, Leforestier C (1978) J Am Chem Soc 100: 2012 (b) Hiberty PC (1981) Int J Quant Chem XIX: 259
21. Harcourt RD (1982) Lecture Notes Chem 30: 1-260 (especially chapter 23)
22. (a) Hirst DM, Linnett JW (1962) J Chem Soc 1035 (b) Hirst DM, Linnett JW (1963) J Chem Soc 1068
23. Coulson CA, Fischer I (1949) Phil Magasine 49: 386
24. Bobrowicz FB, Goddard WA, III (1977) In: Schaefer HF (Ed) Methods of electronic structure theory, Plenum, New York, p 79
25. Cooper DL, Gerratt J, Raimondi M (1987) Adv Chem Phys 69: 319
26. Shaik SS, Bar R (1984) New J Chem 8: 411
27. Shaik SS, Hiberty PC, Ohanessian G, Lefour JM (1985) New J Chem 9: 385
28. Kabbaj OK, Volatron F, Malrieu JP (1988) Chem Phys Lett 147: 353
29. Buncel E, Um IH, Shaik SS, Wolfe S (1988) J Am Chem Soc 110: 1275
30. Kochi JK (1988) Angew Chem Int Ed Engl 27: 1227
31. Shaik SS (1987) J Org Chem 52: 1563
32. (a) Shaik SS (1982) New J Chem 6: 159 (b) Shaik SS, Pross A (1982) J Am Chem Soc 104: 2708
33. Bell RP (1936) Proc R Soc London A 154: 414
34. Brønsted JN (1928) Chem Rev 5: 231
35. Shaik SS, Duzy E, Bartuv A (1990) J Phys Chem 94: 6574
36. Malrieu JP (1986) New J Chem 10: 61
37. Woodward RB, Hoffmann R (1970) The conservation of orbital symmetry, Academic Press, New York, 1970
38. Kabbaj OK, Sini G, Hiberty PC, to be published
39. Cohen D, Bar R, Shaik SS (1986) J Am Chem Soc 108: 231
40. Demolliens A, Eisenstein O, Hiberty PC, Lefour JM, Ohanessian G, Shaik SS, Volatron F (1989) J Am Chem Soc 111: 5623

41. (a) Pellerite MJ, Brauman JI (1983) J Am Chem Soc 105: 2672 (b) Dodd A, Brauman JI (1984) J Am Chem Soc 106: 5356 (c) Barlow SE, van Doren JM, Bierbaum VM (1988) J Am Chem Soc 110: 7240

42. (a) Wolfe S, Mitchell DJ, Schlegel HB (1981) J Am Chem Soc 103: 6794 (b) Mitchell DJ (1981) PhD Thesis, Queen's University, Kingston, Canada (c) Mitchell DJ, Schlegel HB, Shaik SS, Wolfe S (1985) Can J Chem 63: 1642

43. (a) Albery WJ, Kreevoy MM (1978) Adv Phys Org Chem 16: 87 (b) Lewis ES, McLaughlin ML, Douglas TA (1987) In: Harris JM, McManus SP (Eds) Nucleophilicity, American Chemical Society, Washington DC

44. (a) Bertran J (1989) In: Bertran J, Csizmadia IG (Eds) New theoretical concepts for understanding chemical reactions, NATO ASI Series, Kluwer, Dordrecht, Vol C267 (b) Bergsma JP, Gertner BJ, Wilson KR, Hynes JT (1987) J Chem Phys 86: 1356 (c) Gertner BJ, Bergsma JP, Wilson KR, Lee S, Hynes JT (1987) J Chem Phys 85: 1377 (d) Hwang JK, King G, Creighton S, Warshel A (1988) J Am Chem Soc 110: 5297 (e) Van Linde SR, Hase WL (1989) J Am Chem Soc 110: 7240

45. (a) Janousek BK, Brauman JI (1979) In: Bowers MT (Ed) Gas phase ion chemistry, Academic Press, New York, Vol. 2 (b) Chen ECM, Wentworth WE (1975) J Chem Educ 52: 486 (c) Jiordan JC, Moore JH, Tossell JA (1986) Acc Chem Res 19: 281 (d) Olthoff JK (1985): PhD Dissertation, University of Maryland, USA (e) Lindholm E, Li J (1988) J Phys Chem 92: 1713 (f) Jordan KD, Burrow PD (1987) Chem Rev 67: 557 (g) Heinrich N, Koch W, Frenking G (1986) Chem Phys Lett 124: 20 (h) Luke BT, Loew GH, McLean AD (1988) J Am Chem Soc 110: 3396 (i) Hitchcock AP, Tronc M, Modelli A (1989) J Phys Chem 93: 3068

46. Shaik SS (1983) J Am Chem Soc 105: 4359

47. (a) Tucker SC, Truhlar DG (1989) J Phys Chem 93: 8138 (b) Chandrasekhar J, Smith SF, Jorgensen WL (1985) J Am Chem Soc 107: 154

48. Dewar MJS, Healy E (1982) Organometallics 1: 1705

49. Shaik SS, Schlegel HB, Wolfe S (1988) J Chem Soc Chem Commun 1322

50. (a) Sini G (1991) PhD Dissertation, Universite de paris Sud, Orsay, France (b) Kabbaj OK (1989) DEA Thesis, Universite Paul Sabatier, Toulouse, and Universite de paris Sud, Orsay, France

51. (a) Tedder JM Walton JC (1978) Adv Phys Org Chem 16: 51 (b) Tedder JM, Walton JC (1980) Tetrahedron 36: 701 (c) Tedder JM (1982) Angew Chem Int Ed Engl 21: 401

52. Shaik SS, Canadell E (1990) J Am Chem Soc 112: 1446

53. Kitaura K, Morokuma K (1976) Int J Quant Chem 10: 325

54. (a) Wu EC, Rodgers AS (1976) 98: 6112 (b) Kerr JA, Parsonage MJ, Trotman-Dickenson AF (1976) In: The handbook of chemistry and physics, CRC Press, Cleaveland, OH, p F-204-F-220

55. Canadell E, Eisenstein O, Ohanessian G, Poblet JM (1985) J Phys Chem 89: 4856

56. (a) Houk KN, Gandour RW, Strozier RW, Rondan NG, Paquette LA (1979) J Am Chem Soc 101: 6797 (b) Bach RD, Wolber GJ, Schlegel HB (1985) J Am Chem Soc 107: 2837

57. (a) Boal DH, Ozin GA (1971) J Chem Phys 55: 3598 (b) Nelson LY, Pimentel GC (1967) J Chem Phys 47: 3671 (c) Lee YT, Lebreton PR, McDonald JD, Herschbach DR (1969) J Chem Phys 51: 455

58. Shaik SS, Hiberty PC, Lefour JM, Ohanessian G (1987) J Am Chem Soc 109: 363

59. (a) Malrieu JP, Maynau D, Daudey JP (1984) Phys Rev B 30: 1817 (b) Durand P, Malrieu JP (1987) Adv Chem Phys 67: 321

60. Epiotis ND (1984) New J Chem 8: 11

61. Schleyer PvR, Martin JC (1984) Chem Eng News May 28: 4

62. (a) Musher JI (1969) Angew Chem Int Ed Engl 8: 54 (b) Kutzelnigg W (1984) Angew Chem Int Ed Engl 23: 272 (c) Coulson CA (1969) Nature 221: 1106 (d) Martin JC (1983) Science 221: 509

63. (a) McDowell CA, Mitchell KAR, Raghunathan P (1972) J Chem Phys 57: 1699 (b) Clakton TA, Fullam BW, Platt E, Symons MCR (1975) J Chem Soc Dalton Trans 1395 (c) Colussi AJ, Morton JR, Preston KF (1975) J Chem Phys 62: 2004; J Phys Chem 79: 1855

64. Nakamura K, Masaki N, Sato S, Shimokoshi K (1985) J Chem Phys 83: 4504

65. (a) Volatron F, Maitre P, Pelissier M (1990) Chem Phys Lett 166: 49 (b) Volatron F, Demolliens A, Lefour JM, Eisenstein O (1986) Chem Phys Lett 130: 419
66. (a) For H_2S: Turner DW, Baker C, Baker AD, Brundle CR (1970) Molecular photoelectron spectroscopy, Wiley, London (b) For PH_3: Bowers MT (Ed) Gas phase ion chemistry (1979), Academic Press, New York, Vol 2, p 17
67. Hiberty PC, Ohanessian G (1982) J Am Chem Soc 104: 66
68. Katz B, personal communication to Shaik SS
69. Kassab E, Evleth EM (1987) J Am Chem Soc 109: 1653
70. Hajdasz DJ, Squires RR (1986) J Am Chem Soc 108: 3139
71. Reed AE, Schleyer PvR (1987) Chem Phys Lett 133: 553
72. Marcus RA (1964) Annu Rev Phys Chem 15: 155
73. Longuet-Higgins HC, Abrahamson EW (1965) J Am Chem Soc 87: 2045
74. Salem L (1976) Science 191: 822
75. Devaquet A, Sevin A, Bigot B (1978) J Am Chem Soc 100: 2009
76. Robb MA, Bernardi F (1989) In: Bertran J, Csizmadia IG (Ed) New concepts for understanding organic reactions, Kluwer, Dordrecht: NATO ASI series Vol C267

Orbital Interactions and Chemical Reactivity of Metal Particles and Metal Surfaces

R. A. van Santen

E. J. Baerends

Laboratory of Inorganic Chemistry and Catalysis, Technical University of Eindhoven, P.O. Box 513, 5600 MB Eindhoven, The Netherlands
Department of Theoretical Chemistry, Free University Amsterdam, De Boelelaan 7161, 1007 MC Amsterdam

This review of chemical bonding to metal surfaces and small metal particles demonstrates the power of symmetry concepts to predict changes in chemical bonding.

Ab-initio calculations of chemisorption to small particles, as well as semiempirical extended Hückel calculations applied to the study of the reactivity of metal slabs are reviewed.

On small metal particles, classical notions of electron promotion and hybridization are found to apply. The surroundings of a metal atom (ligands in complexes, other metal atoms at surfaces), affect bonding and reactivity through the prehybridization they induce. A factor specific for large particles and surfaces is the required localization of electrons on the atoms involved in the metal surface bond.

At the surface, the bondenergy is found to relate to the grouporbital local density of states at the Fermi level. The use of this concept is extensively discussed and illustrated for chemisorption of CO and dissociation of NO on metal surfaces.

A discussion is given of the current decomposition schemes of bond energies and related concepts (exchange (Pauli-)repulsion, polarization, charge transfer). The role of non-orthogonality of fragment orbitals and of kinetic and potential energy for Pauli repulsion and (orbital)polarization is analyzed.

Numerous examples are discussed to demonstrate the impact of those concepts on chemical bonding theory.

1 Introduction

The increased interest for surfaces and interphases within the past decade has also led to a considerable extension of the quantum chemist's sphere of interest.

Extensive ab-initio calculations have been done on small metal clusters simulating adsorption sites on surfaces. Semiempirical extended Hückel calculations have been applied to study the reactivity of metal slabs and also the techniques of the solid state physicist with the aim to calculate from first principles the interaction between adsorbing molecules and metal surfaces.

As a result, our understanding of the relation between structure and chemical bonding, especially of bonds between metalatoms and adsorbates, has considerably improved.

In this review we will discuss the current state of knowledge in relation to the "classical" frontier orbital concept in terms of interactions between highest occupied molecular orbitals (HOMO) and lowest unoccupied molecular orbitals (LUMO) [1, 2, 3].

If a molecule approaches a metal surface and is at a distance from the surface which is large compared to the spatial extension of the electrondistribution around the atoms, some attraction will be experienced due to induction and van der Waals interactions.

The theory of van der Waals interactions in small atom clusters is well established [4] and has recently also been reviewed for the interaction with metal surfaces [5]. The attractive van der Waals interaction with a metal surface can be considered to be due to the image potentials generated on the metal surface by the fluctuating charge distributions of the interacting molecule.

As long as interaction is weak, the electrons in a metal rapidly adjust to the motions of the electrons in a molecule [6].

Induction forces are generated by the interaction of the induced image potential with the stationary multipoles of the charge distribution of the molecule in the ground state.

Though these contributions to the surface bond will not be considered in this review, one should note that these terms may become important when interactions are weak. They may be of relevance for so-called "precursor" states that are short-lived and are observed in surface molecular beam experiments [7] or postulated in thermal desorption experiments [8].

The Hartree-Fock approximation on which most of our considerations will be based does not include those effects.

Electrostatic long-range forces between metal surfaces and a molecule at a large distance may exist if the surface has become nonhomogeneous, because of the presence of atoms with different configurations (steps or kinks). This is the case for higher Miller index surfaces or if other molecules or atoms have been adsorbed to the surface. Localized dipoles may then exist, leading to long-range Coulombic interactions. The existence of such potentials has been experimentally demonstrated by Geerlings and Los [9] for Li adsorbed on W and has been theoretically calculated for K on Pt [10, 11].

Chemical bonding is considerably affected by the presence of such electrostatic potentials, because the relative position of the adsorbing molecule's molecular orbital energy levels with respect to the metal-surface Fermi level is changed as will be discussed later.

The next terms to consider arise if the atom distances become so small that charge transfer between adsorbate and metal surface becomes possible. The energy cost involved is compensated for by the electrostatic interaction between separated charges in small clusters or by the image potential induced in a metal by the charge on the adsorbing molecule. Distances may still be large compared to the spatial extension of the electron distributions between the atoms. The resulting negative ions sometimes are short-lived and are only observed in molecular beam experiments or at low temperatures. For molecules, these weakly bonded states can be considered to be the precursor states [7, 8] for dissociation. A well-known example is the negative ion of molecular oxygen found at low temperatures at single crystal silver surfaces [12, 13].

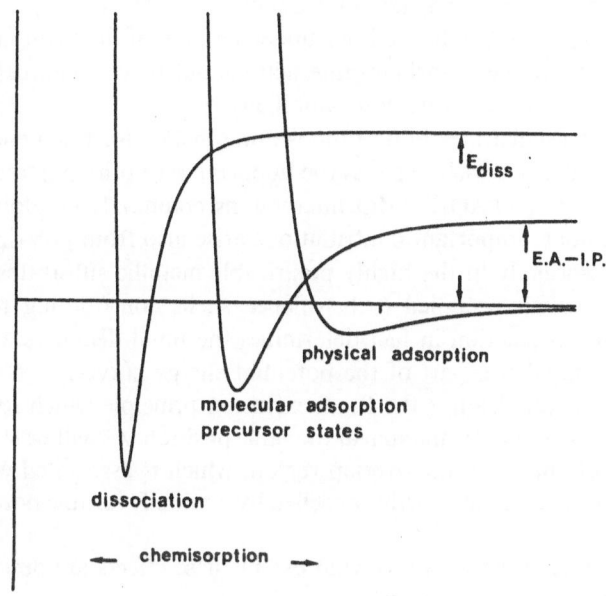

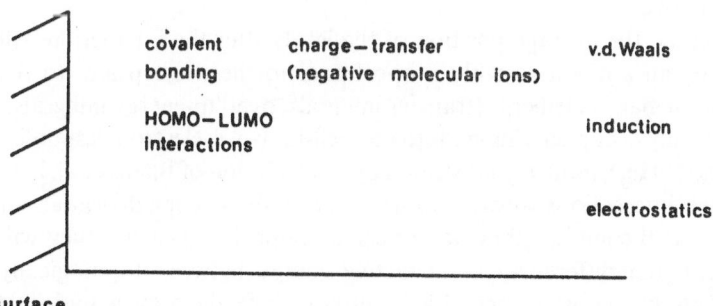

Fig. 1. Types of interactionenergies and potential-energy diagram of a molecule approaching a surface [14]

Alkali atoms become positively charged and are strongly chemisorbed to the metal surface by a potential dominated by the alkali ion-induced image potential.

In this review we will not discuss the interaction regimes mentioned so far, but rather focus on the chemical bonding effects that occur if the distances become short enough for the electrondistributions to overlap. The successive stages of interaction are schematically sketched in Fig. 1.

In the past decade considerable attention has been paid to an analysis of chemical bonding in physically meaningful contributions such as steric repulsion, electrostatic attraction/repulsion, charge transfer, polarization, etc. We may refer to the work by Morokuma et al. [15, 16, 17], Whangbo et al. [18], Bernardi, Bottoni et al. [19] and Stone and Erskine [20]. Applications of such analyses to transition-metal complexes have been carried out by Morokuma et al. [2]), Ziegler [22, 23], Bauschlicher and Bagus [24], and Baerends and Rozendaal [25].

Applications to adsorbates interacting with small clusters of metal atoms are found in the work of Post and Baerends [26] and Bagus et al. [27, 28, 29].

These analyses have considerably enhanced our understanding of the main factors governing metal-ligand (or surface-adsorbate) interactions but they sometimes lead to misunderstanding and are not even without ambiguity.

In the next section, the basic features of decomposition schemes for bond energies will be discussed. In particular, we wish to stress the importance of other interactions than the charge-transfer type HOMO-LUMO interactions commonly employed in frontier orbital considerations. Important contributions arise also from polarization of the interacting units, especially in the highly polarizable metallic substrates, and from repulsive interaction with occupied (sub-)valence shells contributing to the steric repulsion. The latter interaction in fact determines the bond distances as it is responsible for the inner, repulsive, part of the potential energy curves.

The physical origin of the repulsion is the Pauli exclusion principle which forbids the presence of two electrons with the same spin at the same position. As will be shown, this leads to a depletion of charge in the overlap region, which is associated with a strong rise in kinetic energy that is only partly cancelled by a more favorable potential energy.

We will discuss in some detail how and to what extent these effects are described in the simple extended Hückel approximation.

As will be demonstrated, the incorporation of the overlap between the interacting fragment orbitals plays a key role in the extended Hückel analysis, in keeping with the picture sketched above.

If two orbitals interact, the average position of the levels after the interaction will be shifted upwards by an amount roughly proportional to the overlap and to the Hamiltonian interaction matrix elements (transfer integrals, overlap-energy integrals). If the two orbitals are fully occupied, this leads to the well-known 4-electron destabilizing interaction (compare He_2), as nicely illustrated by a calculation of Bagus et al. [30], Fig. 2. Such repulsive interactions always occur between the occupied orbitals of the interacting units. At the surface they are especially important because they will strongly compete with the differences in attractive energy between topologically different sites [3]. This has consequences for reactivity since the way a molecule adsorbs to a metal surface may affect its subsequent reaction mode.

A similar role of changes in kinetic energy is found if the metal is simulated by a

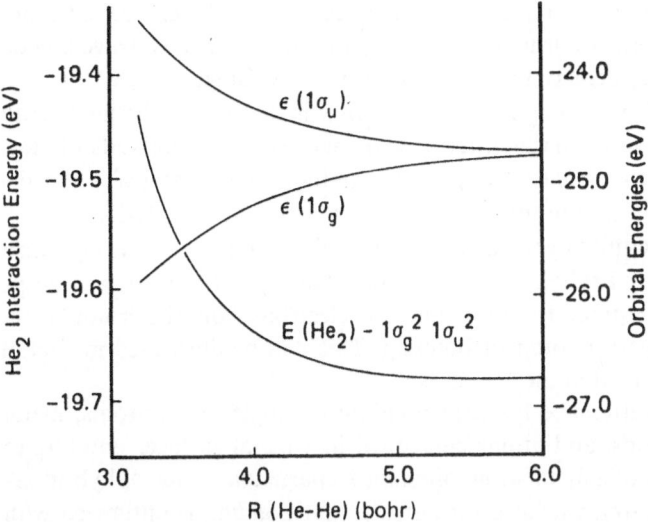

Fig. 2. SCF potential curve for He_2; the interaction energy is shown on the left-hand scale; the orbital energies, $E(1\sigma_g)$ and $E(1\sigma_u)$, as a function of He—He distance, R_0 (He—He); the latter values are shown in the right-hand scale [30]

free electron gas as discussed by Norskov [31] and illustrated in Fig. 3. In the case of chemisorption of atoms, the first-order term to chemical bonding is determined by the change in electron density of the adsorbing atom embedded in the metal surface. To a good approximation, bonding occurs because the atom electron density is replaced by the density of the metal surface. If the atom density is low, an increase in bonding is observed because the negative charge on the embedded atom resulting from the higher metal electron density is stabilized by screening of this charge by the surrounding jellium. This is the earlier discussed induced image potential stabilization of charged adsorbate states. However, if the electron density of the jellium increases,

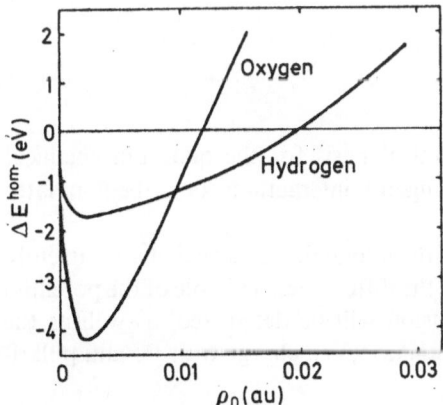

Fig. 3. Energy ΔE^{Hom} of hydrogen and oxygen embedded in infinite jellium as a function of jellium density ϱ_0. The energy zero is the energy of the free atom [16]

the resulting interaction becomes repulsive. A high density on the embedded atom becomes unfavorable because according to free electron theory, the respulsive kinetic energy increases faster with density than the intraatomic exchange energy.

A second topic that will be highlighted in this review is the role of the immediate neighborhood of the metal atom(s) that interact with an incoming ligand (adsorbate): the ancillary ligans in a mononuclear complex (cf. homogeneous catalysis) and the surrounding metal atoms in a semi-infinite lattice (heterogeneous catalysis).

Attractive interactions require hybridization of available orbitals so as to optimize overlap between interacting orbitals. Similar to the Pauling hybridization concept [32], promotion energy is required to redistribute the electrons from the ground state to the state giving the largest amount of bonding. This will be illustrated in Sect. 3 for bonding to isolated transition-metal atoms.

There is an important difference between bonding to single metal atoms, metal atoms surrounded by ligands, and atoms embedded in a metal surface. Bonding to ligands forces the electrons of a metal atom often into a particular mode of hybridization which directs orbitals to a vacant position such that bonding is optimized with a reacting fragment. In such a case little promotion energy may be required.

The metal neighbors of a metal surface atom will have a similar effect on hybridization as the ligands in the organometallic complex [33]. In Sect. 4 we will focus on a special additional feature of the interaction with metal surfaces, namely the energy cost involved due to delocalized electrons [33–37]. In order to bind an electron participating in the chemisorptive bond, it has to decouple from the bonds with the other metal atoms. This localization energy often decreases the bond strength in a complex with respect to that in a free molecule. It also results in weaker bonds between the metal surface atoms after the chemisorptive bond has been formed.

Bonding to a metal surface can also be described in terms of frontier orbital interactions similar as in the cluster case. Symmetry considerations now apply to the coefficients of the metalsurface molecular orbitals at the Fermi elevels that interact with the HOMOs and LUMOs of the adsorbing molecule or atom. Such a theory helps to predict bonding topology of adsorbates, as will be illustrated for bonding of CO to the transition metals.

It will also be used to predict differences in reactivity of transition metal surfaces.

In the last section, differences in reactivity of transition metal surfaces as compared to clusters will be highlighted.

2 Theoretical Intermezzo

In the following, we will discribe the theoretical basis for the quantum chemical expressions used for the analysis of metal-ligand interactions described in later sections.

First, the various decomposition schemes of the interaction energy that are currently used will be briefly reviewed with emphasis on the differences. The role of the potential energy and kinetic energy in the steric repulsion will be delineated, as well as the different ways to treat the frontier orbital interaction, i.e., charge-transfer and polarization type of orbital mixings.

The role of overlap in Pauli repulsion will be stressed. The extended Hückel method will be applied to clusters as well as semi-infinite systems.

Then the relevant theoretical concepts to be used in bonding to semi-infinite systems with a continuous distribution of energy levels, instead of discrete levels as in clusters, will be introduced.

2.1 Analysis of Bonding Energies

Let us consider the interaction between two systems A and B with wave functions Ψ_A and Ψ_B, resp., and energies E_A and E_B, resp.

Although analysis of the bonding energy is not only possible for one-determinantal wave functions, it is most simply carried out and illustrated using such wave functions. We thus take:

$$\Psi_A = |\varphi_1 \ldots \varphi_a| , \qquad \Psi_B = |\phi_1 \ldots \phi_b|$$

where φ_1 to φ_a are the one-electron wave functions of fragment A, and ϕ_1 to ϕ_b are the one-electron wave functions of fragments B. The vertical bars denote antisymmetrization as well as normalization.

As the first step in the bonding process, we consider the wave function Ψ^0 which consists of just the product $\Psi_A\Psi_B$, suitably antisymmetrized and normalized:

$$\Psi^0 = N\hat{A}\{\Psi_A\Psi_B\} = |\varphi_1 \ldots \varphi_a\phi_1 \ldots \phi_b| \tag{1}$$

The total energy is then given by the expression:

$$E^0 = \langle \Psi^0| H |\Psi^0\rangle ,$$

and the bond strength becomes:

$$\Delta E^0 = E^0 - E_A - E_B$$

The energy ΔE^0 computed within this approximation is in general positive (repulsive). It is important to realize that this is not due to electrostatic (Coulomb) interactions between the electrons and nuclei of A with those of B. The electrostatic interaction energy

$$\Delta E_{elstat} = \int \varrho^A(1) \varrho^B(2)/r_{12} \, d\vec{r}_1 \, d\vec{r}_2 + \sum_{\alpha \in A} \sum_{\beta \in B} Z_\alpha Z_\beta/R_{\alpha\beta}$$

$$+ \int V_N^A(1) \varrho^B(1) \, d\vec{r}_1 + \int V_N^B(1) \varrho^A(1) \, d\vec{r}_1$$

where $V_N^A(V_N^B)$ is the nuclear potential of fragment A (resp. B) and $Z_\alpha(Z_\beta)$ are the nuclear charges of A (resp. B), is for neutral systems negative (attractive), except at very short distances. This is simply verified for two atoms with spherically symmetrical charge distributions ϱ^A and ϱ^B, where it is seen to arise from the overlap of the diffuse charge distributions (for instance, the repulsion between ϱ^A and ϱ^B is less than the one between Z_A and Z_B).

What is the origin of the repulsive character of E^0, given that it is not a simple charge superposition effect? Let us first note that the charge distribution corresponding to Ψ^0 is not simply the sum of ϱ^A and ϱ^B. We have to evaluate the electrön density taking the non-orthogonality into account, or alternatively we have to orthogonalize the $\{\phi_i\}$ and $\{\varphi_j\}$ sets (Schmidt, Löwdin or otherwise), which does not change the wave function, but which allows us to write the electron density as the familiar sum of orbital densities.

The density ϱ^0 differs from $\varrho_A + \varrho_B$ in that charge is removed from the overlap region.

Consider the case of one orbital φ_A on A, and ϕ_B on B, with $S = \langle \varphi_A | \phi_B \rangle$. One finds for the antisymmetrized and normalized product function:

$$\Psi^0(1,2) = N\hat{A}\{\varphi_A(1)\,\phi_B(2)\} = \frac{1}{\sqrt{2-2S^2}}\{\varphi_A(1)\,\phi_B(2) - \varphi_A(2)\,\phi_B(1)\}$$

with a corresponding one-electron density distribution:

$$\varrho^0(1) = 2 \int |\psi^0(1,2)|^2 \, d\vec{r}_2$$

$$= \frac{1}{1-S^2}\{|\varphi_A(1)|^2 + |\phi_B(1)|^2 - 2S\varphi_A(1)\,\phi_B(1)\} \tag{2}$$

Expression $\varrho^0(1)$ gives the one-electron distribution of two fragments with overlapping orbitals, without any covalent or other bonding effects present. The changed density $\varrho^0(1)$ is only the result of conservation of charge and the Pauli exclusion principle.

Exactly the same expression results if one orthogonalizes ϕ_B on φ_A to give $\phi_{B'} = (1 - S^2)^{-1/2}(\phi_B - S\varphi_A)$ (Schmidt orthogonalization) which allows to write Ψ^0 as a determinantal wave function with orthogonal orbitals: $\Psi^0(1,2) = |\varphi_A(1)\,\phi_B'(2)|$ and $\varrho^0(1) = |\varphi_A|^2 + |\varphi_B'|^2$.

It is clear from Eq. (2) that the overlap term in ϱ^0 causes depletion of charge in the overlap region. This is demonstrated in Fig. 4, with a contour plot of $\Delta\varrho^0 = \varrho^0 - \varrho_A - \varrho_B$ for the $K^+ - W$ system [38].

The density change $\Delta\varrho^0$ results in changes in both the potential and the kinetic energy. As for the potential energy, the already negative ΔE_{elstat} is changed into an even more negative (attractive) total potential energy as electron density moves to regions closer to the highly attractive nuclei.

It is in fact well known from Ruedenberg's analysis of the chemical bond [39] that building up of a bonddensity that is usually associated with bondformation is not favorable from a potential energy point of view, but has to be judged against the favorable decrease of kinetic energy associated with a more slowly varying density in the internuclear region, änd the unfavorable increase in kinetic energy associated with piling up electron density in the region of low potential energy close to the nucleus.

In keeping with this analysis, it is the latter effect, the rise in kinetic energy, which is the origin of the repulsive character of ΔE^0. This is illustrated by the kinetic and potential energy contributions to ΔE^0 of $K^+ - W$ given in Table 1.

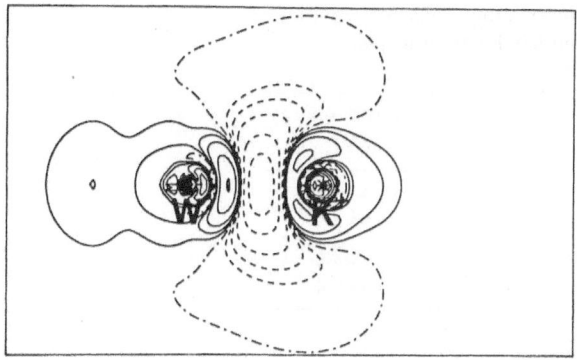

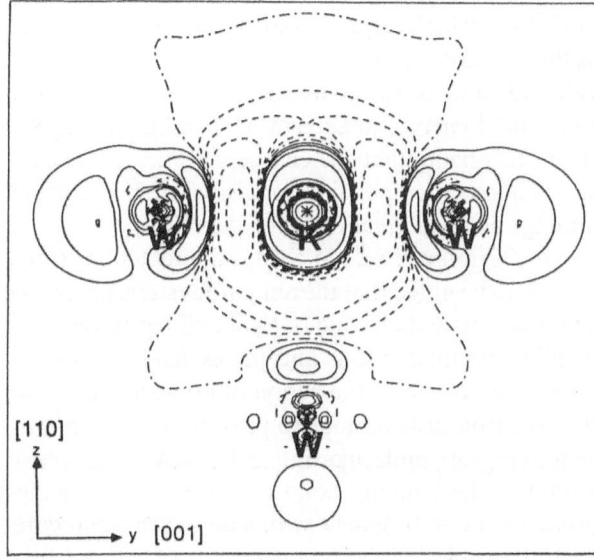

Fig. 4. Contourplot of the change $\Delta\varrho^\circ$ in electron charge density due to the $K^+ - W$ ion-atom (see text). Solid contours indicate $\Delta\varrho^\circ > 0$, dashed contours $\Delta\varrho^\circ < 0$, and dot-dashed contours $\Delta\varrho^\circ = 0$. Contours drawn are $\Delta\varrho^\circ = 0$, $\pm .005$, ± 1.01, ± 0.02, ± 0.05, ± 0.1, ± 0.2, and $\pm 0.5\ e/a_0^3$ [38]

It is possible to qualitatively understand the increase in kinetic energy due to the density change $\Delta\varrho^0$ from the relation between the kinetic energy and the gradient of an orbital $\nabla\psi$, and of the orbital density $\nabla\varrho = \psi^*(\nabla\psi) + (\nabla\psi^*)\,\psi$:

$$E_{kin} = -\frac{1}{2}\int \psi^*\,\nabla^2\psi\;d\tau = \frac{1}{2}\int |\nabla\psi|^2\;d\tau$$

ΔE^0 is denoted in various ways in the literature. If the electrostatic contribution to ΔE^0 is separated, the remainder is commonly referred to as exchange (or Pauli-) repulsion (Fujimoto [40], Morokuma [15]) to indicate that this effect derives from the anti-symmetry requirement on the wave function:

$$\Delta E^0 = \Delta E_{elstat} + \Delta E_{XREP} \tag{3}$$

Table 1. Decomposition of the K^+—W ion-atom exchange repulsions in electronic kinetic energy terms and in Coulomb terms [38]

r (Å)	Kinetic energy (eV)	Coulomb energy (eV)
1.06	635	−436
1.59	251	−211
2.24	66.7	−62.6
3.18	13.1	−14.4

The $\Delta\varrho^0$ plot of Fig. 4 is in fact a clear illustration of the deviation from $\varrho_A + \varrho_B$ induced by the Pauli exclusion principle which requires zero probability of finding two electrons of the same spin at the same position.

In solids, this effect is often referred to as Born repulsion. In the work by Bagus et al. [28], ΔE^0 is called the frozen orbital energy, in agreement with Eq. (1) for Ψ^0. However, one should not infer from this name that the charge densities are frozen, and in particular the terminology "charge superposition repulsion" [29] should be avoided, leading to misunderstandings as to the real cause of the effect.

The exchange repulsion plays a very significant role in chemical interactions (refer to [40] for an extensive discussion). It is to be noted that the ubiquitous steric hindrance effects are not due to Coulombic repulsions but are due to the Pauli repulsion. ΔE^0, comprising both the attractive Coulombic interactions and the exchange repulsion, is appropriately called steric repulsion. Apart from the action of exchange repulsion at the outer edges of the molecular electron distributions to provide steric hindrance and the repulsive part of the potential in, e.g., molecular interactions (Van der Waals complexes), it is also very important in determining bond distances. The exchange repulsion of occupied valence orbitals on one fragment with subvalence (semi-core) orbitals on the other fragment is responsible for the repulsive part of the potential energy curve determining the distance.

When we consider the simple two-orbital model used before, it is clear that the kinetic energy increases due to the orthogonalization of ϕ_B to φ_A to yield ϕ_B', by an amount of approximately $S^2\langle\varphi_A|\,T\,|\varphi_A\rangle$. This rise in kinetic energy is important if φ_A has a high kinetic energy, which is increasingly so for the deep lying doubly occupied core orbitals, and if S is not too small, i.e. if the core orbital is not too contracted. This leads e.g. in the case of CO interacting with a transition metal atom of the third row to the largest contribution to the kinetic repulsion coming from the orthogonalization of the rather bulky C lone pair orbital (5σ) on the 3s, 3p subvalence orbitals.

It has been pointed out in Ref. [26] that this is a short-range, local effect. Therefore, clusters of metal atoms modelling chemisorption by containing at least all the nearest neighbors of an adsorbate yield in general quite reasonable adsorbate surface distances, even if the total chemisorption energy may be quite different.

Apart from the exchange repulsion between valence and semi-core levels, there is of course also exchange repulsion between occupied valence levels. For instance, if the 4s shell is occupied, as it is in most free transition-metal atoms of the third row, there is a very large exchange repulsion with the lone-pair orbitals on ligands [24, 25]. For this reason the effective configuration of the transition-metal atom changes from $3d^n 4s^2$ in the free atom to $3d^{n+2} 4s^0$ in complexes.

In the same way, there is exchange repulsion of lone pair orbitals on adsorbates and the occupied conduction electron levels in a metal. This effect is more important if the conduction electron density is higher. It has been pointed out by Post and Baerends [41] that the much higher conduction electron density of Al (3 valence electrons) compared to Li (1 valence electron) leads to much stronger 5σ/metal-sp exchange repulsion with CO for Al than Li. Bagus et al. [30] have stressed this "σ-repulsion" for CO interacting with metals.

We next turn to the orbital interactions that change the repulsive wave function Ψ^0 in the fully converged Hartree-Fock one-determinantal ground state wave function Ψ_{conv} resulting in covalent bonding. It is common to distinguish between polarization-type interactions, which result from mixing virtual orbitals on A into occupied orbitals on A, and similar for B, and charge-transfer type of interaction consisting of admixture of A virtuals into occupied B levels and vice versa (cf. Fig. 5a). Although the final wave functions is well-defined, as is Ψ^0, the steps in between are not unique. It is immediately evident that if a complete basis is used to describe system A, the sum of occupied and virtual spaces on A necessarily includes the full B space, and, e.g., polarization of A and charge transfer from A to B cannot be distinguished. Apart from this fundamental ambiguity caused by the overlap between the A and B orbital spaces, there is also a dependency of the results of the charge transfer/polarization analysis depending on the way the overlap is treated in practice.

In the original Morokuma scheme [15], the matrix of Hamiltonian interaction elements (Fock matrix) and the overlap matrix in the secular equation:

$$(F - ES) C = 0 \tag{4}$$

are treated simultaneously. For instance, the steric repulsion is evaluated by retaining only matrix elements of both F and S amongst occupied orbitals, and setting all matrix elements connecting an occupied orbital to a virtual empty orbital to zero. This is of course equivalent to the orthogonalization of occupied orbitals discussed before, as diagonalization of F is nothing but a unitary transformation amongst the orthogonalized orbitals. In this way canonical orbitals are obtained for Ψ^0 characterized by their orbital energies.

In Fig. 5b this is shown schematically for the three-orbital system: stabilized bonding and destabilized antibonding combinations are formed. In a one-electron picture the repulsive character of ΔE^0 is reflected in a stronger destabilization of ψ'_B and χ'_B than stabilization of φ'_A (see below).

In order to include, e.g., A to B charge transfer, Morokuma et al. allow, apart from the matrix elements of the Ψ^0 step, also the matrix elements of both F and S among the virtual B orbitals and the matrix elements connecting the virtual B and occupied A orbitals to be non-zero. In a similar way, other contributions such as polarization of B, or charge transfer of B to A, etc. are determined in separate calculations.

In general, the sum of the individual energy contributions differs from the result obtained when allowing all the interactions to take place simultaneously, which is in

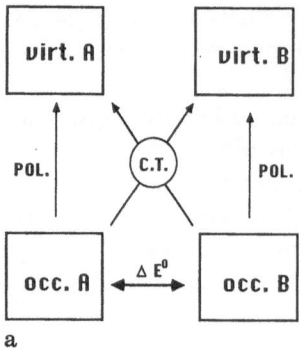

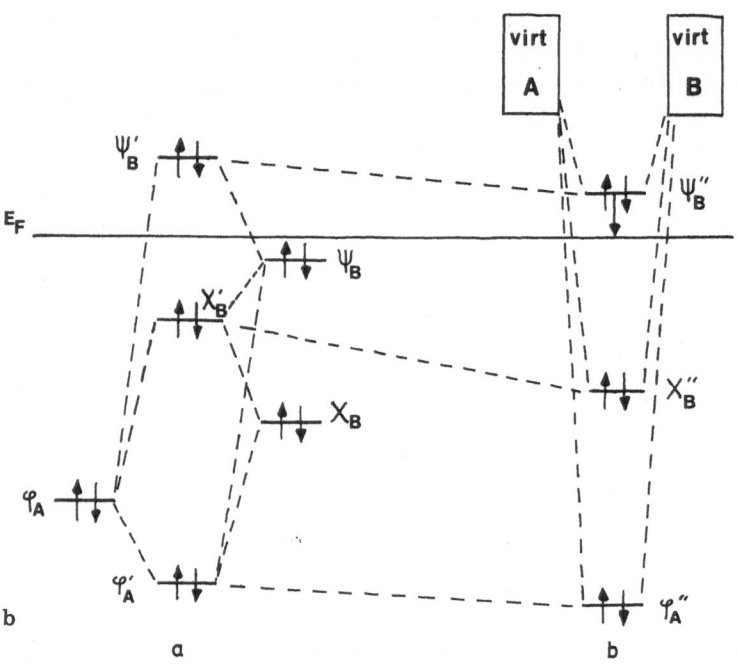

Fig. 5a. Schematic interaction scheme of fragments A and B, indicating polarization and charge transfer interactions. **b** Interaction diagram for the orbital interactions of two interacting units A and B (in the presence of a metal surface). A contains one orbital $\varphi_{B'}$, B contains two orbitals χ_B and ψ_B. a. One electron energies of orbitals forming Ψ^0. b. One electron energies of orbitals forming Ψ_{conv}

accordance which the intuitive concept of synergism of, e.g., donation and back-donation (see for a detailed analysis of synergic effects in metal-ligand bonding Ref [25]).

Bagus et al. have chosen to treat the overlap in a different way. They chose a specific order in which they (Schmidt) orthogonalize the orbitals on all the previous ones, e.g., (occ A) + (occ. B) + (virt. B) + (virt. A). In the Fock matrix in this basis, in which the overlap S matrix is diagonal, blocks are successively allowed to be non-zero. After the Ψ^0 step based on (occ. A) + (occ. B) only, the set (virt. B) is allowed to mix with the set (occ. B). Note that both (occ. B) and (virt. B) now refer to sets Schmidt-orthogonalized on (occ. A). This mixing leads to polarization of B in the field of unmodified A (the set (occ. A), being first in the Schmidt-orthogonalization process, is unchanged). Note that the resulting polarization of B is not only due to an electrostatic field, but is also the result of orthogonaliziation of the full set B on (occ. A). This will be discussed in more detail below. So the polarization concerned is not just classical polarization in an electrostatic field. If the interaction of (occ. B) with (virt. A) is added, also electron donation from B to A is allowed to take place. This process may be continued, for instance, polarization of A in the field of the relaxed B fragment that has been obtained after the previous steps, with relaxed occupied orbitals (occ. B'), may be obtained by allowing (virt. A), now Schmidt-orthogonalized to (occ. B'), to interact with (occ. A). The energy will be lowered in each step, and finally the fully converged SCF·energy will be reached. The procedure is clearly asymmetrical, the steps are consecutive. The dependance of the results on the chosen order in what are called CSOV steps (Constrained Space Orbital Variation) has been discussed by Bauschlicher [42].

Yet another way to decompose the electronic interaction energy, i.e., charge transfer plus polarization energies, is based on symmetry. Only A- and B-orbitals belonging to the same irreducible representation will interact. This is the basis for the widely used distinction between, e.g., σ-donation and π-backdonation. It is indeed possible to write the total electronic interaction energy as the sum of contributions from the irreducible representations. We may write:

$$\Delta E_{int} = E(\Psi_{conv.}) - E(\Psi^0)$$

$$= \frac{1}{2} \sum_{\mu,\nu} P_{\mu\nu}(H_{\mu\nu}^{core} + F_{\mu\nu}) - \frac{1}{2} \sum_{\mu,\nu} P_{\mu\nu}^0(H_{\mu\nu}^{core} + F_{\mu\nu}^0)$$

where H^{core} is the matrix of the purely one-electron operators (kinetic energy and nuclear attraction energy) in Hartree-Fock or the effective core Hamiltonian in semi-emperical schemes, and F is the Fock matrix. As all of the matrices in this expression are symmetry blocked, this means:

$$\Delta E_{int} = \sum_{\Gamma} E_{\Gamma}$$

It has been shown by Ziegler and Rauk [43] that ΔE^{int} may, to a good approximation, be expressed in terms of $\Delta P_{\mu\nu} = P_{\mu\nu} - P_{\mu\nu}^0$ and an effective Fock matrix F^{eff}:

$$\Delta E_{int} = \sum_{\Gamma} \sum_{\mu,\nu} \Delta P_{\mu\nu}(\Gamma) \, F_{\mu\nu}^{eff}(\Gamma)$$

where $F_{\mu\nu}^{eff}(\Gamma)$ is an average over the initial F^0, the final Γ, and an intermediate "transition state" Fock matrix F^{TS}.

The symmetry decomposition is unambiguous. The arbitrariness involved in the distinction between charge transfer and polarization is not present in the symmetry separation.

It should also be noted that the results of a symmetry analysis cannot be compared directly to the results of the previous schemes. A contribution of a given symmetry will in general contain *both* charge transfer and polarization. The symmetry analysis has been introduced by Ziegler [43] and has been applied extensively to transition-metal complexes [22, 23, 25] and to cluster studies of chemisorption [26, 41].

The effect of charge transfer and polarization, i.e., the mixing of virtual orbitals into the levels that have been formed in the steric repulsion step (Ψ^0) is schematically shown in Fig. 5b (to the right). All of the levels φ'_A, χ'_B, and ψ'_B will be stabilized, is symmetry permits, but in particular this will be the case for the high-lying ψ'_B.

We would like to stress that one of the most important results of the polarization is the reduction of Pauli repulsion. An example is the interaction of system A with a low-lying (semi-core) orbital φ_A with system B having a valence s-orbital ψ_B (cf. Fig. 5b) and an additional virtual p-level ψ_{vB}. ψ_{vB} represents a level of the block virt-B in Fig. 5b. If the energy separation between ψ_B and φ_A is large, in particular if the interaction matrix element is negligible compared to the energy separation, the canonical orbital resulting in the Pauli repulsion step from the diagonalization of the Fock matrix in the space of occupied orbitals only (cf. Fig. 5b(a)) will be practically identical to the simple Schmidt-orthogonalized orbital (see Fig. 6):

$$\psi'_B = (\psi_B - S\varphi_A)/\sqrt{(1 - S^2)}$$

As pointed out before, the energy of this orbital will still be raised due to a possibly large kinetic energy of φ_A.

In the subsequent polarization step of allowing admixture of ψ_{vB} into the occupied B-space, we obtain the following coupling matrix element between ψ_{vB} and ψ'_B:

$$\langle \psi_{vB}| H_{eff} |\psi_{B'}\rangle = \frac{1}{\sqrt{(1 - S^2)}} [\langle \psi_{vB}| H_{eff} |\psi_B\rangle - S\langle \psi_{vB}| H_{eff} |\varphi_A\rangle]$$

Here, H_{eff} represents the effective Hamiltonian operator of the one-electron model used.

If the model used is the Hartree-Fock approximation, H_{eff} contains explicitly the field of all the electrons and nuclei in the system. Therefore, the first matrix element in the square brackets represents the coupling of the virtual B orbital to the occupied one because the effective field has changed from the one for isolated B (in which case this matrix element is zero) due to the presence of system A.

System A will exert a direct electrostatic field, particularly if it is not neutral. We are here dealing with the common electric polarization of system B in an external field. [The charge rearrangement occurring in the Pauli repulsion (cf. Fig. 4) of course modifies the field from what it would be if the charges of A and B were simply superimposed].

It is clear, however, that there is another contribution to the polarization which does not have such a simple electrostatic interpretation. This is due to the second matrix element in the square brackets. This part of the polarization has its origin in the orthogonalization or Pauli repulsion, as is evident from the proportionality with the overlap S. It couples the virtual B state to the occupied one through the contamination of the latter with φ_A. It is to be noted that in simple one-electron methods which do not incorporate electrostatic fields due to neighboring moietes (Hückel, extended Hückel), the electric polarization is absent. However, as soon as overlaps are taken into account (extended Hückel), the polarization that relieves the Pauli repulsion will be in effect. Figure 6 illustrates this type of polarization. It can easily be deduced from the sign of the matrix element given above that the admixing of the p-type ψ_{vB} into ψ_B' is such that the original occupied s-type ψ_B hybridizes away from the overlap region so as to alleviate the antibonding with φ_A.

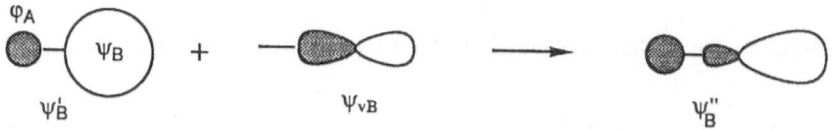

Fig. 6. Polarization due to admixture of virtual p orbital ψ_{vB} into the orthogonalized (onto φ_a) s orbital ψ_B'

The charge rearrangement in this polarization is such that the electron density becomes more smooth. The strong removal of charge from the overlap region and piling up on A and B which is characteristic of the Pauli repulsion is counteracted by a flow of charge back into the overlap region so that the overlap population becomes less negative, a reduction of the net populations of φ_A and ψ_B and an increase of charge at the back of B due to population of ψ_{vB}.

As methods like the extended Hückel are very suitable to obtain insight into the symmetry aspects of orbital interaction, it is important to realize in which way and to what extent the physics of the interaction that has been discussed in this section is embodied in such a method. In Sect. 2.3, we will therefore discuss in some detail, for eseveral model systems, extended Hückel calculations from the point of view developed here.

Another important aspect of interaction of molecules with infinite systems is that in the presence of a Fermi level, as soon as an antibonding level ψ_B' is liftcd above the Fermi energy, the Pauli repulsion will immediately be relieved by deexcitation of the electrons to the Fermi level. In Fig. 5b(b) such a deexcitation is indicated. This is an important difference with finite systems where low-lying empty levels to receive the electrons will not always be present. In that case, considerable repulsion (antibonding interaction) will have to develop before level ψ_B' becomes sufficiently destabilized to lose its electrons to another B orbital. As will be discussed in Sect. 3.1, this situation occurs frequently in interactions of molecules with transition-metal atoms (i.e., fragments), where the ns orbital loses its electrons to the $(n-1)$ d, which corresponds to an excitation in the free atom (promotion energy). At a surface or on clusters, frontier orbital interactions that may lead to bond breaking may meet with a lower activation barrier simply because the opposing repulsive forces develop less strength. We will return to this point in Sect. 3.3.

For finite systems, polarization, i.e., the admixture of virtual orbitals, decreases the Pauli-repulsion effects experienced by the occupied orbitals.

As discussed, the admixture of virtual and occupied orbitals on the same fragment arises mainly from non-orthogonality of the orbitals on different fragments at short distance. Only for charged fragments and if bonding distances become such that overlap of fragment orbitals becomes very small, polarization of the fragments will be found to be due to each others electrostatic field.

2.2 The Extended Hückel Method

In the extended Hückel method [44], molecular orbitals are expanded in a minimum basis set.

The diagonal and non-diagonal matrix elements of the Hamiltonian become:

$$\langle \varphi_i | H | \varphi_j \rangle = \langle \varphi_i | \hat{T}(\vec{r}) - \sum_A \frac{Z_A e^2}{|\vec{R}_A - \vec{r}|} + e^2 \int d\vec{r}' \frac{\varrho(\vec{r}')}{|\vec{r} - \vec{r}'|} - E(\text{exch}, \vec{r}) | \varphi_j \rangle \tag{5}$$

$\hat{T}$ is the kinetic energy operator and $E(\text{exch}, \vec{r})$ the contribution to the exchange energy.

The exchange energy contribution to Eq. (5) is due to the modification of the electron-electron interaction energy because of the requirement that the multielectron wave function most be antisymmetric.

In the extended Hückel method, the repulsion which the electron experiences from electrons on neighboring atoms is supposed to be cancelled by the attraction with the nuclei. In addition, the electron-electron interactions on one atom are averaged or simulated by an electron density-dependent term (iterative extended Hückel).

The first assumption implies that the atoms are considered to be neutral. The total energy is simply the sum of the occupied molecular orbitals.

The diagonal elements of the Hamiltonian matrix become:

$$\langle \varphi_{i,A} | H | \varphi_{i,A} \rangle = \langle \varphi_{i,A} | \hat{T}(r) - \frac{Z_A^{eff}}{|\vec{R}_A - \vec{r}|} | \varphi_{i,A} \rangle \tag{6a}$$

The non-diagonal matrix elements are approximated by:

$$\langle \varphi_{i,A} | H | \varphi_{j,B} \rangle = \frac{1}{2} \langle \varphi_{i,A} | \varphi_{j,B} \rangle \{ \langle \varphi_{i,A} | H | \varphi_{i,A} \rangle + \langle \varphi_{j,B} | H | \varphi_{j,B} \rangle \}$$

$$+ \langle \varphi_{i,A} | \hat{T} | \varphi_{j,B} \rangle - \frac{1}{2} \langle \varphi_{i,A} | \varphi_{j,B} \rangle \{ \langle \varphi_{i,A} | \hat{T} | \varphi_{i,A} \rangle +$$

$$+ \langle \varphi_{i,B} | \hat{T} | \varphi_{j,B} \rangle \} \tag{6b}$$

Equation (6b) is derived from Eq. (6a) by expanding $\varphi_{i,A}$ and $\varphi_{j,B}$ into a complete set of basis functions on the other atom and only retaining the terms containing diagonal Hamiltonian matrix elements. Since this approximation induces a significant error in the kinetic energy term, the kinetic energy term has to be corrected for.

It will be shown later that the second part of Eq. (6b) governs the attractive part of the covalent bond strength.

Equations (6a) and (6b) are often replaced by empirical values.

Anderson [45] corrects for the non-cancellation of nuclear attraction and electron repulsion for neutral atoms by adding in the total energy expression a semi-empirical repulsive expression to the attractive component of the total energy computed as the sum of the energies of the occupied molecular orbitals.

One can incorporate effects due to non-neutrality of charge, by maintaining Eqs. (6a) and (6b) for the matrix elements of the Hamiltonian, but adding to the total energy expression a Madelung potential energy term.

An analysis of modifications of the extended Hückel method which also applies to the recombination of radicals and where changes in electron-electron interaction are important, can be found in Ref [46].

2.3 The Effect of Overlap on the Bond Energy According to the Extended Hückel Method

The polarization effects compensating partly for exchange repulsion in the frozen orbital approximation discussed earlier are also found in the extended Hückel method.

In the extended Hückel method, molecular orbitals are derived by solving the matrix equation:

$$[H - ES] c = 0 \tag{7a}$$

The difference between the solutions of Eq. (7a) and Eq. (7b):

$$[H - EI] c = 0 \tag{7b}$$

with I the unit matrix, will be described with the aim of studying the role of non-orthogonality of the basis functions φ_i. Equation (7b) is used in the Hückel method.

Let us first recall the simple case of a homonuclear diatomic molecule B–B, each atom B with one atomic orbital of energy α and with a coupling matrix element β and overlap S. The molecular orbital energies, corresponding to bonding and anti-bonding combinations $\psi_{\pm} = (\varphi_1 \pm \varphi_2)/\sqrt{(2 \pm 2S)}$, are:

$$\varepsilon_{\pm} = \frac{\alpha - S\beta}{1 - S^2} \pm \frac{\beta - \alpha S}{1 - S^2}$$

Comparing this to the values for $S = 0$, $\varepsilon_{\pm} = \alpha \pm \beta$, it is noted that S has two effects: the average energy of the two levels is raised, and the spacing between the levels is reduced.

The first effect causes, if four electrons are present so that both bonding and anti-bonding levels are filled, a repulsion $= -4S(\beta - \alpha S)/(1 - S^2)$. This is the steric repulsion of all-electron methods. It would clearly be absent here if S is neglected.

If we consider a two-electron bond, i.e., filling of the bonding orbital only, the bond energy appears to be reduced by $-2S(\beta - \alpha S)/(1 - S^2)$ because of the raising of the average level energy and, in addition, it becomes less because β is reduced by $-\alpha S$.

The bondenergy becomes $\Delta E = \{-2S(\beta - \alpha S) + 2(\beta - \alpha S)\}/(1 - S^2) = \dfrac{2(\beta - \alpha S)}{1 + S}$.

These effects of the overlap are general.

In the following, this will be shown for a few simple model systems that can be readily analyzed.

The interaction is described according to the extended Hückel method of the following model systems and configurations:

$$
\begin{array}{ll}
A \ldots B - B & \text{(I)} \\
\quad\; 1 \quad 2 &
\end{array}
$$

$$
\begin{array}{ll}
\quad\; B & \\
A \ldots \mid & \text{(II)} \\
\quad\; B &
\end{array}
$$

$$
\begin{array}{ll}
\quad B\,2 & \\
\quad \mid & \\
A \ldots B\,1 & \text{(III)} \\
\quad \mid & \\
\quad B\,3 &
\end{array}
$$

$$
\begin{array}{ll}
A \ldots \overset{B1}{\diagup}\diagdown & \text{(IV) (symmetric interaction between} \\
\quad\; B2{-}B3 & \text{A and 3 B atoms)}
\end{array}
$$

The fragments B–B, B–B–B, and $\overset{B}{\diagup\diagdown}B{-}B$ are described within the Hückel approximation implying that atomic orbitals $\varphi_B(i)$ are orthogonal. A minimum basis set is assumed, with one atomic orbital per atom of s symmetry, donated by φ_A and $\varphi_B(i)$.

The diagonal matrix elements of the Hamiltonian are supposed to be equal:

$$\langle \varphi_A | H | \varphi_A \rangle = \langle \varphi_B(i) | H | \varphi_B(i) \rangle = \alpha$$

The following parametrizations for the different systems apply:

I.
$$\langle \varphi_A | H | \varphi_B(1) \rangle = \beta' \quad ; \quad \langle \varphi_A | H | \varphi_B(2) \rangle = 0$$
$$\langle \varphi_B(1) | H | \varphi_B(2) \rangle = \beta$$
$$\langle \varphi_A | \varphi_B(1) \rangle = S$$
$$\langle \varphi_A | \varphi_B(2) \rangle = 0$$

II.
$$\langle \varphi_A | H | \varphi_B(1) \rangle = \langle \varphi_A | H | \varphi_B(2) \rangle = \beta'$$
$$\langle \varphi_B(1) | H | \varphi_B(2) \rangle = \beta$$
$$\langle \varphi_A | \varphi_B(1) \rangle = \langle \varphi_A | \varphi_B(2) \rangle = S$$

III. $\langle \varphi_A | \, H \, | \varphi_B(1) \rangle = \beta'$

$\langle \varphi_A | \, \varphi_B(1) \rangle = S$

$\langle \varphi_A | \, \varphi_B(2) \rangle = \langle \varphi_A | \, \varphi_B(3) \rangle = 0$

$\langle \varphi_B(1) | \, H \, | \varphi_B(2) \rangle = \langle \varphi_B(1) | \, H \, | \varphi_B(3) \rangle = \beta$

IV. $\langle \varphi_A | \, H \, | \varphi_B(1) \rangle = \langle \varphi_A | \, H \, | \varphi_B(2) \rangle = \langle \varphi_A | \, H \, | \varphi_B(3) \rangle = \beta'$

$\langle \varphi_A | \, \varphi_B(1) \rangle = \langle \varphi_A | \, \varphi_B(2) \rangle = \langle \varphi_A | \, \varphi_B(3) \rangle = S$

$\langle \varphi_B(1) | \, H \, | \varphi_B(2) \rangle = \langle \varphi_B(1) | \, H \, | \varphi_B(3) \rangle = \langle \varphi_B(2) | \, H \, | \varphi_B(3) \rangle = \beta$

Case I assumes the interaction of a doubly occupied φ_A at energy α with a doubly occupied bonding B–B orbital $\varphi^+ = (\varphi_{B1} + \varphi_{B2})/\sqrt{2}$ at energy $\alpha + \beta$ (see Fig. 7).

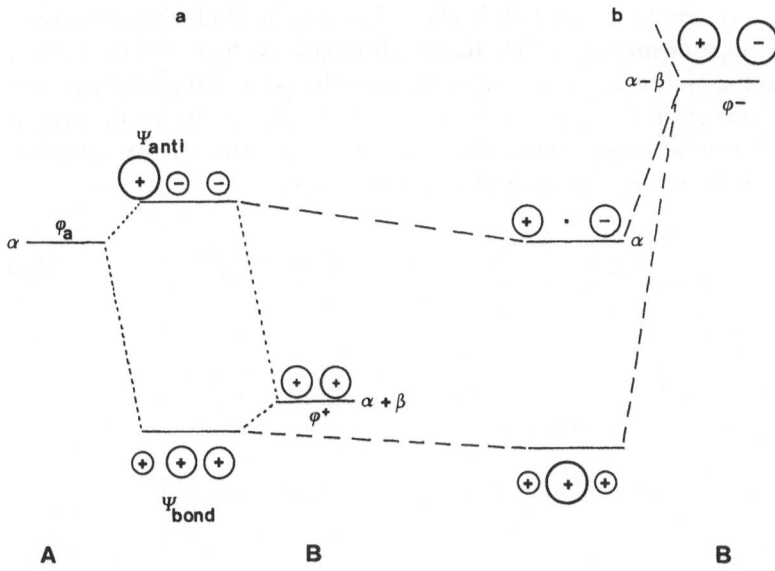

Fig. 7a, b. Interaction of fragment A with orbital φ_a with fragment B having one double occupied orbital φ^+ of s symmetry and an unoccupied orbital φ^- of p symmetry. **a** Interaction between φ_a and occupied orbital φ^+ only. **b** Effect of additional interaction with virtual orbital φ^-

As expected, there is repulsion which is found to be, to lowest order in S, $-2 \left(\beta' - \dfrac{1}{2} (2\alpha + \beta) \, S \right) S$. Evidently, since both the bonding combination:

$$\psi_{bond} = c_A \varphi_A + c_+ \varphi^+$$

and the antibonding combination:

$$\Psi_{anti} = \frac{(C_+ + C_A S/\sqrt{2})}{\sqrt{(1 - S^2/2)}} \varphi_A - \frac{(C_A + C_+ S/\sqrt{2})}{\sqrt{(1 - S^2/2)}} \varphi^+$$

are occupied, the gross populations of φ_A and φ^+ are 2.0.

The charge rearrangements that accompany the repulsion are therefore not visible in the gross populations, but they do show up in the net populations. To second order in S, the net populations of φ_A and φ^+ are $2 + S^2$ and there is a total negative overlap population of $-2S^2$.

It is interesting to consider also the charge distribution within B–B. As the coefficients of B_1 and B_2 are equal to each other in both ψ_{bond} and ψ_{anti} (as depicted in Fig. 7) the total net populations of φ_{B1} and φ_{B2} are equal, viz. $1 + S^2/2$. For φ_{B2} this is also the gross population, but for φ_{B1} there is an overlap population contribution with φ_A which is positive and of first order in S in the bonding orbital ψ_{bond} but negative in ψ_{anti}. The first-order contribution in ψ_{anti} cancels the one in ψ_{bond} and the total overlap population contribution for φ_{B1} is $-S^2$, yielding a gross population of $1 - S^2/2$. The gross and overlap populations clearly fit in with the charge distributions expected from the discussion in Sect. 2.1 for the Pauli-repulsion step, but only if the overlap S is taken into account.

Subsequently, the orbital φ^- on B–B is allowed to mix in. Both charge-transfer- and polarization-type of mixing (which are not distinguished here; see the remark at the end of Sect. 2.1) into ψ_{bond} and ψ_{anti} will cause the ratio of the coefficients of φ_{B1} and φ_{B2} to increase in ψ_{bond} and to decrease in ψ_{anti}. The results for the present case (I) and the other cases may be obtained from full solutions of the secular equations.

For case I the solution for the molecular orbital energies is:

$$E_{1,3} = \frac{\alpha - S\beta'}{1 - S^2} \pm \frac{1}{1 - S^2} \left[(1 - S^2)\beta^2 + (\beta' - \alpha S)^2 \right]^{1/2} \tag{8a}$$

$$= \frac{\alpha - S\beta'}{1 - S^2} \pm \Delta \tag{8b}$$

$$E_2 = \alpha$$

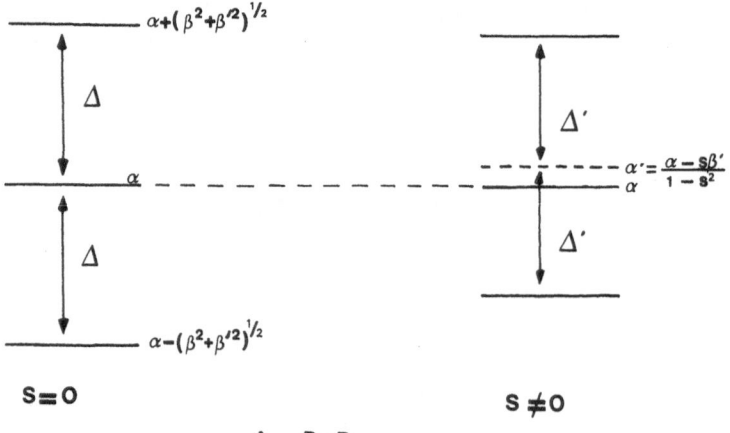

$\alpha + (\beta^2 + \beta'^2)^{1/2}$

Δ

α

Δ

$\alpha - (\beta^2 + \beta'^2)^{1/2}$

S = 0

Δ'

$\alpha' = \frac{\alpha - S\beta'}{1 - S^2}$
α

Δ'

S ≠ 0

A...B—B

Fig. 8. A ... B—B Molecular Orbital Scheme

The solutions for case III are completely analogous, with β replaced by $\beta \sqrt{2}$. Since β' also depends on S, these parameters cannot be chosen independently. Physical solutions to Eq. (8) only exist if:

$$\left| \frac{\alpha - S\beta'}{1 - S^2} \right| < |\alpha| \quad \text{and} \quad \frac{1}{1 - S^2} [(1 - S^2) \beta^2 + (\beta' - \alpha S)^2]^{1/2} < (\beta^2 + \beta'^2)^{1/2}$$

In Fig. 8 the solutions for the molecular orbital energies are compared for $S = 0$ and $S \neq 0$ in Eq. (8).

Figure 8 also reveals that (as always) the condition $S \neq 0$ raises the average position of the molecular levels and decreases the dispersion Δ of the molecular orbital levels. As observed in Eq. (8), the dispersion decrease is of the order αS. So, two effects of S decrease the interaction energy.

a) $|\alpha'| < |\alpha|$, cf. increase of kinetic energy by Pauli repulsion.

b) $|\Delta'| < |\Delta|$, decrease of covalent bonding.

In case I the ratio of the net populations (coefficients squared) on the two B atoms in molecular orbital i now differs from 1. It is given by:

$$\frac{q_B^i(2)}{q_B^i(1)} = \frac{\beta^2}{|\alpha - E_i|^2}$$

It follows that the coefficient of atom 1 in the orbital at energy α, which is ψ_{anti} stabilized by interaction with φ^- (see Fig. 7b), is zero. This is the well-known node in the nonbonding orbital at the central atom of the symmetrical allylic system, which is here shown to occur also if $\beta' \neq \beta$.

Whereas both the net population of $\varphi_{B1}(1)$ and the negative overlap population with φ_A in this orbital thus disappear, the reverse effects take place in the lowest orbital derived from ψ_{bond}.

The precise charge redistribution upon allowing φ^- to interact is subtle. If fragment B contains two electrons, there is polarization of B–B away from A, but the polarized system has stronger bonding (particularly less antibonding) to A. Apart from this, it is interesting to note that since $|\alpha - E_1'| < |\alpha - E_1|$, comparing the cases $S \neq 0$ and $S = 0$ for the full solutions of the three-orbital system, there is a shift in the net population in the lowest orbital towards atom 2 if S differs from zero. At the same time, of course, an overlap population contribution to the gross population of $\varphi_B(1)$ appears when $S \neq 0$.

The molecular orbital energies for cases II and IV are also readily found:

$$E^{\pm} = \frac{\alpha + \dfrac{n}{2} \beta - Z\beta'S}{(1 - ZS^2)} \pm \frac{1}{2(1 - ZS^2)}$$

$$\times [n^2\beta^2 + 4\alpha nZ\beta^2 S - 4nZ\beta\beta'S + 4Z(\beta' - \alpha S)^2]^{1/2} \quad (9)$$

$$n = Z - 1.$$

In case II, $n = 1$, $Z = 2$; In case IV, $n = 2$, $Z = 3$.

In case II the third level has energy $\alpha - \beta$.

In case IV, in addition to levels $E^{\pm}$, two MOs with energy values $\alpha - \beta$ are found. The general result is similar to that found for cases I and III discussed earlier.

The Pauli repulsion term is proportional to Z, the number of atoms on B that are the next neighbor of A.

If molecule B contributes two electrons, the Pauli repulsion energy becomes:

$$E^p_{rep} \approx -ZS(\beta' - \alpha S) \tag{10}$$

Concerning the terms giving dispersion it is noteworthy that again, β' is replaced by $\beta' - \alpha S$ in Eqs. (8a) and (9) if $S \neq 0$.

As a consequence if $S \neq 0$, the first term of Eq. (6b) for the nondiagonal matrix element of the Hamiltonian cancels in the eigenvalue equation and bonding is governed by the second term in Eq. (6b).

This can be easily proven for the general case of an array of equal atoms and s type nonorthogonal atomic orbitals with only nearest neighbor interactions. The expressions for the molecular orbitals are:

$$\psi_k = \frac{1}{\left[\sum_i c_i^{k^2} + 2 \sum_{i<j} c_i^k c_j^k S_{ij}\right]^{1/2}} \sum_j c_j^k \varphi_j \tag{11a}$$

$$\psi_k = \sum_j u_j^k \varphi_j \tag{11b}$$

and for the molecular orbital energy:

$$E_k = \sum_i u_i^{k^2} \alpha_i + 2 \sum_{i<j} u_i^k u_j^k \beta_{ij} \tag{12a}$$

$$= \sum_i \alpha_i \left(u_i^{k^2} + \sum_{j \neq i} u_i^k u_j^k S_{ij}\right) + 2 \sum_{i<j} u_i^k u_j^k \left(\beta_{ij} - \frac{1}{2}(\alpha_i + \alpha_j) S_{ij}\right) \tag{12b}$$

$$= \alpha + 2 \sum_{i<j} u_i^k u_j^k (\beta_{ij} - \alpha S_{ij}) \tag{12c}$$

Equations (12b) and (12c) illustrate the reduction of β_{ij} with $\frac{1}{2}(\alpha_j + \alpha_j) S_{ij}$.

Equation (12c) also shows that the weight of α_i is given by a Mulliken gross population of atomic orbital φ_i in molecular orbital k. For equal α, it becomes 1 by normalization.

Equation (12b) shows the crucial role of the bond-order term to chemical bonding. The bond order of atomic orbitals i and j is given by:

$$P_{ij} = \sum_k^{occ} u_i^k u_j^k$$

For a more general discussion see Ref [46].

Equation (9b) can be used to derive useful expressions for the bond energy as a function of $\bar{\beta}'/\beta$, the ratio of the effective interaction between atoms A and B and that between the B atoms $(\bar{\beta}' = \beta' - \alpha S)$.

If $\bar{\beta}'/\beta \ll 1$, one finds for A symmetrically coordinated to a ring of interacting B atoms.

$$E_{attr} = \frac{-Z\bar{\beta}'^2}{2|\beta|} \qquad (Z > 2) \tag{13}$$

If $\bar{\beta}'/\beta \gg 1$, this contribution becomes:

$$E_{attr} \approx Z^{1/2}\bar{\beta}' \tag{14}$$

For weak bonding $(\bar{\beta}'/\beta \lesssim 1)$, low coordination appears to be favored, since comparison with Eq. (10) shows the much stronger coordination-number dependence of the repulsive term compared to the attractive terms.

We will return to this interesting observation at a later point.

The considerations so far have not taken into account directional effects of atomic orbitals, with angular momentum $l \neq 0$, which also will be discussed later.

The extended Hückel picture of bonding has been shown to be very similar to that found from rigorous first-principle calculations.

Bonding according to first-principle calculations has been interpreted such that explicit inclusion of overlap $(S \neq 0)$ leads to repulsive effects between doubly occupied orbitals. These repulsive effects are reduced if unoccupied orbitals are available to depopulate doubly occupied repulsive orbitals either by polarization of fragments or charge transfer (Fig. 5). The extended Hückel method incorporates both effects, but they appear indirectly. When overlap is taken into account, we have shown that bonding is decreased by an upward shift of the energies of the orthogonalized fragment orbitals with respect to the nonorthogonalized orbitals and by reduction of the dispersive overlap energy integrals. The latter leads to reduction of bonding between partially occupied orbitals, and to repulsion if interaction occurs between doubly occupied orbitals. Clearly, the more unoccupied virtual orbitals are available, the more the repulsive interactions is counteracted. Within the same fragment this interaction is due to nonorthogonality of the unoccupied virtual orbitals and interacting occupied fragment orbitals.

As long as the differences between the discrete levels are large compared to the overlap energy matrix elements, second-order perturbation theory can be used to

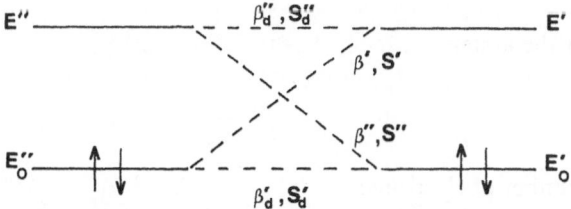

Fig. 9. Interaction energy and scheme according to second order perturbation theory

$$\Delta E = -4\beta'_d \cdot S_{d'} - 2\frac{|\beta' - {}^1\!/_2(E' + E'_0)\,S'|^2}{E' - E''_0}$$

calculate the attractive contribution to the bond energy as in Eq. (13). The repulsive interaction term is calculated analogous as in Eq. (10). The formulae are illustrated in Fig. 9.

In the next section, techniques are discussed to compute the bond energy in case one of the fragments has a continuous energy spectrum, as for a metal surface.

2.4 Embedded Systems

A metal surface is part of a semi-infinite system. Apart from changes in electronic structure and electrostatics, a major difference between a finite and infinite system is, that there is no local conservation of the total number of electrons. The local number of electrons is determined by the requirement that for a system at equilibrium the Fermi-level energy, i.e., the energy of the Highest occupied molecular orbital is the same throughout the system.

The second qualitative difference between a finite and infinite system is that for a finite system, the electron energy spectrum is discrete, whereas for an infinite system, it is continuous erase, be it that localized states may exist. Especially for the valence electron bands of metals with a low electron density, as a consequence, the electron energy spectrum slowly converges with increasing particle size to that of the infinite system.

In semi-infinite systems it is required that the Fermi level does not change and this allows the development of convenient closed expressions for changes in energy due to surface-formation or chemisorption. Such expressions have been derived by Koutecky [47], Grimley [48], Schrieffer [49], and others [50].

According to the extended Hückel emthod, the change in energy is given by:

$$\Delta E = 2 \left\{ \sum_{i}^{occ'} E'_i - \sum_{i}^{occ} E_i \right\} \tag{15}$$

E'_i are the orbital eigenvalues after chemisorption, and E_i the orbital eigenvalues before interaction.

Defining the energy density of states $\varrho(E)$ as:

$$\varrho(E) = \sum_{i} \delta(E - E_i) \tag{16}$$

and using the semi-infiniteness of the system:

$$E_{F'} - E_F \ll E_{F'}$$

as well as conservation of total number of electrons:

$$\int_{-\infty}^{E_{F'}} dE \ \varrho'(E) = \int_{-\infty}^{E_F} dE \ \varrho(E)$$

one finds for the energy change [49]:

$$\Delta E = 2 \int_{-\infty}^{E_F} dE(E - E_F)\, \Delta\varrho(E) \tag{17a}$$

$$= -2 \int_{-\infty}^{E_F} dE\, \Delta n(E) \tag{17b}$$

with $\Delta\varrho(E)$ the change in electron energy density, and $\Delta n(E)$ the change in number of electrons of energy E. Equation (17b) presents the expected result, that changes in energy are related to rearrangements of electrons over energy levels. Using Green's function techniques, elegant practical expressions for $\varrho(E)$ have been developed. In Ref [51], an application of such theories to the problem of CO chemisorption to transsition metals can be found.

Closed expression derived for Eq. (17) are useful, but not very suitable if one wishes to interpret changes in chemical bonding in terms of changes in electrondensity on molecular fragments. For this purpose it is more convenient to return to Eq. (12) for the total energy of a system according to the extended Hückel method:

$$\Delta E = 2\left\{ \sum_i \alpha_i\, \Delta q_i + 2 \sum_{i<i'} \Delta P_{ii'} \left\{ \beta_{ii'} - \frac{1}{2} S_{ii'}(\alpha_i + \alpha_{i'}) \right\} \right\} \tag{18a}$$

$$= 2\left\{ \sum_i \alpha_i\, \Delta q_i + 2 \sum_{i<i'} \Delta P_{ii'}\, \bar{\beta}_{ii'} \right\} \tag{18b}$$

with Δq_i the change in electron density on atomic orbital i:

$$\Delta q_i = \sum_k^{occ'} \left\{ u_i'^{k^2} + \sum_{i'\neq i} u_i'^{k} u_{i'}'^{k} S_{ii'} \right\} +$$

$$- \sum_k^{occ} \left\{ u_i^{k^2} + \sum_{i'\neq i} u_i^k u_{i'}^k S_{ii'} \right\} \tag{19a}$$

$$= q_i' - q_i \tag{19b}$$

and $\Delta P_{ii'}$ the change in bond order between atomic orbitals i and i':

$$\Delta P_{ii'} = \sum_k^{occ'} u_i'^{k} u_{i'}'^{k} - \sum_i^{occ} u_i^k u_{k'}^i \tag{20}$$

$$= P_{ii'}' - P_{ii'}$$

Similar methods as used to calculate Eq. (17) can be used to compute the quantities according to Eqs. (19) and (20).

As already mentioned, the important difference in the electrondistribution of discrete and extended systems is the continuous nature of the electron energy density spectrum. The discrete sums have therefore to be replaced by integrals [51]:

$$q_i = \int_{-\infty}^{E_F} \varrho_{ii}(E) \, dE + \sum_{i' \neq i} P_{ii'} S_{ii'} \tag{20 a}$$

$$P_{ii'} = \int_{-\infty}^{E_F} dE \, \varrho_{ii'}(E) \tag{20 b}$$

and $\varrho_{ii'}(E)$ can be calculated from:

$$\varrho_{ii'}(E) = \frac{1}{\pi} \, \text{Im} \lim_{\varepsilon \to 0} \left[O(E + i\varepsilon) - ES \right]_{ii'}^{-1} \tag{22}$$

The problem posed by embedding a cluster into an extended system is solved once the matrix $O(E)$ is known.

When chemisorption to a metal surface is studied, the matrix $O(E)$ is given by:

$$O = \left[\begin{array}{c|c} H_{ads} & H_{ads,\,latt} \\ \hline H_{latt,\,ads} & O_{latt}(E) \end{array} \right]$$

Equations (22) as well as (17) can be evaluated using the local nature of electron density changes. The electron density changes disappear if one is more than a few atomic distances removed from the absorbate.

The block H_{ads} contains the Hamiltonian matrixelements of the adsorbate and those atoms close to the adsorbate on which there are changes in electrondensity compared to the situation before adsorption.

$O_{lattice}$ contains those atoms that are unperturbed by the adsorbate.

Green's function techniques [48, 50, 52] can be used to reduce the infinite $O_{lattice}$ matrix to a finite one. For instance, a typical diagonal matrix element of O_{latt} has the form [51]:

$$O_{latt_{ij}}(E) = H_{ij} - \sum_{\gamma',\,\bar{\gamma}'} (H_{i\gamma} - ES_{i\gamma}) \, \bar{G}_{\gamma\gamma'}(E) \, (H_{\gamma' j} - ES_{\gamma' j}) \tag{23}$$

$\bar{G}_{\gamma'\gamma''}(E)$ is a Green's function matrix element of the undisturbed lattice, from which interactions with the atoms contained in H_{ads} are excluded.

When chemisorption is modelled by interaction of a hydrogen type adsorbate with a tight-binding (Hückel) s-valence electron metal surface, each metal atom contributes one atomic orbital and the hydrogen atom interacts with one surface atomic orbital. Using the Bethe lattice [54, 55, 56] approximation to calculate the Green's function matrix elements $\bar{G}_{\gamma'\gamma''}(E)$, expressions for $\varrho_{ii'}(E)$ are readily found. and the hydrogen atom interacts with one surface atomic orbital. Using the Bethe lattice [54, 55, 56] approximation to calculate the Green's function matrix elements $\bar{G}_{\gamma'\gamma''}(E)$, expressions for $\varrho_{ii'}(E)$ are readily found.

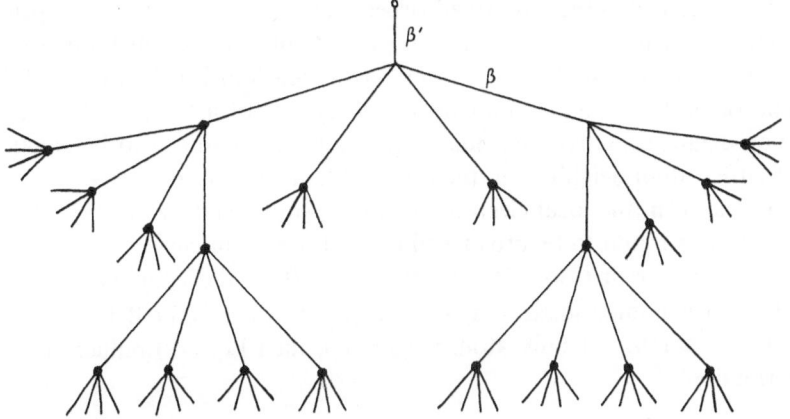

Fig. 10. Bethe lattice and adsorbed atom

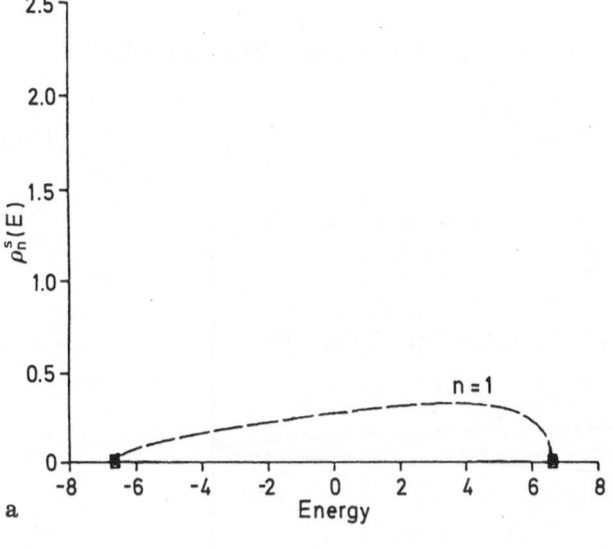

a

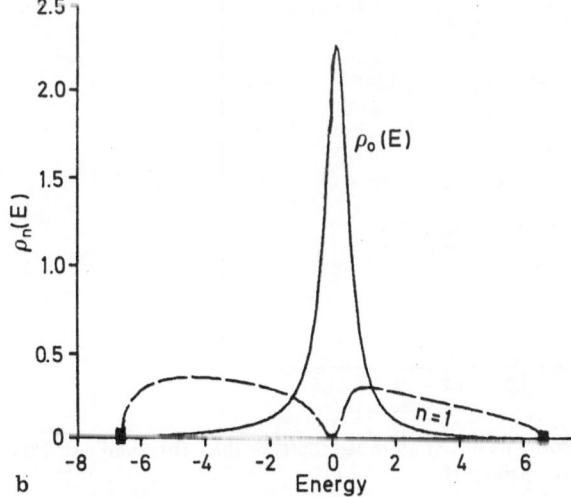

Fig. 11a. ϱ_1: Local electron energy density of states (LDOS) on metal atom to be bonded to hydrogen before adsorption; **b** ϱ_0: LDOS of adsorbed hydrogen atom, ϱ_1: LDOS on metal atom bonded to hydrogen after adsorption

$$\beta' = \beta = -1$$

$$\alpha_0 = \alpha_m$$

b

A Bethe lattice is sketched in Fig. 10. An adsorbed atom is represented by the open circle. Figure 11 b shows calculated electron energy density of states results for hydrogen adsorption. The calculations have been done assuming S in Eq. (22) to be the unit matrix. The Bethe lattice used simulates the (111) face of a f.c.c. crystal. All diagonal and non-diagonal matrix elements, respectively, are assumed to be equal. The number of lattice atom neighbors in the bulk is 12, at the surface it is 9.

As is shown in Fig. 11b, the local electron energy density of states (LDOS) $\varrho_0(E)$ at the hydrogen atom is found to be broadened into a Lorentzian curve.

The LDOS $\varrho_0(E)$ has been extensively studied [49, 57]. If the coupling matrix element β' to the lattice atom is small compared to $2\beta \sqrt{Z}$, which is the Bethe lattice bandwidth ($Z + 1$ is number of bulk atom neighbors), then Eq. (22) reduces to a Lorentzian distribution:

$$\varrho_0(E) = \frac{1}{\pi} \frac{\Gamma(E)}{(\alpha + \Lambda(E) - E)^2 + \Gamma^2(E)} \tag{26a}$$

For weak adsorption, $\Lambda(E)$ is small, and to lowest order, $\Gamma(E)$ is given by:

$$\Gamma = Z'\beta'^2 \bar{\varrho}_1(E = \alpha) = \frac{Z'\beta'^2}{Z^{1/2}|\beta|} \tag{26b}$$

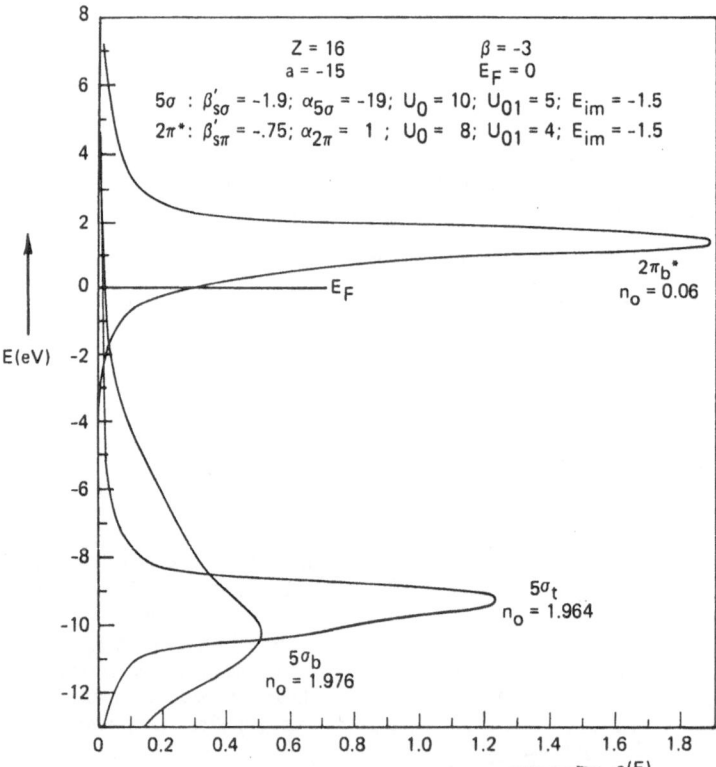

Fig. 12. LDOS $\varrho_0(E)$ for the 5σ and $2\pi^*$ orbitals of CO adsorbed top(t) or three (b) coordinated the (111) surface of a s-band Bethe lattice f.c.c. metal [51]

Z' is the number of adsorbate atom neighbors, $\bar{\varrho}_1(E = \alpha)$ the local density of states of the lattice atom orbital interacting with the adsorbate before adsorption.

The function $\bar{\varrho}_1(E)$ will be discussed later.

Figure 12 illustrates computed adsorbate local density of states (LDOS) using a similar approximation (23) for σ and π symmetry orbitals of cote mono- or three-coordinated to the surface of an s-band model f.c.c. metal. As will be explained later

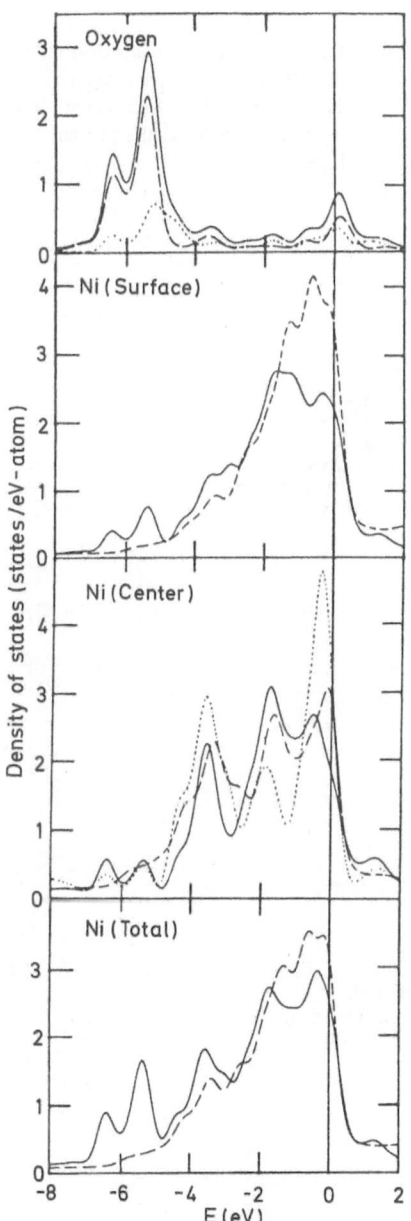

Fig. 13. Total and projected local DOS for three-layer Ni (001) slab (dashed lines) and c (2×2) O on Ni (001) (solid lines) including oxygen e (long dashes) and a_1 (dotted) orbital DOS. Results from ab-initio self-consistent calculations [58]

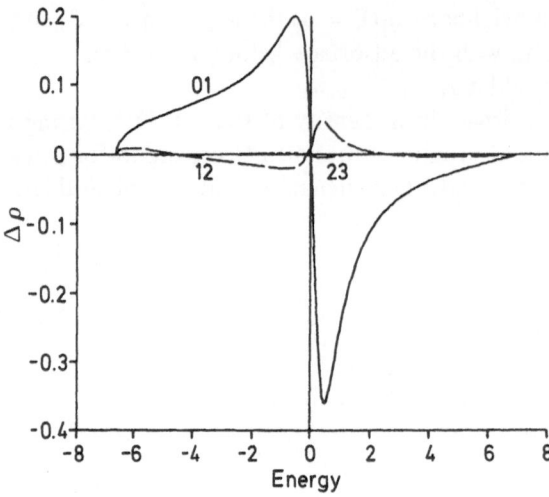

Fig. 14. Bond order electron energy density changes $\Delta\varrho_{01}(E)$, $\Delta\varrho_{12}(E)$, and $\Delta\varrho_{23}(E)$ for the same calculations used to produce results presented in Figs. 11

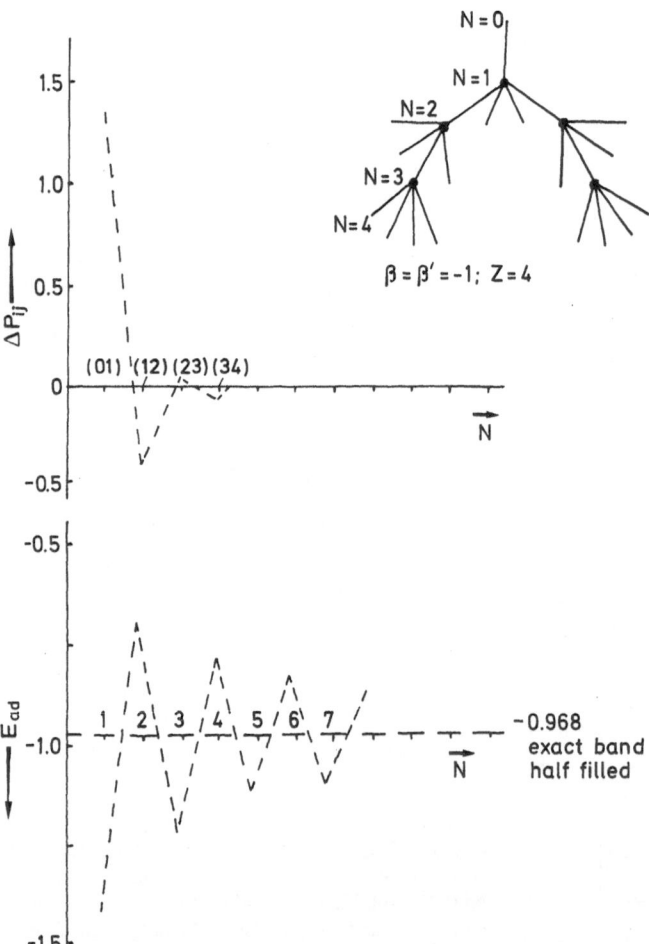

Fig. 15. Changes in bondorder and respective energy contributions as a function of distance to adsorbate. For parameters used, see Ref. [55]

there is only an interaction with π type adsorbate orbitals in bridging coordination sites. An increase in bandwidth of the σ electron density of surface atoms occurs if adsorption increases their coordination number. This is a very general phenomenon and follows from (26b). This is illustrated in Fig. 13 for the change in LDOS of surface Ni atoms when O is adsorbed to it according to an ab initio calculation [58].

The changes of the electron density of Ni surface atoms are shown before and after oxygen adsorption. One observes that the bandwidth of electron density on the surface atoms is less than that of the slab center atoms, which relates to the decreased coordination of the surface-atoms compared to the bulk.

Comparison of $\varrho_1(E)$ in Fig. 11a and 11b also shows a similar broadening of the LDOS of the surface atom in the Bethe lattice calculations.

Because of conservation of density, broadening of the bandwidth implies that at the center of a band, density decreases in order to balance the increase in density at the edges.

Figure 14 shows the bondorder electronenergydensity $\varrho_{01}(E)$ and changes in bondorder electronenergydensities $\Delta\varrho_{12}(E)$, and $\Delta\varrho_{23}(E)$ from the same Bethe lattice calculation done to produce Figs. 11.

In Figs. 11 and 14, indices refer to the coordination shell with respect to the adsorbed atom.

As expected, $\varrho_{01}(E)$ gives a positive bonding contribution at low electron energies, but an antibonding contribution at higher electron energies.

$\Delta\varrho_{12}(E)$ the change in bondorder energy density behaves reverse. After adsorption,

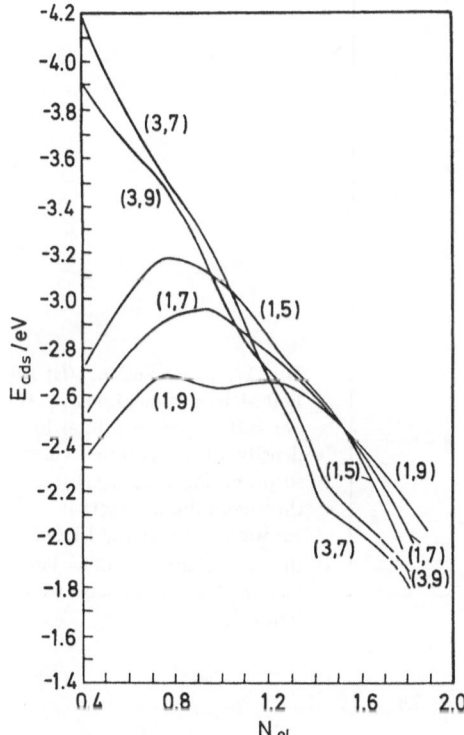

Fig. 16. Interaction energy E_{ads} as a function of $N^e{}_1$ [51]

there is a decrease of the bonding contribution and increase in the antibonding region, $\Delta\varrho_{23}(E)$ is much smaller, but again shows inverse behavior from $\Delta\varrho_{12}(E)$.

Because of coordination of an hydrogen atom to atom 1, there is a decrease in the bondstrength between atom 1 and those in the second coordination shell if the valence electron band is half filled. Respective contributions to the bond energy ΔE calculated on a Bethe lattice simulating a b.c.c. lattice are given in Fig. 15 for the situation that each metal atom contributes one electron.

One observes that there are still significant contributions to the bond energy 2 or 3 coordination shell distances removed from the adsorbate site.

Whether the bond strength increase with the number of neighbors of the atom concerned depends on the position of the Fermi level with respect to the local density of states maximum and the energy dependence of $\varrho(E)$.

Figure 16 illustrates this by presenting the calculated bondstrengths of an hydrogen-type adsorbate to the (111) face of the f.c.c. s-band model metal as a function of the number of valenceband electrons (N_{el}). The same Bethe lattice approximation as discussed earlier has been used. As expected, three-coordinated hydrogen bonds more strongly than mono- or dicoordinated hydrogen atoms at low valence-electron

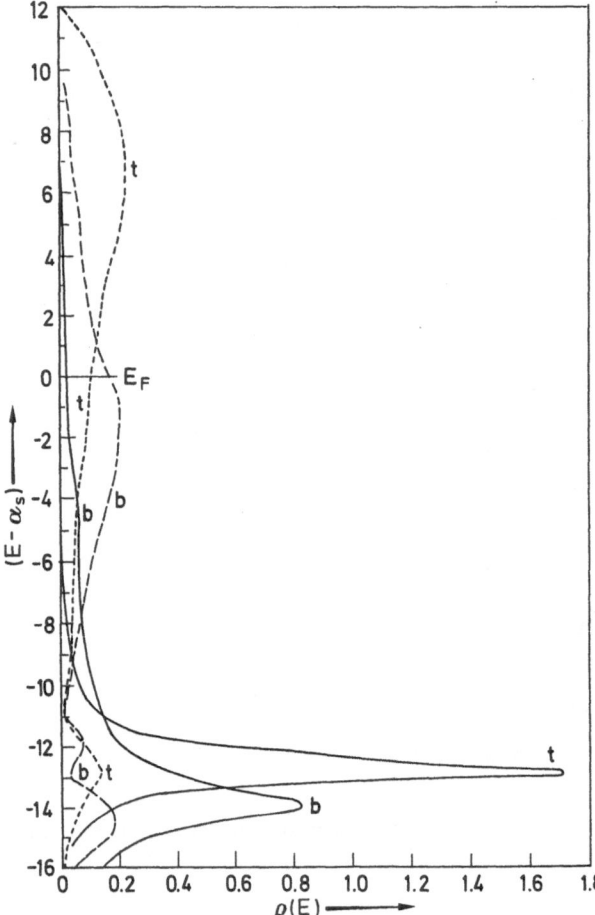

Fig. 17. $\varrho_0(E)$ and $\varrho^s_{t,b}(E)$ for Bethe lattice calculations. $\varrho^s_{t,b}(E)$ are surface group orbital local density of states after chemisorption for atop ($\varrho^s_t(E)$) and three-coordinate ($\varrho^s_b(E)$) adsorption, $\varrho_0(E)$ is the LDOS of the adsorbate orbital. Parameters used are the same as for figure 12.

——— $\varrho_0^{(t,b)}$

- - - - - - $\varrho^s_{(t,b)}$

band filling. This is in line with observations that can be made from Fig. 12. At low valence-electron band filling, more electrons will occupy low energy levels for three-coordinated than for mono-coordinated hydrogen (see also Fig. 17). As a result, according to Eq. (17b), three-coordinated hydrogen is more strongly bonded than monocoordinated hydrogen.

That difference clearly becomes less when the Fermi level increases.

However, the shift from threefold to atop position can only be understood by also considering the LDOS $\varrho_i(E)$ of the surface-fragment orbitals that interact with the adsorbed hydrogen atoms. For atop-adsorbed hydrogen, one has to compute the LDOS $\varrho_1(E)$ of the atomic orbital φ_1 of the metalatom bonded to hydrogen; for three-coordinated hydrogen symmetrically coordinated to three surface metalatoms, it is the LDOS $\varrho_i^S(E)$ of the grouporbital $1/\sqrt{3}\,(\varphi_1 + \varphi_2 + \varphi_3)$.

As is clearly seen in Fig. 17, the metalatom local density of stages $\varrho_i^S(E)$ is high in the antibonding region; the maximum in local density of states being at higher energy for atop coordination than for threefold coordination. The Fermi level in Fig. 17 is chosen such that the valence-electron band filling of the metal lattice atoms equals one electron per metal atom. One observes that more antibonding levels are occupied for threefold coordination (b) than for atop coordination (c). As a result, atop coordination is more favored.

This illustrates that, in general, bandwidth or bandshifts as measured spectroscopically do not correlate with the bondstrength and that deductions on relative bond energies based on approximations to Eq. (18) have to be considered with care.

A very useful approach to estimate bond energies of adsorbates is to use an extension of frontier orbital theory to chemisorption on metal surfaces [33]. This has been discussed extensively elsewhere, so the basic results will be shortly summarized.

We discussed earlier that orbital overlap causes two effects: Pauli repulsion and reduction of energy dispersion.

Equation (15) will now be replaced by the first- and second-order perturbation approximation to it:

$$E = -4 \sum_{\alpha} \bar{\beta}'_\alpha S_\alpha - 2 \left\{ \sum_{\substack{\alpha,\,i \\ E_i > E_F \\ E_\alpha\,occ}} |\bar{\beta}'_{\alpha i}|^2 \frac{1}{E_i - E_\alpha} + \sum_{\substack{\beta,\,j \\ E_j < E_F \\ E_\beta\,unocc}} |\bar{\beta}'_{\beta j}|^2 \frac{1}{E_\beta - E_j} \right\} \quad (24)$$

The first term represents the Pauli repulsion between the fragment orbitals that are allowed to interact. These are sketched in Fig. 18.

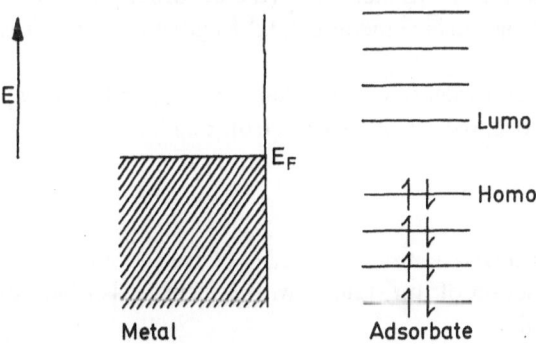

Fig. 18. Schematic illustration of relative position of adsorbate and metal surface orbitals

E_α are the highest occupied molecular orbitals (HOMO) of the adsorbate, E_β are the lowest unoccupied molecular orbitals (LUMO) of the adsorbate, E_i the metal LUMOs, and E_j the metal surface HOMOs.

The second term of Eq. (24) represents electron backdonation from adsorbate to surface, the third electron back donation from metal surface to adsorbate.

The attractive part of Eq. (24) can be partially integrated to give Eq. (25):

$$E_{attr} = -2 \left\{ \sum_\alpha \varrho_\alpha(E_F) \, \bar{\beta}_\alpha'^2 \, \frac{\Delta_\alpha \cdot (1 - P_\alpha)}{-\phi - \dfrac{e^2}{r_\alpha + k_\alpha} - E_\alpha + \Delta_\alpha \cdot (1 - P_\alpha)} + \right.$$

$$\left. + \sum_\beta \varrho_\beta(E_F) \, \bar{\beta}_\beta'^2 \, \frac{\Delta_\beta \cdot P_\beta}{E_\beta - \dfrac{e^2}{4r_\beta + k_\beta} + \phi + \Delta_\beta \cdot P_\beta} \right\} \tag{25}$$

$\varrho_\alpha(E_F)$ is the group-orbital electron density at the Fermi-level energy [59, 60, 61]:

$$\varrho_\alpha(E_F) = \sum_i |\langle \varphi_\alpha | \psi_i \rangle|^2 \, \delta(E_F - E_i)$$

φ_α is a linear combination of surface atomic orbitals, ψ_i a metal surface orbital eigenfunction with corresponding energy E_i.

$$\bar{\beta}_\alpha' = \langle \chi_\alpha | \bar{H}' | \varphi_\alpha \rangle \tag{26}$$

and

$$S_\alpha = \langle \chi_\alpha | \varphi_\alpha \rangle \tag{27}$$

For s-type orbitals:

$$\beta' \sim Z'^{1/2} \beta_0' \tag{28a}$$

β_0' is the overlap energy integral found for atop adsorption, and

$$S_\alpha \sim Z'^{1/2} S_0 \tag{28b}$$

S_0 is the overlap integral found for atop adsorption, with Z' the number of neighbors of the adsorbate orbital. Since H' is totally symmetrix, χ_α (the adsorbate molecular orbital) determines the symmetry of the surface metal-orbital fragment φ_α, which is the group orbital.

Δ_α is the total bandwidth of the metal valence-electron band corresponding to φ_α, and P_α is a measure of the electron occupation of that electron band.

$$P_\alpha = E_F / \Delta_\alpha \tag{29}$$

$-\Phi$ is the work function of the metal surface considered, and the term $-e^2/(4r_\alpha + k_\alpha)$ represents the image potential interaction of ion state α, with effective adsorbate to metal distance r_α and screening k_α [6].

Equation (25) relates the attractive component of the binding energy to:
a. the group orbital local density of states at the surface metal Fermi level,
b. the effective energy difference between adsorbate orbitals and the Fermi level,
c. the surface metal orbital electron occupation,
d. orbital overlap.

This is a remarkable result. Whereas it has been suspected by many authors (see Grimely [62]) that a relation between bondstrength and local density of states at the Fermi level should exist, Eq. (25) states this relation with the modification that the grouporbital local of states should be used. Others [63] have derived related expressions, but ignored the $\varrho(E_F)$ term. Falicov and Somorjai [64] propose to correlate catalytic activity with low-energy local electronic fluctuations in transition metals.

Electronic fluctuations are well known to be related to the local electron density of states at the Fermi levels and the occupied and unoccupied fraction of the valence-electron bands.

Equation (24) appears to quantify this and modifies the importance of the local density of states to that of the grouporbital density of states at the Fermi level.

Figure 19 shows the LDOS of different grouporbitals of the s-electron band for the (111)-face of a f.c.c. crystal calculated in the Bethe lattice approximation.

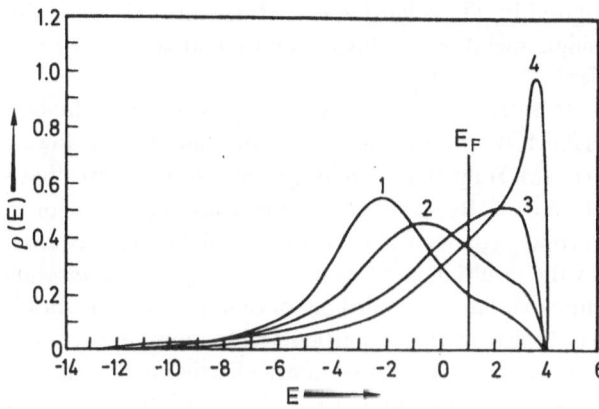

Fig. 19. LDOS $\varrho_i(E)$ of different group orbitals of the electron band for a (111) face of a f.c.c. crystal calculated in the Bethe lattice approximation [51] $i = 1$ σ three coordination; $i = 2$ σ two coordination; $i = 3$ σ atop coordination; $i = 4$ π two coordination

It is observed that at an electron occupation number of one electron per atom and for an adsorbate orbital of σ symmetry:

$$\varrho_{atop}(E_F) > \varrho_{bridge}(E_F)$$

Hence, at this electron bandfilling, already atop adsorption becomes favored as can be seen by comparison with the bond energies presented in Fig. 16.

In agreement with the results discussed earlier at lower band filling, the higher-coordinated hydrogen atoms become more stabilized because of the relative increase in grouporbital local density of states.

3 Bonding to Metal Clusters

In this chapter, the results of ab-initio calculations will be analyzed within the extended Hückel theory conceptual framework. This provides also an opportunity to discuss explicitly effects due to the orbital symmetry of the fragments. The differences in chemical bonding to small and large clusters will be considered here. In the last chapter, the analysis will be extended to semi-infinite lattices. We will discuss studies of H_2 dissociation extensively and comment shortly on the reactivity of methane and ethane. For comparicon, also a discussion of CO chemisorption will be presented.

3.1 Chemisorption and Dissociation (Oxidative Addition) of H_2 to Transition Metal Atoms

Extensive ab-initio calculations on oxidative addition to bare and complexed atoms have appeared during the past years.

Here, we mainly discuss the work of the Siegbahn group [64, 66] and the group of Goddard [66, 67] and Nakarsuji [68]. Other important studies have been done by Kitaura et al. [69] and Noell and Hay [70]. The work up to 1985 has been excellently reviewed by Dedieu [71]. The elementary interactions playing a role in oxidative addition have been recently analyzed by JT Saillard and R Hoffmann [47], who also compared H_2 dissociation on single metal atom clusters and metal surfaces.

Three main effects play a role:

As expected on the basis of frontier orbital theory, the *symmetries* of highest occupied molecular orbitals (HOMO) on the one fragment, and empty lowest unoccupied molecular orbitals (LUMO) on the other fragment and vice versa have to match. Sometimes they do not match. By *promotion* of electrons in a fragment to unoccupied orbitals, orbitals of proper symmetry may become available. The required promotion energy is paid back by the resulting increase in bond energy. *Hybridization* of the orbitals to allow for optimum overlap is the third factor of importance. Hybridization of the orbitals in order to optimize overlap in particular directions also requires promotion of electrons, so promotion and hybridization are closely interconnected. Such effects are very general. We will discuss them here for H_2 dissociation as well as for CO chemisorption. Adsorption and dissociation of H_2 to a Ni atom has been thoroughly studied by Blomberg and Siegbahn [64] using CASSCF (complete active space SCF) and contracted CI calculations. The results are summarized in Fig. 20. The calculated ground state of Ni is $3d^8 4s^2$ with the spins in a triplet state (experimentally the $^1D(d^9s^1)$ state is 0.03 eV lower). The H_2 molecule approaches the Ni atom symmetrically.

Because of the small overlap, the triplet state Ni 3d electrons interact only weakly with H_2. The doubly occupied Ni 4s orbital has a repulsive interaction with the doubly occupied H_2 σ orbital, so bond formation is symmetry-forbidden. This results in a large activation energy for hydrogen addition (38 kcal/mole). In NiH$_2$, the angle ϕ between the NiH bonds is $180°$, agreeing with sp hybridization between the doubly occupied 4s and empty 4p orbital of Ni. Addition of H_2 appears to be energetically neutral. The activation energy for H_2 addition becomes significantly decreased if

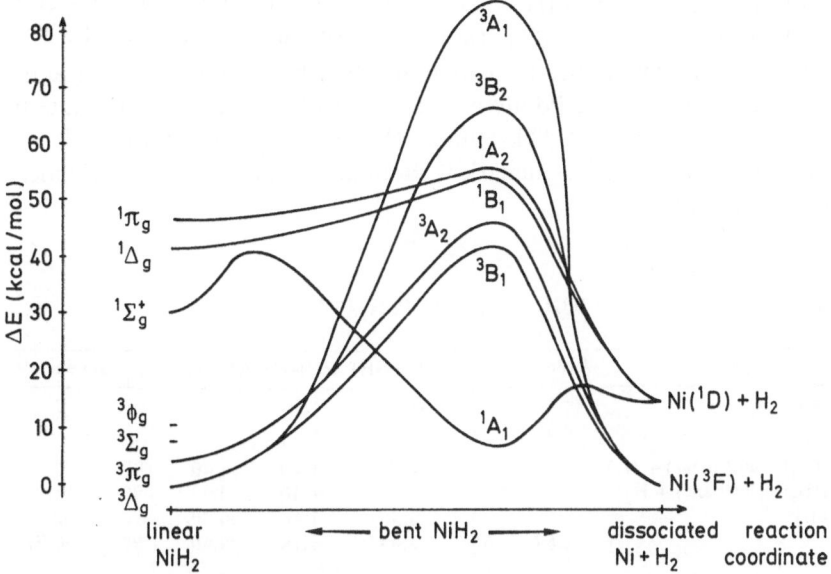

Fig. 20. Dissociation curves for triplet and singulet states of NiH_2. The energies are relative to H_2 and the ^3F state of nickel [64]

promotion of electrons from 4s to 3d orbitals can occur. The promotion energy cost is 15 kcal/mol, but reaction between the Ni d^9s^1 (^1D) and H_2 is now symmetry-allowed. Electrons can be donated from the doubly occupied H_2 σ orbital into the now partially occupied Ni 4s orbital and an antisymmetric Ni d orbital can backdonate into the empty H_2 σ^* orbital. As a result, the overall activation energy for H_2 addition decreases to 15 kcal/mol and an endothermic quasi-stable $^1A_1$ NiH_2 state is found to exist at an energy of 8 kcal/mol above the triplet state. The configuration of the $^1A_1$ NiH_2 state is now bent, with an angle ϕ equal to 50°. The $^1A_1$ state can be approximately described as a Ni atom in the d^9 state and hybridization between d_{xy} and s orbital (Fig. 21). The predicted angle would however be 90°, indicating that the H_2 molecule has not yet completely dissociated. It should be noted that H_2 readily dissociates on a Ni metal surface, without any activation energy. The reason for this difference will be discussed later.

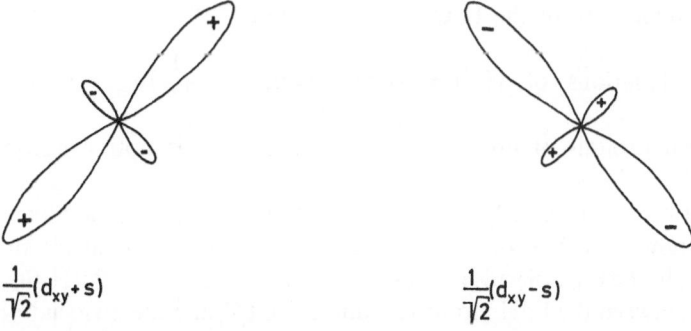

$$\frac{1}{\sqrt{2}}(d_{xy}+s)\qquad\qquad\frac{1}{\sqrt{2}}(d_{xy}-s)$$

Fig. 21. Hybridized d_{xy} and s orbitals

Addition of H_2 to Pd and a $Pd(H_2O)_2$ complex using similar methods has been studied by UB Brandemark et al. [65]. Table 2 [65] summarizes their results. This study demonstrates the importance of prehybridization by the presence of ligands. We first consider addition to the Pd atom. Promotion of the Pd atom $4d^{10}$ state to a $4d^9s^1$ state is calculated to cost 40 kcal/mol. The ground state of Pd is $4d^{10}$. So, if only the 4d orbital is involved, symmetrical addition of H_2 is symmetry-forbidden, similarly to Ni in the d^8s^2 state.

Table 2. Geometries and energies for PdH_2 and $(H_2O)_2$ PdH_2 [65]

Geometry	$R(Pd-H)$	θ	$R(Pd-HH)$	$R(H-H)$	$R(Pd-O)$	ϕ	rel energy
Asymptote (Pd + H_2)	∞		∞	1.40			0
PdH_2	3.65	23	3.58	1.46			−5.5
Asymptote (($H_2O)_2$ Pd(linear) + H_2)	∞		∞	1.40	4.40	180	0
Asymptote (($H_2O)_2$ Pd(bent) + H_2)	∞		∞	1.40	4.40	90	15.0
PdH_2 geometry	3.65	23	3.58	1.46	4.40	90	−2.0
Equilibrium	2.89	65	2.43	3.14	4.40	90	−6.9

Distances in au and angles in degrees. Energies in kcal/mol.

H_2 addition to Pd is exothermic by −5.5 kcal/mol.

(Using SAC and SAC-CI methods, Nakatsuji et al. [68] report a value of 15 kcal/mol indicating perhaps that the energy differences calculated are more comparative than quantitative). The angle ϕ between the two PdH bonds equals 23° and H_2 is not dissociated. Bonding occurs because of donation of electrons from the H_2 σ orbital into the empty Pd 5s orbital and backdonation from an antisymmetric Pd 4d orbital into the antibonding H_2 σ^* orbital. The Pd configuration remains, however, close to Pd d^{10}. The calculations on $Pd(H_2O)_2$ were done with an angle of 90° between the H_2O molecules. The important characteristic of the H_2O ligands (and the PR_3 ligands to be discussed below) is the presence of a low-lying occupied σ (lone-pair donor) orbital. The d_{xy} orbital mixes into the antisymmetric combination of these σ ligand orbitals in a bonding (stabilizing) fashion, and is itself destabilized by antibonding interaction. The antibonding interaction with the σ lone pairs is alleviated by admixture of a 5 p_y Pd orbital which has the effect of hybridizing the d_{xy} away from the σ lone pairs and towards the vacant coordination site. The 5s is less destabilized by interaction with the symmetry combination of the H_2O lone-pair orbitals. It leads to bonding and antibonding combinations of H_2O σ orbitals with the $\frac{1}{\sqrt{2}}(d_{xy} + s)$ and $\frac{1}{\sqrt{2}}(d_{xy} - s)$ hybridized orbitals of Pd. As a consequence, the promotion energy from the $4d^{10}$ configuration is lowered. The computed difference between the $^1S(d^{10})$ and $^1D(d^9s^1)$ state is now only 20.4 kcal/mol instead of 40 kcal/mol found for the bare Pd atom. As a result, the H_2 molecule can dissociate when reacted with $Pd(H_2O)_2$. The resulting angle ϕ between the PdH atoms is found to be 65° and the dissociation energy as −6.0 kcal/mol, with respect to linear $Pd(H_2O)_2$. The difference in energy between bent and linear $Pd(H_2O)_2$ is 14 kcal/mol, linear $Pd(H_2O)_2$ being more stable. Its electronic structure is very similar to that of the linear $Pt(PH_3)_2$ to be discussed later. Addition of H_2 is accompanied by moving the H_2O molecules from the linear to bent configurations. In this process, the Pd atom becomes excited to the d^9 configuration.

This demonstrates that ligand addition to an atom may increase the interaction energy with a reacting fragment. The ligands promote electrons in the complex for optimal bonding. Vacant hybridized orbitals become directed towards empty ligand position(s).

As will be shown later, similar effects also occur on surfaces, explaining partly the changed reactivity of surface atoms compared to free atoms [33].

Low and Goddard [66, 67] also studied the interaction of H_2, methane, and ethane with Pd and the interaction of H_2 with the $Pt(PH_3)_2$ complex using ab-initio methods.

Representative results for Pd and Pt are presented in Figs. 22 and 23 resp. Starting with dissociated H_2, CH_4, and C_2H_6 on Pd (not stable according to Blomberg et al. [64], they calculate the activation energies for dissociation. They find an increasing barrier for reductive elimination moving from H_2 to CH_4 to C_2H_6.

The orbital density plots show the hybridized nature of the bonding Pd atom orbitals. Reaction of H_2 with the $Pt(PH_3)_2$ complex is very similar as that with $Pd(H_2O)_2$. As in the case of $Pd(H_2O)_2$ the free $Pt(PH_3)_2$ molecule is linear. Whereas the groundstate of Pt is $5d^96s^1$, the Pt d orbital occupation in $Pt(PH_3)_2$ is $5d^{10}$. This is because the doubly occupied phosphine σ orbitals form bonding as well as antibonding orbitals with sp hybrids. The four phosphine electrons are accomodated in the bonding orbitals, the single s electron is placed in a low d orbital rather than in an antibonding ligand-metal (sp) orbital. An alternative way of viewing this process is that promotion of the 6s electron into the 5d orbital of Pt reduces the repulsive interaction between the two doubly occupied phosphine orbitals and the Pt s orbital. Oxidative addition of H_2 decreases the angle between the phosphine groups to $100°$, enabling hybridization of the Pt 6s and $5d_{xy}$ orbital with a resulting angle between the PtH bonds of $80°$. This is very similar to bonding to the bent $Pd(H_2O)_2$ complex. The activation energy to dissociation is 2.34 kcal/mol, the dissociation energy of 18 kcal/mol is exothermic. Noell and Hay report similar but quantitatively different results [70]. The same holds for the work by Kitaura et al. [69]. (cf., the comparison made by Dedieu [71]).

Again we observe the favorable effect of surrounding the reacting metal atom with ligands. Now, the rotated PH_3 groups prepromote the electrons in Pt so as to give hybridized orbitals of favorable orientation.

Bonding with a Pt atom complex exceeds that with Pd because of the spatial extension of the Pt orbitals. Compare the respective bond energies of the hydrides (PtH = 83 kcal/mol, PdH = 76 kcal/mol, NiH = 60 kcal/mol) [69].

Figure 23 shows the computed generalized valence-bond (GVB) orbitals for the $H_2Pt(PH_3)_2$ complex. Figure 23b shows the GVB orbitals in PtH. Note that the 6p orbitals do not participate in the PtH bond, which can be considered as a linear combination of a hybridized $Pt(5d_z 2, 6s)$ bond and H(s) orbital, but they do participate in the Pt–H bonds in $H_2Pt(PH_3)_2$. The picture that derives for bonding to the last members of the transition-metal series in the periodic system is very clear. Upon dissociation of H_2, localized bonds are formed between two (s, d_{xy}) metal orbitals

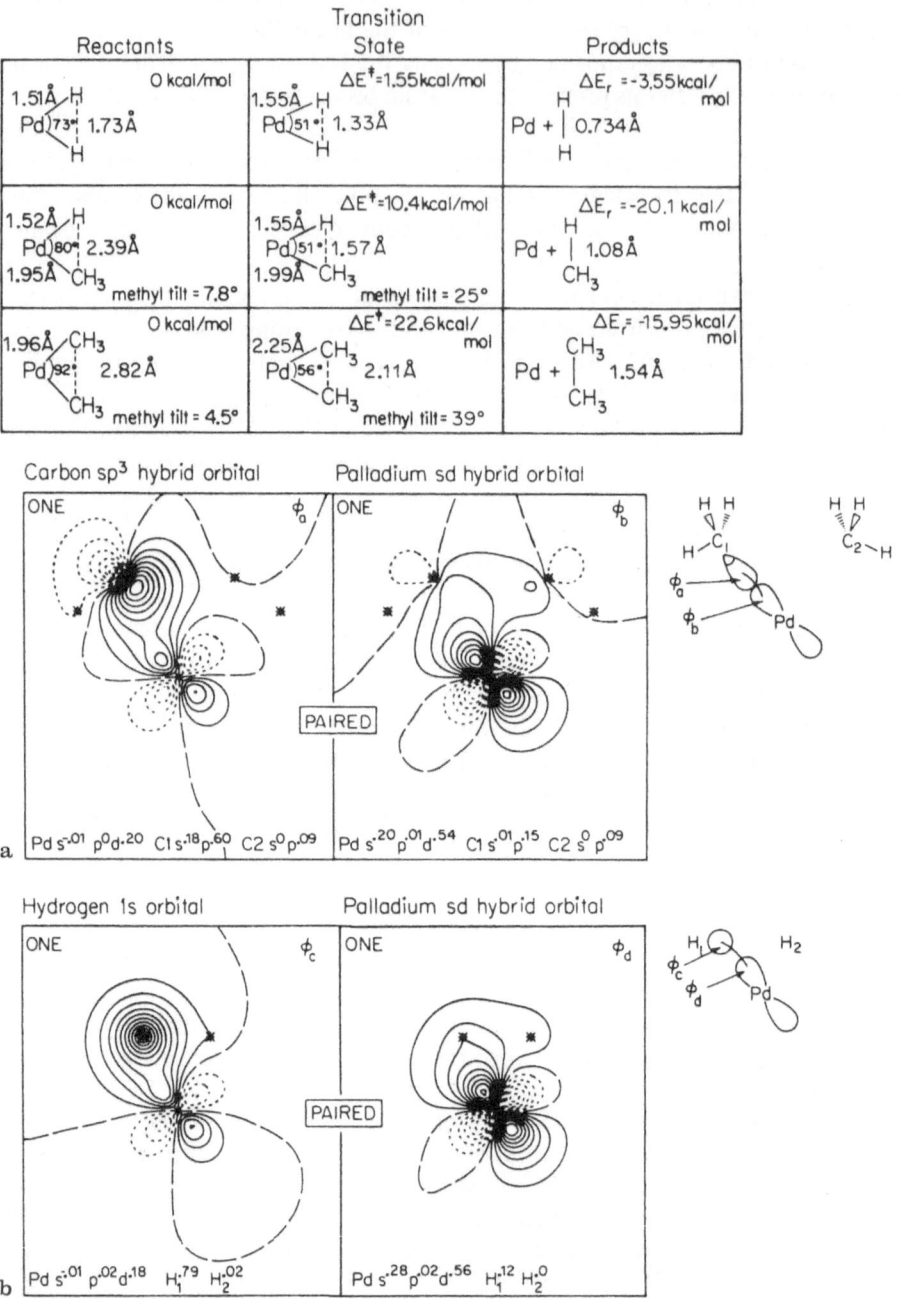

Fig. 22a. Geometries and energetics for the reactions $PdH_2 \rightarrow Pd + H_2$, $PdH(CH_3) \rightarrow Pd + CH_4$, and $Pd(CH_3)_2 \rightarrow Pd + C_2H_6$. The angle between the Pd—C bond and the vector from the C atom to the center of mass of the methyl hydrogen atoms is defined to be the methyl tilt [67]; **b** GVB orbital for the Pd—C bonds at the transition state for the reaction $Pd(CH_3)_2 \rightarrow Pd + C_2H_6$ and the GVB orbitals for the PdH bond for the reaction $PdH_2 \rightarrow Pd + H_2$. The Mulliken populations are listed with each orbital to show the hybridization of each orbital [67]

and the hydrogen atom orbitals, resulting in orbital energies low compared to the d-atomic orbitals (Fig. 24b).

The theoretical angle between the M-H bonds is 90° and two electrons reside in each of the bonding M-H orbitals. The bondstrength of a M-H bond is large, between 80 and 60 kcal/H atom and decreases from Pt to Pd to Ni. The metal atom part not involved in bonding is in a M^{2+} state, in line with the notion of oxidative addition

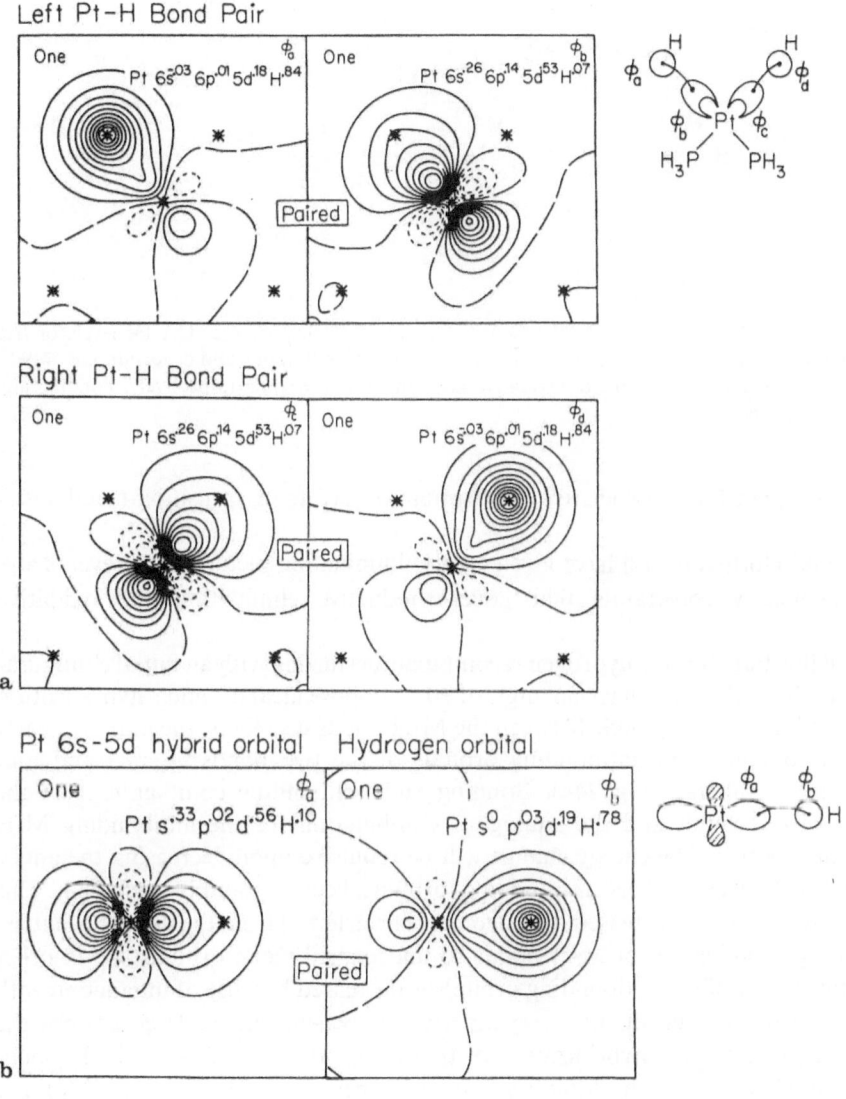

Fig. 23a. GVB orbitals for the Pt—H bonds of H_2 $Pt(CH_3)_2$ at equilibrium. Hybridization for each singly occupied GVB orbital is shown on each plot. Each contour represents a change of 0.05 in amplitude. Solid lines represent positive amplitude. Asteriks represent positions of atoms [66]; **b** GVB orbitals for diatomic PtH [66]

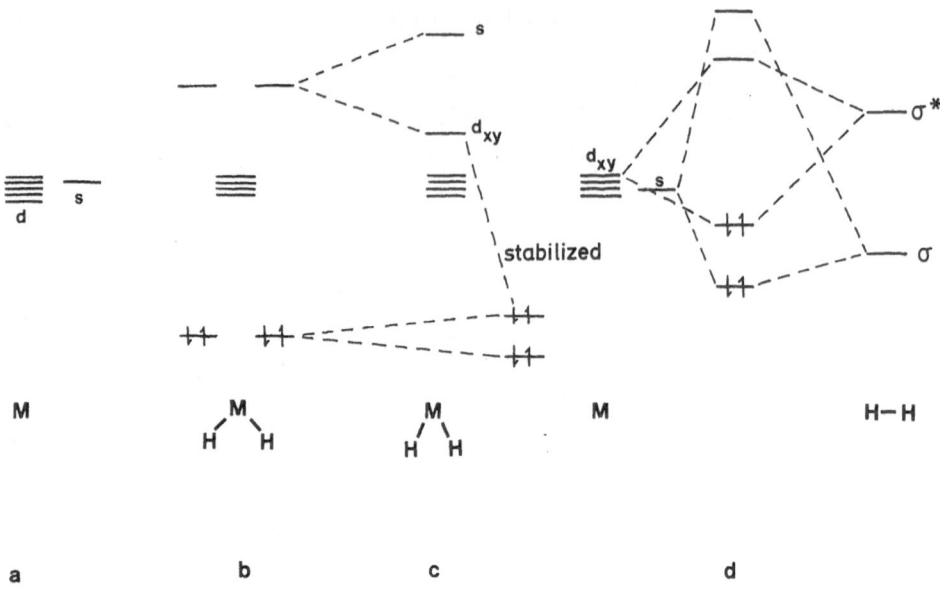

Fig. 24a–d. Schematic electronic orbital scheme of reductive elimination. **a.** Orbital levels of free metal atom. **b.** Dissociated H_2 bonded to a metal atom (hybridization of s and d_{xy} orbital), $\varphi = 90°$. **c.** Reductive elimination; $\varphi < 90°$. **d.** Oxidative addition; σ and σ^* are, respectively, bonding and antibonding H_2 orbitals

commonly applied to dissociative adsorption of H_2 to a transition-metal atom complex.

Berke and Hoffmann [72] have given a very illuminating picture of the events and orbital symmetry constraints that govern reductive elimination and oxidative addition.

We will illustrate this for hydrogen recombination starting with an initial configuration where the M-H bonds have an angle of 90°. Suppose ideal d_{xy} and s hybridization on the metal atom. If the angle between the M-H bonds decreases, the initially nearly degenerate bonding and antibonding orbitals of the two bonds interact and four new orbitals result (see Fig. 24c). Bonding and antibonding combinations of the bonding M-H orbitals and the analogous combination for the antibonding M-H orbitals are formed. The energy change will be repulsive upon decreasing the angle, because bonding as well as antibonding orbitals become doubly occupied. This repulsive interaction is decreased if electrons are transferred from the doubly occupied antibonding combination of M-H bonds to unoccupied metal orbitals. Or in other words, the energy of the antibonding orbitals is decreased because of interaction with an unoccupied d orbital of proper symmetry. Decreasing the angle ϕ between the M-H bonds changes the hybridization of the metal orbital part of the M-H bond. So at an angle $\phi < 90°$, the orbital scheme becomes as sketched in Fig. (24c). An empty d orbital of d_{xy} symmetry is required on the metal atom. In summary apart from this orbital three orbitals are of importance in the recombination of the hydrogen atoms: a symmetric low-lying orbital leading to the bonding H_2 σ orbital, an antisymmetric doubly occupied orbital leading to the σ^* orbital of H_2, and an s-type metal orbital that is pushed away to high energy. The antisymmetric orbital is initially occupied and results in repulsion with decreasing angle ϕ. The energy of the anti-symmetric orbital is decreased if it can interact with a suitable d-orbital of the same symmetry. Whether such an orbital is empty and energetically available usually depends on the ligand configuration around the metal atom and the metal atom electron occupation. In our case of a free metal atom, such an orbital is provided by the d_{xy} orbital. As a result, when the hydrogen atom distance decreases, more electrons are donated in the empty d-orbital and at infinite distance from each other, a neutral metal atom and H_2 molecular appear (reductive elimination).

The increased activation energies found for the reductive elimination of methane and ethane derive from additional repulsive interactions between doubly occupied bonding C-H bonds with the M-H bond or the C-H bond of the molecule fragment. Such repulsive effects have also been described by Zheng, Apelaig and Hoffmann [73] between methyl hydrogen atoms and surface metal atoms.

3.2 Reaction of H_2 with Transition Metal Clusters

A recent ab-initio study of H_2 dissociation on Pd_2 [68] and an older study on the reaction of H_2 on Ni exist [74]. Addition of H_2 in a symmetrical way parallel to the metal dimer is considered. Compared to bonding to a metal atom, the main additional feature of a dimer is that antibonding combinations of occupied s or d orbitals on the different atoms, if occupied by electrons, provide additional possibilities for inter-action with unoccupied orbitals of π symmetry and that dissociated atoms can bind to different metal atoms.

Nakatsuji et al. [68] report a complexation energy of -12 kcal/mol and an H-H distance of 0.9 Å and a Pd-Pd distance equal to 2.8 Å for H_2 to Pd_2 using the CAS-MS-SCF method. An adsorption energy of -15 kcal/mol for H_2 molecule complexation to a Pd atom and a PdH bond strength of 54 kcal/mol is calculated by the same method. Upon dissociation, an activation energy of 3.4 kcal/mol and

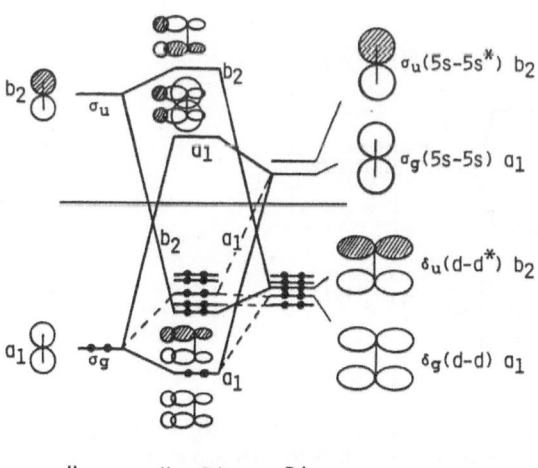

Fig. 25. Schematic correlation diagram for the interaction of H_2 and Pd_2 [68]

an additional energy gain of -2.2 kcal/mol is found, implying a PdH bond strength of 59 kcal/gat H. The H-H distance after dissociation becomes 2.1 Å. According to this calculation, first a precursor state of molecular H_2 is formed that consecutively converts to dissociated hydrogen, with each hydrogen atom attached to a different Pd atom. Bonding in the Pd_2H_2 molecule is schematically sketched in Fig. 25. In the Pd_2 molecule, the bonding as well as antibonding s orbitals are unoccupied and the symmetric as well as antisymmetric d-orbital combinations are occupied (implying weak bonding in the Pd_2 molecule, $E_{diss}^{exp} = 17$ kcal/mol).

According to HOMO-LUMO considerations, addition of H_2 to Pd_2 is clearly symmetry-allowed. As a result, exothermic dissociation of H_2 on Pd_2 becomes possible, whereas on an isolated Pd atom the dissociation reaction does not occur.

Bonding of Pd atoms in a dimer has decreased the promotion energy from 4d to 5s which is the cause of the strong repulsion of H_2 with the Pd atom.

Whereas no calculation of dissociation of H_2 on Cu_2 is available, clearly dissociation as well as addition of the hydrogen molecule becomes symmetry-forbidden. The a_1 orbital that consists of the symmetrical combination of metal s orbitals, empty in Pd_2, becomes occupied in Cu_2.

We will return to the reactivity of Cu_2 in a discussion of the reactivity of Co. Melius et al. [74] studied dissociation of H_2 parallel to the Ni_2 dimer.

The major difference between Ni_2 and Pd_2 is that in Ni_2 the d valence orbitals remain partly empty. Therefore, in Ni_2 the bonding orbital formed from the two

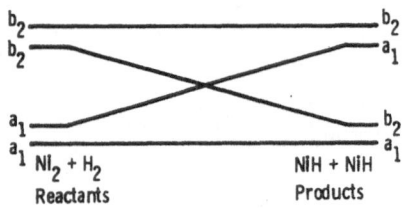

a   Orbital Correlation Diagram

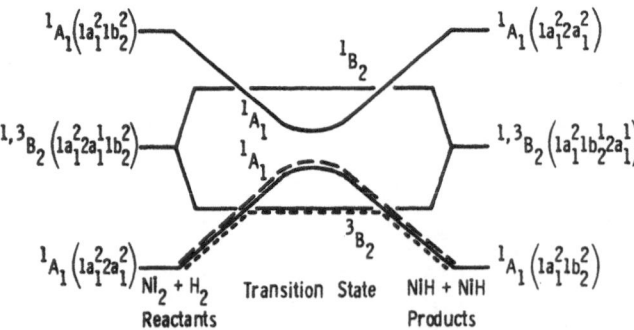

b   State Correlation Diagram

Fig. 26a and b. Orbital and state correlation diagrams for the reaction $H_2 + Ni_2 \rightarrow 2\,NiH$. The long-dashed curve indicates the reaction path for which the valence symmetry is conserved. The short-dashed curve indicates the reaction path for which the valence symmetry changes ($^1A_1 \rightarrow {}^3B_2 \rightarrow {}^1A_1$) [74]

Ni 4s atomic orbitals is occupied. Ignoring the interaction with the Ni d orbitals, the dissociation reaction of H_2 to Ni_2 would be clearly Woodward-Hoffmann-forbidden and a large activation energy for dissociation is expected. The orbital and state correlation diagrams for the dissociation reaction is given in Fig. 26. Total energies are given in Table 3. As observed, the calculated activation energy is 21 kcal/mol, which is comparable to that found for the Ni atom.

Table 3. Ni_2H_2 energies [74]

Method	State	R_1 (0.74 A)	R_2 (1.06 A)	R_3 (2.49 A)
HF (1 cf)	$^1A_1$	0.0 [a]	1.57	0.15
HF (1 cf)	$^3B_2$	−0.73	0.03	1.80
MCSCF (12 cf)	$^1A_1$	0.0 [a]	0.96	0.18
MCSCF (8 cf)	$^3B_2$	0.09	0.69	2.49

[a] Reference point energy has been set equal to zero. The total energy of the $^1A_1$ state at R_1 is −39.33 eV (HF) and −40.66 eV (MCSCF).

Siegbahn et al. [75] studied H_2 dissociation on a Ni_{13} cluster simulating the Ni(100) face using similar techniques as applied to the NiH_2 systems discussed earlier. However, to make the computational problem tractable, in some cases modified effective core potentials (MEP) had to be used excluding 3d orbitals to take part in the dissociation process. Dissociation atop of a Ni atom was, however, calculated without restricting the degree of freedom corresponding to that of the Ni 3d atom electrons. Dissociation in a bridging configuration again could only be done with frozen 3d electrons.

Table 4a shows for atop dissociation that in order to lower the activation energy for dissociation, covalent interaction with the transition-metal d electrons is a necessity. This indicates the importance of the metal to adsorbate backdonation interaction involving the unoccupied H_2 σ^* level. Since s orbitals are totally symmetric, only interaction with d orbitals can provide this stabilizing interaction. If d electrons are not allowed to interact, an activation energy of 48 kcal/mol is calculated. The interaction with the d electrons lowers the activation energy for dissociation to 4–5 kcal/mol, which compares with 15 kcal/mol for the isolated Ni atom. Whereas no complete dissociation occurs on a Ni atom and H_2 addition is thermodynamically neutral, on the Ni cluster H_2 dissociates exothermally with a dissociation energy of 2 kcal/mole.

Note that in contrast to the activation energy, the interaction with the d electrons is not essential to compute proper values for the Ni–H interaction, resulting in Ni–H bonds of 53 kcal/gat, which is close to the experimental value.

This agrees with our conclusions [37] based on an analysis of experimental data, and those of Upton and Goddard [76] to be discussed later.

MEP calculation with frozen d orbitals results in an activation energy of 28.5 kcal/mol for H_2 dissociation in a bridgeing configuration (Table 4b). This compares with 48 kcal/mol for atop dissociation using the same frozen d-orbital approximation.

The difference derives because in the bridge site of the Ni_{13} cluster backdonation of electrons into the antibonding σ^* H_2 level becomes possible by interaction with a populated antisymmetric combination of 4s Ni atom orbitals. In the bridgeing configuration, antisymmetric 4s Ni atomic orbital combinations become populated.

Table 4a. Reaction energies (kcal/mol) and geometries for on-top dissociation of H_2 on Ni_{13} [75]

	Energetics			
	Transition state		Adsorbed	
	SCF	CI[a]	SCF	CI[a]
12 MEP	66.4	4.4(17.0)	29.2	−2.0(10.5)
All MEP	57.3	48.0(50.3)	−4.7	−3.8(−4.5)

	Geometries					
	Transition state					
	R_1		R_{H-H}		R_{Ni-H}	
	SCF	CI	SCF	CI	SCF	CI
12 MEP	2.66	2.51	1.95	2.37	2.83	2.78
All MEP	2.53	2.53	2.47	2.55	2.82	2.83
	Adsorbed					
	R_1		R_{H-H}		R_{Ni-H}	
	SCF	CI	SCF	CI	SCF	CI
12 MEP	2.04	2.07	4.51	3.89	3.04	2.84
All MEP	1.91	1.92	4.87	4.88	3.10	3.10

[a] Values within parentheses do not contain Davidson's correction

Table 4b. Reaction energies and geometries of the bridge site dissociation of H_2 toward the on-top site on Ni_{14} [75]

	Energetic			
	Transition state		Adsorbed	
	SCF	CI[a]	SCF	CI[a]
All MEP	4.10	28.5(31.6)	−32.8	−33.8(−33.8)

	Geometries					
	Transition state					
	R_1		R_{H-H}		R_{Ni-H}	
	SCF	CI	SCF	CI	SCF	CI
All MEP	2.77	2.79	2.70	2.79	2.95	2.95
	Adsorbed					
	R_1		R_{H-H}[b]		R_{Ni-H}	
	SCF	CI	SCF	CI	SCF	CI
All MEP	1.92	1.92	4.71	4.71	3.10	3.10

[a] Values within parentheses do not contain Davidson's correction
[b] The R_{H-H} was not optimized; the H's were positioned at the bridge site

These are empty for the Ni_2 molecule, but in the Ni_{13} cluster they become partially occupied because of the interaction with the s orbitals of the surrounding Ni atoms.

The dissociation energy for H_2 dissociation resulting in two H atoms sharing the same Ni atom is -2 kcal/mol. If no Ni atoms are shared, the dissociated energy increases significantly to -34 kcal/mol, because now the unfavorable interaction between two Ni–H bonds on the same Ni atom is absent.

Upton and Goddard [76] studied hydrogen-atom adsorption to Ni_{20} and Ni_{28} atom clusters, also in the MEP approximation for the Ni 3d electrons.

We will discuss their results for Ni_{20} clusters.

These results are very relevant also to our discussion of the interaction of adsorbing molecules to semi-infinite metal lattices. Table 5 shows bond strengths and frequencies as a function of hydrogen-atom coordination. One observes increases in bond strength with hydrogen coordination number. Secondly, bonding is strongest to those nickel atoms that have the largest number of nickel metal atom neighbors. The last result is unexpected. Since delocalization of electrons is expected to increase with increasing number of metal atom neighbors, one would expect a decrease in bond strength of an hydrogen atom bonded to a Ni atom if the number of Ni atom neighbors increases [33]. The very different result found derives from the strong asymmetry of the LDOS of the electron energy density in the face-centered cubic lattices discussed earlier (see Fig. 19) and depends strongly on band filling. At intermediate electron-band filling, increased delocalization dominates, but at the electron-band edges this inverts. This behavior can be completely understood from considerations based on the group orbital density of states at the Fermi level [33, 51]. Figure 27 shows the computed spectrum of Ni_{20} and $N_{20}H$ molecular orbitals. Note that hydrogen-type orbitals systematically shift to lower energy with increasing hydrogen-atom coordination number. These orbitals clearly are of the bonding type. Antibonding orbitals are

Table 5. Bond parameters for H binding sites [76]

| Site | Description | | Bond length (Å) | | Vibrational frequency (meV) | Chemisorption energy (eV) |
	Surface	Ligancy of H[a]	$R_\perp$[b]	R_e[c]		
B	⟨001⟩	1(7)	1.50	1.50	283	1.56
F	⟨110⟩	1(5)	1.49	1.49	280	1.43
C	⟨112⟩	1(7)	1.49	1.49	(231)	(1.00)
A	⟨001⟩	2(7)	0.99	1.59	177	2.73
G	⟨001⟩	2(5)	0.99	1.59	(173)	(2.17)
E	⟨110⟩	2(6)	0.93	1.55	(161)	(1.56)
H	⟨112⟩	2(5)	0.96	1.57	176	2.43
I	⟨111⟩	3(5, 5) hcp	0.78	1.63	155	3.21
D	⟨111⟩	3(6, 7) fcc	0.79	1.64	(131)	(2.12)
J	⟨001⟩	4(7, 5)	0.30	1.78	73	3.04

[a] In parentheses is the number of nearest neighbors for the Ni atoms(s) binding site. Where nonequivalent surface atoms are present, values are given for each type
[b] Optimum distance from H to the plane representing the surface
[c] Distance from H to nearest neighbor Ni atoms

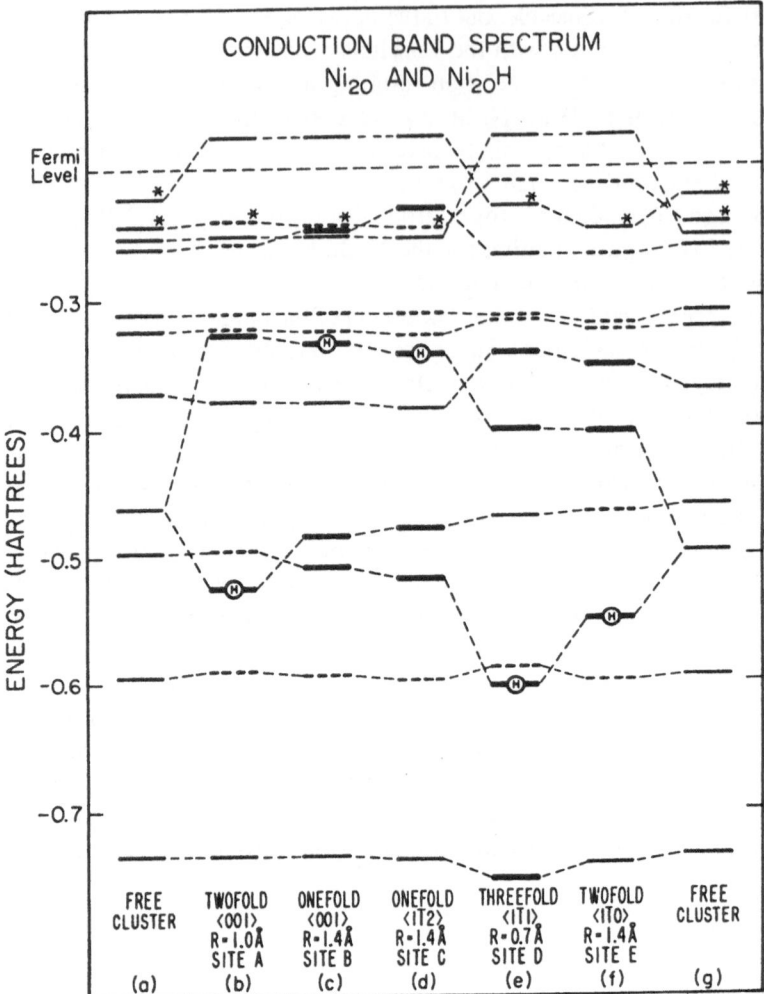

Fig. 27. Spectrum of states for Ni_{20} and $Ni_{20}H$ clusters. Levels connected by light dashed lines are of similar orbital character. Heavy lines indicate levels interacting significantly with the H atom. The H on a level indicates the orbital with maximum H character. Dashed levels are for orbitals unable to interact with the H atom by symmetry. Asteriks indicate simply occupied orbital levels. The d levels are spatially localized and have been replaced by an effective potential. The position of the Fermi level is approximate and is used primarily to distinguish between occupied and unoccupied levels [76]

shifted above the Fermi level (and become empty). According to Upton and Goddard, the bond strengths calculated correlate with the coefficients of the interacting atomic orbitals of molecular orbitals close to the Fermi level. In other words, there appears to be a relation with the local density of states of the group orbital at the Fermi level, as suggested by Eq. (28).

In summary, symmetry considerations based on frontier orbital theory enable a good understanding of H_2 dissociation on transition-metal clusters. For a single atom, electron promotion energy and hybridization are important variables. In diatomics, the relative position of d versus s electrons determines whether the bonding symmetric s orbital built from s-atomic orbitals is occupied. Occupation of this orbital usually implies large repulsive effects. These can be decreased by promotion to empty d orbitals. In the molecule, this promotion energy is usually less than that of the free atom. The dissociation energy of H_2 and the activation energy for H_2 dissociation is also less than that of the free atom, because now the dissociated atoms can bond to different atoms. Dependent on whether the σ orbital is occupied or not, increasing the cluster size will change the repulsive interaction originating from the interaction with this orbital. In addition, antibonding σ^* surface orbitals become populated, which favors metal electron backdonation into antibonding empty adsorbate orbitals.

The ionization energies of the metal clusters may decrease sharply for metals with a filled d band (e.g., Cu) enhancing backdonation further. This will be discussed more extensively in the following section. If the particle size increases further, delocalization increases and the interaction with adsorbates tends to decrease.

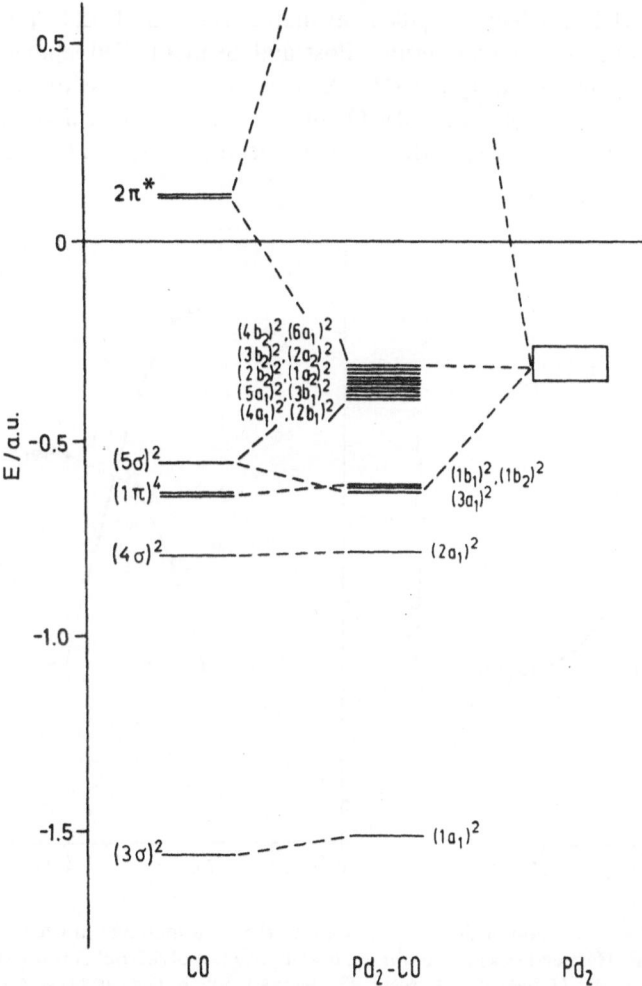

Fig. 28. Schematic MO diagram for the $^1A_1$ ground state of Pd_2—CO. The notation of the MOs in Pd_2—CO concerns the valence MOs only [77]

3.3 Cluster Chemisorption Models of CO Adsorption

Pacchioni and Koutecky [77] studied the interaction of CO with several Pd clusters using multireference doubly excited configuration interaction (MRD CI) procedures.

The schematic interaction scheme presented in Fig. 28 is very similar to that derived for the interaction with H_2 discussed earlier. The doubly occupied 5σ CO orbital interacts strongly with the empty symmetric σ orbital of Pd_2 consisting mainly of the Pd 5s atomic orbitals. This is a stabilizing interaction and electrons are donated to the metal.

The $2\pi^*$ level interacts with the occupied antisymmetric Pd atomic d orbital combination, resulting in a shift upwards for the CO $2\pi^*$ levels and a bonding stabilization of the Pd d-orbitals.

The bond strength increases from 6 kcal/mol for Pd-CO, and 13.2 kcal/mol for Pd_2CO, to 17.1 kcal/mol for Pd_3CO. So it increases with the number of metal neighbors of CO.

A recent study by Andzelm and Salahub [78] is available using the Local Spin Density method, with qualitatively similar results. Post and Baerends [79] studied chemisorption of CO to Cu clusters using the HFS-X_α method. Of interest to our discussion are their results for the interaction of CO with Cu_2. In their calculation, CO approaches Cu_2 in a symmetrical way with its axis perpendicular to the Cu-Cu

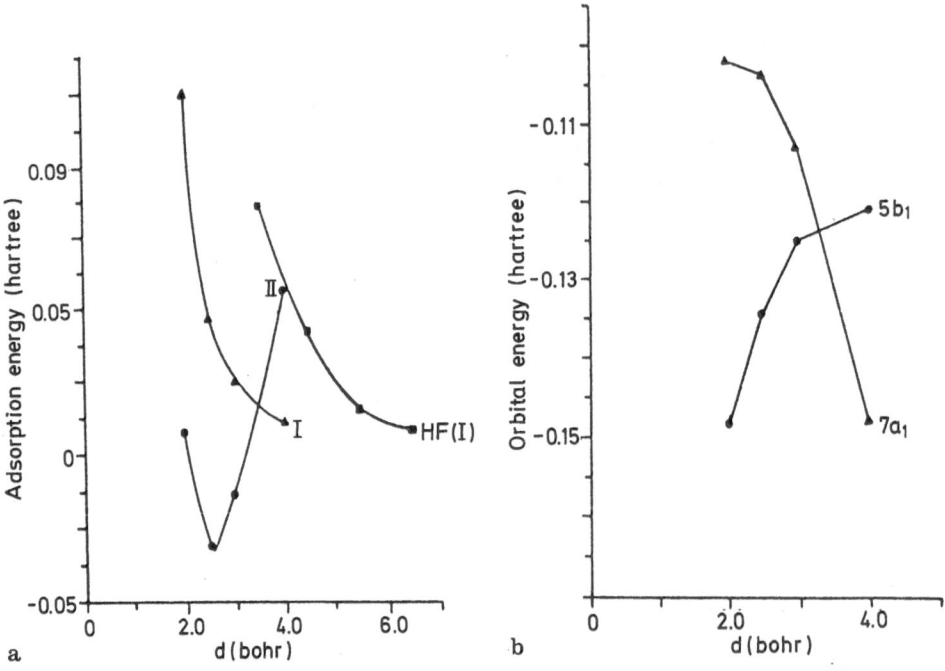

Fig. 29a. "Adsorption energy" as a function of the CO height above the band midpoint in Cu_2 for configurations I and II. The HF (Hartree-Fock) curve (for a Cu—Cu distance of 4.2 Bohr) is taken from: Kaleveld EW (1981) PhD thesis, University of Utrecht, The Netherlands; **b.** The Cu_2 (4s + 4s), CO $5\sigma(7a_1)$ and CO $2\pi^*$ ($5b_1$) one-electron energies as a function of CO height above the bridged position in Cu_2 [79]

axis. Since the Cu atom has one valence electron more than the Pd atom, the symmetric molecular orbital of Cu_2 built from the Cu 4s atomic orbitals is doubly occupied.

The HOMO of CO is its 5σ orbital. This orbital is also doubly occupied resulting in repulsive interaction with the Cu_2 σ orbital. This repulsion is not compensated by the electron backdonation interaction with the unoccupied CO $2\pi^*$ level (curve I, Fig. 29). Promotion of electrons from the Cu_2 $(4s+4s)\sigma$ to the empty Cu_2 $(4s-4s)\sigma^*$ molecular orbital, changes the repulsive interaction curve into an attractive one (curve II, Fig. 29a). A HOMO of σ^* symmetry becomes available for backdonation into the CO $2\pi^*$ orbital, and an unoccupied LUMO of σ symmetry becomes available to bonding with the doubly occupied 5σ CO orbital.

The most important effect of the configuration change from $(4s+4s)$ doubly occupied to $(4s-4s)$ doubly occupied is the disappearance of the 4-electron de-stabilizing interaction (exchange or Pauli repulsion) of the 5σ (carbon lone pair) orbital of CO with the $(4s+4s)$ metal orbital.

In particular, in large clusters such configuration changes almost always occur. The reason is that for an adsorbate like CO, with a pronounced lone pair orbital pointing toward the metal, the Pauli repulsion is large because of the large overlap of the lone pair orbital with metal orbitals. This point has also been stressed by Bagus et al. [30].

In a simple one-electron picture, the Pauli repulsion is due to the occupation of the destabilized antibonding orbitals (cf. Sect. 2).

In chemisorption to a metal, however, these antibonding orbitals will not remain occupied but they will become unoccupied as soon as they are "pushed" above the Fermi level. This corresponds to the configuration change in the metal cluster referred to above.

The implication is that the steric repulsion (caused by Pauli repulsion) calculated using the "frozen configuration" wave function Ψ^0, introduced in Sect. 2a, is often relieved in the same way as in Cu_2, by an electron transfer out of a repulsive σ orbital into a π orbital that can backdonate into the π^* of CO. The σ orbital that caused the repulsion with the CO 5σ due to a large overlap, will now for the same reason be a good acceptor orbital.

Therefore, large steric repulsion (or frozen orbital repulsion) does not imply that there will be no σ donation. It does, however, require a configuration change to make the σ donation possible.

We may clarify the situation by referring back to Fig. 5, which may serve to represent a number of interacting orbitals of σ symmetry. The Pauli repulsion corresponds to the destabilization of ψ_B, a typical substrate level close to the Fermi surface, and of χ_B, representing a manifold of levels well below E_F. As discussed in Sect. 2, the overlap region between φ_A and ψ_B and χ_B will be depleted of electron charge in Ψ^0, with a concomitant lowering of the electrostatic energy and rise in kinetic energy.

Subsequently, there is the energy lowering from this situation to the converged wave function Ψ_{conv}, which arises from the orbital interactions of σ symmetry depicted in Fig. 5 and which may therefore be called σ bonding.

It consists of two types of electron rearrangement.

In the first place, the virtual orbitals will mix into the destabilized $\psi_{B'}$ and $\chi_{B''}$. This results in the more stable $\psi_{B''}$ and $\chi_{B'''}$. In terms of electron rearrangements, this step consists of polarization and charge transfer.

In the second place, there is the depopulation of the $\psi_{B''}$ level if it remains above the Fermi level, in spite of the stabilization by the virtuals. It is, as often, somewhat hard to distinguish charge transfer and polarization in this last effect.

If B represents the metal (cluster), $\psi_{B''}$ will have predominantly metal character as will the level at the Fermi surface receiving the electrons. This means that this is mainly a rearrangement of electrons within the cluster, i.e., polarization of the metal.

As discussed earlier, one has to remember that the polarization discussed here is due to non-orthogonality of fragment orbitals and not due to the presence of an electrostatic field.

Still, the occurrence of polarization is at the same time an indication of some charge transfer. This is most clearly seen (see Ref [26]) when a configuration change is first achieved in the bare cluster so that ψ_B becomes empty, corresponding to the cluster polarization, and next the charge-transfer (HOMO-LUMO) type of interaction between φ_A and ψ_B is allowed to occur (σ donation).

A very interesting application and verification of these effects is provided in the work of Raatz and Salahub [80]. These authors studied the change in the magnetism of Ni upon chemisorption of CO. In agreement with the decrease of the saturation magnetization of transition-metal particles upon chemisorption found experimentally, they observe a decrease of the magnetic moments of the Ni atoms in their clusters near to the adsorption site.

The explanation is provided by the depletion of antibonding levels as discussed above. Part of the $d\sigma$ density of states is pushed above the Fermi level through anti-bonding interaction with CO (cf. $\psi_{B'}$, resp., $\psi_{B''}$). This triggers a highly spin-dependent rearrangement of the d electrons. A 'hole' is created in the $d\sigma$ manifold which is compensated by an increased d density of states in other symmetries. As there are no majority-spin holes, the increase is in the minority-spin levels, leading to a reduction of the total net up-spin density. The authors noted that this magnetic effect is intimately related to the Ni–CO bonding, in the sense that the emptying of the antibonding levels relieves the Pauli repulsion which would otherwise result.

The effects discussed here are fairly large and will be found to occur at the SCF level in any reasonably accurate electronic structure method. It has, for instance, been shown by Baerends and Rozendaal [25] in a detailed study on $Cr(CO)_6$ that the Hartree-Fock and X_α models yield completely analogous pictures of the importance of the various contributions (steric repulsion, σ bonding, π backbonding) to metal-carbonyl bonding.

Overall bonding is somewhat stronger in X_α than in Hartree-Fock owing to a stronger π bonding.

On the other hand, Bagus et al. [30] noted, upon comparing their Hartree-Fock results for CO interacting with an $Al_5(5, 0)$ cluster to previous X_α results for the same system obtained by Post and Baerends [41], that there was a huge discrepancy of ~ 5 eV in the calculated adsorption energies.

Whereas the X_α calculation resulted in a bonding energy of -1.9 eV for CO atop of the central Al atom of the Al(5,0) cluster, at 3.7 Bohr distance, the Hartree-Fock calculation yielded an antibonding contribution by 2.5 eV at 3.5 Bohr, a difference of 4.4 eV.

Although X_α is known to give stronger bonding, this difference is too large to be attributed to a difference in the model (X_α or Hartree-Fock) used.

We have, therefore, repeated the Hartree-Fock calculations by Bagus et al. [30] using exactly the same Gaussian basis and the same geometry [81]. We do indeed reproduce the antibonding of 2.5 eV quoted in Ref [30], but only if we freeze in the Al_5–CO system the electron configuration of the Al_5 part at the $(3a_1)^2 (1b_1)^2 (2e)^3$ ground state configuration of the bare Al_5 cluster. However, it had been noted in the X_α calculation that the highest occupied a_1 orbital of the Al_5 cluster $(3a_1)$ has considerable $3p_z$ character on the central Al and, as a consequence, a sizable overlap with the CO 5σ orbital. As noted above for the $(4s + 4s)$ orbital of Cu_2, and as has been found in many clusters [26], also in this case depopulation of the $3a_1$ orbital was found to occur as it was shifted above the Fermi level by the antibonding interaction with the Co 5σ orbital.

We have verified that precisely the same effect occurs in the Hartree-Fock calculation: changing the configuration to $(3a_1)^0 (1b_1)^2 (2e)^4 (1b_2)^1$ lowers the energy of the Al_5–CO system by 3 eV. CO is, therefore, bound in Hartree-Fock to the Al_5 cluster by 0.5 eV. This is still considerably less than the 1.9 eV in the X_α calculation, but this difference is more in line with the usual difference between Hartree-Fock and X_α (or local density) calculations for CO interacting with metal clusters and for bond energies in general [83].

This example shows that it is incorrect to use unchanged cluster electronic configurations in chemisorption calculations. Most importantly, however, it shows that a gratifyingly similar picture of the importance of such effects as (relieve of) Pauli repulsion, configuration changes, polarization of the metal substrate, etc. emerge from such different electronic structure methods as Hartree-Fock and X_α (LSD).

Although the energy decompositions discussed so far provide a rather detailed understanding of the interaction of adsorbates with clusters, it remains an open question how accurately the cluster calculations mimick chemisorption. Among the many studies of the effect of cluster size, there are two on the Cu_n/CO system which used energy decomposition to consider the variation of the various energy contributions with cluster size and shape (Post and Baerends [26], and Hermann, Bagus, Nelin [84].

The conclusions of these two studies are virtually identical. There are a number of properties of the CO/Cu system which converge fairly rapidly with cluster size, such as distance to the surface, vibration frequency, bond lengthening of CO, decrease of CO vibration frequency. The notable exception is the bonding energy of the adsorbate to the metal cluster, which shows strong variation with cluster size (cf. also Ref [78]).

Also the relative energies of different adsorption sites on clusters are different from those found on semi-infinite lattices, the hollow site, for instance, being strongly preferred in calculations for CO on Cu clusters, whereas at surfaces, CO is found to adsorb atop.

Therefore, we conclude this section with a short discussion of the most important differences of the clusters compared with extended systems.

It has been noted [26, 84] that the chemisorption energy obtained with a small cluster depends on the "accidental" position of the important frontier orbitals of $\sigma(A_1)$ and $\pi(E)$ symmetry.

If, for instance, the a_1 frontier orbital is empty, or so little below E_Γ that it becomes unoccupied, the cluster is a good σ acceptor.

On the other hand, if an a_1 orbital that strongly overlaps the CO 5σ orbital (or the adsorbate lone-pair orbital in general) remains occupied, the cluster is not a good σ acceptor and there is strong Pauli repulsion.

It is clear from the results in Refs [26, 84] that, in particular, this repulsive term, which is large, varies considerably between the clusters. The positions of the crucial levels in the clusters have been termed "accidental" as they are determined, in small clusters, by the shape and size of the clusters.

The same arguments apply, mutatis mutandis, to the bonding contributions. Concentrating on the cluster orbitals originating from the metal s atomic orbitals, we note that the antisymmetric combinations important for π bonding will be occupied or unoccupied depending on the shape of the cluster and of course on the electron count of the metal. For instance, triangular or larger clusters of one-electron sp-metals (Cu, alkalis) populate such orbitals. As demonstrated in Ref [26], such clusters have a strong interaction with CO in high coordination sites, with significant back-donation into the $2\pi^*$ orbitals of CO.

In the next section it will be shown that these effects are much smaller on extended lattices because electrons become delocalized. The effective population of anti-symmetric "group orbitals" present in the cluster mentioned above, becomes much smaller in an extended lattice. The π interaction is therefore smaller on the extended substrate. The reverse may also happen. In metals that have too low an electron count to populate antisymmetric combinations of s atomic orbitals in small clusters (the d orbitals will be populated instead), emyedding in an extended metallic lattice will populate such antisymmetric group orbitals to some extent. We already discussed such a situation for the Ni_{13} cluster. In such cases, interaction with π-type orbitals will be enhanced in the extended lattice as compared to the small cluster.

As discussed in the theoretical intermezzo, repulsion due to inferaction of doubly occupied orbitals is proportional to the number of neighbours.

Calculations by Hermann et al. [85] confirm this for the interaction of NH_3 with an Al_{10} cluster, assuming $^3A_2$ symmetry (promotion absent). Whereas an overall bonding interaction is found for ammonia in the atop site, a strong repulsion is found in the threefold position (Fig. 30).

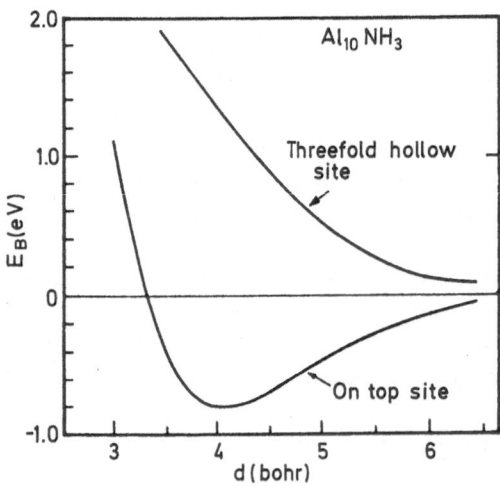

Fig. 30. Binding curves of the Al_{10} (7,3) NH_3 (on-top site), and the Al_{10} (3,7) NH_3 (threefold hollow site) clusters [85]

The more favorable backdonation into empty π-type orbitals of ammonium favoring the threefold position does not compensate for the strong repulsive effects experienced by the doubly occupied ammonia σ orbital in that position. This is understandable since the π acceptor orbitals of NH_3 are at very high energy. Even for CO, however, with low lying π acceptor levels, the topsite is preferred and the hollow site only weakly bonding [41].

4 Chemisorption to Metal Surfaces

General features of the chemical bond to a metal surface were discribed in Sect. 2.

Here, we will focus on those aspects of the surface chemical bond that sensitively depend on the distribution of the electrons over valence orbitals with a different angular distribution.

Studies of Hoffmann and Anderson [86, 87] as well as Baetzold [88] apply the extended Hückel method to the calculation of the electron distribution of molecules interacting with metal slabs.

We will use approximations to the extended Hückel method to derive a useful and convenient description of the local density of states of a metal. The methods developed in Sect. 2d are suitable to derive basic features of the surface chemical bond.

This is especially a suitable approach if one wishes to discuss the role of orbital symmetry in surface reactions.

Early work by Bond [89] and Weinberg and Merrill [90] used the orientation of d orbitals at a surface and Goodenough's [91] band theory to study the interaction of molecules and atoms with a surface.

A similar approach will be used here [37], but one based on more recent electron band models.

Trends will be discussed in the bond strengths and reactivities of CO and NO chemisorbed to different faces of group VIII transition metals with varying valence d electron count.

Of particular interest are the differences in coordination of CO to the surfaces of different metals.

At low surface coverage, CO adsorbs atop on the most dense faces of Pt [92], Rh [93], Co [94], and Cu [95], but it adsorbs in a bridge coordination on the (111) face of Ni [96].

We also intend to indicate the reasons for the differences in face dependence of the reactivity comparing different Pt faces to that of Rh.

Finally, the opposing trends in CO bond strength and H bond strength found for elements in the last row of group VIII metals will be explained [37].

The work-function dependence according to Eq. (28) results in similar effects of the electrostatic field on chemisorption as found from first-principle calculations [11], effective medium theory [10], or adapted extended Hückel theory [55].

Lowering the effective ionization potential enhances electron backdonation between metal and adsorbate. The authors have discussed this effect extensively elsewhere

[51]. Also earlier we discussed the use of Eq. (27) to interpret coadsorption effects [51].

The changes of the surface group-orbital local density of states at the Fermi level (see Sect. 2d for definition), computed in the presence of coadsorbates (e.g., S) [97] can be inserted into Eq. (28) to compute the resulting energy changes.

The following simplified model of the electronic structure of the transition-metal surface is very useful to discuss the elementary interactions playing a role in the formation of the surface-adsorbate chemical bond.

In the bulk, a f.c.c. metal atom has 12 nearest neighbors. As sketched in Fig. 31, the three d_{xy}, d_{yz}, d_{zx}, x orbitals each have 4 nearest neighbors which are not shared.

This leads to a symmetric electron density of states, threefold degenerate in the three perpendicular planes.

The d_{z^2} and $d_{x^2-y^2}$ orbitals have nearest neighbors at a $\sqrt{2}$ larger distance than the d_{xy}, d_{yz} and d_{zx} atomic orbitals. The d_{z^2} and $d_{x^2-y^2}$ orbitals will form a twofold degenerate electron band of much narrower bandwidth than the d_{xy}, d_{yz}, and d_{zx} orbitals.

Since the overlap between the s and p atomic orbitals is much larger than between the d atomic orbitals, they will form a broad band, usually overlapping the much smaller d valence-electron energy band.

This is sketched in Fig. 32 for the s and d bands.

For a f.c.c. crystal, the s valence-electron band is strongly asymmetric as discussed earlier [51], the d valence-electron bands are symmetric as long as only interactions between nearest neighbors are considered and the overlap matrix is assumed to be diagonal.

For group VIII metals, a good approximation gives 1 electron per atom in the s valence-electron band and varying electron number in the d valence-electron band.

Because the d valence-electron bands are nearly completely filled for the metals Ni, Pd, or Pt, one expects holes in the d valence-electron band to have considerable d_{xy}, d_{yz}, and d_{zx} and little d_{z^2} and $d_{x^2-y^2}$ character.

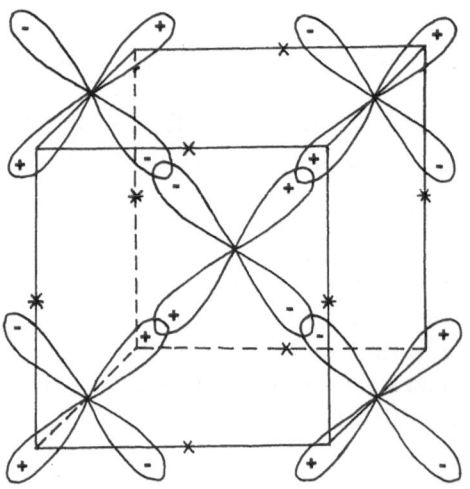

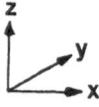

Fig. 31. d_{xy} orbitals in f.c.c. crystal

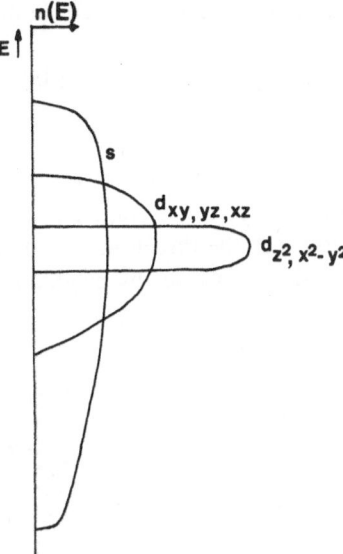

Fig. 32. Band structure (schematic) of f.c.c. metal [51]

The picture of the valence-electron structure presented is, of course, highly schematized; nonetheless, extended Hückel calculation on slabs indicate the general validity of the schematic electron distribution derived earlier [98]. In Fig. 33, 34, and 35, the respective changes at the (111), (100), and (110) surfaces of the d_{xy}, d_{yz}, and d_{zx} valence-electron bonds are sketched.

In the sense of the second-order perturbation theory expression for the energy, only those orbitals have to be considered, because for the group VIII metals in the last columns of the periodic system they are the only ones to have a finite density of states at E_F.

At the (111) surface, each of the d_{xy}, d_{yz}, and d_{zx} surface orbitals looses one neighbor. The resulting orbital configuration is sketched in Fig. 33a.

As pointed out by Kahn and Salem [99], the three degenerate dangling surface orbitals will rehyvridize according to the local symmetry of the surface atoms.

As a result, two degenerate asymmetric surface orbitals and one symmetric surface orbital is formed:

$$\phi_1^{as} = \frac{1}{\sqrt{2}} (\phi_{d_{xy}} - \phi_{d_{yz}}); \quad \phi_2^{as} = \frac{1}{\sqrt{6}} (\phi_{d_{xy}} + \phi_{d_{yz}} - 2\phi_{d_{zx}})$$

$$\phi_3^{s} = \frac{1}{\sqrt{3}} (\phi_{d_{xy}} + \phi_{d_{yz}} + \phi_{d_{zx}})$$

The original degeneracy is lifted by the presence of the metal surface and the resulting surface symmetry electron density of states is sketched in Fig. 33c (the figure is the result of a Bethe lattice approximation calculation). It is essential to consider this lifting of degeneracy and the creation of asymmetric orbital combinations, since some authors, e.g., Banholzer et al. [100], have erroneously ignored this. As a result, their relations between symmetry of surface orbitals and adsorbate are based on an incorrect use of surface orbital analysis.

As is seen in Fig. 33c, the maximum in LDOS of the asymmetric LDOS ($d\pi$) is at higher energy than that of the symmetric part ($d\sigma$). After hybridization, the surface orbitals can be represented as sketched in Fig. 33b.

Interaction between lobes originating from different surface atoms, but directed towards the missing atom, can also be considered and will lead to symmetric and asymmetric combination of atomic orbitals directed towards the threefold position.

At the (100) surface, the d_{yz} orbital in the (100) plane does not lose any neighbors.

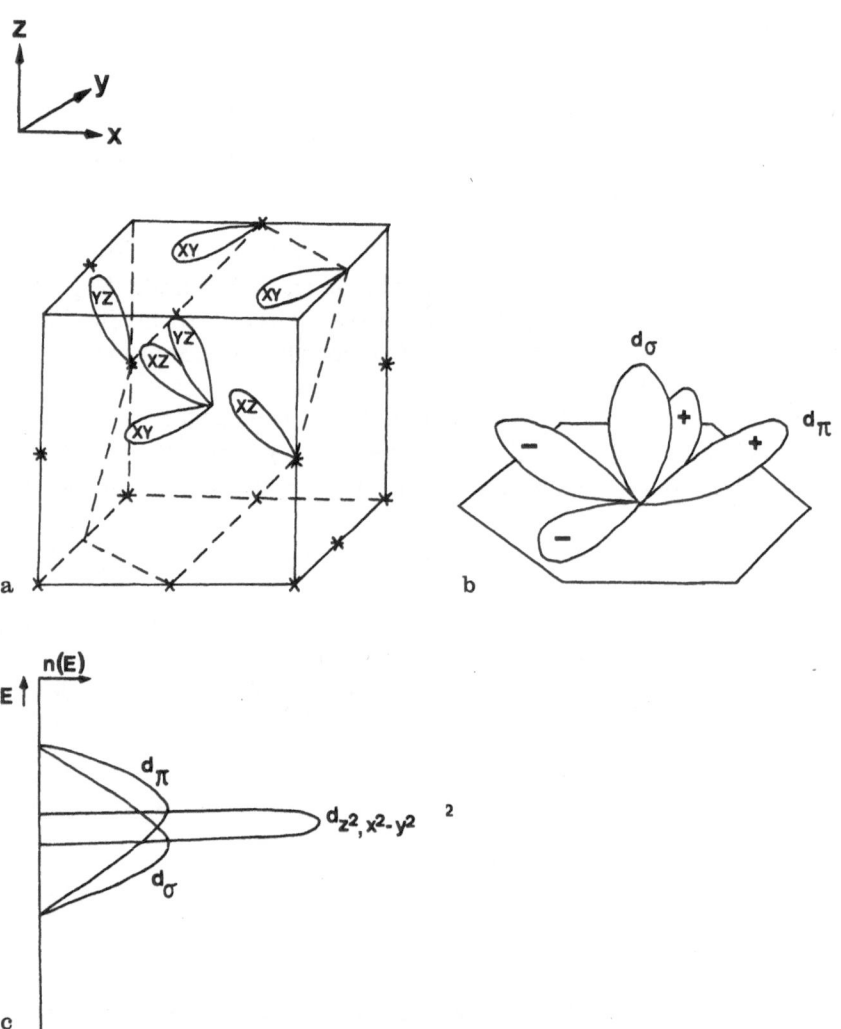

Fig. 33a–c. d-valence electron distribution at the (111) surface. **a.** The out-of-plane lobes of the degenerate d_{xy}, d_{yz}, and d_{zx} atomic orbitals [37]. **b.** The linear combinations of the plane lobes of the d_{xy}, d_{yz}, and d_{zx} atomic orbitals symmetry adapted to the (111) surface. **c.** Scheme of surface d-electron density of states

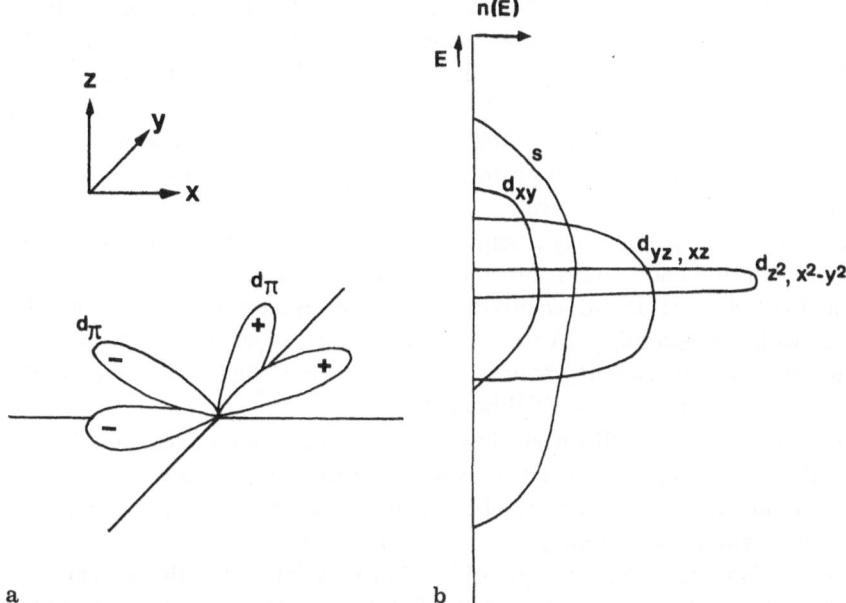

Fig. 34a, b. d-valence electron distribution at the (100) surface, **a.** d_{xz} and d_{yz} lobes. **b.** Schematic sketch of surface electron density of states [37]

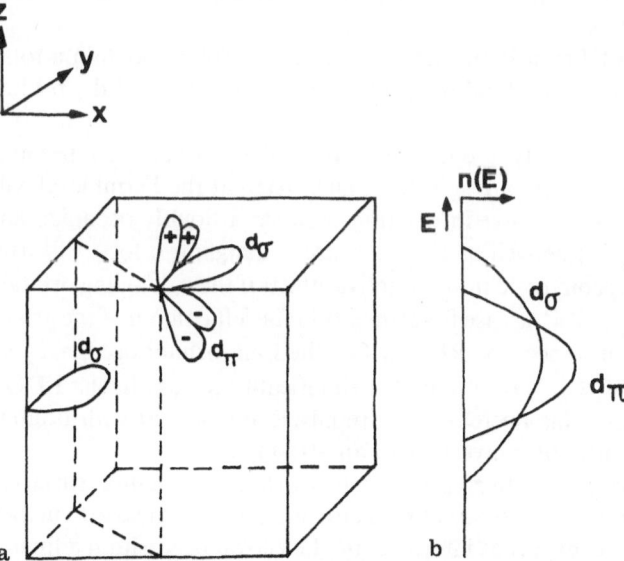

Fig. 35a, b. d-valence electron distribution at the (110) surface. **a.** d_{xy}, d_{yz}, and d_{zx} lobes. **b.** Schematic d-valence electrondensity of states

The d_{xy} and d_{zx} orbitals each lose two neighbors. As a result, the local density of states $\varrho_i(E)$ at the surface is split into two bands, a broad band corresponding to the d_{yz} orbitals and two more narrow bands d_{xy}, d_{zx}, with a higher electron occupation (see Fig. 34).

At the (110) surface, two different atoms are generated, atoms a and b (Fig. 35). Atom a loses one neighbor, resulting in a dangling orbital σ symmetric into the direction perpendicular to the surface.

The edge orbitals of t_{2g} symmetry are missing 5 neighboring orbitals. This generates one σ-dangling orbital and two π symmetry orbitals, as sketched in Fig. 35.

To summarize, at the (111) surface orbitals of σ as well as π symmetry are generated with a bandwith corresponding to three neighbors, instead of four as in the bulk At the (100) surface, only dangling bonds of π symmetry are generated, with a smaller bandwidth corresponding to two neighbors.

The edge atoms of the (110) surface have one σ and two π-type dangling bonds. The σ dangling bond has a broader bandwidth than the π-type bond.

Let us first discuss the consequences of this orbital scheme for the interaction with an orbital of σ symmetry (H atoms, 5σ orbital of CO, etc.).

In the atop adsorbed state, a σ-type orbital will only interact with the (111) and (110) surface d orbitals, because at the (100) face no d orbital of σ symmetry with a finite density of state at the Fermi level is available.

In this approximation the contribution of the interacting d_{z^2} and $d_{x^2-y^2}$ orbitals that are doubly occupied, but positioned below E_F is ignored.

Since the σ-dangling bond bandwidth at the (111) face and (110) face is comparable, the interaction will be the same.

So, a σ adsorbate orbital with one electron in its orbital will bind with comparable energy to the (111) as well as (110) d orbitals in the atop configuration. There will be no interaction with the d orbitals at the (100) face.

A σ orbital with two electrons in its orbital will have a repulsive interaction with other doubly occupied orbitals.

At the end of the group VIII transition-metal series, the d orbitals contain a total of 9 electrons, this results in a 5/6 orbital occupation of the d_{xy}, d_{yz}, and d_{zx} orbital bands.

The orbital occupation of the σ-type dangling bond will be even higher because of its narrower surface bandwith. As a result, the LDOS $\varrho_i(E)$ at the Fermi level will be low, resulting in an overall repulsive interaction between a doubly occupied adsorbate σ orbital and the (111) as well as (110) surface σ dangling d lobe. Clearly, these repulsive effects may be converted into attractive effects if the σ dangling orbitals become depleted of electrons, as is the case for elements in the left column of the group VIII transition metals. For instance, for Rh and Co, the bulk orbital occupancy of the d_{xy}, d_{yz}, and d_{zx} orbitals is 1/2, resulting in a significant increase in the LDOS $\varrho_i(E)$ at E_F for those metals. This decreases the repulsive interaction with doubly occupied σ orbitals significantly or converts it to an attractive one.

π-Type d surface orbitals interact with π-type adsorbate orbitals on all three surfaces. Since the bandwidth of the π-type d orbitals at the (110) and (100) surfaces are smaller than at the (111) face, one may expect at intermediate d-electron occupation a higher LDOS and electron occupation for those orbitals at the (110) or (100) surfaces than the (111) face.

So one expects on Pt, with a nearly completely filled d-valence electron band that adsorbed molecules such as NO and CO in the atop position favor binding to the (100) face, since then repulsion with the adsorbate orbitals is minimized and the attractive contribution dominates.

Clearly, this will be only the case if interaction with the d valence electrons dominates, since because of symmetry the adsorbate π electrons do not interact with the s valence-band electrons in this position.

In bridging or higher coordinated positions the π adsorbate orbitals will interact with antisymmetric combinations of surface metal s orbitals, hence this interaction with π adsorbate orbitals tends to favor bridge coordination.

Thus, the bonding interaction with the d valence electrons increases if the d occupancy decreases from 9 to 8 electrons. The d valence-electron occupancy then shifts to the center of d valence bands, where the LDOS at the E_F is maximum.

On Pt, CO adsorbs atop, but on Ni, CO adsorbs in bridging coordination. The work function of Pt is larger than that of Ni, 5.65 and 5.15 (ev) [100], respectively, so that the contribution of $2\pi^*$ donation is least on the Pt surface.

As discussed in Sect. 2, the 5σ orbital of CO has the largest attractive interaction with the s valence electrons in the atop position, since the LDOS at the E_F is highest in that position.

The interaction between the 5σ electrons and the σ d valence dangling-bond electrons is repulsive. The repulsive effect is proportional to the number of neighbors, this favors also atop adsorption.

It has been found for NO, that it dissociates more rapidly on the 100 surface than on the (111) and (110) surfaces of Pt [100].

On all three surfaces, there is a favorable interaction of π symmetry. As discussed, the repulsive interaction with the σ orbitals will be least at the (100) surface. Comparing π symmetric interactions, they will be favored on the (100) surface compared to the (110) surface because of the higher degree of coordinative unsaturation at the (110) face.

This may explain the results of Nieuwenhuis et al. [101] for Rh, with one electron less than Pt, showing that the (110) face becomes more reactive with respect to the (100) surface than was the case for Pt.

The interaction of CO with Ni, Pd, and Pt increases from Ni to Pt. This indicates also an increasing contribution of the adsorbate to metal donation interaction to chemical bonding and domination of the σ donating term over the $2\pi^*$ backdonating term. The increasing repulsive interaction with the d valence electrons tends to favor also the atop position for CO adsorbed to Pt.

Because of the decrease in work function, $2\pi^*$ backdonation increases from Pt to Ni. This is reflected in the favored bridge-coordinated adsorption site of CO to Ni and Pd.

In contrast to CO, the bond strength of hydrogen atoms increase from Pt to Ni. Whereas bonding in the hydride molecule is larger in PtH than in NiH, at the surface bonding to Ni is stronger [37]. Part of this change in sequence is due to prehybridization of Ni at the Ni metal surface compared to that of the Ni atom (bonding to the Ni atom requires electron promotion).

The same holds, of course, for Pd.

The orbital occupancy of the hydrogen atoms is half. The work function decrease from Pt to Pd to Ni will favor bonding to the Ni surface if electron backdonation dominates chemical bonding.

As demonstrated in the discussion on oxidative addition, the metal hydride bond gives an excess negative charge on the adsorbed H atom.

Comparing the bond strength in the M–H molecule and that of an H atom to the metal surface, one finds that bonding of the metal surface of Pt is significantly decreased [37]. This reflects the importance of the electron localization term in the energy to the bond strength of the adsorbate metal surface bond [33, 37].

The *localization* energy of an electron is the energy required to decouple a metal electron from the metal lattice valence-electron band and localize it on a surface atom, so that it can bind to an adsorbate. If a metal valence electron band is half filled, this will result in a decrease of the metal adsorbate bond strength. The localization energy is not present in a single metal atom, where, on the other hand, the electron promotion energy is an important energy term.

As discussed, CO chemisorption to Pt in the atop position is ascribed to the dominance of the interaction with CO 5σ electrons.

On Ir, Rh, and Co the depletion of the d valence-electron band enhances the local density of states of the d electrons at the Fermi level, and it appears that the interaction with the doubly occupied σ orbitals may become attractive. Whereas on the Ni surface backdonation into the $2\pi^*$ orbitals is favored over σ donation and, as a consequence, bridge adsorption is favored, on Co the increased interaction with the d valence electrons enhances the $2\pi^*$ backdonation in the atop position with respect to the bridge position, changing the balance to the atop position. The same happens on Rh compared to Pd.

On Pt, the high work function decreasing backdonation into the $2\pi^*$ orbitals and the spatial extension of the d orbitals resulting in a large repulsion help to favor the atop configuration.

On Ir, the interaction with the d band will be significantly enhanced and some increase in $2\pi^*$ backdonation is expected. Again the atop position is expected to be favored.

Comparing Cu with Ni, the attractive part of the bond energy should become more dominated by $2\pi^*$ backdonation, because of the further decreased work function.

The total bond strength, however, decreases because of the loss of d valence-electron density at the Fermi level.

All of the d valence-electron orbitals now are doubly occupied and their interaction with the 5σ orbitals becomes exclusively repulsive. Since repulsive effects are lowest for minimum coordination, atop adsorption results [31].

We will conclude this section with a short discussion of the effect of explicitly including orbital overlap on the LDOS $\varrho_i(E)$ of adsorbate orbitals. Figure 36 [31] compares the LDOS of the CO 5σ and $2\pi^*$ orbitals of CO chemisorbed on Pt, as well as the computed metal valence d and s surface local density of states.

These are results of Bethe lattice approximation to the extended Hückel method calculations

One observes very clearly the earlier-discussed loss of electron density in the bonding

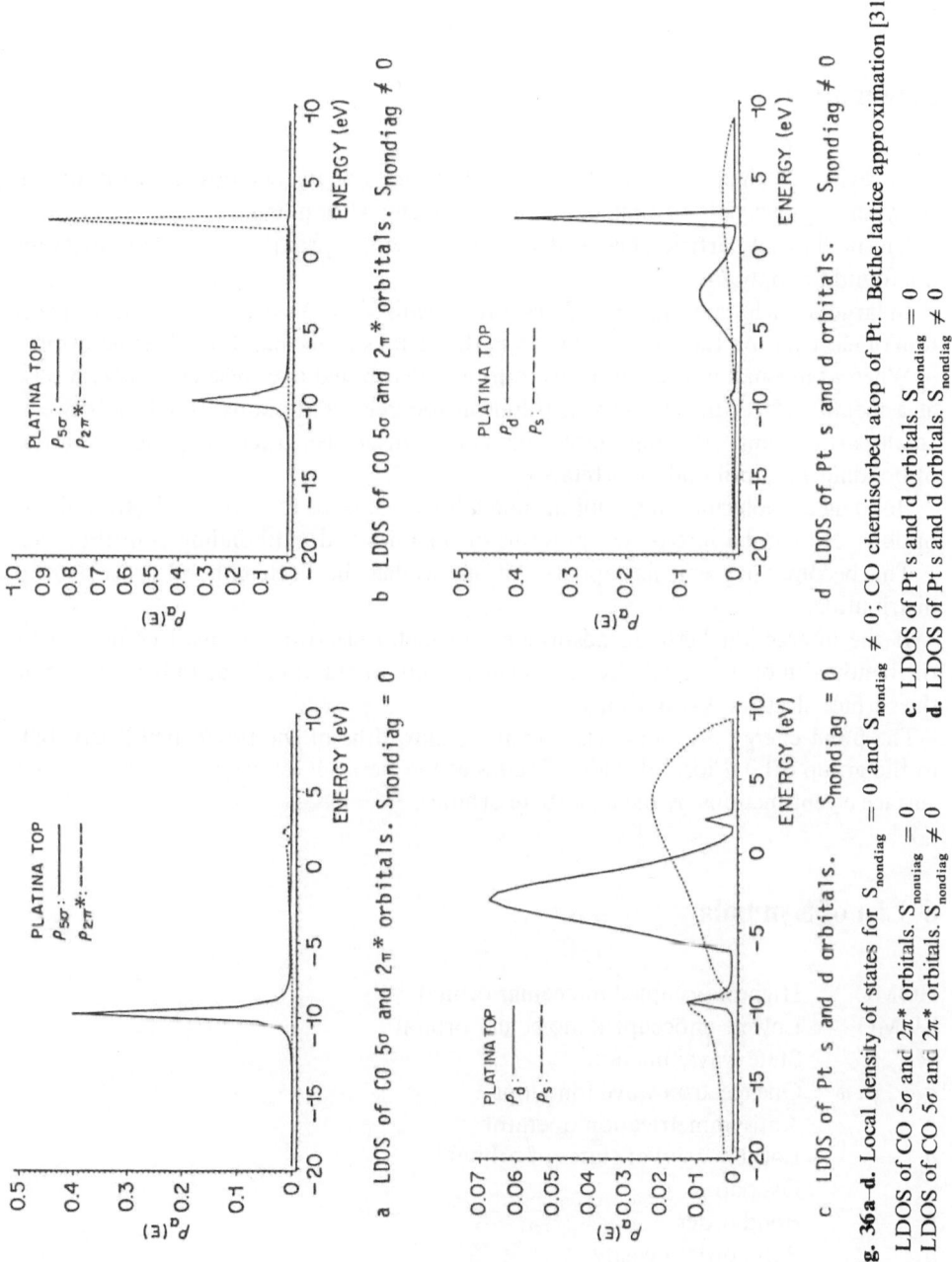

Fig. 36a–d. Local density of states for $S_{nondiag} = 0$ and $S_{nondiag} \neq 0$; CO chemisorbed atop of Pt. Bethe lattice approximation [31].

a. LDOS of CO 5σ and $2\pi^*$ orbitals. $S_{nondiag} = 0$ **c.** LDOS of Pt s and d orbitals. $S_{nondiag} = 0$
b. LDOS of CO 5σ and $2\pi^*$ orbitals. $S_{nondiag} \neq 0$ **d.** LDOS of Pt s and d orbitals. $S_{nondiag} \neq 0$

region and the increase of electron density in the antibonding region if the overlap matrix is assumed to be nondiagonal.

This is completely in line with the electron polarization effects discussed earlier due to non-orthogonality of the interaction fragment orbitals.

5 Summary

This review of chemical bonding to metal surfaces again demonstrates the power of symmetry concepts to predict changes in chemical bonding.

On small metal particles, classical notions of electron promotion and hybridization are found to apply.

In large particles and metal surfaces, prehybridized atoms occur, but now localization of electrons on the atoms involved in the metal surface bond becomes necessary.

Whereas in small particles the electrons are distributed over discrete energy levels, at a metal surface this electron distribution becomes continuous. Bonding between small particles and adsorbing molecules can be discussed in terms of the formation of bonding and antibonding orbitals.

Bonding of molecules to a semi-infinite lattice also leads to changed electron distributions that can be interpreted in terms of bonding and antibonding contributions.

This becomes in particular apparent if one studies the electron bond order density distributions.

If the interaction between adsorbate and metal electrons is small compared to the bandwidth of the metal electron valence band, interaction leads to broadening of the surface electron distributions.

The bond energy does not relate to the bandwidths of the broadened levels, but to the group orbital local density of states at the Fermi level projected out from the surface eigenfunctions by the adsorbate orbitals.

6 List of Symbols

HOMO	Highest occupied molecular orbital
LUMO	Lowest unoccupied molecular orbital
Ψ	State wave function
$\phi, \chi, \psi, \varphi$	One electron wave functions
$\hat{A}$	Antisymmetrization operator
$\varrho_{i,i}$	Local density of states of orbital i
S	Overlap
$P_{\mu, \nu}$	Bond order
$\varrho_{\mu, \nu}$	Bond order density
$F_{\mu\nu}$	Fock matrix
H	Hamiltonian operator
T	Kinetic energy operator
Z_A	Nuclear charge of atom A

$E_{(exch, s)}$	Electron exchange potential
$\Delta_{x, REP}$	Exchange repulsion energy
c_i^k	Coefficient of atomic orbitals i in molecular orbital k
u_i^k	Normalized coefficients of atomic orbitals in molecular orbital k
α_i	One electron energy of atomic orbital i
β_{ij}	Overlap energy between atomic orbitals i and j
Δ	Dispersion energy
Z	Number of atom neighbors
E_i	Molecular orbital eigenvalue
$\Delta n(E)$	Change in number of electrons of energy E
q_i	Charge on atom i
$G_{\gamma, \gamma'}$	Green's function
$\Gamma(E)$	Linewidth function
$\Lambda(E)$	Lineshift function
P_α	Orbital occupation fraction of electron band α
Δ_α	Bandwidth of electron band α
E_F	Fermi level
k_d	Screening length of orbital
σ	Symmetric orbital
σ^*	Antisymmetric orbital
π	Symmetric p-type orbital
π^*	Antisymmetric p-type orbital
ϕ	Angle

7 References

1. Woodward RB, Hoffman R (1961) Ang Chem Int Ed Engl 8: 781; Hoffman R (1982) Ang Chem Int Ed Engl Science 21: 711
2. Fukui K (1982) Science 218: 747
3. Pearson RG (1971) Acc Chem Res 4: 152
4. van der Avoird A, Wormer PES, Mulder F, Berns RM (1986) Top Curr Chem 93: 1
5. March NH (1986) Chemical Bonds outside metal surfaces, Plenum
6. van Santen RA (1985) J Chem Phys 83: 6039
7. Grunze M, Kreuzer HJ (eds) (1986) Kinetics of Interface Reactions, Springer Series in Surface Sciences, vol. 8 Springer, Heidelberg—Berlin—New York
8. Campbell CT (1986) Surf Sci Lett 173: 641
9. Geerlings JJC, Los J (1984) Phys Lett A 102: 204
10. Norskov JK, Holloway S, Lang ND (1984) Surf Sci 137: 65
11. Wimmer E, Fu CL, Freeman AJ (1985) Phys Rev Lett 55: 2618
12. Backx C, de Groot CPM, Biloen P (1981) Surf Sci 104: 300
13. Barteau MA, Madix RJ (1980) Surf Sci 97: 101
14. Norskov JK, Besenbacher F (1987) J Less Common Metals 130: 475
15. Kitaura K, Morokuma K (1976) Int. J Quantum Chem 10: 325
16. Morokuma K (1977) Acc Chem Res 10: 244
17. Kitaura K, Sakaki S, Morokuma K (1981) Inorg Chem 20: 2292
18. Wolfe S, Mitchell DJ, Whangbo MH (1978) J Am Chem Soc 100: 1936
19. Bernardi F, Bottoni A, Mangini A, Tonachini G (1981) J Molec Struct (Theochem) 86: 163
20. Stone AJ, Erskine RW (1980) J Am Chem Soc 102: 7185
21. Noell JO, Morokuma K (1979) Inorg Chem 18: 2774

22. Ziegler T (1985) Inorg Chem 24: 1547
23. Ziegler T, Tschinke V, Becke A (1987) J Am Chem Soc 109: 1351
24. Bauschlicher CW, Bagus PS (1984) J Chem Phys 81: 5889
25. Baerends EJ, Rozendaal A (1986) in: Veillard A (ed) Quantum Chemistry: The Challenge of Transition Metals and Coordination Chemistry, Nato, ASI series, Reidel, Dordrecht, p 159
26. Post D, Baerends EJ (1983) J Chem Phys 78: 5663
27. Bauschlicher Jr. CW, Bagus PS, Nelin CJ, Roos BO (1986) J Chem Phys 85(1): 354
28. Bagus PS, Hermann K, Bauschlicher CW (1984) J Chem Phys 80: 4378
29. Hermann K, Bagus PS, Nerlin CJ (1987) Phys Rev B35: 9467
30. Bagus PS, Nerlin CJ, Bauschlicher ChW (1983) Phys Rev B28: 5423
31. Norskov JK, Lang ND (1980) Phys Rev B21: 2131
31. van Santen RA (1988) J Mol Struct 173: 157
32. Pauling L (1931) J Am Soc 53: 1367; see also Coulson A (1961) Valence, Oxford University Press
33. van Santen RA (1987) Progr Surf Sci 25: 253
34. Newns PM (1969) Phys Rev 178: 1123
35. Grimely TB, Torrini M (1973) J Phys C6: 868
36. Schrieffer JR (1972) J Vac Sci Techn 9: 561; Einstein TL, Schrieffer JR (1973) Phys Rev B7: 3627
37. van Santen RA (1982) Recl Trav Chem Pays-Bas 101: 121
38. van den Hoek PJ, Tenner AD, Kleyn AW, Baerends EJ (1986) Phys Rev B34: 5030
39. Feinberg MJ, Ruedenberg K (1971) J Chem Phys 54: 1495
40. Fujimoto H, Osamura J, Minato T (1978) J Am Chem Soc 100: 2954
41. Post D, Baerends EJ (1982) Surf Sci 116: 177
42. Bauschlicher CW (1986) Chem Phys 106: 391
43. Ziegler T, Rauk A (1977) Theoret Chim Acta 46: 1
44. Hoffmann R (1963) J Chem Phys 39: 1397; Hoffmann R, Lipscomb WN (1962) J Chem Phys 36: 2179; Hoffmann R, Lipscomb WN (1962) J Chem Phys 37: 2872; Albright ThA, Burdett JR, Whangbo MH (1985) Orbital Interactions in Chemistry, Wiley, New York
45. Anderson AB (1975) J Chem Phys 62: 1187
46. van Santen RA (1979) J Chem Phys 71(1): 163
47. Koutecky J (1958) Trans Far Soc 54: 1038; Koutecky J (1957) Phys Rev 108: 13
48. Grimley TB, Pisani C (1974) J Phys C7: 2831; Pisani C (1978) Phys Rev B17: 3134
49. Schrieffer JR (1974) in: Dynamic Aspects of Surface Physics, Goddman FO (ed) Proc Int School of Physics "Enrico Fermi", Course XVII
50. van Santen RA, Toneman LH (1977) Int J Quantum Chem 12, suppl 2: 83
51. van Santen RA (1987) J Chem Soc Far Trans 1, 83: 1915
52. Callaway J (1971) Phys Rev B83: 2556
53. Seel M, Del Re G, Ladik J (1982) J Comp Chem 3: 451
54. Haydock R, Heine V, Kelly MG (1975) J Phys C: Solid State Phys 8: 259
55. van Santen RA (1984) Proc 8th Int Catal Conf Berlin, pp 97; Dechema: J Phys C, Solid State Phys (1982) 15: L513
56. Khanra BC (1983) Chem Phys Lett 96: 76
57. Grimley TB (1976) CRC Crit Rev in Solid State Sci, pp 239
58. Wang CS, Freeman AJ (1979) Phys Rev B19: 4930
59. Kelly MG (1974) Surf Sci 43: 587
57. Grimley TB (1976) CRC Crit Rev in Solid State Sci, pp 239
58. Wang CS, Freeman AJ (1979) Phys Rev B19: 4930
59. Kelly MG (1974) Surf Sci 43: 587
60. Gadzuk GW (1976) NATO Adv Study Inst Ser, Phys Ser B, vol 16, Plenum, NY
61. Salem L, Elliott R (1983) J Mol Structure 93: 75
62. Grimley TB (1086) Phil Trans Roy Soc (London) A318: 135
63. Shustorovich E (1982) Solid State Comm 44: 567; Shustorovich E, Baetzold RC, Muetterties EL (1983) J Phys Chem 87: 1100; Shustorovich E (1984) J Phys Chem 88: 1927
64. Blomberg MRA, Siegbahn PEM (1983) J Chem Phys 78: 5689
65. Brandemark UB, Blomberg MRA, Petterson LGM, Siegbahn PEM (1984) J Phys Chem 88: 4617
66. Low JJ, Goddard WA (1984) J Am Chem Soc 106: 6928
67. Low JJ, Goddard WA (1084) J Am Chem Soc 106: 8321

68. Nakatsuji H, Hada M, Yonezawa T (1987) J Am Chem Soc 109: 1902
69. Kitaura K, Obara S, Morokuma K (1981) J Am Chem Soc 103: 2891
70. Noell JO, Hay PJ (1984) J Am Chem Soc 106: 6928
71. Dedieu A (1985) in: Gielen M (ed) Topics in Organometallic Chemistry, Vol. 1, pp 1, Freund Publishing House, London
72. Berke H, Hoffmann R (1978) J Am Chem Soc 100: 7224
73. Zheng Ch, Hoffmann R (1988) J Am Chem Soc 110: 749
74. Melius CF, Moskowitz JW, Mortola AP, Baillie MB, Ratner MA (1976) Surf Sci 59: 279
75. Siegbahn PEM, Blomberg MRA, Bauschlicher Jr, CW (1984) J Chem Phys, 81(4): 2103
76. Upton ThH, Goddard WA (1981) CRC Crit Rev in Solid State and Mat Sci, pp 261
77. Pachioni G, Koutecky J (1987) J Chem Phys 78(9): 5663
78. Andzelm MJ, Salahub DR (1986) Int J Quantum Chem 29: 1091
79. Post D, Baerends EJ (1983) J Chem Phys 78(9): 5663
80. Raatz F, Salahub DR (1986) Surf Sci 176: 219
81. Baerends EJ, Siebbeles L, unpublished
82. Koutecky J, Hanke U, Fantucci P, Bonacic-Koutecky V, Papierouska-Kamienski D (1986) Surf Sci 165: 161
83. Painter GS, Averill FW (1987) B35: 7713
84. Hermann K, Bagus PS, Nelin C (1987) Phys Rev B35: 9467
85. Hermann K, Bagus PS, Bauschlicher CW (1985) Phys Rev B31: 6371
86. Sung S, Hoffmann R (1984) J Am Chem Soc 107: 2006; Sung S, Hoffmann R, Thiel P (1986) J Phys Chem 90: 1380; Hoffmann R, Zeng C (1982) in: Veillard (ed) Quantum chemistry: The challenge of transition metals and coordination chemistry, Nato, ASI series, Reidel, Dordrecht
87. Ray NK, Anderson AB (1982) Surf Sci 119: 35; Ray NK, Anderson (1983) Surf Sci 125: 803; Anderson AB, Awad MK (1985) J Am Chem Soc 107: 7854; Mechandru SP, Anderson AB, Ross PN (1986) J Catal 100: 210; Mechandru SP, Anderson AB (1986) Surf Sci 169: L281; Anderson AB, Grime RW, Hory SP (1987) J Phys Chem 91: 4245
88. Baetzold RJ (1983) J Am Chem Soc 105: 4271
89. Bond GC (1966) Disc Far Soc 41: 200
90. Weinberg WH, Merrill RP (1975) J Catal 40: 268
91. Goodenough JB (1963) Magnetism and the chemical bond, J Wiley, NY
92. Freitsheim H, Ibach H, Lehwald S (1977) Appl Phys 13: 147; Baro AM, Ibach H (1979) J Chem Phys 71: 4812
93. Dubois LH, Somorjai GA (1980) Surf Sci 91: 514
94. Backx C, unpublished
95. Hollins P, Pritchard J (1985) Progr Surf Sci 19: 275
96. Anderson S (1976) Solid State Comm 20: 229; Erley W, Wagner H, Ibach H (1979) Surf Sci 80: 612
97. Maclaren JM, Pendry JB, Joyner RW, Mechan P (1986) Surf Sci 175: 263; Maclaren JM, Pendry JB, Joyner RW (1986) Surf Sci L80: 165
98. Fassaert DJM, Verbeek H, van de Avoird A (1972) Surf Sci 9: 501
99. Kahn O, Salem L (1977) in: Proc of the 6th International Congress on Catalysis, London, 1: 101 (The Chemical Society, London)
100. Banholzer WF, Park PO, Mak KM, Masel RI (1983) Surf Sci 128: 176; Park PO, Banholzer WF, Masel RI (1983) Surf Sci 19: 145; Park PO, Banholzer WF, Masel RI (1985) Surf Sci 155: 341 and 653
101. Hendrickx HACM, Jongenelis APJM, Niewenhuys BE (1985) Surf Sci 154: 503; Hendrickx HACM, Nieuwenhuys BE (1986) Surf Sci 175: 185

Intermolecular Forces and the Properties of Molecular Solids

Ad van der Avoird

Institute of Theoretical Chemistry, University of Nijmegen, Toernooiveld, 6525 ED Nijmegen, The Netherlands

Quantum chemical *ab initio* calculations have reached the stage where they can yield detailed quantitative information about the interaction potentials between molecules, in particular about the orientational dependence of these potentials and, in the case of open-shell molecules, also about their spin dependence. In order to use the intermolecular potentials in the calculation of aggregate properties and, thereby, to test and improve these potentials, it is necessary to express them in analytic form. Moreover, the question is important whether the intermolecular interactions are additive. These topics are addressed in the first part of this paper.

The second part deals with the actual calculation of the spectra and the bulk properties of molecular solids. The (lattice dynamics) methods to perform such calculations are outlined and it is demonstrated that a complete *ab initio* treatment of "simple" molecular crystals yields new and interesting insight in the behavior of these solids. Finally, it is illustrated how lattice dynamics calculations based on empirical atom-atom potentials are useful for the interpretation of the measurements on more complex molecular crystals.

1 Intermolecular Potentials

1.1 Introduction

The notion of intermolecular potentials is based on separability at two different levels. The Born-Oppenheimer separation between electronic and nuclear motions prescribes the use of the electronic energy surface as the potential energy for the nuclear motions. The nuclear motions can be separated into internal molecular motions, i.e. molecular vibrations, and external motions, i.e. (relative) translations and rotations of whole molecules. The latter separation follows from the shape of the potential energy surface, which is determined by the nature of the interactions involved. Molecules are kept together by "chemical", mainly covalent, bonds between the atoms, which, for neutral molecules, are considerably stronger than the intermolecular interactions. For molecular ions the intermolecular Coulomb interaction energies are equally large as the intramolecular covalent binding energies, but even in this case the steep distance dependence and strong directionality of the covalent bonds make the potential energy surface depend most sensitively on the internal molecular coordinates. So there is a clear separation between internal molecular coordinates and external ones. Molecules are recognizable by their electronic and vibrational spectra; the intermolecular interactions cause (slight) modifications of these spectra (line shifts, splittings, broadening). This separation becomes less distinct for larger molecules which are often flexible in some of their internal coordinates. The motions along those specific coordinates will be strongly influenced by intermolecular interactions and coupled to the overall motions of the molecules.

Thus, in general, for interactions varying from strong Coulombic between molecular ions, through intermediate between polar molecules (hydrogen bonding), to weak Van der Waals interactions, the central concept to describe these interactions is the intermolecular potential energy surface (short: intermolecular potential). This potential depends on the distances between the centers of mass of the molecules and on the (relative) molecular orientations (the orientational dependence is often called anisotropy). It also depends on the internal molecular coordinates. For flexible molecules, in particular, the choice of convenient internal coordinates and the definition of axes which fix the molecular orientations are not trivial, however. (This is related to the separation between internal molecular vibrations and overall rotations, which is an approximate one.) The dependence of the intermolecular potential on the internal coordinates is important for many energy transfer and relaxation processes. In the rigid-molecule approximation this dependence is neglected. Further it is worth noting that, besides the interaction energy, also other observable properties of molecular systems such as (transition) dipole moments, polarizabilities, etc., can be described by scalar, vector or tensor functions [1] which depend on the external and internal molecular coordinates.

1.2 Contributions to the Intermolecular Potential; ab initio Calculations

A survey of the state of the art in this field is given in several recent reviews [2–6]. So only the most important points are indicated here.

The interactions between molecules are usually divided into long-range interactions and short-range interactions. In the long range, i.e. when the charge clouds of the interacting molecules do not overlap, the interaction energy can be formally obtained by standard Rayleigh-Schrödinger perturbation theory. The perturbation, which is the intermolecular interaction operator, can be expanded as a multipole series in powers of R^{-1}, where R is the distance between the centers of mass of the molecules. The first-order energy is the electrostatic multipole-multipole interaction energy. The second-order energy contains the induction (multipole-induced multipole) energy and the, non-classical, dispersion energy. Extensive formulas, using cartesian as well as spherical tensors for the multipole operators, are given in the preceding chapter by Stone [7]. For molecular ions the electrostatic and induction interactions are strongly dominant. For polar, e.g. hydrogen bonded, molecules the electrostatic interactions are still the most important contribution, while the induction and dispersion energies are comparable. For apolar molecules, i.e. molecules with small dipole moments, the dispersion energy becomes the most important (attractive) long range interaction. As it is evident from the formulas in the preceding chapter [7], the long range interactions are completely determined by the permanent multipole moments and the, static as well as frequency-dependent, multipole polarizabilities of the monomers. All these molecular properties can be calculated *ab initio*, although especially the calculation of accurate frequency-dependent polarizabilities including the effects of electron correlation [8, 9] is not yet a routine job.

Since the molecular charge clouds have exponential tails, there is always some overlap between them. The effects of this overlap are twofold. Penetration causes the exact electrostatic interaction between continuous, overlapping charge clouds to deviate from its representation by a multipole series. This is correctly included in the Rayleigh-Schrödinger perturbation theory if one avoids the expansion of the electrostatic interaction operator [5, 6]. Not included in the standard perturbation theory are the exchange effects, which arise from the antisymmetrization of the overall electronic wave functions, required by the Pauli postulate. Both penetration and exchange effects modify the interaction energy in all orders of perturbation theory. They have been explicitly calculated in first and second order for very small systems (He, H_2), by means of symmetry-adapted perturbation theory [10, 11]. In practice, they become important near the Van der Waals minimum and they are mostly calculated in first order only, by means of a Heitler-London type formula, which already gives the dominant exchange repulsion. Even then, the inclusion of the intramolecular electron correlation in such a calculation is not easy [5, 12]. The effects of penetration and exchange on the second-order induction and dispersion energy, which are much less important than the first-order exchange repulsion, are often taken into account by the use of (semi-empirical) damping functions [13–15].

Most popular in the *ab initio* calculation of intermolecular potentials is the so-called supermolecule method, because it allows the use of standard computer programs for electronic structure calculations. This method automatically includes all the electrostatic, penetration and exchange effects. If the calculations are performed at the SCF (self-consistent field) level the induction effects are included, too, but the dispersion energy is not. The latter, which is an intermolecular electron correlation effect, can be obtained by configuration interaction (CI), coupled cluster (CC) calculations or many-body perturbation theory (MBPT). These calculations are all plagued by basis set superposition errors (BSSE) [2–6], however, which are mostly of the same magnitude as the intermolecular interaction energy. For small molecules these errors can now be nearly avoided at the SCF level, but not when the electron correlation is included. Moreover, the CI methods, which are the most generally applicable, are suffering from the lack of size-consistency [3, 5]. Further experimentation with these methods will be required, in order to ensure that they will produce reliable results. In principle, they are very attractive because they yield complete interaction potentials, over the full range of distances.

Most of the current work on intermolecular interaction potentials is concerned with closed-shell molecules, but it is worth noting that the interactions between open-shell molecules are especially interesting. As a direct consequence of the relation between the spin and the permutation symmetry of electronic wave functions [16], different couplings between the non-zero spin states of interacting open-shell monomers will lead to different exchange interactions. In other words, interacting open-shell molecules possess a manifold of intermolecular potential energy surfaces, one surface for each total spin state. The splitting between these surfaces is caused by exchange interactions. Some of these potential surfaces may correspond to chemical bonding, just as it occurs between open-shell atoms. A very weak bond of this type seems to be present [17, 18] in the singlet state of $(NO)_2$. In the $(O_2)_2$ dimer, on the other hand, the singlet, triplet and quintet state are all showing a net, although different, exchange repulsion [19] between the triplet O_2 molecules. So here one finds a very interesting system with a Van der Waals interaction potential which is spin-dependent.

1.3 Analytic Forms of Intermolecular Potentials

The knowledge of (*ab initio*) intermolecular potentials opens the way to (the calculation of) many observable properties for systems consisting of few molecules up to bulk systems. In the first category are the spectra and thermodynamic stability of Van der Waals molecules [1, 4, 6, 20] and molecular beam elastic and inelastic, total and state-to-state, scattering cross sections [21–23]. In the second category are various bulk gas as well as condensed matter properties. Measured gas phase properties [24, 25] which depend directly on the intermolecular potential are second virial coefficients, viscosity and diffusion coefficients, thermal conductivity, sound absorption, pressure broadening of spectral lines, nuclear magnetic relaxation and depolarized Rayleigh scattering. Additional information is obtained from the effects of electric and magnetic fields on the transport properties (Senftleben-Beenakker effects). In the condensed phases one may calculate (by liquid state theory) or simulate (by Monte Carlo or Molecular Dynamics methods) the behavior of liquids [26], or study the stability and lattice vibrations of molecular solids [27]. Especially the latter subject will be treated in Sect. 2 of this paper, in relation to various experimental data for molecular crystals. On the other hand, all the measured data may be used, and have actually been used in several examples, to construct or improve (semi-)-empirical intermolecular potentials.

In all these studies it is practical to write the intermolecular potential in some analytic form or another. Thus it can be easily calculated for many different distances and orientations of the molecules, as required in Monte Carlo simulations, for example. Specific forms will be convenient for scattering, liquid state or lattice dynamics calculations. Moreover, if the potential is to be improved through such studies, this form must contain a limited number of variable parameters. In practice, two basic forms can be recognized, a spherical expansion and an (isotropic) atom-atom or site-site potential.

1.3.1 Spherical Expansion of Intermolecular Potentials

In this expansion the orientational dependence (anisotropy) of the intermolecular potential between two molecules A and B is explicitly expressed in a (complete) basis of symmetry-adapted free-rotor functions, defined as:

$$G_{M_A N_A}^{(L_A)}(\Omega_A) = \sum_{K_A} a_{K_A N_A}^{(L_A)} D_{M_A K_A}^{(L_A)}(\Omega_A)^* \tag{1}$$

for molecule A, and analogously for molecule B. The functions $D_{MK}^{(L)}(\Omega)$ are Wigner rotation matrix elements [28]. The Euler angles $\Omega_A = (\alpha_A, \beta_A, \gamma_A)$ and $\Omega_B = (\alpha_B, \beta_B, \gamma_B)$ describe the orientations of the molecules with respect to an arbitrary coordinate frame. The coefficients $a_{KN}^{(L)}$ are determined by the totally symmetric irreducible representations of the point groups of the molecules A and B, in their equilibrium structures. For linear molecules one has $a_{KN}^{(L)} = \delta_{K0}$ and the symmetry-adapted functions are simply spherical harmonics in the Racah normalization [28]:

$$G_M^{(L)}(\Omega) = D_{M0}^{(L)}(\alpha, \beta, \gamma)^*$$

$$= C_M^{(L)}(\beta, \alpha)$$

$$= \left[\frac{4\pi}{2L+1}\right]^{1/2} Y_M^{(L)}(\beta, \alpha) . \tag{2}$$

For tetrahedral molecules (point group T) the symmetry adapted functions are listed in Table 1. In this case the index N must be used to label the two functions for $L = 6, 9$ or 10. The direction of the vector $\mathbf{R}$, which connects the center of mass of molecule A to that of molecule B, is given by the polar angles (Θ, Φ). The intermolecular potential can be expressed as [1]:

$$V(\mathbf{R}, \mathbf{q}_A, \mathbf{q}_B, \Omega_A, \Omega_B)$$

$$= \sum_{L_A N_A L_B N_B L} v_{L_A L_B L}^{N_A N_B}(R, \mathbf{q}_A, \mathbf{q}_B) A_{L_A L_B L}^{N_A N_B}(\Omega_A, \Omega_B, \Theta, \Phi), \tag{3}$$

in terms of the complete orthogonal set of angular functions:

$$A_{L_A L_B L}^{N_A N_B}(\Omega_A, \Omega_B, \Theta, \Phi) =$$

$$\sum_{M_A M_B M} \begin{pmatrix} L_A & L_B & L \\ M_A & M_B & M \end{pmatrix} G_{M_A N_A}^{(L_A)}(\Omega_A) G_{M_B N_B}^{(L_B)}(\Omega_B) C_M^{(L)}(\Theta, \Phi). \tag{4}$$

Table 1. Tetrahedral rotation functions (symmetry group T)^a

l	Function

0    $G_0^{(0)} = D_{0,0}^{(0)*}$

3    $G_m^{(3)} = -\frac{1}{2} i \sqrt{2} [D_{m,2}^{(3)*} - D_{m,-2}^{(3)*}]$

4    $G_m^{(4)} = \frac{1}{2} \sqrt{\frac{7}{3}} D_{m,0}^{(4)*} + \frac{1}{2} \sqrt{\frac{5}{6}} [D_{m,4}^{(4)*} + D_{m,-4}^{(4)*}]$

6    $G_{m,1}^{(6)} = \frac{1}{4} \sqrt{2} D_{m,0}^{(6)*} - \frac{1}{4} \sqrt{7} [D_{m,4}^{(6)*} + D_{m,-4}^{(6)*}]$

       $G_{m,2}^{(6)} = \frac{1}{4} \sqrt{\frac{11}{2}} [D_{m,2}^{(6)*} + D_{m,-2}^{(6)*}] - \frac{1}{4} \sqrt{\frac{5}{2}} [D_{m,6}^{(6)*} + D_{m,-6}^{(6)*}]$

7    $G_m^{(7)} = -\frac{1}{4} i \sqrt{\frac{13}{3}} [D_{m,2}^{(7)*} - D_{m,-2}^{(7)*}] - \frac{1}{4} i \sqrt{\frac{11}{3}} [D_{m,6}^{(7)*} - D_{m,-6}^{(7)*}]$

8    $G_{m,1}^{(8)} = \frac{1}{8} \sqrt{33} D_{m,0}^{(8)*} + \frac{1}{4} \sqrt{\frac{7}{6}} [D_{m,4}^{(8)*} + D_{m,-4}^{(8)*}] + \frac{1}{8} \sqrt{\frac{65}{6}} [D_{m,8}^{(8)*} + D_{m,-8}^{(8)*}]$

9    $G_{m,1}^{(9)} = -\frac{1}{4} i \sqrt{\frac{3}{2}} [D_{m,2}^{(9)*} - D_{m,-2}^{(9)*}] + \frac{1}{4} i \sqrt{\frac{13}{2}} [D_{m,6}^{(9)*} - D_{m,-6}^{(9)*}]$

       $G_{m,2}^{(9)} = -\frac{1}{4} i \sqrt{\frac{17}{3}} [D_{m,4}^{(9)*} - D_{m,-4}^{(9)*}] + \frac{1}{4} i \sqrt{\frac{7}{3}} [D_{m,8}^{(9)*} - D_{m,-8}^{(9)*}]$

10    $G_{m,1}^{(10)} = \frac{1}{8} \sqrt{\frac{65}{6}} D_{m,0}^{(10)*} - \frac{1}{8} \sqrt{11} [D_{m,4}^{(10)*} + D_{m,-4}^{(10)*}] - \frac{1}{16} \sqrt{\frac{187}{3}} [D_{m,8}^{(10)*} + D_{m,-8}^{(10)*}]$

       $G_{m,2}^{(10)} = \frac{1}{16} \sqrt{\frac{247}{3}} [D_{m,2}^{(10)*} + D_{m,-2}^{(10)*}] + \frac{1}{16} \sqrt{\frac{19}{6}} [D_{m,6}^{(10)*} + D_{m,-6}^{(10)*}] - \frac{1}{16} \sqrt{\frac{85}{2}} [D_{m,10}^{(10)*} + D_{m,-10}^{(10)*}]$

^{a} The z axis of the molecule-fixed system is a two-fold axis; the [1,1,1] axis is a threefold axis

These functions $A_{L_A L_B L}^{N_A N_B}$ are very similar to the functions $\bar{S}_{l_1 l_2 j}^{k_1 k_2}$ defined by Stone in the preceding chapter [7], the expression in large round brackets is a 3-j symbol [28]. The expansion coefficients $v_{L_A L_B L}^{N_A N_B}(R, \boldsymbol{q}_A, \boldsymbol{q}_B)$ depend on the distance R between the molecules and on the internal molecular coordinates, symbolized by $\boldsymbol{q}_A$ and $\boldsymbol{q}_B$. Any intermolecular pair potential can be expressed to any desired accuracy when a sufficient number of these expansion coefficients is given. For molecules with well-defined equilibrium structures (sometimes called nearly-rigid molecules) it is convenient to make a Taylor expansion of the expansion coefficients about the equilibrium values of the internal coordinates $\boldsymbol{q}_A$ and $\boldsymbol{q}_B$. In practice, this has only been done yet for very simple systems [29]. Mostly one has used the rigid-molecule approximation, i.e. replaced $\boldsymbol{q}_A$ and $\boldsymbol{q}_B$ by their equilibrium values, or one has assumed a Born-Oppenheimer separation between the internal and external molecular coordinates and replaced $\boldsymbol{q}_A$ and $\boldsymbol{q}_B$ by their vibrationally averaged values [30].

All the long-range contributions to the interaction potential, defined in Sect. 1.2, depend on the distance R as power series in R^{-1}, the short range overlap (penetration and exchange) effects depend exponentially on R. It is shown in the preceding chapter [7] that the first order electrostatic interactions automatically adopt the form of Eq. (3); the expansion coefficients $v_{L_A L_B L}^{N_A N_B}(R, q_A, q_B)$ are products of multipole moments $Q_{N_A}^{(L_A)}(q_A)$ and $Q_{N_B}^{(L_B)}(q_B)$ and powers R^{-L-1} of the distance, for $L = L_A + L_B$ (for $L \neq L_A + L_B$ they vanish). Also the second-order long-range contributions (induction and dispersion) can be directly expressed in the form of Eq. (3), if coupled polarizabilities $\alpha_{N_A}^{(l_A l'_A)\,l_A}$ and $\alpha_{N_B}^{(l_B l'_B)\,l_B}$ are used [8] instead of the uncoupled ones $(\alpha_{lkl'k'})$ defined in the preceding chapter [7]. So by means of *ab initio* calculated multipole moments and frequency-dependent coupled polarizabilities one can directly obtain the long range form of the expansion coefficients $v_{L_A L_B L}^{N_A N_B}$, i.e. electrostatic, induction and dispersion coefficients C_n which multiply the various R^{-n} contributions. For the *ab initio* calculation of the short range behavior of the coefficients $v_{L_A L_B L}^{N_A N_B}$ a numerical (Gauss quadrature) integration procedure has been devised in

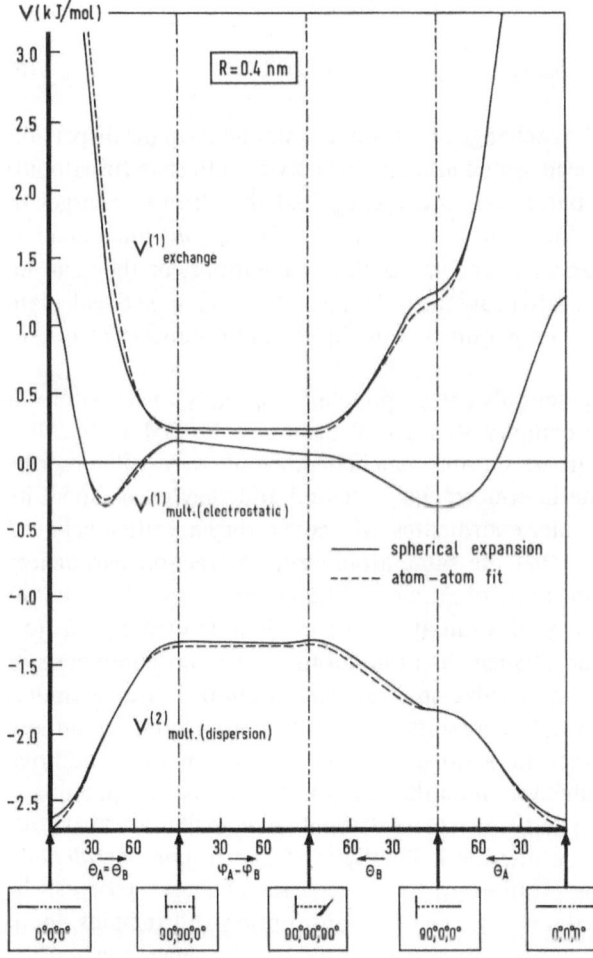

Fig. 1. Orientational dependence of different long-range (multipole) and short-range (exchange and penetration) contributions to the $N_2 - N_2$ potential [31]

our group [1, 31]. It requires the *ab initio* calculation of short range interaction energies on an angular grid of quadrature points. This procedure has been applied to linear molecules (the N_2–N_2 [31, 32] and O_2–O_2 [19] potentials) and to atom-molecule systems such as Ar-NH_3. For non-linear molecule dimers it is still too expensive. The anisotropy of the different long range and short range contributions to the N_2-N_2 potential is illustrated in Fig. 1.

1.3.2 Atom-Atom or Site-Site Potentials

A widespread approximation of the intermolecular potential between two molecules A and B is to write it as a sum of atom-atom potentials, which are assumed to be isotropic, i.e. to depend only on the distance R_{ab} between the atoms a in molecule A and the atoms b in molecule B:

$$V_{AB} = \sum_{a \in A} \sum_{b \in B} v_{ab}(R_{ab}) .$$ (5)

The atom-atom potential functions are mostly written as:

$$v_{ab}(R_{ab}) = a_{ab} \exp(-b_{ab}R_{ab}) - c_{ab}R_{ab}^{-6} + q_a q_b R_{ab}^{-1} ,$$ (6)

with the first term representing the exchange repulsion, the second term the dispersion attraction and the last term the electrostatic interactions between effective (fractional) atomic charges q_a and q_b. The parameters a_{ab}, b_{ab}, c_{ab} and the effective charges q_a and q_b can be obtained [31, 33] by fitting the potential in Eq. (5) to an *ab initio* potential V_{AB} calculated for various distances R and molecular orientations, or they can be determined empirically [34]. In the latter case the potential in Eq. (5) is used to calculate various measured properties and the parameters in Eq. (6) are optimized to obtain a best fit to these properties.

The reason that atom-atom potentials are so popular, especially in the study of condensed phases [34] and more complex Van der Waals molecules [35], is that they contain few parameters and can be cheaply calculated, while they still describe (implicitly) the anisotropy of the intermolecular potential and they even model its dependence on the internal molecular coordinates. Moreover, they are often believed to be transferable, which implies that the same atom-atom interaction parameters in Eq. (6) can be used for the same types of atoms in different molecules. One should realize, however, that the accuracy of atom-atom potentials is limited by Eq. (5). Further inaccuracies are introduced when the atom-atom interaction parameters in Eq. (6) are transferred from one molecular environment to another. Furthermore, Eq. (6) does not include a term which represents the induction interactions and there is the intrinsic problem that these interactions are inherently not pairwise additive (see Sect. 1.4). Numerical experimentation on the C_2H_4–C_2H_4 and N_2–N_2 potentials, for example, has taught us [31, 33] that even when sufficient *ab initio* data are available, so that the terms in Eq. (6) can be fitted individually to the corresponding *ab initio* contributions and, moreover, the positions of the force centers for each term can be optimized, the average error in the best fit of each contribution still remains about 10%. Since the different contributions to the potential partly cancel each other and the errors for specific molecular orientations are considerably larger, it will be clear that even atom-atom potentials with shifted force centers (called site-site potentials) are of limited value for the description of the detailed anisotropy of the intermolecular potential surface. This has been confirmed by results on solid nitrogen, see Sect. 2. In some cases one has tried to overcome these problems by including several point charges per atom for the electrostatic interactions and by adopting atomic or bond polarizabilities for the description of induction forces [7, 36, 37].

1.3.3 Other Analytic Potentials; Relations

Besides the two basic forms just described, which are used the most in the literature, other expressions for intermolecular potentials have been proposed. Some potentials are given, for instance, by analytic forms of the generalized Lennard-Jones type [38]:

$$V_{AB}(R, \Omega) = 4\varepsilon(\Omega)\left[\left(\frac{R}{\varrho(\Omega)}\right)^{-12} - \left(\frac{R}{\varrho(\Omega)}\right)^{-6}\right] \tag{7}$$

or by exponential functions for the short range repulsion [39]:

$$V_{AB}^{short}(R, \Omega) = \exp\left[-a(\Omega)(R - \varrho(\Omega))\right], \tag{8}$$

in which the well depth $\varepsilon(\Omega)$ and the non-linear "range parameters" $\varrho(\Omega)$ and "slope parameters" $a(\Omega)$ are assumed to be functions of the molecular orientations Ω_A and Ω_B. Such functions may be expanded [39] in terms of the angular basis functions of Eq. (4), which then occur non-linearly in the potential. An advantage of this procedure is [39] that these expansions need only very few terms.

Another way of expressing the intermolecular potential, which is intermediate between the two basic forms, is treated extensively in the preceding chapter [7]. Molecules are divided in segments, which may be individual atoms or groups of atoms. The inaccuracy inherent in the isotropic atom-atom model is avoided by assuming anisotropic potentials between the segments. which have the same form as Eq. (3). The expansion in Eq. (3) should converge more rapidly for the segment interactions than it does for the interactions between the whole molecules. In particular, it has been illustrated by Stone [7] that the long range multipole interaction series converges for smaller distances and with fewer multipole moments for segments than for molecules. The arbitrariness in dividing the multipole moments and multipole polarizabilities over the segments can be used to improve this convergence. Such ideas will be useful in particular for larger molecules. The transferability of segment-segment interaction potentials and their extension to induction and dispersion forces are still open to investigation.

Atom-atom or segment-segment potentials, and potentials of the form of Eqs. (7) or (8), usually contain fewer terms and fewer parameters than the spherical expansion of the intermolecular potential given by Eq. (3). This is useful if the parameters have to be varied, for instance, to improve the agreement with experimental data. On the other hand, the spherical expansion is convenient for several applications of the potential, such as scattering calculations or lattice dynamics calculations by special methods (see Sect. 2). It is always possible [1] to calculate the spherical expansion coefficients, see Eq. (3), of any known intermolecular potential by numerical quadrature methods. Atom-atom potentials of the form of Eq. (6) can also be transformed analytically into a spherical expansion; the formulas required are given in Ref. [27]. Also segment-segment potentials of the type of Eq. (3) can be analytically converted to a molecular expansion of the same type [7, 39].

1.3.4 Spin-dependent Potentials Between Open-Shell Molecules

The exchange interactions between open-shell molecules, which lead to different potential energy surfaces for different spin-couplings between the molecules (see Sect. 1.2), can be represented in the form of a Heisenberg Hamiltonian [19]:

$$V^{\text{spin-dependent}}(R, q_A, q_B, \Omega_A, \Omega_B, S_A, S_B) = -2J(R, q_A, q_B, \Omega_A, \Omega_B) S_A \cdot S_B,$$

$$(9)$$

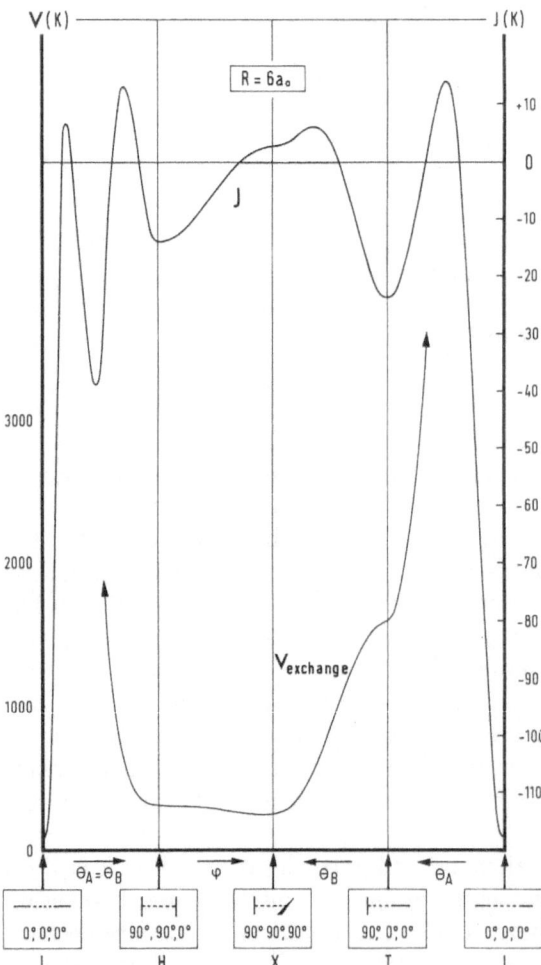

Fig. 2. Orientational dependence of the exchange contributions to the spin-dependent O_2-O_2 potential [19]. V_{exchange} represents the spin-averaged exchange repulsion between the triplet O_2 molecules; the Heisenberg Hamiltonian of Eq. (9) with the coupling function J represents the exchange splitting between the singlet, triplet and quintet potential surfaces

where S_A and S_B are the total electron spin operators of the interacting monomers A and B. Higher terms in the scalar product $(S_A \cdot S_B)$ will be needed in general [16, 40], when the overlap between the molecular charge clouds becomes appreciable, i.e. for distances much closer than the Van der Waals minimum. The Heisenberg coupling parameter J is a scalar function, just as the spin-independent potential, and it can be expanded via Eq. (3). In the case which has been investigated in our group, the O_2-O_2 potential, J appears to be extremely dependent on the molecular orientations, so that rather high terms in Eq. (3) are important. Even the sign of J changes, i.e. the exchange coupling between the triplet O_2 molecules alternates between ferromagnetic and antiferromagnetic, when the molecules are rotated about certain axes, see Fig. 2. This is related to the nodal planes of the antibonding π_g molecular orbitals which form the open shell. Other terms which may contribute to the spin-dependent potentials between open-shell molecules are spin-orbit and spin-spin interactions, intra- as well as inter-molecular [1, 41]. Such terms are anisotropic in two respects: they depend on the orientations of the molecular axes attached to the nuclear framework, and on the orientations of the molecular spin momenta S_A and S_B. Only when the intramolecular spin-orbit and spin-spin couplings are much stronger than the intermolecular exchange interactions, the monomer spin momenta are fixed with respect to the molecular axes; in general, and for O_2-O_2 in particular, they are not. All the spin-dependent terms in the O_2-O_2 potential have been analytically expressed in Refs. [19] and [41].

1.4 Intermolecular Interactions in Condensed Matter; Additivity

So far we have been concerned only with the interaction potential for a pair of molecules A and B. In bulk calculations it is mostly assumed that the total potential energy can be written as a sum of pairwise intermolecular potentials. In the gas phase, where simultaneous collisions of more than two molecules are relatively rare, this approximation is fairly realistic for most properties. In the condensed phases each molecule is surrounded by several others at Van der Waals distances, however, and one has to worry about the importance of three- and more-body interactions. This question can be addressed most clearly by considering the basic interaction mechanisms. The first-order electrostatic interactions are exactly pairwise additive, whether they are calculated in the multipole expansion or between continuous charge clouds. The first-order exchange effects are not pairwise additive. The dominant three-body exchange contributions are roughly proportional to the overlap product $S_{AB}S_{BC}S_{AC}$, while the dominant pairwise exchange energy between three molecules A, B and C increases with $S_{AB}^2 + S_{BC}^2 + S_{AC}^2$. For Van der Waals distances between the molecules the overlap integrals S_{AB}, S_{BC} and S_{AC} are of the order of a few hundredths, and so the first-order exchange non-additivity is also of the order of a few percent. At high pressures this non-additivity will be more important, however, since the overlap integrals increase rapidly with decreasing intermolecular distances.

The second-order dispersion interaction is exactly pairwise additive when it is calculated (as usual) in the multipole approximation, but not when penetration and exchange are included. The second-order penetration and exchange effects are of minor importance, however, and so they do not play a great role in the non-additivity, in general. An important non-additive interaction between non-polar molecules is the third-order dispersion energy, the dominant term of which is called the Axilrod-Teller or triple-dipole dispersion interaction [42]. This long range term which is proportional to $R_{AB}^{-3}R_{BC}^{-3}R_{AC}^{-3}$ and the first-order exchange term which is the dominant short-range three-body interaction, have been studied in detail in rare gas solids [43–45]. At normal pressures the three-body contributions to the cohesion energy are estimated to be of the order of 10%, at high pressures they may be more important. For specific distances (pressures) the Axilrod-Teller and the first-order exchange three-body interactions may nearly cancel; in that case, the small second-order exchange effects dominate the non-additivity [46].

The interaction contribution which lacks pairwise additivity completely is the second-order induction energy. This is easily understood since the energy of polarization of a molecule A in the field of two other molecules B and C is given by $-\frac{1}{2}\alpha_A F^2 = -\frac{1}{2}\alpha_A(F_{BA} + F_{CA}) \cdot (F_{BA} + F_{CA})$, where α_A is the polarizability of molecule A and F_{BA} and F_{CA} are the electric field vectors at the center of molecule A, originating from the (permanent and induced) multipole moments of the molecules B and C. The three-body part of this interaction energy, $-\alpha_A F_{BA} \cdot F_{CA}$, is generally of similar size as the pairwise contribution, $-\frac{1}{2}\alpha_A(F_{BA}^2 + F_{CA}^2)$. In systems with polar molecules and, *a fortiori*, in ionic systems where the induction energy is relatively important, the total interaction energy is therefore highly non-additive. Mainly responsible for this is the second-order induction energy, but the higher order induction contributions are important too. These can be understood as the (non-additive) polarization energies of the molecules in the fields produced by induced moments on other molecules. The induced moments are proportional to the fields at these molecules, which again originate from permanent or induced multipole moments on other molecules, etc. A fully self-consistent treatment of the local fields and the induced moments on all the molecules in the system, which is related to the dielectric theory of macroscopic bodies [47], is necessary to include all these effects. Such a theory should also take into account the higher multipole moments and field gradients.

In practice, it is often assumed in calculations or simulations, even of ionic systems, that the interaction energy is pairwise additive. Some justification of this assumption may be given by the fact that in bulk systems the molecules are more or less symmetrically surrounded, so that the local fields at the molecular sites partly cancel and the induced (lower) multipole moments are small (considerably smaller than expected by considering the fields from individual neighbours). However, even in the most perfect crystal of high symmetry this holds only when all the molecules occupy their equilibrium positions, but not when they are (dynamically) displaced. Simplified models have been devised to account for (non-additive) polarization effects in dynamical calculations, such as the shell model for ionic crystals [48].

2 Dynamics and Properties of Molecular Solids

2.1 Introduction

In molecular solids the molecules cannot move around freely, but they are trapped in relatively deep potential wells, caused by the intermolecular potential. In these wells they can vibrate and since the vibrations of individual molecules are coupled, again by the intermolecular potential, one obtains collective vibrations of all the molecules in the solid, called lattice vibrations or phonons. Phonons associated with the center of mass motions of the molecules are called translational phonons, phonons associated with their hindered rotations or librations are called librons. The degree of hindrance of the rotations may vary. If the molecules have well-defined equilibrium orientations and perform small amplitude librations about these, one speaks about ordered phases. If the molecular rotations are nearly free or if the molecules can oscillate in several orientational pockets and easily jump between these pockets, then the solid is called orientationally disordered or plastic. Several molecular solids may occur in each of these phases, depending on the temperature and pressure; they undergo order/disorder phase transitions. Also the intramolecular vibrations are coupled by the intermolecular potential, via its dependence on the internal coordinates. The excitations of the solid associated with such vibrations are called vibrational excitons or vibrons.

We restrict the attention to periodic solids, molecular crystals. The excitations are characterized by wave vectors q, that lie in the first Brillouin zone of the lattice considered. These excitations are not necessarily pure translational phonons, librons or vibrons, in general they will be mixed. Much experimental information has been collected about such excitations, by infrared and Raman spectroscopy and, in particular, by inelastic neutron scattering. Due to the optical selection rules infrared and Raman spectra can only probe the $q = 0$ excitations. By neutron scattering one can excite states of any given q and thus measure the complete dispersion (wave vector dependence) of the phonon and vibron frequencies.

The vibrations in molecular crystals, the lattice vibrations in particular, are important for the macroscopic properties. First, one has to consider the zero-point vibrations, i.e. the energy difference between the quantum mechanical vibrational ground state of the crystal and the minimum of its potential energy. Especially in the weakly bound Van der Waals crystals formed with apolar molecules, the zero-point motions affect the cohesion energy (i.e. the crystal sublimation energy, extrapolated to $T = 0$ K) to a non-negligible extent. Secondly, the characteristic excitation energies associated with the lattice vibrations are of the same order of magnitude as $k_B T$ (k_B being the Boltzmann constant) in most experimental circumstances. The excited states are thermally populated and, thus, contribute to the properties of the system. In other words, entropy effects are essential. For the thermodynamic stability and for studying various processes that may occur, one should consider the Helmholtz (F) or Gibbs (G) free energy rather than the energy (E). The phase transition between the ordered and disordered phases of the same material (sometimes called orientational melting of the material) may serve as an example.

Usually the ordered phase occurs at low temperature, the plastic phase at higher temperature. The stability of the ordered phase is due to a favourable packing of the molecules, such that the ground-state energy (E) of the solid is as low as possible. The excitation energies of the molecules oscillating in deep (orientational) pockets are relatively high. The thermodynamic stability of the disordered phase at higher temperatures is caused by its lower excitation energies, which are typically those of a hindered rotor. For a given temperature, the excited states of the disordered phase will be more populated, and its entropy (S) will be larger. Above the phase transition temperature T, which can be calculated by comparison of the free energy $F = E - TS$ of the two phases, this will be sufficient to overcome the difference in energy E between the disordered phase and the ordered phase. It will be clear from this example that in order to give a reasonable account of any phase transition, one must accurately calculate the ground-state energies (including the zero-point motions) and the excitation energies of the phases involved. The comparison between calculated and observed phase transition temperatures (or pressures) provides a sensitive test, both for the intermolecular potential and for the method used to describe the lattice vibrations.

An overview of the current lattice dynamics methods is given in Sect. 2.2 (see also Ref. [27]). In practically all cases, up to now, these methods have used semi-empirical intermolecular potentials, mostly of the atom-atom type. The parameters occurring in these potentials are usually fitted to the properties of interest, such as the lattice structure, the cohesion energy and the phonon frequencies. This procedure hides the flaws which are present in the intermolecular potentials as well as in the lattice dynamics method. In studies of solid N_2 [49–53], solid O_2 [41, 54–56] and solid H_2 [57] *ab initio* potentials have been used, however, which contain detailed information on the anisotropy of the potential and, in the case of O_2, also on its spin-dependence. Illustrative results of these studies are described in Sect. 2.3. The final Sect. 2.4, shows some typical phenomena occurring in more complex molecular crystals, such as phonon-vibron mixing, the dispersion and shifts of vibron bands and the effects of isotopic substitution, i.e. changes of nuclear masses, on the lattice- and internal vibrations. These phenomena are illustrated by results obtained on solid tetra-cyano-ethene [58] and on several chlorinated-benzene crystals [59, 60].

2.2 Lattice Dynamics Methods

2.2.1 The Crystal Hamiltonian

Any quantum mechanical lattice dynamics method has to start by giving the crystal Hamiltonian. For a molecular crystal the dynamical coordinates are: the center of mass displacements u_A of the molecules A from their equilibrium positions, the molecular orientations Ω_A and the vibrational coordinates $Q_A = \{Q_{i_A}\}$ of the internal molecular modes i_A. The crystal Hamiltonian can be written as:

$$H = \sum_A [H_A(Q_A) + T(u_A) + L(\Omega_A)] + \frac{1}{2} \sum_{A,B}' V_{AB}(u_A, \Omega_A, Q_A; u_B, \Omega_B, Q_B).$$

$$(10)$$

The pair potential V_{AB} in this equation is the intermolecular potential, defined in Eq. (3), for example. It is now written in terms of displacement coordinates, however. The translational displacements u_A are related to the (instantaneous and equilibrium) vectors R: $R = R_{eq} + u_B - u_A$, and the vibrational displacements Q_A are related to the internal coordinates q_A: $Q_A = q_A - q_{A, eq}$. Since the remainder of this chapter will be dealing only with crystals of apolar molecules (Van der Waals solids), it has been assumed here that the three- and more-body terms in the potential energy may be neglected. Further, it has been assumed that the vibrations of the free molecules are decoupled from their rotations (i.e. centrifugal distortion and coriolis effects are neglected). If the free-molecule vibrations are harmonic, then:

$$H_A(Q_A) = T(Q_A) + V_A^{intra}(Q_A)$$

$$= \sum_{i_A} \left(\frac{1}{2} P_{i_A}^2 + \frac{1}{2} \omega_{i_A}^2 Q_{i_A}^2 \right), \tag{11}$$

where the summation runs over all the normal modes i_A of the free molecule A, with normal coordinates Q_{i_A}, conjugate momenta P_{i_A} and frequencies ω_{i_A}. The kinetic energy operators for the center of mass vibrations and the (rigid) rotations of the molecules are, respectively:

$$T(u_A) = -\frac{\hbar^2}{2M_A} \Delta(u_A) \tag{12}$$

and:

$$L(\Omega_A) = \frac{J_a^2(\Omega_A)}{2I_a} + \frac{J_b^2(\Omega_A)}{2I_c} + \frac{J_c^2(\Omega_A)}{2I_c}. \tag{13}$$

The operator $\Delta(u_A)$ is the Laplacian, M_A is the molecular mass, and $J_a(\Omega_A)$, $J_b(\Omega_A)$, $J_c(\Omega_A)$ are the components of the angular momentum operator with respect to the principal axes a, b and c of the molecule A; I_a, I_b and I_c are the principal values of the molecular inertia tensor (for the equilibrium structure).

The intermolecular potential V_{AB}, as it is given in Eq. (3) for example, does not depend explicitly on the (external) molecular displacements u_A or on the (internal) normal coordinates Q_A, as required by Eq. (10). The atom-atom potential in Eq. (5) does not even depend explicitly on the molecular orientations Ω_A. All these dependencies have to be brought out, by expansion and transformation of the potentials in Eq. (3) and Eq. (5), before these can actually be used in lattice dynamics calculations. The way this is performed depends on the lattice dynamics method chosen (see below). If one is not interested in the internal molecular vibrations, the free-molecule Hamiltonians H_A may be omitted from Eq. (10) and the potential V_{AB} may be averaged over the molecular vibrational states. The effective potential V_{AB} thus obtained no longer depends on the coordinates Q_A and Q_B.

The excitation spectra of molecular crystals and also their thermodynamic functions can be determined, in principle, by diagonalization of the crystal Hamiltonian in Eq. (10). In practice one has to make further approximations. The standard lattice dynamics method is the harmonic method, which is described in many books and reviews [61–64]. Improvements of this method, which are outlined below, can be made by perturbation theory [64, 65] or they can be based on the following thermodynamic variation principle, also called the Gibbs-Bogoliubov inequality:

$$F_{\text{var}} = F_0 + \langle H - H_0 \rangle_0 \geq F. \tag{14}$$

The quantity F is the Helmholtz free energy ($\beta^{-1} = k_B T$):

$$F = -\beta^{-1} \ln(\text{Tr}[\exp(-\beta H)]), \tag{15}$$

associated with the "exact" Hamiltonian H, and F_0 is associated in the same way with an arbitrary approximate Hamiltonian H_0. The thermodynamic expectation value in Eq. (14) also refers to this approximate Hamiltonian:

$$\langle H - H_0 \rangle_0 = \frac{\text{Tr}[(H - H_0)\exp(-\beta H_0)]}{\text{Tr}[\exp(-\beta H_0)]}. \tag{16}$$

A proof of Eq. (14) is given in several places [27, 66–68]. The advantage of this equation is that it can be used to systematically improve the approximate Hamiltonian H_0 and the free energy F_{var}, by optimizing a set of variable parameters or functions contained in H_0. Specific forms chosen for H_0 lead to the Self-Consistent Phonon (SCP) method, the Mean-Field (MF) method and the Time-Dependent Hartree (TDH) or Random-Phase Approximation (RPA).

2.2.2 The Harmonic Method and the Self-Consistent Phonon Method

The basis for the harmonic model is the assumption that the molecules in the solid have well-defined equilibrium positions and orientations, so that one can make a Taylor expansion of the potential about these positions and orientations and truncate after the quadratic terms. It is not always realized that the rotational part of the kinetic energy, Eq. (13), must be approximated too, see Ref. [27]. The dynamical coordinates for the molecular rotations in the harmonic model are linearized angular displacements $\Delta\boldsymbol{\Omega} = (\Delta\Omega_x, \Delta\Omega_y, \Delta\Omega_z)$. Together with the center of mass displacements $\boldsymbol{u} = (u_x, u_y, u_z)$ and the normal coordinates $\{Q_i; i = 1, \dots, 3n - 6\}$ of the intramolecular vibrations, they form a set of $3n$ displacement coordinates for each molecule in the crystal:

$$Q_{j_A} = \begin{cases} u_{x_A}, u_{y_A}, u_{z_A} & \text{for } j = 1, 2, 3 \\ \Delta\Omega_{x_A}, \Delta\Omega_{y_A}, \Delta\Omega_{z_A} & \text{for } j = 4, 5, 6 \\ Q_{i_A} & \text{for } j = i + 6 = 7, \dots, 3n \end{cases} \tag{17}$$

where n is the number of atoms per molecule. The harmonic crystal Hamiltonian reads:

$$H = \frac{1}{2} \sum_A \boldsymbol{P}_A^+ \boldsymbol{G}_A \boldsymbol{P}_A + \frac{1}{2} \sum_{A,B} \boldsymbol{Q}_A^+ \boldsymbol{F}_{AB} \boldsymbol{Q}_B. \tag{18}$$

The column vector Q_A contains the displacement coordinates from Eq. (17), the column vector P_A their conjugate momenta. The matrix G_A is a generalized inverse inertia tensor, which contains the inverse molecular mass M_A^{-1}, the inverse molecular inertia tensor I_A^{-1} and the $3n - 6$ dimensional unit matrix on the diagonal. The force constant matrix contains two-particle blocks:

$$F_{AB} = \left(\frac{\partial^2 V_{AB}}{\partial Q_A \, \partial Q_B} \right)_{eq}, \tag{19}$$

which depend on the intermolecular potential V_{AB}, and one-particle blocks F_{AA} which depend on the intramolecular force-field as well as on the intermolecular potential. The effect of the intramolecular force-field is contained already in the normal coordinates $\{Q_{i_A}; i_A = 1, \ldots, 3n - 6\}$ of the free molecule, see Eq. (11). If these are used as part of the molecular displacement coordinates, as in Eq. (17), then the blocks F_{AA} can simply be written as

$$(F_{AA})_{jj'} = F_{j_A j'_A} = \delta_{j_A j'_A} \omega_{j_A}^2 + F_{j_A j'_A}^{inter}, \tag{20}$$

with $\omega_{j_A} = 0$ for $j_A = 1, \ldots, 6$ and $\omega_{j_A} = \omega_{i_A}$ for $j_A = i_A + 6 = 7, \ldots, 3n$. The intermolecular contributions to F_{AA} depend on the intermolecular potential V_{AB}; they are given in terms of the elements F_{AB} by the conditions that the Hamiltonian of Eq. (18) must be invariant under overall translations and rotations of the crystal [64]. They can also be obtained by direct differentiation of the intermolecular potential $V^{inter} = \frac{1}{2} \sum'_{A,B} V_{AB}$ with respect to the displacement coordinates Q_A.

The most convenient way to calculate the force constants F_{AB} and F_{AA}^{inter} depends on the analytic form of the intermolecular potential V_{AB}. If this potential is given in terms of the molecular orientations Ω_A and Ω_B and the center of mass distances R, as in Eq. (3), then it is most efficient to differentiate V_{AB} directly with respect to the molecular displacements u_A and $\Delta\Omega_A$. If the potential V_{AB} is given as an atom-atom potential, as in Eq. (5), then one may differentiate V_{AB} with respect to atomic displacements (applying the formulas given in Ref. [70], for example) and use the relation between the molecular displacement coordinates in Eq. (17) and the atomic displacements. In transforming the atomic derivatives one must take into account that the rotational molecular displacements are curvilinear [59, 69, 70].

The harmonic Hamiltonian in Eq. (18) can be diagonalized exactly. First, one introduces Fourier-transformed displacement coordinates:

$$Q_\alpha(q) = \frac{1}{\sqrt{N}} \sum_m \exp(-i q \cdot R_m) Q_A \tag{21}$$

and their conjugate momenta $P_\alpha(q)$. A molecule A is assumed to lie in the unit cell with origin R_m and to belong to sublattice α, so we write $A = \{m, \alpha\}$, and N is the number of unit cells in the lattice. After this transformation the Hamiltonian in Eq. (18) separates into a set of commuting Hamiltonians, one for every wave vector q:

$$H(q) = \frac{1}{2} \sum_\alpha P_\alpha^\dagger(q) \, G_\alpha P_\alpha(q) + \frac{1}{2} \sum_{\alpha, \beta} Q_\alpha^+(q) \, F_{\alpha\beta}(q) \, Q_\beta(q), \tag{22}$$

where G_α is the generalized inverse inertia tensor G_A, which is the same for every molecule A in sublattice α, and $F_{\alpha\beta}(q)$ is the Fourier transformed force constant matrix:

$$F_{\alpha\beta}(q) = \sum_m F_{\{0,\,\alpha\}\,\{m,\,\beta\}} \exp\left(iq \cdot R_m\right). \tag{23}$$

The latter matrix is often called the dynamical matrix. The transformation coefficients $e_\alpha(q)$ which diagonalize the Hamiltonian can be found from the following generalized eigenvalue problem for the dynamical matrix:

$$\sum_\beta F_{\alpha\beta}(q)\, e_\beta(q) = G_\alpha^{-1} e_\alpha(q)\, \omega(q)^2 \,. \tag{24}$$

The eigenfrequencies $\omega_k(q)$ are the (internal as well as external) vibrational excitation frequencies of the crystal in the harmonic approximation. The eigenvectors $e_{\alpha k}(q)$, which express the crystal normal modes in terms of the displacement coordinates $Q_\alpha(q)$, are usually called the polarization vectors.

The macroscopic properties of the crystal can be derived from the partition function, which reads in the harmonic approximation:

$$\begin{aligned} Z &= \mathrm{Tr}\left[\exp\left(-\beta H\right)\right] \\ &= \mathrm{Tr}\left[\prod_q \exp\left(-\beta H(q)\right)\right] \\ &= \prod_q \prod_k \left[2 \sinh\left(\frac{1}{2}\beta\hbar\omega_k(q)\right)\right]^{-1}. \end{aligned} \tag{25}$$

As found in any textbook on statistical thermodynamics one can write, for instance, for the Helmholtz free energy:

$$\begin{aligned} F &= -\beta^{-1} \ln Z \\ &= \sum_q \sum_k \beta^{-1} \ln\left[2 \sinh\left(\frac{1}{2}\beta\hbar\omega_k(q)\right)\right], \end{aligned} \tag{26}$$

for the entropy:

$$S = -\left(\frac{\partial F}{\partial T}\right)_V, \tag{27}$$

for the energy:

$$E = F + TS\,, \tag{28}$$

for the pressure:

$$p = -\left(\frac{\partial F}{\partial V}\right)_T, \tag{29}$$

for the specific heat:

$$c_v = T \left(\frac{\partial S}{\partial T} \right)_V = -T \left(\frac{\partial^2 F}{\partial T^2} \right)_V \tag{30}$$

$$c_p = T \left(\frac{\partial S}{\partial T} \right)_p = c_v + \alpha_p^2 \frac{TV}{\varkappa_T}, \tag{31}$$

for the thermal expansion coefficient:

$$\alpha_p = \frac{1}{V} \left(\frac{\partial V}{\partial T} \right)_p, \tag{32}$$

for the compressibility:

$$\varkappa_T = -\frac{1}{V} \left(\frac{\partial V}{\partial p} \right)_T, \tag{33}$$

etc.

The anharmonic terms, i.e. the cubic and higher terms in the displacement expansion of the intermolecular potential and the rotational kinetic energy terms, which are neglected in the harmonic Hamiltonian, can be considered as perturbations. They affect the vibrational excitations of the crystal in two ways: they shift the excitation frequencies and they lead to finite lifetimes of the excited states, which are visible as spectral line broadening. By means of anharmonic perturbation theory based on a Green's function approach [64, 65] it is possible to calculate the frequency shifts, as well as the line widths.

Another technique to obtain the effects of the anharmonic terms on the excitation frequencies and the properties of molecular crystals is the Self-Consistent Phonon (SCP) method [71]. This method is based on the thermodynamic variation principle, Eq. (14), for the exact Hamiltonian given in Eq. (10), with the internal coordinates Q_A not explicitly considered. As the approximate Hamiltonian H_0 one takes the harmonic Hamiltonian of Eq. (18). The force constants F_{AB} in Eq. (18) are not calculated at the equilibrium positions and orientations of the molecules as in Eq. (19), however. Instead, they are considered as variational parameters, to be optimized by minimization of the Helmholtz free energy according to Eq. (14). The optimized force constants are found to be the thermodynamic (and thus temperature dependent) averages of the second derivatives of the potential over the (harmonic) lattice vibrations:

$$F_{AB} = \left\langle \frac{\partial^2 V_{AB}}{\partial Q_A \partial Q_B} \right\rangle_0 \tag{34}$$

and the result for the optimized free energy is:

$$F = \sum_q \sum_k \beta^{-1} \ln \left[2 \sinh \left(\frac{1}{2} \beta \hbar \omega_k(q) \right) \right]$$

$$+ \frac{1}{2} \sum_{A, B}{}' \langle V_{AB}(u_A, \Omega_A; u_B, \Omega_B) \rangle_0$$

$$- \frac{1}{4} \sum_q \sum_k \hbar \omega_k(q) \coth \left(\frac{1}{2} \beta \hbar \omega_k(q) \right). \tag{35}$$

The thermodynamic averages of the force constants and the potential over harmonic oscillator states are evaluated by first calculating the displacement-displacement correlation matrix:

$$D_{AB} = \langle Q_A Q_B^+ \rangle_0 , \tag{36}$$

including the diagonal blocks D_{AA} and D_{BB}. The inverse of this matrix determines the width (i.e. the exponents) of a Gaussian distribution function $\varrho_{AB}(Q_A, Q_B)$ over which the quantities of interest can be integrated. The nature of this procedure implies that the SCP method takes into account only the anharmonic terms of even power in the displacements Q_A. The SCP method was originally formulated for atomic (rare gas) crystals [71]; its first extension to molecular crystals [72] was still written in terms of atomic displacements and it was restricted to the use of atom-atom potentials, Eq. (5). In Ref. [27] the SCP method is formulated more generally for molecular crystals with the explicit inclusion of the librational displacements $\Delta \Omega_A$ and the possibility to use potentials as in Eqs. (3), (7) and (8) that depend directly on the orientational coordinates Ω_A.

In the harmonic method and its extensions it is always assumed that the amplitudes of the molecular vibrations about their equilibrium positions and orientations remain small. It will be illustrated by several examples in Sect. 2.3 that this is often not realistic for molecular crystals. In the plastic phases there is not even a well-defined equilibrium orientation of the molecules, but also in the ordered phases the librational amplitudes may become substantial. The motions in such systems have been studied by classical methods, in particular the Molecular Dynamics method and the Monte Carlo method [73]. The advantage of these methods is that they can also be applied to study liquids, and the melting of solids, and other systems (glasses, solutions, mixed crystals) which have lost translational periodicity. Large amplitude motions in molecular crystals can also be studied quantum mechanically, however, by the methods described below.

2.2.3 The Mean-Field Method, the Random-Phase Approximation and the Time-Dependent Hartree Method

Just as the SCP method, the Mean Field (MF) method [74, 75] is based on the thermodynamic variation principle, Eq. (14). This time, however, the approximate Hamiltonian H_0 is chosen to be the sum of effective single-particle Hamiltonians:

$$H_0 = \sum_A H_A^{MF}(u_A, \Omega_A) . \tag{37}$$

Substitution of this form, together with Eq. (10), into F_{var} in Eq. (14) and minimization of the latter quantity yields the result that the optimized effective single-particle Hamiltonian is precisely the MF Hamiltonian:

$$H_A^{MF}(\boldsymbol{u}_A, \boldsymbol{\Omega}_A) = T(\boldsymbol{u}_A) + L(\boldsymbol{\Omega}_A) + \sum_{B \neq A} \langle V_{AB}(\boldsymbol{u}_A, \boldsymbol{\Omega}_A; \boldsymbol{u}_B, \boldsymbol{\Omega}_B) \rangle_B . \tag{38}$$

The thermodynamic (temperature dependent) average of the potential V_{AB} has to be taken with respect to the Hamiltonians $H_B^{MF}(\boldsymbol{u}_B, \boldsymbol{\Omega}_B)$, and so the calculation of the MF Hamiltonians has to proceed self-consistently. In principle, the Hamiltonians H_A^{MF} may be different for all molecules A. In practice, one can relate them by all the symmetry operations in the (adopted) space group of the crystal. The expansion of the potential V_{AB} in the symmetry-adapted functions of Eq. (1), as given by Eqs. (3) and (4), is very convenient for this purpose, since the transformation of these functions in Eq. (1) under the rotations $\hat{R}(\boldsymbol{\Omega}_g)$ contained in the space group follows directly from the relation [28]:

$$\hat{R}(\boldsymbol{\Omega}_g) \, D_{MK}^{(L)}(\boldsymbol{\Omega})^* = \sum_{M'} D_{M'K}(\boldsymbol{\Omega})^* \, D_{M'M}(\boldsymbol{\Omega}_g) . \tag{39}$$

In practice the calculation of the MF Hamiltonians is performed in a finite basis [27, 49–51]. In every cycle of the self-consistent procedure, the MF Hamiltonians are diagonalized in this basis,

$$H_A^{MF} |\psi_{k_A}\rangle = \varepsilon_{k_A} |\psi_{k_A}\rangle , \tag{40}$$

and the thermodynamic average of any operator X is simply:

$$\langle X \rangle_A = \frac{\text{Tr}\,[X \exp(-\beta H_A^{MF})]}{\text{Tr}\,[\exp(-\beta H_A^{MF})]} = \frac{\sum\limits_{k_A} \langle \psi_{k_A}| X | \psi_{k_A}\rangle \exp(-\beta \varepsilon_{k_A})}{\sum\limits_{k_A} \exp(-\beta \varepsilon_{k_A})} . \tag{41}$$

The average potentials $\langle V_{AB} \rangle_B$ yield the MF Hamiltonians of Eq. (38) for the next cycle, until self-consistency is reached. An additional simplification may be obtained by a further separation between the center of mass vibrations of the molecules and their orientational motions:

$$H_A^{MF}(\boldsymbol{u}_A, \boldsymbol{\Omega}_A) = H_A^T(\boldsymbol{u}_A) + H_A^L(\boldsymbol{\Omega}_A) . \tag{42}$$

This yields two "particles" for every molecule, one translating and one rotating, whose motions have to be solved self-consistently. The correlation between these motions is recovered at the RPA or TDH level (see below).

With the potential V_{AB} expanded as in Eqs. (3) and (4), a convenient basis [27, 49] for the orientational motions of the molecules consists of (symmetric top) free-rotor functions $D_{m_1 k_1}^{(l_1)}(\boldsymbol{\Omega}_A)^*$. For linear molecules the latter functions are simply spherical harmonics $C_{m_1}^{(l_1)}(\beta_A, \alpha_A)$. This basis is appropriate, of course, when the molecules undergo nearly free rotations, as in solid hydrogen. But also in the ordered phases of solid nitrogen and solid oxygen, for example, a set of spherical harmonics with quantum numbers up to $l_1 = 12$ yields well-converged (orientationally localized) MF states for the librations of the molecules. For homonuclear diatomic molecules only the basis functions with even l_1 or those with odd l_1 have to be included, depending on the nuclear spin (ortho/para) states.

A suitable basis for the center of mass vibrations of the molecules consists of the three-dimensional (spherical) harmonic oscillator functions $\chi_{l_2 m_2}^{(n)}(u_A)$. The number of such functions to be used depends on the anharmonicity of the potential. For the details about the expansion of the potential in Eqs. (3) and (4) in terms of the displacements u_A and u_B, including the higher (anharmonic) terms, and about the calculation of the matrix elements, we refer to Refs. [27, 51].

From the converged MF Hamiltonians it is possible again to calculate all the thermodynamic functions of the crystal. The free energy can be found from Eq. (14), and other quantities follow from it:

$$F = -k_B T \sum_A \ln(Z_A^{MF}) - \frac{1}{2} \sum_{A,B}' \langle V_{AB} \rangle_{A,B} \tag{43}$$

$$S = -\frac{\partial F}{\partial T} = k_B \sum_A \ln(Z_A^{MF}) + T^{-1} \sum_A \langle H_A^{MF} \rangle_A \tag{44}$$

$$E = F + TS = \sum_A \langle H_A^{MF} \rangle_A - \frac{1}{2} \sum_{A,B}' \langle V_{AB} \rangle_{A,B} \tag{45}$$

with the Mean-Field partition function:

$$Z_A^{MF} = \sum_{k_A} \exp(-\beta \varepsilon_{k_A}). \tag{46}$$

The excitations of the crystal that originate from the MF model may correspond with strongly anharmonic translational vibrations or librations of the molecules; they may even correspond with hindered- or free-rotor states. They remain single particle excitations, however, which do not show any dispersion (i.e. wave vector dependence) in their frequencies. The simplest manner to obtain this dispersion is by the so-called Exciton Model or Tamm-Dancoff Approximation [76]. From the crystal ground state, which is a product of (known) MF states:

$$|\Psi_{\{k\}}\rangle = \prod_{k_A} |\psi_{k_A}\rangle, \tag{47}$$

with all $k_A = 0$, excited states are constructed by promoting a single molecule to a state with $k_A \neq 0$. Excitations are allowed to travel through the crystal with wave vector q by defining a set of crystal excitation operators:

$$E_{k\alpha}^+(q) = \frac{1}{\sqrt{N}} \sum_m \exp(iq \cdot R_m) E_{k_A}^+, \tag{48}$$

which are Fourier-transformed single-particle excitation operators [cf. Eq. (21); it is assumed again that molecule A lies in the unit cell with origin R_m and belongs to sublattice α]. In the Exciton Model the crystal Hamiltonian of Eq. (10) is diagonalized between the states created by the excitation operators of Eq. (48). Although this model does yield dispersion and performs reasonably well for the collective intramolecular vibrations (vibrons) in molecular crystals, it is not satisfactory for the lattice vibrations (phonons). The reason is that the invariance of the crystal Hamiltonian of Eq. (10) under arbitrary translations of the whole crystal is lost in the Exciton Model. As a consequence, the frequency of the acoustic phonon branches will not vanish in the limit of infinitely long wavelength ($q = 0$), as it should.

The Random-Phase Approximation (RPA) [77, 78], which is hardly more complicated than the Exciton Model, behaves correctly in this respect. It uses the same excitation operators and the corresponding de-excitation operators and it writes the crystal Hamiltonian as:

$$H - H_0 = \sum_{A,B} \sum_{k_A k'_B} c_{k_A k'_B} E^+_{k_A} E_{k'_B} + \sum_{A,B}' \sum_{k_A k'_B} [d_{k_A k'_B} E^+_{k_A} E^+_{k'_B} + d^*_{k_A k'_B} E_{k_A} E_{k'_B}].$$

(49)

The coefficients $c_{k_A k_B}$ are matrix elements of the Hamiltonian of Eq. (10) between singly excited MF states of Eq. (47) and the coefficients $d_{k_A k' B}$ are matrix elements between the MF ground state and the doubly excited states. The following formulas are easily derived for these coefficients (for real MF functions ψ_{k_A}):

$$c_{k_A k'_B} = \chi_{k_A k'_B} + \Phi_{k_A k'_B}$$

$$d_{k_A k'_B} = \Phi_{k_A k'_B},$$

(50)

where the diagonal matrix χ contains the MF excitation energies:

$$\chi_{k_A k'_B} = \delta_{kk'} \delta_{AB} (\varepsilon_{k_A} - \varepsilon_{0_A})$$

(51)

and the elements of the matrices Φ_{AB} are defined by:

$$(\Phi_{AB})_{kk'} = \Phi_{k_A k'_B} = \langle \psi_{k_A} \psi_{0_B} | V_{AB} | \psi_{0_A} \psi_{k'_B} \rangle.$$

(52)

The first term in this RPA Hamiltonian of Eq. (49) is the Exciton Model Hamiltonian; the additional terms ensure translational invariance. From Eq. (50) it follows that the construction of the RPA Hamiltonian is equally simple as that of the Exciton Model Hamiltonian. Also its diagonalization is not very difficult. First it can be split into separate Hamiltonians $H(q)$ for every wave vector q by the Fourier transformation of Eq. (48) [cf. Eqs. (21) and (22)]. The excitation operators $a^+_p(q)$ which diagonalize the Hamiltonian are then written as:

$$a^+_p(q) = \sum_\alpha ' \sum_k [x^p_{k\alpha}(q) E^+_{k\alpha}(q) + y^p_{k\alpha}(q) E_{k\alpha}(-q)]$$

(53)

and it can be shown that the coefficients $x_{k\alpha}^p(q)$ and $y_{k\alpha}^p(q)$ must be obtained from the following matrix equation:

$$\begin{pmatrix} \chi + \Phi(q) & -\Phi(q) \\ \Phi(q) & -\chi - \Phi(q) \end{pmatrix} \begin{pmatrix} x(q) \\ y(q) \end{pmatrix} = \begin{pmatrix} x(q) \\ y(q) \end{pmatrix} \omega(q) . \tag{54}$$

The matrix χ is given by Eq. (51) and the blocks of the matrices $\Phi(q)$ are defined by:

$$\Phi_{\alpha\beta}(q) = \sum_m \Phi_{\{0, \alpha\}\{m, \beta\}} \exp(iq \cdot R_m), \tag{55}$$

in terms of the matrices Φ_{AB} from Eq. (52) with $A = \{0, \alpha\}$ and $B = \{m, \beta\}$. It is obvious that the matrix $\Phi(q)$ given by Eqs. (55) and (52) bears some similarity to the dynamical matrix $F(q)$ in the harmonic model, defined in Eqs. (23) and (19). In fact, if the Hamiltonian of Eq. (10) would be exactly harmonic, the RPA model would render the exact solution.

When the MF model is further separated by Eq. (42) the MF excitations are either translational (T) or librational (L). The indices k and k' which label the rows and columns of the matrix $\Phi(q)$ run over both types of excitations. The elements $\Phi^{TT}(q)$ and $\Phi^{LL}(q)$ correlate the translational and rotational vibrations of the individual molecules; the resulting lattice vibrations are the translational phonons and the librons. These are coupled by the elements $\Phi^{TL}(q)$ and $\Phi^{LT}(q)$, which contain the translation-rotation coupling terms present in the intermolecular potential, cf. Eq. (3). The frequencies $\omega_p(q)$ obtained by solving the eigenvalue equation in Eq. (54) are the (vibrational) excitation frequencies of the crystal, which in general correspond with mixed translational-phonon/libron modes.

The Time-Dependent Hartree (TDH) method [79, 80] is a generalization of the RPA formalism. In RPA only the MF excitations from the ground state $|0_A\rangle$ to the excited states $|k_A\rangle$ are taken into account. In TDH it is assumed that the excited states are thermally populated and that excitations to higher excited states can occur as well as de-excitations to lower states. The matrix equation to be solved for the TDH excitation frequencies $\omega_p(q)$ can be written in the same form as Eq. (51). The matrices χ and $\Phi(q)$ have larger dimensions, however. The diagonal matrix χ is given by:

$$\chi_{k_1 k_2 A, \, k_1' k_2' B} = \delta_{k_1 k_1'} \delta_{k_2 k_2'} \delta_{AB}(\varepsilon_{k_1 A} - \varepsilon_{k_2 A}) \tag{56}$$

and the matrix $\Phi(q)$ is defined by Eq. (55) with:

$$\begin{aligned} (\Phi_{AB})_{k_1 k_2, \, k_1' k_2'} &= \Phi_{k_1 k_2 A, \, k_1' k_2' B} \\ &= (\varrho_{k_2 A} - \varrho_{k_1 A}) \langle \psi_{k_1 A} \psi_{k_2' B} | V_{AB} | \psi_{k_2 A} \psi_{k_1' B} \rangle . \end{aligned} \tag{57}$$

The excited state populations are given by:

$$\varrho_{k_A} = \frac{\exp(-\beta\varepsilon_{k_A})}{\sum\limits_{k_A} \exp(-\beta\varepsilon_{k_A})} . \tag{58}$$

It is easily checked that TDH is equivalent to RPA at $T = 0$ K, because only the ground state is populated then. At higher temperatures the TDH model is expected to work better than the RPA formalism.

From the structure of the eigenvalue equation of Eq. (54) it follows [79] that the eigenvalues $\omega_p(q)$ must be purely real or purely imaginary. The real frequencies are obtained in pairs with the same absolute values but different signs; they represent the excitations and corresponding de-excitations of the crystal. If one or more frequencies are imaginary, it can be proved [27, 50, 79] that the self-consistent solutions of the MF equations, which have been used to construct the excitation and de-excitation operators, do not correspond to a minimum of the Helmholtz free energy F_{var}, see Eq. (14), but rather to a saddle point. By choosing a different MF solution, which corresponds to a different symmetry of the unit cell for instance, it will be possible to obtain a more stable solution. An example of this situation occurs in the orientationally disordered β-phase of solid nitrogen [50], see Sect. 2.3. It is similar to the occurrence of soft-mode behavior in the harmonic model: if one or more frequencies in the harmonic model become zero or imaginary, the equilibrium structure chosen for the crystal does no longer correspond to a minimum of the energy. The experimental observation of certain modes becoming soft is a sign for the occurrence of a phase transition.

The thermodynamic functions of the crystal are not uniquely defined by the RPA or the TDH method. Both methods produce the excitation energies $\omega_p(q)$ and the corresponding excitation operators of Eq. (53). They do not produce an explicit ground state wave function, however. Implicity it appears from the occurrence of MF de-excitation operators in Eq. (53) that the ground state in RPA and TDH must contain excited MF states. From the form of the RPA Hamiltonian, Eq. (49), it follows that doubly excited MF states are admixed into the MF ground state; the corresponding energy lowering is the correlation energy. By making specific assumptions about the ground state wave function, which also fixes the excited states through the excitation operators of Eq. (53), it is possible to define the thermodynamic functions at the RPA or TDH level [53].

2.2.4 Spin Waves

In crystals composed of open-shell molecules such as solid oxygen, see Sect. 2.3, the electronic spin momenta provide an extra degree of freedom. In the ground state these spins will be coupled, mainly by the Heisenberg term of Eq. (9) in the spin-dependent intermolecular potential, to a ferromagnetic or to some kind of antiferromagnetic order. The collective excitations of the spin system are called spin waves or magnons. Just as the translational phonons and librons, these magnetic excitations can be observed by infrared and Raman spectroscopy and by inelastic neutron scattering and they must obey certain symmetry selection rules [56]. Their frequencies are typically of the order of a few cm^{-1}, so they contribute substantially to the thermodynamic properties of the solid. One of the obvious signs of the importance of magnons is the observed temperature variation of the specific heat in solid oxygen, for example.

Since the orientations of the molecular spin momenta are quantized, it is natural to treat the spin waves by a quantum mechanical theory. This theory follows closely the methods outlined in the previous subsection. First, the ground state of the spin system and the single-particle excitations are obtained by MF theory, in a basis of $2S + 1$ spin-functions per molecule (if S is its total spin quantum number). Next, the collective spin excitations (magnons) are calculated by the RPA or TDH method. When the translational phonons and the librons have also been obtained by these methods, it is relatively easy to make an integrated treatment [41] of the spin waves and the lattice vibrations. The "dynamical matrix" $\boldsymbol{\Phi}(\boldsymbol{q})$, see Eqs. (55) and (52) and the text below these equations, will contain elements $\boldsymbol{\Phi}^{TS}(\boldsymbol{q})$ and $\boldsymbol{\Phi}^{LS}(\boldsymbol{q})$ that couple the translational phonons and the librons to the magnons. In principle, this may lead to the occurrence of mixed vibrational and magnetic excitations. Whether this coupling is actually important depends on the form of the spin-dependent intermolecular potential. It has been observed in Sect. 1.3, for example, that the spin-dependent Heisenberg term of Eq. (9) is extremely dependent on the orientations $\boldsymbol{\Omega}_A$ and $\boldsymbol{\Omega}_B$ of the molecules, see Fig. 2, and also on their translational displacements (which affect the intermolecular distance R). In the subsection on solid oxygen some consequences of these effects will be illustrated.

2.3 "Simple" Molecular Crystals; ab initio Treatment

2.3.1 Solid Nitrogen; Orientational Order/Disorder

Solid N_2 is one of the simplest molecular crystals and, as such, it has been subject to many studies, experimental as well as theoretical. It is still of fundamental interest, since it has ordered phases as well as disordered phases and the nature of the orientational motions in the latter phases (nearly-free rotations, precessions or orientational jumps) could not yet be established experimentally [81]. Even in the ordered phases the amplitudes of the molecular librations are not really small, however. At normal pressure the ordered α-phase is stable below 35.6 K; above this temperature, up to the melting point at 63 K, one finds the disordered β-phase. At higher pressure the latter phase remains stable even at room temperature, but several other phases are

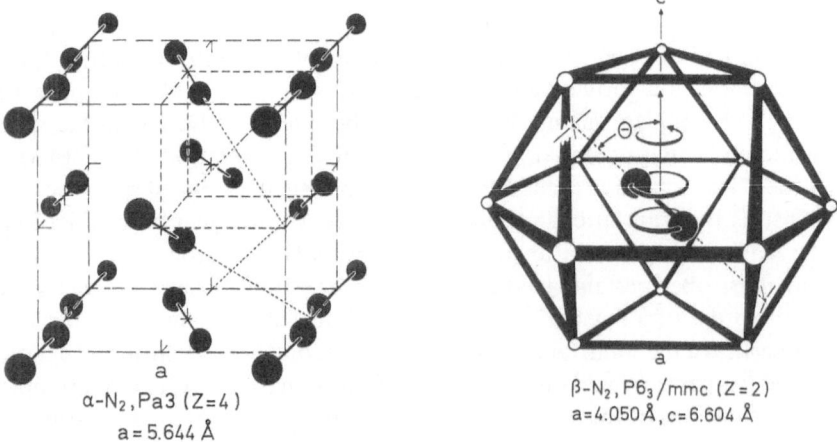

α-N_2,Pa3 (Z=4)
a = 5.644 Å

β-N_2, P6$_3$/mmc (Z=2)
a=4.050 Å, c=6.604 Å

Fig. 3. Crystal structures of α- and β-nitrogen [81]

known also. The structure of the α and β phases is shown in Fig. 3. The results prior to 1976 have been collected by Scott [81]; more recent studies are cited in Ref. [27]. All the lattice dynamics methods mentioned in Sect. 2.3 have been applied to solid nitrogen, using empirical atom-atom potentials or simply quadrupole-quadrupole interactions, as well as detailed anisotropic potentials from *ab initio* calculations. Here we concentrate on the latter.

The *ab initio* potentials used in solid nitrogen are from Refs. [31] and [32]. They have been respresented by a spherical expansion, Eq. (3), with coefficients $v_{L_A L_B L}(R)$ up to $L_A = 6$ and $L_B = 6$ inclusive, which describe the anisotropic short-range repulsion, the multipole-multipole interactions and the anisotropic dispersion interactions. They have also been fitted by a site-site model potential, Eq. (5), with force centers shifted away from the atoms, optimized for each interaction contribution. In the most advanced lattice dynamics model used, the TDH or RPA model, the librations are expanded in spherical harmonics up to $l = 12$ and the translational vibrations in harmonic oscillator functions up to $n = 4$, inclusive.

Results for the translational-phonon, libron and mixed mode frequencies in the ordered α-phase are shown in Table 2, for various points in the Brillouin zone. Phonon

Table 2. Lattice frequencies in α-N_2 (in cm^{-1})

		Experiment [82]	Semi-empirical harmonic [83]	*Ab initio* harmonic [84]	SCP [84]	RPA [51]
Lattice constant a (Å)		5.644	5.644	5.611	5.796	5.699
$\Gamma(0,0,0)$						
Librations	E_g	32.3	37.5	42.4	41.1	31.0
	T_g	36.3	47.7	52.9	50.7	41.0
	T_g	59.7	75.2	77.7	73.7	68.0
Translational vibrations	A_u	46.8	45.9	52.8	49.2	47.2
	T_u	48.4	47.7	52.6	49.0	48.8
	E_u	54.0	54.0	58.9	54.1	55.6
	T_u	69.4	69.5	78.8	73.3	73.1
$M\left(\dfrac{\pi}{a}, \dfrac{\pi}{a}, 0\right)$						
Mixed	M_{12}	27.8	29.6	34.9	32.7	27.6
	M_{12}	37.9	40.6	46.4	43.8	39.1
	M_{12}	46.8	51.8	59.1	55.8	50.2
	M_{12}	54.9	59.0	64.4	60.4	59.1
	M_{12}	62.5	66.4	72.3	67.6	66.5
$R\left(\dfrac{\pi}{a}, \dfrac{\pi}{a}, \dfrac{\pi}{a}\right)$						
Translational vibrations	R_1^-	33.9	34.4	37.1	34.7	34.4
	R_{23}^-	34.7	35.7	39.2	36.5	35.8
	R_{23}^-	68.6	68.3	77.6	72.3	72.3
Librations	R_1^+	43.6	50.7	58.1	55.2	47.9
	R_{23}^+	47.2	57.8	61.0	58.4	50.8
rms deviation of librational frequencies			10.6	14.8	12.2	5.0
rms deviation of translational frequencies			0.6	6.3	2.4	2.1
rms deviation of all lattice frequencies			6.1	10.4	7.6	3.4

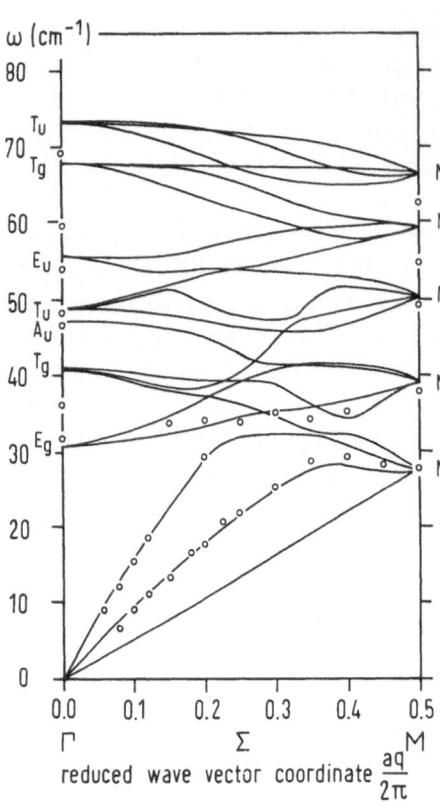

Fig. 4. Calculated (TDH) dispersion curves in α-nitrogen for (mixed) phonon modes propagating along the [1,1,0] direction, from Ref. [51]. The circles correspond with inelastic neutron scattering data [82], measured at $T = 15$ K

dispersion curves are displayed in Fig. 4. Given the fact that the potential has not been adjusted (as the empirical potentials usually are, cf. the second column of Table 2), the agreement with the experimental frequencies is very satisfactory. Also the macroscopic properties of α-N$_2$ calculated with the *ab initio* potentials agree well with the experimental data, see Table 3 and Fig. 5. It is clear from Table 2 that especially the librational modes are improved by the RPA model. The mean amplitude of the zero-point librations in α-N$_2$ appears to be 16°. It has been found recently in our group [87] that the correct description of the anisotropy of the potential by the spherical expansion of Eq. (3) used in the RPA model is mainly responsible for this improvement.

Table 3. Macroscopic properties of α-N$_2$

		Experiment	Calculated (*ab initio*)
		Ref. [85]	(RPA) [53]
Elastic constants	C_{11}	29.0	28.5 ± 0.6
(kbar)	C_{12}	20.0	22.0 ± 0.6
	C_{44}	13.5	13.3 ± 0.6
		Ref. [86]	(MF) [32]
Compressibility	$\varkappa_T$	4.6	4.69

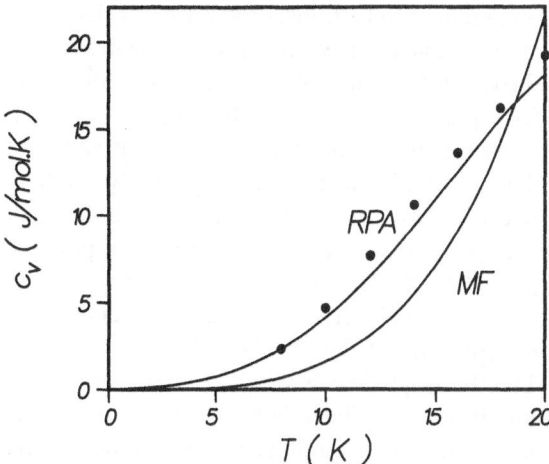

Fig. 5. Specific heat of α-nitrogen. The dots represent the experimental values [81], the curves have been calculated in Ref. [53]

Apparently, the site-site potential used in the harmonic and SCP calculations, which is fitted to the same *ab initio* data represented by the spherical expansion, has lost substantial accuracy in the anisotropy.

Lattice dynamics studies of the disordered β-phase are more scarce because, obviously, the standard harmonic method and the SCP method cannot be applied to this phase (although in some studies the harmonic method has still been used for the translational phonons, while neglecting the anisotropy of the potential.) Most calculations on β-N_2 have been made by classical Monte Carlo or Molecular Dynamics methods, using semiempirical atom-atom or quadrupole-quadrupole potentials. In our group [50, 52] we have investigated the motions in β-N_2 and the α — β order/disorder phase transition by means of the MF, RPA and TDH methods, using the same spherically expanded anisotropic *ab initio* potential which yields accurate properties for α-N_2.

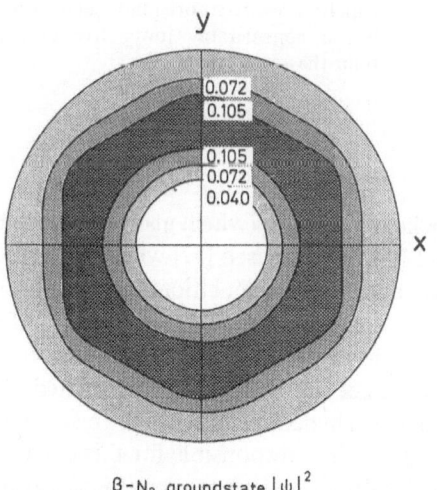

Fig. 6. Orientational probability distribution of the N_2 molecules in the disordered β-phase of solid nitrogen, from Mean Field calculations [50]. This result, which would correspond to a nearly free precession of the molecules about the z-axis with a large spread in the precession angle, is thermodynamically unstable

If all the molecules in the β-phase are assumed to be translationally equivalent, the MF orientational distribution corresponds to a nearly free precession of the molecular axes about the crystal c-axis, with a broad distribution of the precession angle, see Fig. 6. Some of the TDH frequencies corresponding to this MF solution are imaginary, however, and so the distribution shown in Fig. 6 does not represent a stable solution for the β-phase. The structure found to be stable, with a considerably lower free energy F, is a solution in which the two molecules in the unit cell of the hexagonal β-phase have found different (not in shape, but in direction) orientational pockets in which they perform rather localized librations. Thus they avoid to a large extent the steric hindrance between nearest-neighbor molecules, which would occur in the "precession" model from Fig. 6. The frequencies of the librations localized in these pockets, as calculated [50] by the TDH method, agree fairly well with the experimental spectra (which show very broad bands, however). The symmetry of the calculated stable solution, with two different molecules in the unit cell which have specific orientations, is much lower than the symmetry of the crystal observed [81] by X-ray and neutron diffraction studies and by nuclear magnetic and quadrupole resonance (NMR and NQR). It has to be assumed that every N_2 molecule can librate in six orientational pockets, with different but equivalent directions, see Fig. 7, and that

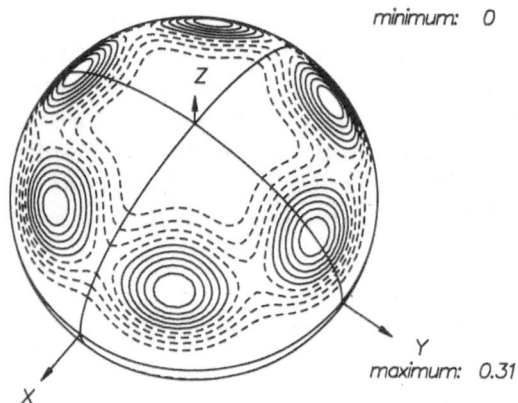

minimum: 0

maximum: 0.31

Fig. 7. Orientational probability distribution of the N_2 molecules in β-nitrogen, calculated in Refs. [50, 52]. This result, which represents six orientational pockets, has a considerably lower free energy than the result in Fig. 6

the molecules jump occasionally from one pocket to another (which also explains the spectral line broadening). A theoretical model has been devised [52] which takes the orientational pockets obtained from the MF and TDH calculations as a starting point, and uses the same anisotropic *ab initio* potential. This model has yielded the orientational distribution of the molecular axes displayed in Fig. 7. It appears from this model that the occupancy of the different pocket states is strongly correlated for nearest neighbors, see Fig. 8. It has not been analyzed in detail which of the anisotropic terms present in the intermolecular potential are mainly responsible for this correlation, but it will be clear, see Fig. 1, that steric hindrance (exchange repulsion) effects play an important role.

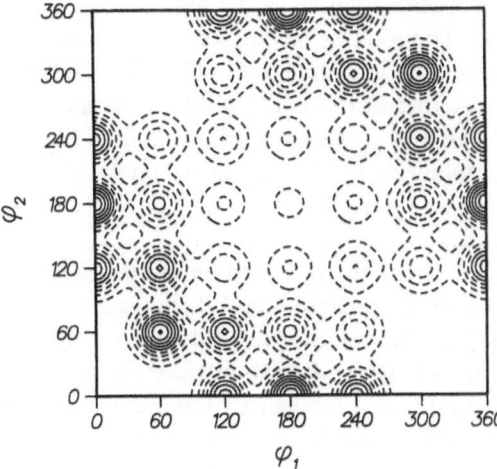

Fig. 8. Orientational correlation function corresponding with the pocket states in Fig. 7 for two nearest-neighbor molecules in β-nitrogen, from Ref. [52]

2.3.2 Solid Oxygen; A Magnetic Molecular Crystal

O_2, with its $^3\Sigma_g^-$ ground state, is one of the few stable molecules with a nonvanishing electronic spin momentum. This makes oxygen a most interesting molecular solid, which has received much attention [88]. Different kinds of magnetic order occur, in addition to orientational order and disorder. At normal pressure one finds already three different phases: the α phase, between 0 K and 23.8 K, the β-phase, between 23.8 K and 43.8 K, and the γ-phase between 43.8 K and the melting point at 54.4 K. The α and β phases are orientationally ordered, see Fig. 9, the γ-phase is plastic. The α-phase is antiferromagnetic with the (average) spin momenta lying along the crystal

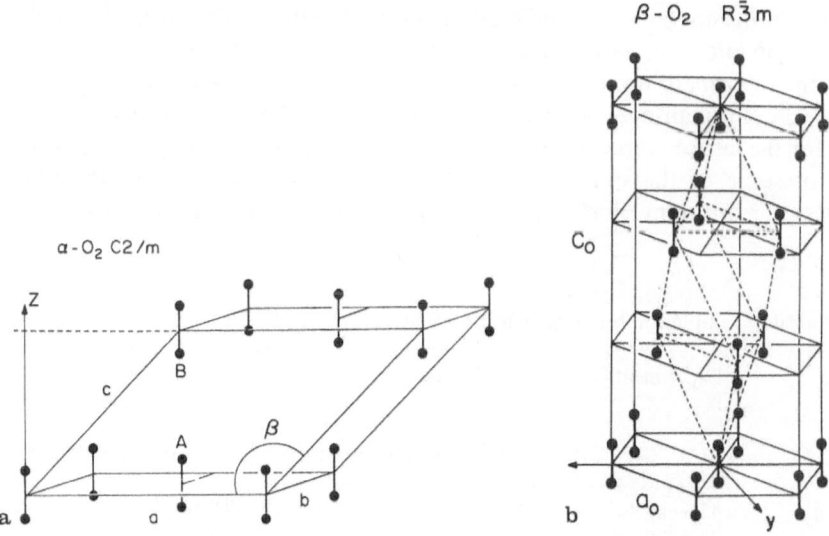

Fig. 9. Crystal structures of α- and β-oxygen [88]

b-axis, in two collinear sublattices with opposite spins. It is the only homogeneous system known which is antiferromagnetic. The β-phase has no long-range magnetic order, but it has short-range antiferromagnetic order with a three-sublattice 120° spin arrangement. The γ-phase is paramagnetic, just as the liquid phase. Especially the magneto-elastic (i.e. magnetic and structural) $\alpha - \beta$ phase transition has been studied in detail. In the β-phase the molecules are closely packed in a hexagonal layered lattice, see Fig. 9. In the monoclinic α-phase the hexagonal structure is slightly distorted, in order to achieve a more favorable antiferromagnetic order.

Most calculations on solid oxygen refer either to its lattice vibrations or to its magnetic properties. The lattice vibrations of the ordered α and β phases have been treated by the harmonic method [89–91], usually without regard to the magnetic coupling. It could not be explained by any of these calculations why the observed $q = 0$ libron peak, which is degenerate in β-O_2 at a frequency of about 50 cm^{-1}, is split by as much as 36 cm^{-1} in the α-phase [92]. Since the structural distortion at the $\beta - \alpha$ phase transition is small, none of the calculations predicts a splitting larger than 10 cm^{-1} (even when the empirical potentials were adjusted). The magnetic properties of α-O_2 have been studied on a pure spin system, by MF theory or spin-wave (RPA) methods, on the basis of a phenomenological spin-Hamiltonian with adaptable parameters. Also here it is found that the adjustment of these parameters could not explain all the observed quantities (magnetic susceptibilities, magnon frequencies, Néel temperature, spin-flop field) simultaneously [88].

We have obtained an anisotropic and spin-dependent O_2-O_2 potential from *ab initio* calculations [19, 41]. The exchange interactions between the open-shell molecules have been represented by a Heisenberg Hamiltonian as in Eq. (9), and the smaller spin-orbit and spin-spin coupling terms have been included too. The complete potential, including the Heisenberg exchange (J) and spin-coupling parameters, has been brought into the form of a spherical expansion as in Eq. (3). Thus, it could be used in an integrated lattice-dynamics/spin-wave study [41] by the RPA method. It appears that the actual mixing between the lattice modes and the spin waves is of limited importance; it occurs only at specific points in the Brillouin zone, near the avoided crossings of the phonon and magnon branches. The effect of a combined treatment is mainly the renormalization of the spin-coupling parameters by averaging the spin-dependent Hamiltonian over the lattice vibrations, and the creation of an effective potential for the lattice vibrations by averaging the Hamiltonian over the spin states. Still, the presence of the spin-averaged Heisenberg term is crucial for the libron frequencies; it is this term which explains the large splitting of the libron frequency at

Table 4. Optical ($q = 0$) libron frequencies in α and β-O_2

		Experiment [92]	Calculated (*ab initio*) [41]	
			Including Heisenberg term (9)	Neglecting Heisenberg term (9)
α-O_2:	B_g	42.6 cm^{-1}	39.9 cm^{-1}	38.9 cm^{-1}
	A_g	74.2	72.2	50.7
β-O_2:	E_g	48.0	53.6	42.9

the $\beta - \alpha$ phase transition, see Table 4. This is due to the extremely strong anisotropy of the Heisenberg coupling parameter J, see Fig. 2. Also the magnon frequencies in α-O_2 calculated from the renormalized spin-dependent *ab initio* Hamiltonian, which contains no adjustable parameters, agree fairly well with the infrared and Raman spectra, see Table 5. The calculated Néel temperature of α-oxygen, $T_N = 49.5$ K, and spin-flop field, 7.1 Tesla (experimentally 7.5 ± 0.5 Tesla), are reasonable too, but the magnetic susceptibility, obtained at the MF level, is considerably too high [54]. Further it has been predicted by the *ab initio* calculations (but not yet observed) that the presence of strong external magnetic fields will shift the $\alpha - \beta$ phase transition temperature and the libron frequencies in the α and β phases, to a measurable extent [54].

Table 5. Optical ($q = 0$) magnon frequencies in α-O_2

Experiment [88]	Calculated (*ab initio*) [41]
6.4 cm^{-1}	6.7 cm^{-1}
27.5	22.2

2.3.3 Solid Hydrogen; A Quantum Crystal

At first sight it may seem that the simplest molecular crystal is solid hydrogen, but this is not so because of strong quantum effects. H_2 molecules are very light and, therefore, their translational vibrations have large amplitudes and are rather anharmonic. Most exceptional are the orientational motions, however. The rotational constant of H_2 is extremely large, $B = 59$ cm^{-1}, and the anisotropy of the potential, which is mainly due to quadrupole-quadrupole interactions is much smaller (between two nearest neighbors at the equilibrium distance it is about 5 cm^{-1}). So, in the solid at normal pressure (melting point about 14 K), there is very little mixing between the free-rotor states of the H_2 molecules. In para-H_2 and ortho-D_2 the ground state has $l = 0$ and the crystal looks like an atomic (rare gas) crystal with hcp (hexagonal closed packed) structure. The first excited $l = 2$ free-rotor state lies at 354 cm^{-1} (179 cm^{-1} for D_2). In the solid, this fivefold degenerate state is split and broadened into a band by the anisotropic (quadrupole-quadrupole) interactions. Because of this dominant free-rotor character, the collective excitations to the $l = 2$ band are called rotons (rather than librons). Only at very high pressures, which correspond with nearest-neighbor distances that are smaller by a factor of 1.8, a substantial admixture of the $l = 2$ wave functions into the ground state $l = 0$ function will occur and the molecules will become orientationally ordered (with large amplitude zero-point librations). In ortho-D_2 the order/disorder phase transition has been observed at 27.8 GPa, in para-H_2 it lies probably around 100 GPa. For the same reasons the situation in ortho-H_2 and para-D_2 is completely different. Here the ground state of the molecules has $l = 1$. This threefold degenerate state is split by the interactions in the solid and the molecules order into states which have $m = 0$ with respect to the four [1, 1, 1] body diagonals of an fcc (face centered cubic) lattice. Thus, ortho-H_2 and para-D_2 are orientationally ordered already at low pressure. The $m = \pm 1$ states lie not far above the ground state, however, and so the libron frequencies are very small and the order/disorder phase transition occurs at very low temperature (2.8 K for ortho-H_2 and 3.8 K for para-D_2). All these interesting properties have been described in reviews [93, 94] and in a book [95] devoted especially to solid hydrogen.

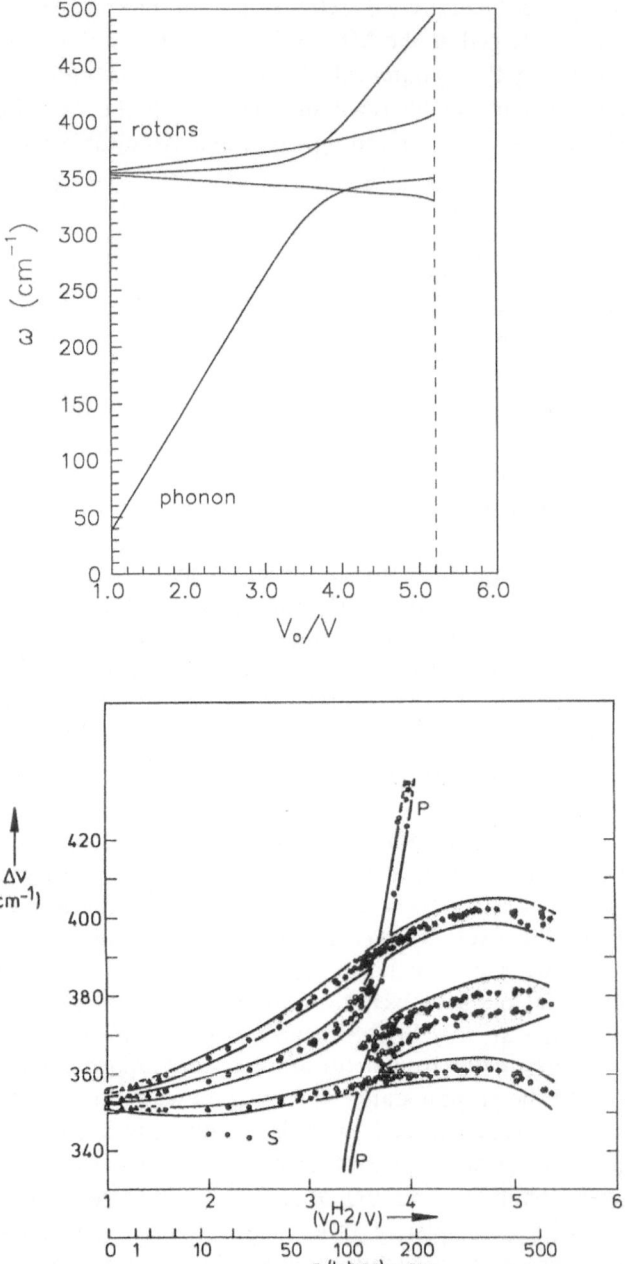

Fig. 10. Pressure dependence of the Raman-active phonon and roton frequencies in solid para-hydrogen, as calculated [57] and measured [100]

Up to now, lattice dynamics calculations have addressed the translational phonons and the rotons separately. For the translational phonons one has used an isotropic (i.e. orientationally averaged) H_2–H_2 potential. The SCP method has been applied to the anharmonic vibrations [96], but it appeared to be necessary to introduce an approximate Jastrov function into this method (with one adjustable parameter) in order to obtain realistic results. The roton frequencies, and their softening at higher pressures (smaller volume) which precedes the disorder/order phase transition, have been calculated by the MF [97] and RPA [98] methods. Only quadrupole-quadrupole interactions were taken into account, and translation-rotation coupling was neglected.

Recently we have introduced the *ab initio* H_2–H_2 potential of Meyer and Schäfer [99] into our RPA and TDH programs. The full anisotropy of this potential (mainly due to quadrupole-quadrupole coupling) was retained and the anharmonic expansion of the potential in the translational displacements was truncated only after the sixth power. In this anharmonic potential the vibrational wave functions were calculated explicitly with the inclusion of higher harmonic oscillator basis functions. Thus, it appeared no longer necessary to use the adjustable Jastrov function. Translation-rotation coupling was included by the RPA method. The phonon and roton dispersion curves calculated [57] for para-H_2 are shown in Fig. 10. At normal pressure the translational phonon frequencies are much lower than the roton frequencies. At high pressure the phonon and roton branches show an avoided crossing, however, and phonon-roton mixing becomes important (see Fig. 10). The pressure and temperature of the order/disorder phase transition can be calculated, either by comparison of the free energy F for the two phases, or by studying the softening of the roton/libron modes.

2.4 Phenomena in More Complex Molecular Crystals

In collaboration with experimental groups, we have recently studied some chlorinated-benzene crystals, 1,2,4,5-tetrachlorobenzene (TCB) [59] and 1,4-dichlorobenzene (DCB) [60], as well as solid tetracyanoethene (TCNE) [58]. In these studies we have used empirical atom-atom potentials, of exp-6 type [see Eq. (6)], which we have supplemented with the Coulomb interactions between fractional atomic charges. Lattice dynamics calculations have been performed by the harmonic method, with inclusion of intramolecular vibrations [70], see Eqs. (17) to (24). The normal modes of the free molecules have been calculated from empirical Valence Force Fields, using the standard **GF**-matrix method [101, 102]. The results of these calculations are used here to illustrate some phenomena occurring in more complex molecular crystals. These phenomena are well known; the numerical results show their quantitative importance, in some specific systems.

2.4.1 Phonon-Vibron Mixing

When the molecules get larger and more flexible, the separation between the intramolecular vibrations and the lattice vibrations becomes less strict. Some of the intramolecular modes have relatively low frequencies and large amplitudes and they will mix with the lattice modes. A clear example of such mixing occurs in solid TCNE.

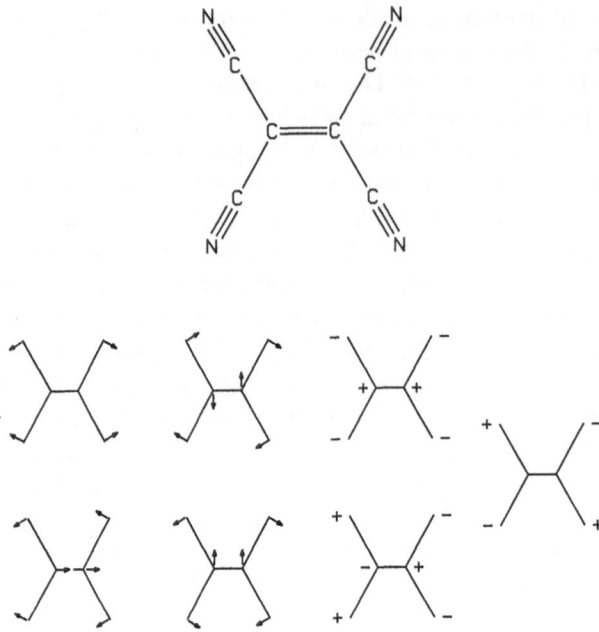

scissoring   rocking   wagging   torsion

Fig. 11. Low frequency vibrations of the free TCNE molecule

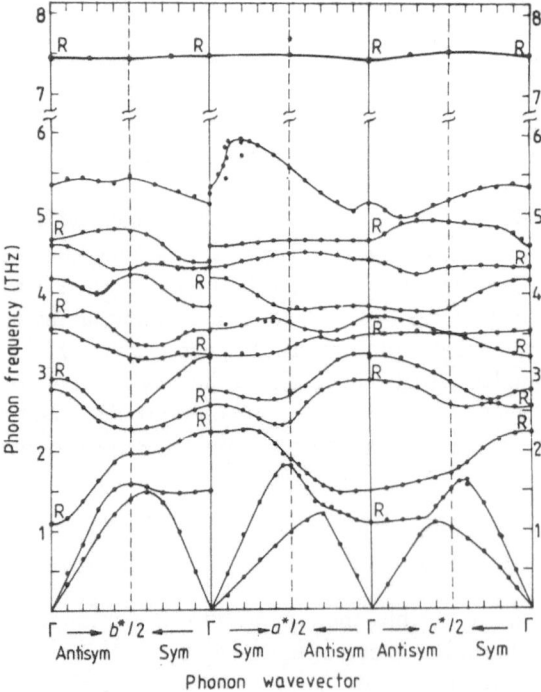

Fig. 12. Phonon-vibron dispersion curves in solid TCNE from inelastic neutron scattering [103] (1 THz = 33.36 cm^{-1}). The modes below 5 THz are strongly mixed [58, 103]

Although the TCNE molecule, which is a flat conjugated π-electron system, is rather rigid, it has several low lying vibrations. These vibrations correspond with the collective motions (torsion, out-of-plane wagging, rocking and scissoring) of the $C-C \equiv N$ "legs" (see Fig. 11) and they have frequencies between 80 and 250 cm^{-1}. Since the frequencies of the highest lattice vibrations reach up to about 120 cm^{-1}, it is understandable that the phonon dispersion curves measured by inelastic neutron scattering ([103], see Fig. 12) cannot be explained by lattice dynamics calculations based on the rigid-molecule model. If the mixing between the lowest intramolecular modes and the lattice modes is included, the calculated phonon-vibron dispersion curves agree well with the experimental data, in general. For several modes in solid TCNE this mixing is about fifty-fifty [58, 103].

2.4.2 Vibron Band Structure; Davydov and Site Splitting

Even when the intramolecular vibrations do not mix with the phonons, they are coupled by the intermolecular potential, through its dependence on the internal molecular coordinates. This coupling determines the band structure of the vibrational excitons or vibrons in molecular crystals, which can be studied by inelastic neutron scattering (see Fig. 12). In the case of TCB crystals the vibron band structure has been observed recently by laser phosphorescence [59, 104]. This is a rather special achievement since, normally, optical techniques probe only the $q = 0$ excitations. It is related to the long lifetime of triplet electronic excitons in TCB, which are scattered into different q states. By phosphorescent de-excitation of $q \neq 0$ triplet excitons to the electronic ground state it is possible to detect the $q \neq 0$ vibrons, with intensities controlled by the (very low) temperature and by the phosphorescence delay time.

Generally, the vibrations with the lowest frequencies and the largest amplitudes have the largest vibron band widths. It appears that there are strong and seemingly irregular variations of the vibron band widths in molecular solids, however, from a few tenths of a wave number to about 30 cm^{-1}. This is determined by the nature of the intramolecular vibrations involved and by the intermolecular coupling mechanism. Each of the terms in the intermolecular potential can contribute to this coupling, the exchange repulsion, the dispersion attraction, as well as the electrostatic interactions [cf. Eq. (6)]. From lattice dynamics calculations on different modifications of solid TCB (the α and β phases) [59] and solid DCB (the α, β and γ phases) [60] it follows that either the electrostatic interactions or the exchange repulsion terms are dominant, with a smaller, compensating, contribution from the dispersion attraction, in the latter case. This depends on the type of intramolecular vibrations and on the crystal structure. When the available empirical exp-6 atom-atom potentials were supplemented with fractional atomic charges, in order to represent the electrostatic interactions, the variations calculated in the vibron band widths [59] agree well with the experimental observations. Also the nature of the vibrational excitons, quasi one-dimensional for several vibrations in TCB, three-dimensional in other cases, emerges well from the calculations.

Particular attention has been given to the vibron band around 5.5 THz (180 cm^{-1}) in solid TCNE which shows "anomalous" dispersion [58, 103] along the a^*-direction in the Brillouin zone (see Fig. 12). It follows from calculations that the phonon-vibron mixing which is strikingly present for some of the lower vibrations in TCNE (as explained in the preceding subsection), does not occur for this particular vibron. It is a pure vibron, associated with the out-of-plane wagging mode of the TCNE molecules. This mode is strongly infrared-active and the "anomalous" dispersion appears[58] to be caused by the resonant dipole-dipole coupling between the molecules. A model has been developed [58] with fractional atomic charges that represent the *ab initio* calculated quadrupole moment of TCNE, as well as the transition-dipole moment of the out-of-plane wagging mode. The use of this model in lattice dynamics calculations and the evaluation of the long range electrostatic lattice sums in the dynamical matrix, which automatically takes into account the resonant transition-dipole interactions, nicely reproduces the "anomalous" vibron dispersion curve, both in shape and in amplitude (band width ≈ 30 cm^{-1}).

When the unit cell of the crystal contains more than one molecule, it is possible to observe the same type of couplings that lead to the vibron band structure even in regular optical (infrared and Raman) spectroscopy. In principle, the peaks corresponding with the $q = 0$ vibrons can be split by two mechanisms. The off-diagonal elements F_{AB} of Eq. (19) in the dynamical matrix of Eq. (23), which couple the different molecules in the unit cell, lead to the so-called Davydov or factor-group splittings. When these molecules, say A and B, are not symmetry related there may also be a difference in the diagonal elements F_{AA} and F_{BB}. This difference is called the site or crystal-field splitting. Both these effects have been calculated and observed by Raman spectroscopy in solid TCB [59] and DCB [60]. In the high temperature β-phase of TCB the two molecules in the monoclinic unit cell are related by a glide plane and one finds almost no (Davydov) splitting. In the low temperature triclinic α-phase of TCB this symmetry relation is lost; the observed (site) splitting is much larger. In α- and γ-DCB one observes and calculates somewhat larger Davydov splittings than in β-TCB. Either the electrostatic or the exchange repulsion terms in the intermolecular potential appear to be mainly responsible for these splittings. The splitting of the vibron bands and the vibron band widths are not necessarily dominated by the same terms, however.

2.4.3 Isotope Substitution; Effects of Random Disorder

When in a molecular solid all the atoms of a certain type are replaced by a different isotope, the frequencies of the vibrational excitations are simply shifted. So, for instance, when ethene crystals are grown from C_2D_4 rather than C_2H_4, all the translational phonon frequencies decrease by a factor of $\sqrt{32/28}$, the librations about the $C=C$ axis by a factor of $\sqrt{2}$ and the librations about the other axes by different but fixed amounts. The actual shifts found in the lattice frequencies can be used to determine the nature of the lattice modes.

The situation is much more complex, and also more interesting, when the crystals are composed of isotopically mixed substances. Examples are the chlorinated-benzene crystals recently investigated [59, 60], which contain ^{35}Cl and ^{37}Cl atoms in the natural abundance ratio of 3:1. In 1,2,4,5-tetrachlorobenzene (TCB) this leads to seven different molecular species with different distributions of ^{35}Cl and ^{37}Cl; in 1,4-dichlorobenzene (DCB) there are three such species. In TCB and DCB crystals these seven or three species occur randomly distributed over the lattice. Theoretical models used to describe the band structure of crystals with random disorder [105] are the Virtual Crystal Approximation (VCA), the Average T-matrix Approximation (ATA) and the Coherent Potential Approximation (CPA). The most sophisticated of these, the CPA model, describes two extreme cases. One extreme is the amalgamated band limit, which occurs when the difference between the single-particle excitation frequencies of the different species is small compared with the coupling matrix elements between the particles. According to the CPA model, this yields practically the same (single-)band structure as a homogeneous crystal composed of the average species. For vibrations this is the species with identical atoms of average mass. Such a (hypothetical) crystal is called the virtual crystal; the VCA model applies just to this case. The other extreme occurs when the difference between the single-particle excitation frequencies is large compared with the interparticle coupling. Separated bands occur in that case, which are centered around the excitation frequencies of the different particles. These correspond with the excitations remaining localized in clusters of the same species; the size of such clusters shows a broad distribution. In the intermediate case, which has been simulated and treated theoretically in simple two-component model systems [62, 106], the separated bands start to amalgamate and the resulting band shape becomes very complex.

The phonons in mixed molecular crystals [107] occur practically always in the amalgamated band limit. The vibrons may occur in either limit, and these limits may be found in the same crystal, for different intramolecular modes. This is clearly illustrated by the lattice dynamics calculations and the observed Raman spectra of TCB and DCB [59, 60]. The 312 cm^{-1} B_{3g} mode in TCB, which is a C—Cl stretch vibration, is shifted by about 2 cm^{-1} with each substitution of one ^{37}Cl atom by a ^{35}Cl atom. The band width of this vibration in α-TCB and β-TCB is about 0.5 cm^{-1}, which implies that the nearest-neighbor coupling elements are only about 0.1 cm^{-1}. So, this vibra-

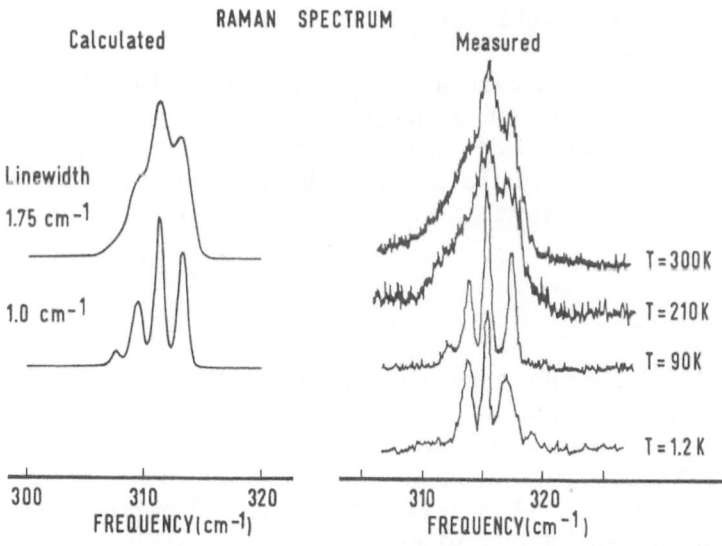

Fig. 13. Calculated and measured ^{35}Cl/^{37}Cl isotope splitting of a C—Cl stretch vibron band in solid TCB [59]

tion should obviously occur in the separated band limit. In the Raman spectra the seven separated bands are indeed observed, with the correct (calculated) frequency splitting and intensity ratios according to the natural abundance (see Fig. 13). A very similar situation occurs for the 328 cm^{-1} A_g C—Cl stretch mode in α-DCB and γ-DCB. The three separated bands are observed in the Raman spectra with the calculated splittings of 4 cm^{-1} and the correct intensity ratio (9:6:1). In β-DCB the situation is different, however. The width calculated for the vibron band around 328 cm^{-1} (more than 10 cm^{-1}) is much larger than the corresponding band widths in α-DCB and γ-DCB (which are less than 1 cm^{-1}). So in β-DCB the nearest-neighbor coupling matrix elements in the dynamical matrix for this mode are comparable with the isotope shifts in the free-molecule vibration frequency. The structure of the peak around 328 cm^{-1} in the observed Raman spectrum of β-DCB is rather complex indeed [60].

Another phenomenon which is related to the occurrence of random disorder in molecular solids with mixed isotopes is the observation of "symmetry forbidden" transitions. Also this phenomenon is nicely exemplified in TCB and DCB. For all the phases of these solids the space groups of the observed X-ray structures contain a center of inversion, which coincides with the molecular center of mass. Therefore, all the gerade vibrations of the molecules will be Raman active also in the solid and the ungerade vibrations (though not all) will be infrared active. In crystals with random disorder this symmetry should only be interpreted as an average symmetry, however, which corresponds to the virtual crystal. The actual symmetry at every specific site will be lower. Those intramolecular vibrations in TCB and DCB which involve the Cl-atoms will be most sensitive to this symmetry lowering. Some of the "ungerade" vibrations of this type have indeed been observed in the Raman spectra. It will be clear that especially those molecular species which lack an inversion center (DCB molecules with one ^{35}Cl and one ^{37}Cl atom, for example) will contribute to the intensity of the "forbidden" transition. Since these transitions are weak anyway, it may be expected that, in the separated band limit, only the bands corresponding to these asymmetric species will be visible. Also this prediction agrees with the experimental observations [59, 60].

Summarizing, we conclude that the examples of TCNE, TCB and DCB discussed in this section are very useful to illustrate several "textbook phenomena". The effects of these phenomena have been observed by inelastic neutron scattering and by optical spectroscopy, but the interpretation of the spectra and the actual recognition of these effects leans strongly on lattice dynamics calculations.

Acknowledgement: An always stimulating co-operation with Paul Wormer in the calculation of intermolecular potentials and other matters is gratefully acknowledged. The results used to illustrate Sect. 2 are taken from lattice dynamics work in Nijmegen by Tadeusz Wasiutynski, Tadeusz Luty, Wim Briels, Tonek Jansen, Tom van den Berg and Wilfred Janssen. The study of chlorinated-benzene crystals was made in a fruitful collaboration with the experimentalists Ad Jongenelis and Jan Schmidt (Leiden). Paul Wormer and Tom van den Berg are thanked for critically reading this manuscript.

3 References

1. van der Avoird A, Wormer PES, Mulder F, Berns RM (1980) Topics Curr. Chem. 93: 1
2. Kaplan IG (1986) Theory of molecular interactions. North Holland, Amsterdam
3. van Lenthe JH, van Duijneveldt-van de Rijdt JCGM, van Duijeneveldt FB (1987) Advan. Chem. Phys. 69: 521
4. Hobza P, Zahradnik R (1988) Intermolecular complexes. Elsevier, Amsterdam
5. Chałasiński G, Gutowski M (1988) Chem. Rev. 88: 943
6. Buckingham AD, Fowler PW, Hutson JM (1988) Chem. Rev. 88: 963
7. Stone AJ (1989) in: Maksić ZB (ed) Theoretical models of chemical bonding. Springer, Berlin, part 4, chapter 6
8. Rijks W, Wormer PES (1989) J. Chem. Phys. 90: 6507, and references therein
9. Rijks W, van Heeringen M, Wormer PES (1989) J. Chem. Phys. 90: 6501
10. Chipman DM, Hirschfelder JO (1973) J. Chem. Phys. 59: 2830, 2838
11. Jeziorski B, Kołos W (1977) Int. J. Quantum Chem. 12 Suppl. 1: 91
12. Rijks W, Gerritsen M, Wormer PES (1989) Mol. Phys. 66: 929
13. Douketis C, Scoles G, Marchetti S, Zen M, Thakkar AJ (1982) J. Chem. Phys. 76: 3057
14. Tang KT, Toennies JP (1984) J. Chem. Phys. 80: 3726
15. Knowles PJ, Meath WJ (1987) Mol. Phys. 60: 1143
16. Matsen FA, Klein DJ, Foyt DC (1971) J. Phys. Chem. 75: 1866
17. Western CM, Langridge-Smith PRR, Howard BJ (1981) Mol. Phys. 44: 145
18. Brechignac Ph, De Benedictis S, Halberstadt N, Whitaker BJ (1985) J. Chem. Phys. 83: 2064
19. Wormer PES, van der Avoird A, (1984) J. Chem. Phys. 81: 1929
20. Miller RE (1988) Science 240: 447
21. Faubel M (1983) Advan. At. Mol. Phys. 19: 345
22. Buck U, Huisken F, Schleusener J (1980) J. Chem. Phys. 72: 1512
23. Bergmann K, Hefter U, Witt J (1980) J. Chem. Phys. 72: 4777
24. Hirschfelder JO, Curtiss CF, Bird RB (1964) Molecular theory of gases and liquids. Wiley, New York
25. Maitland GC, Rigby M, Smith EB, Wakeham WA (1981) Intermolekular forces. Clarendon, Oxford
26. Gray CG, Gubbins KE (1984) Theory of molecular fluids. Clarendon, Oxford
27. Briels WJ, Jansen APJ, van der Avoird A (1986) Adv. Quantum Chem. 18: 131
28. Brink DM, Satchler GR (1975) Angular momentum. Clarendon, Oxford
29. Le Roy RJ, Hutson JM (1987) J. Chem. Phys. 86: 837
30. Le Roy RJ, Carley JS (1980) Advan. Chem. Phys. 42: 353
31. Berns RM, van der Avoird A (1980) J. Chem. Phys. 72: 6107
32. van der Avoird A, Wormer PES, Jansen APJ (1986) J. Chem. Phys. 84: 1629
33. Wasiutynski T, van der Avoird A, Berns RM (1978) J. Chem. Phys. 69: 5288
34. Pertsin AJ, Kitaigorodsky AI (1987) The atom-atom potential method for organic molecular solids. Springer, Berlin
35. Hair SR, Beswick JA, Janda KC (1988) J. Chem. Phys. 89: 3970
36. Claverie P (1978) in: Pullman B (ed) Intermolecular interactions: From diatomics to biopolymers, Wiley, New York, p 69
37. Rullman JAC, van Duijnen PTh (1988) Mol. Phys. 63: 451
38. Gay JG, Berne BJ (1981) J. Chem. Phys. 74: 3316
39. Stone AJ, Price SL (1988) J. Phys. Chem. 92: 3325
40. Fuchikama N, Block R (1982) Physica B 112: 369
41. Jansen APJ, van der Avoird A (1987) J. Chem. Phys. 86: 3583
42. Margenau H, Kestner NR (1971) Theory of intermolecular forces, 2nd edn. Pergamon, New York
43. Meath WJ, Aziz RA (1984) Mol. Phys. 52: 225
44. Loubeyre P (1987) Phys. Rev. Lett. 58: 1857
45. Bulski M (1989) in: Polian et al. (ed) Simple molecular systems at very high density. Plenum, New York

46. Bulski M, Chałasiński (1987) J. Chem. Phys. 86: 937
47. Born M, Huang K (1954) Dynamical theory of crystal lattices. Clarendon, Oxford
48. Cochran W (1971) CRC Critical reviews in solid state science 2: 1
49. Jansen APJ, Briels WJ, van der Avoird A (1984) J. Chem. Phys. 81: 3648
50. van der Avoird A, Briels WJ, Jansen APJ (1984) J. Chem. Phys. 81: 3658
51. Briels WJ, Jansen APJ, van der Avoird A (1984) J. Chem. Phys. 81: 4118
52. Jansen APJ (1988) J. Chem. Phys. 88: 1914
53. Jansen APJ, Schoorl R (1988) Phys. Rev. B38: 11711
54. Jansen APJ, van der Avoird A (1987) J. Chem. Phys. 86: 3597
55. Jansen APJ (1986) Phys. Rev. B33: 6352
56. Jansen APJ (1988) J. Phys. C: Solid State Phys. 21: 4221
57. Janssen WBJM, van der Avoird A (1990) Phys. Rev. B in press
58. van den Berg THM, van der Avoird A (1989) J. Phys. Condensed Matter, 1: 4047
59. Jongenelis APJM, van den Berg THM, Jansen APJ, Schmidt J, van der Avoird A (1988) J. Chem. Phys. 89: 4023
60. Jongenelis APJM, van der Berg THM, Schmidt J, van der Avoird A (1989) J. Phys. Condensed Matter, 1: 5051
61. Maradudin AA, Vosko SH (1968) Rev. Mod. Phys. 40: 1
62. Maradudin AA, Montroll EW, Weiss GH, Ipatova P (1971) Theory of lattice dynamics in the harmonic approximation, Academic Press, New York
63. Maradudin AA (1974) in: Horton GK, Maradudin AA (ed) Dynamical properties of solids, Vol. 1, North Holland, Amsterdam, p 1
64. Califano S, Schettino V, Neto N (1981) Lattice dynamics of molecular crystals, Lecture notes in Chemistry, Vol. 26, Springer, Berlin
65. Barron THK, Klein ML (1974) in: Horton GK, Maradudin AA (ed) Dynamical properties of solids, Vol. 1, North Holland, Amsterdam, p 391
66. Girardeau MD, Mazo RM (1973) Advan. Chem. Phys. 24: 187
67. Feynman RP (1972) Statistical mechanics, Benjamin, Reading, Massachusetts
68. Löwdin PO (1988) Int. J. Quantum Chem. Symp. 22: 337
69. Neto N, Kirin D (1979) Chem. Phys. 44: 245
70. Taddei G, Bonadeo H, Marzocchi MP, Califano S (1973) J. Chem. Phys. 58: 966
71. Werthamer NR (1976) in: Klein ML, Venables J (ed) Rare gas solids, Vol. 1, Academic Press, London
72. Wasiutynski T (1976) Phys. Status Solidi B76: 175
73. Hansen JP, McDonald IR (1976) Theory of simple liquids, Academic Press, New York
74. Kirkwood JG (1940) J. Chem. Phys. 8: 205
75. James HM, Keenan TA (1959) J. Chem. Phys. 31: 12
76. Raich JC (1972) J. Chem. Phys. 56: 2395
77. Raich JC, Etters RD (1968) Phys. Rev. 168, 425
78. Dunmore PV (1972) J. Chem. Phys. 57: 3348
79. Fredkin DR, Werthamer NR (1965) Phys. Rev. A138: 1527
80. Hüller A (1974) Phys. Rev. B10: 4403
81. Scott TA (1976) Phys. Rep. 27: 89
82. Kjems JK, Dolling G (1975) Phys. Rev. B11: 1639
83. Raich JC, Gillis NS (1977) J. Chem. Phys. 66: 846
84. Luty T, van der Avoird A, Berns RM (1980) J. Chem. Phys. 73: 5305
85. Kjems JK, Dolling G (1981) Phys. Rev. B24: 2967
86. Heberlein DC, Adams ED, Scott TA (1970) J. Low Temp. Phys. 2: 449
87. van den Berg THM, Bongers MMG, van der Avoird A (1990) to be published
88. DeFotis GC (1981) Phys. Rev. B23: 4714 and references therein
89. Kobashi K, Klein ML, Chandrasekharan V (1979) J. Chem. Phys. 71: 843
90. Etters RD, Helmy AA, Kobashi K (1983) Phys. Rev. B28: 2166
91. Kuchta B (1985) Chem. Phys. 95: 391
92. Bier KD, Jodl HJ (1984) J. Chem. Phys. 81: 1192
93. Silvera IF (1980) Rev. Mod. Phys. 52: 393
94. Silvera IF (1989) in: Polian et al. (ed) Simple molecular systems at very high density, Plenum, New York

95. van Kranendonk J (1985) Solid hydrogen, Plenum, New York
96. Klein ML, Koehler R (1970) J. Phys. C3: L102
97. England W, Raich JC, Etters RD (1976) J. Low Temp. Phys. 22: 213
98. Lagendijk A, Silvera IF (1981) Phys. Lett. 84A: 28
99. Schäfer J, Köhler W (1989) Z. Physik D13: 217
100. Silvera IF, Wijngaarden RJ (1981) Phys. Rev. Lett. 47: 39
101. Wilson EB, Decius JC, Cross PC (1955) Molecular vibrations, McGraw-Hill, New York
102. Califano S (1976) Vibrational states, Wiley, London
103. Chaplot SL, Mierzejewski A, Pawley GS, Lefebvre J, Luty T (1983) J. Phys. C: Solid State Phys. 16: 625
104. Abramson EH, Jongenelis APJM, Schmidt J (1987) J. Chem. Phys. 87: 3719
105. Kopelman R (1976) in: Lim EC (ed) Excited states, Vol. 2, Academic Press, New York, p 33
106. Economou EN (1983) Green's functions in quantum physics, 2nd ed., Springer, Berlin, Chap. 7
107. Bellows JC, Prasad PN, Monberg EM, Kopelman R (1978) Chem. Phys. Lett. 54: 439

Theoretical Evaluation of Solvent Effects

O. Tapia

Department of Physical Chemistry, University of Uppsala, Box 532, S-75121 UPPSALA, Sweden

The theory of solvent-effects and some of its applications are overviewed. The generalized selfconsistent reaction field (SCRF)theory has been used to give a unified approach to quantum chemical calculations of subsystems embedded in a given milieu. The statistical mechanical theory of projected equations of motion has been briefly described. This theory underlies applications of molecular dynamics simulations to the study of solvent and thermal bath effects on carefully defined subsystem of interest. The relationship between different approaches used so far to calculate solvent effects and the general SCRF has been established. Recent work using the continuum approach to model the surrounding media is overviewed. Monte Carlo and molecular dynamics studies of solvent effects on molecular properties and chemical reactions together with simulations of solvent effects on protein structure and dynamics are reviewed.

1 Introduction

Solvent effects have been recognized as important and pervasive phenomena in physics, chemistry and biology [1–7]. Recent developments in experimental and computer sciences [3, 4] and technology have lead to spatially detailed and time resolved descriptions of a host of significant phenomena occurring in condensed phases and bioenvironments thereby raising serious challenges to theoreticians of solvent effects. Two theoretical approaches have evolved more or less independently for treating such phenomena: solvation theory and solvent-effects theory. The former targets direct evaluation of the solvation free energies and excitation energies; this domain has recently been reviewed [5] and will not be examined here. While in solvent-effects theory the attention is focused on the changes induced by the medium onto the electronic structure and molecular properties of the solute system; the effects are measured with respect to the properties of the solute in vauo; the approach has chemical connotations that makes it suitable for studying chemical systems at a microscopic level [6, 7]. This subject will be the main concern of this chapter.

The theory of solvent effects has been progressing along two different, albeit convergent, lines. In one of them, emphasis is put onto quantum mechanical aspects in representing the solute (or system of interest), while the surrounding medium is modeled with more or less sophisticated approaches. In the other one, the representation of the solute and surrounding medium is made with the help of classical statistical mechanical techniques, e.g. Monte Carlo (MC) or molecular dynamics (MD) procedures; quantum mechanics, when it is employed, serves to generate information on intermolecular potentials, charge distributions or other relevant molecular information. In this paper, these two aspects are reviewed: quantum and classical statistical mechanical methods are examined. The quantum mechanical theory of solvent effects on the electronic structure is presented from a unified viewpoint: the generalized self-consistent reaction field (GSCRF) theory. While classical statistical mechanics of projected Liouvillians describing the system of interest coupled to a thermal bath [8, 9] is summarized. Such a formulation is conceptually adequate to treat solvent-effects [10] in conjunction with the GSCRF theory. Practical calculations of solvent effects with computer simulations are overviewed.

The paper is organized as follows. In Sect. 2, quantum theories of solvent effects are examined. There, a generalized self-consistent reaction field approach is described. The theory is based on a microscopic representation of the surrounding medium, that overcomes limitations of preceding formulations and affords a scheme well adapted to perform integrated quantum/statistical mechanical molecular dynamics or Monte Carlo simulations. Thereafter, a statistical mechanical analysis of the GSCRF theory is presented. The resulting equation can be solved following two different methods: a microscopic approach and a dielectric one. The former is discussed first; in the dipolar approximation, the inhomogeneous SCRF [7] method is recovered; the relationships to other methods are established. A general dielectric formulation that follows the lines discussed in Ref. [7] is examined; this theory contains previously proposed schemes based in the cavity approach as particular cases. Recent chemical work based upon continuum and quasi-continuum models is reviewed.

In Sect. 3, the fundamentals of classical statixtical mechanics underlying applications of molecular dynamics simulations are presented. Projected equations of motion, that are essential for describing solvent effects are sketched; the equation of motion for an effective probability distribution function describing a subsystem of interest coupled to a thermal bath is given. A generalized Langevin equation is obtained for the dynamical variables of the subsystem. The active site model approach is described; recent applications made both for liquid phase and biomolecules simulations are discussed.

In Sect. 4, an overview of solvent effects evaluation with the techniques described in the preceding sections is presented. MC and MD studies on molecular properties are overviewed. Of particular interest are recent developments in the simulation of solvent effects on chemical reactions. The focus is in computer simulations. Analytical models for describing chemical reactions have been thoroughly discussed by Hynes [11] and will not be examined here.

The final sections, Sect. 5, contains the conclusions and trends.

2 Quantum Theory of Solvent Effects

2.1 Generalized Selfconsistent Reaction Field Theory

A general effective Hamiltonian describing the solute system interacting with the surrounding medium held at a fixed configuration can be constructed by replacing the total electronic interaction Hamiltonian, that is previously averaged over the exact solvent wavefunction, by a solvent electrostatic potential coupled to the solute charge density operator; the solvent potential fulfils a Poisson's equation. For a molecule, or any well defined subsystem as for instance a model active site in an enzyme, the solute and surrounding medium electronic Hamiltonians, $\hat{H}_s(r_s, R_s)$ and $\hat{H}_m(r_m, R_m)$, respectively, and the interaction operator $\hat{V}_{sm} = V(r_s, r_m, R_s, R_m)$ are well known [7, 12–14]. The latter, in the Coulomb gauge [14], is the electrostatic interaction between all charges in the system and can be written with the help of charge density operators $(\hat{\Omega})$ as follows:

$$\hat{V}_{ms} = \int dr \int dr' \, \hat{\Omega}_s(r) \, T(r - r') \, \hat{\Omega}_m(r') \tag{1}$$

where $T(r - r') = 1/|r - r'|$ is the Coulomb kernel [15]; $\hat{\Omega}_s(r) = -\Sigma_i \delta(r - r_i) + \Sigma_{si} Z_{si} \delta(r - R_{si})$; with a similar, albeit more cumbersome, expressions for the solvent charge density operator; r_i stands for the i-th electron position vector operator; R_{si} is the position vector of the solute si-th nuclei; Z_{si} its corresponding nuclear charge; $\delta(r)$ is Dirac's delta distribution and dr is a volume element in real space (R^3). The electron coordinate operators for each subsystem r_s and r_m and the nuclear vector position coordinates of each subsystem R_s and R_m are embedded in the definitions of the charge density operator.

The construction of an effective Hamiltonian $\hat{H}_s$ implies separability of the total wavefunction Ψ into a product of wavefunctions for each subsystem. The product must be antisymmetrized as electrons are indistinguishable. For closed shell electronic systems, exchange interactions due to the Pauli exclusion principle, which are included in the antisymmetrization of the total wave function Ψ are responsible for molecular shapes. Explicit antisymmetrization of the total wavefunction is avoided by including the resultant repulsive forces as interatomic potential functions in the configurational space. Analogously, the instantaneous electron dipole-dipole correlations between the subsystems are included as standard van der Waals potentials in the statistical mechanical treatment. Under these assumptions, the product wavefunction can be given a Hartree form: $\Psi \approx \Psi_s(r_s; R_s, R_m) \Psi_m(r_m; R_m, R_s)$, where each wavefunction in the product is computed at a fixed nuclear configuration $X = (R_m, R_s)$. The theory is therefore placed in the Born-Oppenheimer framework.

The interaction Hamiltonian describes now the coupling of the solute charge density operator with the electrostatic potential created by the surroundings at fix X: $V_m(r) = \int dr' \langle \Psi_m | T(r - r') \hat{\Omega}_m(r') | \Psi_m \rangle$. This expression can be cast into: $\int dr' T(r - r') \langle \Psi_m | \hat{\Omega}_m(r') | \Psi_m \rangle = \int dr' T(r - r') \Omega_m(r')$. The potential $V_m(r)$ fulfils Poisson's equation: $\nabla_r^2 V_m(r) = \Omega_m(r)$. The effective Hamiltonian acquires the simple form:

$$\hat{H}_s(r_s; X) = \hat{H}_s(r_s; R_s) + \int dr \hat{\Omega}_s(r) V_m(r; X) \tag{2}$$

For each nuclear configuration, the solute wavefunction Ψ_s and the effective energy $E_s(X)$ are obtained as a solution of the effective Schrödinger equation:

$$\hat{H}_s | \Psi_s \rangle = E_s(X) | \Psi_s \rangle \tag{3}$$

Eq. (2) is not very useful unless the solvent charge density can be given a tractable form. This goal can be attained if one realizes that the solute can be taken as a classical external electrostatic source to the surrounding medium. For this approximation to be accurate, the solute wavefunction must be fairly well localized in the volume assigned to the solute system; overlap with the surrounding medium must be minimal. This condition can not be applied to solute-solvent charge transfer interactions unless the solute includes the charge transfer partners.

Let $\Omega_m^0(r)$ be the charge density of the surrounding medium in absence of the solute field. The total solvent charge density is obtained by adding to $\Omega_m^0(r)$ the polarization charge density $\Omega'(r)$ set up by the solute electric field $e(r)$. This latter density is the divergence of the polarization vector $p(r)$ [16]:

$$\Omega_m(r) = \Omega_m^0(r) - \nabla_r \cdot p(r) \tag{4}$$

where ∇_r is the gradient operator at r. Note that $p = 0$ at the boundaries and outside the macroscopic volume occupied by the solvent; p is the dipole moment density in the medium. The effective Hamiltonian (2) acquires an implicit non-linear structure via the polarization density term:

$$\hat{H}_s = \hat{H}_s + \int dr \hat{\Omega}_s(r) [V_m^0(r) + \int dr' T(r - r') \cdot p(r')] \tag{5}$$

where $T(r - r') = \nabla_r T(r - r')$ is a unit electric field at r' produced by a unit charge at r. The term $T(r - r') \cdot p(r') dr'$ is the differential reaction field (RF) potential

$(d\Pi(\mathbf{r}))$ at the point $\mathbf{r}$ in the volume occupied by the solute. The total RF at $\mathbf{r}$, $\Pi(\mathbf{r})$, is obtained after integration over the $\mathbf{r}'$-variable. V_m^0 is the potential acting on the quantum system that is generated by the surrounding medium charge density. $V_m^0(\mathbf{r})$ and $\Pi(\mathbf{r})$ must now be operationally determined.

Surrounding Potential. $V_m^0(\mathbf{r})$ is obtained from model charge densities representing the atoms or molecules in the solvent or surrounding medium. There are several model charge densities currently available. For proteins, Warme-Scheraga effective charges for the amino acid residues have been used in studies of enzyme catalyzed reactions [7]. Current MD or MC program packages [17, 18] have model point charges to calculate electrostatic interactions. They result from fittings of empirical data [17–20]. Potential functions for water based on ab initio calculations have also been produced [21]. The construction of intermolecular potentials is still a very active field [22].

Reaction Field Potential. The polarization vector can be obtained from model solvent response functions without resorting to a multipolar expansion. In this way, the effective Hamiltonian can be cast as a functional of the solute charge density, thereby expliciting the non-linear structure of Eq. (5).

Let $\mathbf{X}(\mathbf{r})$ be the static response tensor to the external electric field $\mathbf{e}(\mathbf{r})$. For the present case the field is produced by the solute charge density $\Omega_s(\mathbf{r}') = \langle \Psi_s| \hat{\Omega}_s(\mathbf{r}') |\Psi_v \rangle$ and the field generated is given by:

$$\mathbf{e}(\mathbf{r}) = -\nabla_r \int d\mathbf{r}' \, \Omega_s(\mathbf{r}') \, T(\mathbf{r} - \mathbf{r}') = -\int d\mathbf{r}' \, \Omega_s(\mathbf{r}') \, \mathbf{T}(\mathbf{r} - \mathbf{r}') \tag{6}$$

The polarization density is given then by:

$$\mathbf{p}(\mathbf{r}) = \mathbf{X}(\mathbf{r}) \cdot [\mathbf{e}(\mathbf{r}) + \int d\mathbf{r}' \, \mathbf{T}(\mathbf{r} - \mathbf{r}') \cdot \mathbf{p}(\mathbf{r}')] \tag{7}$$

where $\mathbf{T}(\mathbf{r} - \mathbf{r}') = \nabla_r T(\mathbf{r} - \mathbf{r}') = \nabla_r \nabla_r T(\mathbf{r} - \mathbf{r}')$ is the dipole-dipole interaction tensor. The integration can be carried out over the whole volume occupied by the surrouding medium while avoiding the singularity by an appropriate cutoff; in practical schemes, this is never a real problem [7]. Equation (7) is iteratively solved and, by introducing the surrounding medium response tensor $\mathbf{C}(\mathbf{r}, \mathbf{r}')$ defined by:

$$\mathbf{C}(\mathbf{r}, \mathbf{r}') = \mathbf{X}(\mathbf{r}) \cdot [\mathbf{1}\delta(\mathbf{r} - \mathbf{r}') + \mathbf{T}(\mathbf{r} - \mathbf{r}') \cdot \mathbf{X}(\mathbf{r}') +$$
$$\int d\mathbf{r}'' \, \mathbf{T}(\mathbf{r} - \mathbf{r}'') \cdot \mathbf{X}(\mathbf{r}'') \cdot \mathbf{T}(\mathbf{r}'' - \mathbf{r}') \cdot \mathbf{X}(\mathbf{r}') \times ...] \tag{8}$$

where $\mathbf{1}$ is the unit tensor, the polarization density is cast into a compact form resembling the one used in the electrodynamics with spatial dispersion [7]:

$$\mathbf{p}(\mathbf{r}) = \int d\mathbf{r}' \, \mathbf{C}(\mathbf{r}, \mathbf{r}') \cdot \mathbf{e}(\mathbf{r}') \tag{9}$$

This equation emphasizes the non local response of the medium towards the solute (external) field.

The reaction field contribution to the interaction Hamiltonian is given by:

$$\hat{V}_{RF} = \int d\mathbf{r} \, \hat{\Omega}_s(\mathbf{r}) \int d\mathbf{r}' \int d\mathbf{r}'' \int d\mathbf{r}''' \, \mathbf{T}(\mathbf{r} - \mathbf{r}'') \cdot$$
$$\mathbf{C}(\mathbf{r}'' - \mathbf{r}''') \cdot \mathbf{T}(\mathbf{r}''' - \mathbf{r}') \, \Omega_s(\mathbf{r}') \tag{10}$$

the integration variables $\mathbf{r}$ and $\mathbf{r}'$ are points associated only with the solute system, while integration over $\mathbf{r}''$ and $\mathbf{r}'''$ are points associated only with the solvent system, thus, defining now a RF kernel function, $K(\mathbf{r}, \mathbf{r}') = \int d\mathbf{r}'' \int d\mathbf{r}''' \, T(\mathbf{r} - \mathbf{r}'') \cdot C(\mathbf{r}'' - \mathbf{r}''') \cdot T(\mathbf{r}''' - \mathbf{r}')$ the effective Hamiltonian of Eq. (5) is cast now into its final form by using Eq. (8) and (9):

$$\hat{H}_s = \hat{H}_s + \int d\mathbf{r} \, \hat{\Omega}_s(\mathbf{r}) \, [V_m^0(\mathbf{r}) + \int d\mathbf{r}' \, K(\mathbf{r}, \mathbf{r}') \, \Omega_s(\mathbf{r}')] \tag{11}$$

The solution of Eq. (2) with this Hamiltonian has to be made selfconsistently, irrespective of the quantum chemical approach used. The total energy of the solute, W_s, is obtained by adding to $E_s(\mathbf{X})$ the energy required to polarize the surrounding medium:

$$W_s = -(1/2) \int d\mathbf{r} \langle \Psi_s | \, \hat{\Omega}_s(\mathbf{r}) \, | \Psi_s \rangle \int d\mathbf{r}' \, K(\mathbf{r}, \mathbf{r}') \, \Omega_s(\mathbf{r}') + E_s(\mathbf{X}) \tag{12}$$

at the end of the selfconsistent calculation.

The general form of the RF attained in the present theory, namely,

$$\Pi(\mathbf{r}) = \int d\mathbf{r}' \, K(\mathbf{r}, \mathbf{r}') \, \Omega_s(\mathbf{r}') , \tag{13}$$

shows an explicit functionality on the solute charge density.

The theory described thus far is well suited for designing computation schemes coupled to MC or MD simulations of the surrounding medium. MC and MD provide information on the global configuration $\mathbf{X}$. Once $\mathbf{X}$ is known, $V_m^0(\mathbf{r})$ and $\mathbf{C}$ can be evaluated provided a solvent charge density and external field susceptibility are given. These problems have already been solved in the framework of the inhomogeneous SCRF theory [7, 23]. At the quantum chemical level, matrix elements over one-electron operators are required. But there is more in this approach as different types of theories can be obtained for varied choices of $\mathbf{C}$-tensors. This is the point where a microscopic approach to the surrounding medium can be used. Another possibility is open if the dielectric theory is employed instead. The latter procedure allows for derivations of the continuum models as particular cases of a generalized dielectric theory.

2.2 Statistically Significant GSCRF Theory

The connection with standard solvent effects theories must be made after a statistical mechanical averaging over the solvent configurations has been made. Temperature enters here in a natural manner.

As the solute-solvent interaction Hamiltonian does not depend upon particles momenta, the statistical averaging is carried out over the configurational space; it will be designated by angular brackets: $\langle \dots \rangle$. A fixed molecular reference frame is assigned to the solute by using, for instance, its nuclear equilibrium configuration $\mathbf{R}_s$. Following Kirkwood, an averaging over the whole solvent configuration is performed while the solute is kept at a fixed orientation defined by the Euler angles of a fixed molecular frame [6]. This averaging is designated by the angular brackets with a subindex m to indicate this fact.

Polarization density. Let us consider first the statistically averaged polarization density:

$$\langle \mathbf{p}(\mathbf{r}; X) \rangle_m = \int d\mathbf{r}' \langle \mathbf{C}(\mathbf{r}, \mathbf{r}'; X) \cdot \mathbf{e}(\mathbf{r}'; X) \rangle_m \tag{14}$$

and introduce the statistically averaged response tensor $\langle \mathbf{C}(\mathbf{r}, \mathbf{r}'; X) \rangle_m = \langle \mathbf{C} \rangle_m$ and electric field $\langle \mathbf{e}(\mathbf{r}'; X) \rangle_m = \langle \mathbf{e} \rangle_m$. At thermodynamic equilibrium (in Feynman's sense, namely, all fast things have happened while the slow ones have not [24]) the fluctuation correlation vector function:

$$\langle \delta \mathbf{C} \cdot \delta \mathbf{e} \rangle_m = \langle (\mathbf{C}(\mathbf{r}, \mathbf{r}'; X) - \langle \mathbf{C} \rangle_m) \cdot (\mathbf{e}(\mathbf{r}'; X) - \langle \mathbf{e} \rangle_m) \rangle_m \tag{15}$$

fades away and Eq. (14) can be uncoupled:

$$\langle \mathbf{p}(\mathbf{r}; X) \rangle_m = \int d\mathbf{r}' \langle \mathbf{C}(\mathbf{r}, \mathbf{r}'; X) \rangle_m \cdot \langle \mathbf{e}(\mathbf{r}'; X) \rangle_m \tag{16}$$

This relationship indicates that, at thermal equilibrium, the surrounding medium will react to the averaged solute electric field. Note that if we average over the orientation of the solute, the rhs of (14) cancels out. The theory described in this paragraph is of a mean field type.

The statistically averaged reaction field susceptibility $\langle K(\mathbf{r}, \mathbf{r}') \rangle_m$ can also be uncoupled following a reasoning similar to the one above. Thus,

$$\langle K(\mathbf{r}, \mathbf{r}') \rangle_m = \int d\mathbf{r}'' \int d\mathbf{r}''' \, \mathbf{T}(\mathbf{r} - \mathbf{r}'') \cdot \langle \mathbf{C}(\mathbf{r}'', \mathbf{r}''') \rangle_m \cdot \mathbf{T}(\mathbf{r}''' - \mathbf{r}) \tag{17}$$

and the RF becomes a response of an statistically averaged susceptibility tensor $\langle \mathbf{C}(\mathbf{r}'', \mathbf{r}''') \rangle_m$.

Solvent potential. The averaged solvent electrostatic field, $\langle V_m^0(\mathbf{r}) \rangle_m$ is important for inhomogeneous media, such as enzymes, membranes, miscelles and crystalline environments systems. Due to the existence of strong correlations, such a field does not cancel out. This factor becomes an important contribution to solvent effects at a microscopic level. In a study of non-rigid molecules in solution, Sese et al. [25] constructed a $\langle V_m^0(\mathbf{r}) \rangle_m$ by using the solute-solvent atom-atom radial distribution function. Electrostatic interactions in three-dimensional solids were treated by Angyan and Silvi [26] in their self-consistent Madelung potential approach; such a procedure can be traced back to a calculation of $\langle V_m^0(\mathbf{r}) \rangle_m$. An earlier application of the ISCRF theory to the study of proton mechanisms in crystals of hydronium perchlorate both $\langle V_m^0(\mathbf{r}) \rangle_m = V_m^0(\mathbf{r}; \langle X \rangle)$ and the RF potential [27] were computed. Another example was provided by the calculation of the electric field produced at the active site of alcohol dehydrogenase by the protein atoms surrounding it [23].

Averaged GSCRF Equation. In the mean field approach, a statistically significant effective Hamiltonian is obtained by m-average ($\langle \dots \rangle_m$) Eq. (11):

$$\mathscr{H}_s = \{\hat{H}_s + \int d\mathbf{r} \, \hat{\Omega}_s(\mathbf{r}) \, [\langle V_m^0(\mathbf{r}) \rangle_m + \int d\mathbf{r}' \langle K(\mathbf{r}, \mathbf{r}') \rangle_m \, \langle \Omega_s(\mathbf{r}') \rangle_m] \tag{18}$$

The key step here for constructing a simple SCRF Hamiltonian is to replace the m-averaged solute charge density ($\langle\Omega_s(\mathbf{r}')\rangle_m$) by the expectation value of the solute charge density operator with respect to the wavefunction Ψ_s, i.e. $\langle\Omega_s(\mathbf{r}')\rangle_m = \langle\Psi_s|\hat{\Omega}_s(\mathbf{r}')|\Psi_s\rangle$. The solute wavefunction is now an ensemble representative object that is obtained as solution of the effective Schrödinger equation with Hamiltonian Eq. (19):

$$\mathscr{H}_s = \{\hat{H}_s + \int d\mathbf{r}\,\hat{\Omega}_s(\mathbf{r})\,[\langle V_m^0(\mathbf{r})\rangle_m + \int d\mathbf{r}'\langle K(\mathbf{r},\mathbf{r}')\rangle_m\langle\Psi_s|\Omega_s(\mathbf{r}')|\Psi_s\rangle\} \tag{19}$$

this Hamiltonian represents a system in thermal equilibrium with respect to all the degrees of freedom except the solute orientation. This assumption is congruent to the mean field approach. Under this hypothesis, the effective Schrödinger equation has the same form as in the generalized SCRF theory presented in the preceding section. The interpretation being fundamentally different in both cases.

An explicit formulation of Hamiltonian (19) is obtained if $\langle K(\mathbf{r},\mathbf{r}')\rangle_m$ is replaced by Eq. (17):

$$\mathscr{H}_s = \{\hat{H}_s + \int d\mathbf{r}\,\hat{\Omega}_s(\mathbf{r})\,[\langle V_m^0(\mathbf{r};\mathbf{X})\rangle_m +$$
$$\int d\mathbf{r}'\int d\mathbf{r}''\int d\mathbf{r}'''\,\mathbf{T}(\mathbf{r}-\mathbf{r}'')\cdot\langle\mathbf{C}(\mathbf{r}'',\mathbf{r}''';\mathbf{X})\rangle_m\cdot$$
$$\mathbf{T}(\mathbf{r}'''-\mathbf{r}')\,\langle\Psi_s|\hat{\Omega}_s(\mathbf{r}')|\Psi_s\rangle\} \tag{20}$$

where the dependence of both the charge density and the averaged response tensor is explicitly shown. This is the fundamental Hamiltonian of the statistically significant GSCRF approach.

The generalized SCRF theory can be used now in several different ways. One makes use of the knowledge provided by MD and/or MC statistical mechanical simulation procedures on the statistical distribution of $\mathbf{X}$. Thus, for systems having a well defined average structure, as for instance a native enzyme, $\langle\mathbf{C}(\mathbf{r}''-\mathbf{r}''';\mathbf{X})\rangle_m = \mathbf{C}(\mathbf{r}''-\mathbf{r}''';\langle\mathbf{X}\rangle_m)$. It is also possible to replace the theory into the framework of the electrodynamics of dielectric materials. In this case, $\langle\mathbf{C}(\mathbf{r}''-\mathbf{r}''';\mathbf{X})\rangle_m$ is represented with the help of the static permitivity tensor. In the next section, the microscopic approach is examined first and thereafter the dielectric approach.

2.3 Microscopic Approach

In the microscopic approach, the theory is completed with a recipe to calculate $\mathbf{C}$. Since we are handing the response to the electronic motion, the high frequency polarizability must be chosen. The ansatz adapted to the present context is given by equation:

$$\mathbf{X}(\mathbf{r}) = \Sigma_m\alpha_m\delta(\mathbf{r}-\mathbf{R}_m) \tag{21}$$

where the summation is carried over the atoms or molecules of the surrounding medium; α_m is the static polarizability tensor of the m-th species in the medium [7, 23, 27].

This form is particularly useful if the solute system can undergo a chemical reaction. Both factors in the RF depend on the global nuclear configuration **X**. The nuclear configuration(s) can be obtained either by sampling the space with MD or MC procedures, or it can be obtained, for particular systems, from X-ray and/or neutron diffraction methods. In crystals and protein surroundings an average structure can be defined [23, 26, 27]. The fluctuations around such structures can be sampled with MC or MD procedures. In this manner, structural fluctuation effects on the solute electronic wavefunction, and associated properties can be subjected to numerical calculations.

2.4 GSCRF Theory in the Dielectric Approach

In the electrodynamics of dielectrics, and electric induction vector $\mathbf{D}(r)$ and a dielectric polarization vector $\mathbf{P}(r)$ are introduced for describing the fields acting in the medium. Correspondences between these objects and those coming from the GSCRF theory are established below.

The averaged polarization density $\langle \mathbf{p}(\mathbf{r}) \rangle_m$ corresponds, by construction, to $\mathbf{P}(\mathbf{r})$. The assigment of $\langle \mathbf{e}(\mathbf{r}) \rangle_m$ is more delicate. In standard electrodynamics, the external sources are not perturbed by the dielectric medium. If such were the case here, there would be no effective Schrödinger equation including reaction fields; first order perturbation theory would be enough to calculate interaction energies between the solute and the medium. As $\langle \mathbf{e}(\mathbf{r}) \rangle_m$ results from a selfconsistency process, the polarization effects are built into the "external" charge density. Thus, it is more natural to assign this field to the electric induction $\mathbf{D}(\mathbf{r})$ when the selfconsistent wave function is used to calculate it. Thus, at selfconsistency, Eq. (7) becomes a relationship between the polarization density and the electric induction:

$$\mathbf{P}(\mathbf{r}) = \int d\mathbf{r}' \langle \mathbf{C}(\mathbf{r}, \mathbf{r}') \rangle_m \cdot \mathbf{D}(\mathbf{r}').$$

The statistical mechanical averaging of **C** introduces correlations related to all molecular motions in the solvent that are allowed to relax until attaining thermal equilibrium. It is natural to seek a function of the static permitivity tensor $\varepsilon(\mathbf{r}, \mathbf{r}')$ to represent the averaged response function. By writing the averaged susceptibility tensor as:

$$\langle \mathbf{C}(\mathbf{r}, \mathbf{r}') \rangle_m = \mathbf{1}\delta(\mathbf{r} - \mathbf{r}') - \varepsilon^{-1}(\mathbf{r}, \mathbf{r}') \tag{22}$$

where $\varepsilon^{-1}(\mathbf{r}, \mathbf{r}')$ is the inverse of the static permitivity tensor, the statistical mechanical averaged polarization density now satisfies standard electrostatic equations for media having spatial dispersion, namely,

$$\mathbf{P}(\mathbf{r}) = \mathbf{D}(\mathbf{r}) - \int d\mathbf{r}' \, \varepsilon^{-1}(\mathbf{r}, \mathbf{r}') \cdot \mathbf{D}(\mathbf{r}') \tag{23}$$

Note that the last term corresponds to the electric field of the source as if it were in vacuo: $\mathbf{E}(\mathbf{r}) = \int d\mathbf{r}' \, \varepsilon^{-1}(\mathbf{r}, \mathbf{r}') \cdot \mathbf{D}(\mathbf{r}')$. Thus, the standard electrostatic relationship between the displacement and external field is recovered: $\mathbf{D}(\mathbf{r}) = \mathbf{E}(\mathbf{r}) + \mathbf{P}(\mathbf{r})$. The internal consistency of the present theory is therefore ensured.

The generalized selfconsistent reaction field equation in the dielectric approach now reads:

$$\{\hat{H}_s + \int d\mathbf{r}\, \hat{\Omega}_s(\mathbf{r})\, [\langle V_m^0(\mathbf{r})\rangle_m + \int d\mathbf{r}'\, \mathbf{T}(\mathbf{r} - \mathbf{r}') \int d\mathbf{r}''\, [\mathbf{1}\, \delta(\mathbf{r}' - \mathbf{r}'')]\, \mathbf{D}(\mathbf{r}'')]\} |\Phi_s\rangle$$
$$= U_s(\mathbf{R}_s) |\Phi_s\rangle \tag{24}$$

For homogeneous environments only, $\langle V_m^0(\mathbf{r})\rangle_m$ averages out to zero. For liquids if a very structures solvation shell remains around the solute, it is convenient to introduce this region (cybotactic region) into the definition of the solute and treat the surroundings as homogeneous.

Hamiltonian Eq. (24) has the most general form among those proposed in the literature to represent solvent effects in the dielectric approach. As a matter of fact, Hamiltonian Eq. (24) contains those as special cases.

2.4.1 Continuum Approach

In the continuum approach to the surrounding medium one has by definition, $\langle V_m^0(\mathbf{r})\rangle_m = 0$. Medium effects are therefore presented by a reaction field term in Eq. (24). Three types of environment can be represented in this framework: i) an anisotropic medium without spatial dispersion, where the permitivity tensor is defined with the ansatz: $\boldsymbol{\varepsilon}(\mathbf{r} - \mathbf{r}') = \boldsymbol{\varepsilon}(\mathbf{r})\, \delta(\mathbf{r} - \mathbf{r}')$, that leads to a distance dependent dielectric system dependance; ii) an isotropic medium which is characterized by $\boldsymbol{\varepsilon}(\mathbf{r} - \mathbf{r}') = \varepsilon(\mathbf{r})\mathbf{1}\delta(\mathbf{r} - \mathbf{r}')$; iii) a homogeneous and isotropic medium, the permitivity tensor is the unit tensor multiplied by the static dielectric constant ε_{st}. Thus the effective Schrödinger equation for each case is obtained from Eq. (24) after integration of the $\mathbf{r}'$-variable with the corresponding ansatz for the permitivity tensor.

2.4.2 Cavity Models

The cavity immersed in a continuum dielectric continues to attract attention from quantum chemists trying to incorporate solvent effects in their in vacuo calculations [28–41]. Cavity models in a continuum dielectric follow in a simple manner by directly modelling $\langle \mathbf{C}(\mathbf{r}, \mathbf{r}')\rangle_m$.

Let $\mathbf{r}^c$ be the cavity radius and V' the sample colume where the spherical cavity containing the solute has been cut out. For a homogeneous isotropic dielectric Eq. (13) can be written as: $\langle \mathbf{C}(\mathbf{r}'', \mathbf{r}''')\rangle_m = (1 - \varepsilon^{-1})\, \Theta(\mathbf{r}'' - \mathbf{r}^c)\, \Theta(\mathbf{r}''' - \mathbf{r}^c)\mathbf{1}\, \delta(\mathbf{r}'' - \mathbf{r}''')$, where $\Theta(\mathbf{r}'' - \mathbf{r}^c)$ is the Heasivide step function. The RF potential can then be cast into:

$$\Pi(\mathbf{r}) = (1 - \varepsilon^{-1}) \int d\mathbf{r}''\, \mathbf{T}(\mathbf{r} - \mathbf{r}'')\, \cdot$$
$$\Theta(\mathbf{r}'' - \mathbf{r}^c) \int d\mathbf{r}'''\, \mathbf{1}\, \Theta(\mathbf{r}''' - \mathbf{r}^c)\, \delta(\mathbf{r}'' - \mathbf{r}''')\, \cdot \mathbf{D}(\mathbf{r}''') \tag{25}$$

The inner integral can be carried out strightforwardly as r″ and r‴ run outside the cavity, thus, $\Pi(\mathbf{r}) = (1 - \varepsilon^{-1}) \int d\mathbf{r}'' \, \Theta(\mathbf{r}'' - \mathbf{r}^c) \, \mathbf{T}(\mathbf{r} - \mathbf{r}'') \cdot \mathbf{D}(\mathbf{r}'')$ and introducing the quantum mechanical definition of the solute field, one finally gets:

$$\Pi(\mathbf{r}) = (1 - \varepsilon^{-1}) \int d\mathbf{r}' [\int d\mathbf{r}'' \, \Theta(\mathbf{r}'' - \mathbf{r}^c) \, \mathbf{T}(\mathbf{r} - \mathbf{r}'') \cdot \mathbf{T}(\mathbf{r}'' - \mathbf{r}')] \, \Omega_s(\mathbf{r}')$$

(26)

This equation contains the basic model used by M. Newton in his pioneering calculations of the solvated electron [7]. If the charge density is replaced by a classical unit charge at the origin of the sphere, the RF potential obtained after integration of Eq. (26) corresponds to Born's model for a metalized sphere immersed in an isotropic continuum.

The model using spherical harmonics expansions for the RF potential can be derived from Eq. (26) by introducing spherical boundary conditions. The procedure has already been outlined by this author [6] and will not be repeated here.

Semi-continuum models. In this type of approach, the first solvation shell is represented in the supermolecule and, consequently, enters into the quantum chemical description. Basically, the radius of the sphere embedded in the continuum dielectric is much larger than for the desolvated solute. This model has been used in several occasions, e.g. solvated electron, electron transfer in solution.

Recently, two developments have taken place in this field: one concerns the quantum chemical level used to solve open shell electronic structures [37], the other one is an extention of the homogeneous model to treat an anisotropic surrounding medium [38].

The level of electronic structure theory used by Mikkelsen et al. [37] is given by the multiconfigurational (MC) selfconsistent field (SCF) where the wavefunction is fully optimized with respect to all variational parameters; these include both orbital and configurations. The main deficiency of standard SCF ab initio procedures, namely, lack of correlation effects, is overcome in this MCSCF approach. The level of solvent-effects theory is the standard spherical cavity immersed in a continuum dielectric; an early formalism proposed by Rinaldi and Rivail was used (see Ref. [6] for an extensive analysis).

The contribution by Hoshi et al. addresses the formulation of a theory for the estimation of a molecular electronic structure surrounded by an anisotropic medium [38]. It is assumed there, that the medium surrounding the solute system is composed of more than two polarizable dielectrics with different dielectric constants; the different dielectric regions make contact at each other through arbitrary shape boundaries. As usual, the solute charge distribution interacts with the dielectrics via a reaction field.

In a different vein, Claverie and coworkers [52] have made a significant contribution to improve the continuum model used for calculation of solvation thermodynamics quantities of a molecule embedded in a cavity formed by the intersecting van der Waals spheres of the solute in a polarizable medium.

At a more elementary level, Gersten and Sapse have investigated solvent effect through the use of an extended Born equation [43]. Honig and Rashin [44] have also contributed with an empirical method to correct Born's results.

An analysis of discrete and continuum dielectric models as they have been applied to calculate protonation energies in solution has been presented by Rullman and van Duijnen [45]. The model proposed by these authors combines a discrete molecular description of the first two or three solvation layers with a continuum description of the bulk solvent. The solute is described quantum chemically. The accuracy of the results depends on the dielectric model as well as on the details of the electrostatic potential and on inductive interactions.

The simple virtual charge model discussed by Constanciel and Tapia [6] has been developed into an extended generalized Born (EGB) approach. Different approximations have been proposed. Constanciel [40] has analyzed the theoretical basis used as foundations for empirical reaction field approximations through the continuum model to the surrounding medium. Artifacts in the EGB scheme have been clearly identified. The new approximate formulation proposed derives from an exact integral equation of classical electrostatics following a well defined procedure. It is shown there how the wavefunction of solvated species imbedded in cavities formed by interlocking sphere in a polarizable continuum can be computed.

3 Classical Statistical Mechanical Scheme

The dynamics in the X-space, namely, the configurational space, is normally assumed to be driven at the classical mechanical level. Once the quantum mechanics of electronic motion has been solved, the nuclei are submitted to intra- and inter-molecular potentials. The total Hamiltonian H is written as a sum of molecular Hamiltonians (H_m and H_s) and the intermolecular interactions H_{ms}:

$$H(X, P) = H_m(P_m, R_m) + H_s(P_s, R_s) + H_{ms}(R_m, R_s) \qquad (25)$$

where, P is the linear momentum in cartesian coordinates of the ensemble of particles; P_m and P_s identify the momenta associated to each subsystem. The solute subsystem, in the classical statistical mechanical treatment, may contain a number of molecules larger than those entering in the quantum subsystem treated in the preceding section. To simplify notation, and without loss of generality the solvent molecules are taken as atoms. Then, $H_m(P_m, R_m) = \Sigma_i[H(P_{m_i}) + \Sigma'_j V(R_{m_i}, R_{m_j})]$; where R_{m_i} is the position vector and P_{m_i} is the canonically conjugated momentum of the i-th solvent atom; $H(P_{m_i})$ is the kinetic energy. The atoms' interact via the total charge densities; $V(H_{m_s}, H_{m_j})$ is the interatomic potential between solvent atoms.

The solute is a polyatomic system characterized by the intramolecular potential $V(R_s)$. The solute Hamiltonian in the laboratory coordinate frame can be cast into a matrix form as follows:

$$H_s = (1/2) P_s^+ \cdot M^{-1} \cdot P_s + V(R_s) \qquad (26)$$

$\mathbf{M}$ is the mass-diagonal matrix, $\mathbf{M}^{-1}$ its inverse; $\mathbf{P}_s^+$ is the row vector obtained by hermitian conjugation of the column vector $\mathbf{P}_s$. The potential $V(\mathbf{R}_s)$ represents the force field associated to the solute system.

The solute-solvent interaction H_{ms} is represented as a pairwise potential energy function:

$$H_{ms} = \Sigma_i^{N_m} \Sigma_j^{N_s} V(|\mathbf{R}_{m_i} - \mathbf{R}_{s_j}|) \tag{27}$$

where N_m and N_s are the number of atoms in the solvent and solute, respectively. As discussed in section 2, $V(|\mathbf{R}_{m_i} - \mathbf{R}_{s_j}|)$ contains exchange repulsive terms, van der Waals attractive interactions and electrostatic terms. The dependence of this potential function with the internal geometry of the solute is neglected in standard treatments. It is usually assumed that the solute is found at its equilibrium nuclear configuration when $V(|\mathbf{R}_{m_i} - \mathbf{R}_{s_j}|)$ is either calculated from ab initio quantum chemical calculations, or fitted from experimental data.

The time evolution of the dynamical variables is controlled by the Liouvillian superoperator $i\hat{L}$. Let $\mathbf{A} = (A_1, A_2, ...)$ be a row vector gathering the dynamical variables of our system. According to classical statistical mechanics, the equation of motion is given by the Liouville equation

$$dA(t)/dt = \dot{A}(t) = i\hat{L}(t) A(t) \tag{28}$$

where the Liouvillian operator is given by:

$$i\hat{L}(t) = \Sigma_k (\partial H/\partial \mathbf{P}_k) \, \partial[]/\partial \mathbf{R}_k - \Sigma_k (\partial H/\partial \mathbf{R}_k) \, \partial[]/\partial \mathbf{P}_k \tag{29}$$

the sum over k goes over the 3N degrees of freedom including both the solute and its surroundings. From this definition, it is obvious that the Liouvillian can be written in a way similar to the Hamiltonian, i.e. $i\hat{L} = i\hat{L}_m + i\hat{L}_s + i\hat{L}_{ms}$. The interaction Liouvillian acquires a simple form:

$$i\hat{L}_{ms} = -\Sigma_i \Sigma_j (\partial V(|\mathbf{R}_{m_i} - \mathbf{R}_{s_j}|)/\partial \mathbf{R}_{mi} \, \partial[]/\partial \mathbf{P}_{sj} -$$
$$\Sigma_j \{\Sigma_s \, \partial V(|\mathbf{R}_{m_i} - \mathbf{R}_{s_j}|)/\partial \mathbf{R}_{mj}\} \, \partial[]/\partial \mathbf{P}_{mi} \tag{30}$$

describing the coupling between the solvent and solute at time t in terms of forces.

In molecular dynamics simulations, the fundamental dynamical variables are the coordinates and canonically conjugated momenta. The time evolution of the probability distribution function $f(T)$, where $T = (\mathbf{P}_{m1}, \mathbf{P}_{m2}, ..., \mathbf{R}_{m1}, \mathbf{R}_{m2}, ..., \mathbf{P}_{s1}, \mathbf{P}_{s2}, ..., \mathbf{R}_{s1}, \mathbf{R}_{s2}, ...)$ is a point in phase space at time t, is controlled by the Liouville Eq.:

$$\partial f/\partial t = -i\hat{L}f(T, t) \tag{31}$$

The formal solution of this equation is $f(t) = \exp(-i\hat{L}t) f(T, 0)$. The initial condition is taken to be an ensemble in thermal equilibrium which, for a number of situations of chemical interest, is represented by the canonical distribution:

$$f(T, 0) = f(T_0) = \exp(-\beta H)/\int dT_0 \exp(-\beta H) \tag{32}$$

with $\beta = 1/k_B T$; k_B stands for the Boltzman constant and T is the absolute temperature. This distribution function is used to carry out the statistical averages ($\langle \dots \rangle$) in what follows.

3.1 Projected Equation of Motion

For a theory of solvent effects on a given subsystem, it is important to define a reduced (effective) distribution function $f_s(T_s)$ for the system of interest. This can be achieved by using the theory of projection operators. The operator projecting onto the dynamical variable space of the subsystem of interest can be defined as: $f_s(T_s, t) = f_m(T_m, 0) \int dT_m f(T, t) = \hat{P}f$; the density for the subsystem m is obtained as: $f_m(T_m, t) = (1 - \hat{P}) f(T, t)$. Multiplying Eq. (31), in turn, by $\hat{P}$ and the projection operator for the orthogonal complement $\hat{Q} = \hat{1} - \hat{P}$, two differential equations are obtained. Solving the one for $f_m(T_m, t)$ first, and introducing this solution into the one for $f_s(T_s, t)$, the density of the subsystem s can be obtained as a solution of the master equation:

$$\partial f_s(T_s, t)/\partial t = iPLPf(T, t) + iPLe^{iQL^t}f(T, 0) + \int du\, iPLe^{iQLu}QLPf(T, t - u) \tag{33}$$

This equation describes the time evolution of the projected density $f_s(T_s, t)$. The first term in (33) evolves in the subspace of the s-system; the second term acts as force produced by the evolution in the m-system which has a projection onto the s-system; the last term has a non-Markovian character, as can be seen from the memory kernel $iPLe^{iQLu}QLP$. The non-Markovian character of Eq. (33) is a consequence of the procedure of variable contraction.

As a result of the partitioning accomplished above, an effective Liouvillian can be written,

$$i\hat{\mathscr{L}}_s = i\hat{L}_s + i\langle \hat{L}_{ms} \rangle_m + i\hat{L}_g \tag{34}$$

where $i\langle \hat{L}_{ms} \rangle_m$ describes the conservative force effects between both subsystems, and $i\hat{L}_g$ stands for the non-conservative effects. This latter factor is neglected in standard treatments.

It is a commonly encountered situation to find that the solvent molecules in a neighborhood of the solute are strongly perturbed by it. The subsystem of interest is therefore defined so as to include sufficient solvent molecules into the simulation. In this way, the correlations between this redefined subsystem and the reminder may be weak enough to neglect them. The effective Liouvillian $i\hat{\mathscr{L}}_s$ describes this latter type of subsystem of interest.

The projection operator technique can also be applied to the dynamical variables of the system. In order to proceed, it is necesary to define a scalar product among linearly independent dynamical variables. One useful definition is via correlation functions between two variables, say **A** and **B**:

$$\langle \mathbf{A}(t) \, \mathbf{B}(0) \rangle = \int dT_0 \, f(T_0) \, \mathbf{A}(t) \, \mathbf{B}^+(0) = (\mathbf{A}(t), \mathbf{B}^+(0)) \tag{35}$$

After introducing a projection operator for the variables of interest, the time evolution of the momentum variables for the subsystem can be writen as:

$$\dot{\mathbf{P}}_s = (i\hat{L}\mathbf{P}_s(0), \, \mathbf{P}_s^+(0)) \, (\mathbf{P}_s(0), \, \mathbf{P}_s^+(0))^{-1} \, \mathbf{P}(t) + \mathbf{F}_s(t) + \int_0^t dt' \, \Phi(t - t') \cdot \mathbf{P}(t') \tag{36}$$

which is the analog of Eq. (33), since the first term is nothing but $i\hat{P}\hat{L}\mathbf{P}\mathbf{P}(t)$, the second $\mathbf{F}_s(t) = i\hat{P}\hat{L}\hat{Q} \, e^{i\hat{Q}\hat{L}t} \, \mathbf{P}(0)$ and, the third, is another way of writing the non-Markovian effects. It can be shown that $\Phi(t)$ is the autocorrelation matrix of the stochastic force $\mathbf{F}_s(t)$, namely,

$$\Phi_s(t) = (\mathbf{F}_s(t), \, \mathbf{F}_s^+(0)) \, (\mathbf{P}_s(0), \, \mathbf{P}_s^+(0))^{-1} \tag{37}$$

which is the second fluctuation-dissipation theorem. The time integral of $\Phi_s(t)$ is a generalized friction function: $\gamma(t) = \int dt' \, \Phi(t - t')$.

Approximate time evolution equation for the system of interest that are used in molecular dynamics simulations can now be obtained. If the total Liouvillian is replaced by $i\mathscr{L}_s$ in Eq. (36) and use is made of Eqs. (29) and (30), a generalized Langevin equation follows:

$$\hat{\mathbf{P}}_s = -\partial V(\mathbf{R}_s)/\partial \mathbf{R}_s - \partial \langle H_{ms} \rangle_m/\partial \mathbf{R}_s + \mathbf{F}_s(t) + \int dt' \, \Phi(t - t') \cdot \mathbf{P}_s(t') \tag{38}$$

The first term describes the subsystem force field; the second one is important if the medium surrounding the subsystem of interest is structured, such a situation is met in an enzyme's active site simulation (see below), otherwise, this term cancels out. The stochastic forces and memory term are subject to particular modelizations.

Force Fields. The potential energy function, $V(\mathbf{R}_s)$, describes the interactions between the atoms in the system of interest. For a protein or nucleic acids, the potential is composed of terms representing covalent bond stretching; bond angle bending; quadratic dihedral bending which takes care of out-of-plane and out-of-tetrahedral configuration motion; sinusoidal dihedral torsion; interatomic repulsive forces and van der Waals attractive interactions which make up for the Lennard-Jones potential used in normal liquid simulations; and Coulomb interactions. Empirical parameter sets have been developed, and are now embodied in Rehovot (M. Levitt), Harvard (CHARMM) [17] and Groningen (GROMOS) [18] computer programs.

Water-water potential functions have been reported from different laboratories. The procedure leads either to analytic functions fitted to empirical data, e.g. SPC (single point charge) [19] and TIPS (transferable intermolecular potential functions) family [20], or to ab initio quantum chemical calculations, e.g. Clementi's MCY potential [21]. At this point, it is interesting to examine the performance of the MCY potential. This water model has been extensively used in the literature. Rice and coworkers have tested the accuracy of water-water potentials [46–49]. Predictions of densities, lattice energies and lattice geometries of the proton ordered ices have been used to test this potential. The MCY potential predicts the correct ice lattice structure but not the density. Frequency dependent properties have also been examined. As pointed out by Beveridge [50] excellent agreement is obtained with experiment on the oxygen-oxygen radial distribution function. The model fails to give reasonable pressure. In a future release of this potential announced by Clementi, this drawback is overcome. Three-body interactions were computed and used in MC simulations by Clementi and Corongiu [51]. Later on fourth-body interactions were included [52]. Improvements in the oxygen-oxygen radial distribution function and enthalpy have been reported. Computer simulations with the MCY potential of the dielectric constant of water have been reported by Neumann [53].

3.2 Active Site Models

The theory of solvent effects is constructed on a spatially localized model for the events of interest. By such techniques, as was illustrated in the quantum mechanical section, only a small part of the system is induced in the explicit simulation and the effects of the remainder of the system are treated implicitly. Such is the spirit in which Eq. (38) is constructed.

For liquids, Adelman, Karplus, McCammon and coworkers [54–58] have proposed a method for simulating a localized region and replacing distant atoms by a suitably constructed boundary, including a stochastic heat bath. Warshel's surface constraint compressible dipole model pioneered this type of approach [59]. Basically, the system is partitioned into a reactant subsystem and a boundary or reservoir region. The former is further parted into a reaction region and a buffer one. The molecules in the buffer are treated as Langevin particles and, in the reaction region, they are simulated by standard molecular dynamics. The reservoir provides a static force field that helps insure that correct structural and dynamic properties will be maintained within the reaction zone. This type of approach has been extended to treat active site dynamics of enzymes by several authors [60–62].

4 Evaluation of Solvent Effects

Progress in computer technology has profoundly influenced the field of solvent effects. From small solutes up to proteins in water, a number of studies have been reported in the literature.

4.1 Solvent Effects on Molecular Properties

4.1.1 Monte Carlo Studies

Monte Carlo simulation techniques have been extensively used to study solvent effects on molecular properties and equilibrium points. Jorgensen has summarized theoretical work of condensed-phase effects on conformational equilibria [63].

Monte Carlo studies of a dilute aqueous solution of benzene have been reported by Beveridge and coworkers [50] and by Linse et al. [64]. Intermolecular pairwise potential functions determined from quantum mechanical calculations were used; for water-water interactions both groups have used pairwise Matsuoka-Clementi-Yoshimine potential. Interesting solvation patterns around benzene have been found.

Solvent effects on the relative energies for the planar and perpendicular allyl cations in liquid hydrogen fluoride have been studied with MC simulations. The intermolecular potential functions describing the solute-solvent interactions have been obtained from ab initio molecular orbital calculations with a 4-31G basis set. The conformers were represented by different charge distributions but the same Lennard-Jones parameters. The solvent-solvent interactions were described by a function of the TIPS form inclusing one Lennard-Jones term and three charged sites for each hydrogen fluoride monomer [65]. Significant differences in solvation are detected. The more localized (perpendicular) conformer is found to be better solvated in agreement with traditional notions of ion solvation.

Clementi and coworkers have been deriving intermolecular potentials from ab initio computations and using these potentials in statistical mechanical computer experiments on pure solvents and solutions. The goal being the derivation of an ordered set of approximations leading to an increasingly realistic description of water and aqueous solutions. Extensive work on solvation of biomolecules has been carried out [66–69] by this group.

Free energy simulations using the Metropolis MC method and a coupling parameter approach with umbrella sampling have been performed for a number of systems going from liquid water to chemical reaction in solution. Applications to the study of liquid water, hydrophobic interactions and solvent effects on conformational stability have been reported by Beveridge et al. [70]. The effect of hydration on the torsional energy surface of butane using statistical perturbation theory has been studied by Jorgensen and Buckner [71]. The methodology therein proposed is shown to yield results with high precision and to have significant advantages over umbrella sampling.

4.1.2 Molecular Dynamics Studies

Dynamics of proteins and nucleic acids and their solvent surroundings have been reviewed recently by McCammon and Harvey [72]. Solvent effects on biochemical solutes studied by MD techniques have also been discussed by Jorgensen. In this section recent studies on protein solvation are examined.

Only a few MD simulations of proteins in aqueous solution have been carried out. Such studies provide data for examining conformational differences between crystal structures and solution structures [18, 73–77]. Thus, Ahlström et al. have studied Parvalbumine in vacuo and in aqueous solution. Parvalbumine is a Ca^{+2} binding protein. Simulations in vacuo of Ca^2-free protein were also reported. Considerable structural changes were detected relative to the initial coordinate in the two in vacuo simulations. Water surroundings, although they help to maintain the structure in the neighborhood of the crystallographic structure a little better, are not sufficient to keep a correct structure. It is not clear from the analysis presented by these authors, whether their results are mere artifacts deriving from the potential function used or if the results are a fair representation of the protein behavior in solution.

Simulations of BPTI (bovine pancreatic trypsin inhibitor) in van der Waals solvents have been reported [74, 75], the density and molecular size were chosen to simulate those of water. More realistic water representations were used in further simulations [18, 76, 77]. Avian pancreatic polypeptide hormone in crystal and in aqueous solution has been reported by Kruger [78]. These studies tend to indicate that the calculations in vacuo represent fairly correctly the motion of the protein core, while exposed sidechains react more strongly to solvent effects.

Collective motions of secondary and supersecondary structures in proteins are important for understanding functionality. A unique collective motion has been detected by molecular dynamics dynamics simulations of the carboxy terminal fragment (CTF) of the L7/L12 ribosomal protein [79]. In the crystal state, the unit cell embodies eight units of CTF. A dimer has been proposed as a functional significant structure. MD simulations of the dimer [80] and of CTF immersed in a bath of 2352 SPC water molecules carried out in our laboratory, have shown that protein collective motions are not damped down. For this particular case, the simulations in vacuo are good enough for reproducing structural features and the dynamical behavior of the whole protein.

As the number of systems studied so far is rather small, general conclusions concerning model water effects on structural and dynamic properties of proteins cannot be drawn.

4.2 Solvent Effects on Chemical Reactions

One of the most spectacular application of solvent effects evaluation by computer simulation techniques is the study of nucleophilic addition and the bimolecular nucleophilic substitution S_N2 reactions.

The S_N2 reaction, $X^- + RY \rightarrow XR + Y^-$, has been simulated with MC equilibrium calculations by Jorgensen and coworkers [81, 82]. The procedure used by these authors involves three steps: i) the lowest energy reaction path is determined for the in vacuo system by using ab initio molecular orbital calculations; ii) intermolecular potential functions are obtained to describe the interactions between the substrate and a solvent molecule; these potentials depend on the internal structure of the substrate; iii) MC simulations are carried out to determine the free energy profile for the reaction in solution. This is a difficult computational task since importance sampling methods are required to explore all the values of the reaction coordinate. A similar technique was used by Madura and Jorgensen [83] in simulating the nucleophilic addition of hydroxide ion to formaldehyde in the gas phase and in aqueous solution.

Molecular dynamics simulations of the S_N2 type reaction were reported by Bergsma et al. [84]. The technique is used to explore the role of polar solvent dynamics and configurations in modulating the reaction trajectoris and the ratio between the true value of the rate constant k and the one obtained from the transition state theory k^{TST}. Important results concerning solvent effects on the dynamics are obtained.

In a more chemical vein, Tapia and coworkers [85–87] have studied hydration effects for the rate limiting step of the acid catalyzed rearrangement of α-acetylenic alcohol to α,β-unsaturated carbonyl compounds. Water intervenes in the reaction mechanism. Ab initio MO studies of the energy hypersurface were carried out at a 4-31G basis set level [86]. The topography of this energy hypersurface presents two solvation sites for the protonated alcohol and two for the protonated allenol; the solvation minima of the protonated alcohol are connected via a saddle point whose structure is the reactant in the rate limiting step(RLS) model; an analogous situation is found for the protonated allenol where the saddle point is now the product in the RLS model. MC simulations at stationary points on the energy hypersurface show that the solvation sites are occupied by solvent water molecules, and and due to solvent caging effects, the reactant and product of the RLS in solution become stable species. Mechanistically, the transition state for the RLS derives from the solvated reactant by the jump of one solvent molecule towards its nucleophilic center. As the unrelaxed solvation shell is less efficient than the equilibrium one, the TS for this reaction is better described as a poorly solvated saddle point structure; relaxation of this shell opens the channel leading to final products. This set of studies illustrate the difficulties simulation methods encounter when water takes an active role in the molecular mechanism.

4.3 Integrated Quantum/Statistical Studies

The key issue of incorporating solvent effects in the quantum mechanical calculation has not been solved satisfactorily in MC and molecular dynamics studies overviewed above. Warshel's empirical valence bond approach, van Duijnen's direct reaction field method, and Tapia's ISCRF theory, by including these solvent effects, are steps forward in this direction. Although the key theoretical issue cannot be considered satisfactorily solved, the applications made are most interesting.

The energetics and dynamics of the S_N2 class of reactions in aqueous solution have been studied by a combination of the empirical valence-bond method and a free perturbation technique. The solvent is represented by a surface constrained all-atom solvent (SCAAS) model, and many-body interactions are taken into account with a solvent parameter set that includes atomic polarizabilities [88]. In this work, activation free energies for the $X^- + CH_3Z \rightarrow XCH_3 + Z^-$ reactions are computed and the general relationship between the reaction free energies and the solvent contribution to the activation free energies are examined. The dynamical aspects of the S_N2 charge transfer reaction are explored by propagating trajectories downhill from the transition state by using linear response theory. The simulations suggest that solvent fluctuations play a central role in driving the system towards the transition state; the relaxation time for the reactive fluctuations is determined by both the polarization time of the solute dipole moment and the dielectric relaxation time of the solvent. Effects of solute-solvent coupling and solvent saturation effects on solvation dynamics of charge transfer reactions were pursed in another work by Hwang et al. [89].

Evaluation of catalytic free energies in genetically modified proteins was made by Warshel et al. [90]. A combination of the EVB method and free energy perturbation approach similar to the one described above was applied to study activity of genetically modified enzymes: trypsin and subtilisin. The importance of dynamics aspects was shown by using autocorrelation functions of the protein reaction field on the reacting substrate. This procedure shows the enormous power simulation techniques have to help predict important biological changes in enzyme function.

Earlier theoretical work on charge-relay catalysis in the function of serine proteases has been critically evaluated by Schowen [91]. The catalytic steps for acylation and hydrolysis of a model ester by chymotrypsin has been studied with the ISCRF theory. The results are in good qualitative agreement with experimental facts. Since the semiempirical MO method was not calibrated to reproduce in vacuo properties, no quantitative agreement was reached [92].

The mechanism of liver alcohol dehydrogenase (LADH) has been extensively studied. For a recent overview the reader is referred to Ref. [93]. Reaction field effects on the transition structure of model hydride transfer systems have been calculated at ab initio 4-31G basis set level [93, 94]. The active site of enzymes are usually assumed to be designed to receive molecules in the transition state for the reaction they catalyze. This special sort of surrounding medium effects has been computationally documented recently [95]. From the reaction geodesic passing through the transition state for hybride transfer in the pyridium cation/methanolate model system, only the TS-structure could be fitted into the LADH active site. The normal mode analysis carried out on the TS showed an excellent agreement with isotopic substitution experiments [95]. Reaction field calculations on this model systems have also been performed. For an overview of biomolecular interactions the reader is referred to Ref. [96].

5 Conclusions and Prospects

The theory of solvent-effects and some of its applications have been overviewed. The generalized self-consistent reaction field theory has been used to give a unified approach to quantum chemical calculations of subsystems embedded in a given milieu. The statistical mechanical theory of projected equation of motion has been briefly described. This theory underlies applications of molecular dynamics simulations to the study of solvent and thermal bath effects on carefully defined subsystems of interest. The relationship between different approaches used so far to calculate solvent effects and the general approach advocated by this reviewer has been established. Applications to molecular properties in a time independent framework have been presented.

The applications described in this chapter have been selected to illustrate the power of the solvent-effects approach to biochemical and physicochemical problems. The techniques are being now used in other fields. Thus, molecular dynamics simulations are being used to study solvent effects on static and dynamic properties of linear and star polymer [97, 98]. A generalized Langevin equation has been used to represent solvent around a polymer with stochastic forces; the solution of this equation agrees well with corresponding MD simulations [99].

Non-equilibrium solvent effects have now become the subject of intensive experimental investigations. Different blends of time-resolved spectroscopies are contributing to the knowledge of dynamical processes in liquids, solutions and bioenvironments [100, 101]. The electronic properties of the solute system become affected by fluctuations of the solvent configuration $X(t)$. The response of the solvent to a newly created charge or dipole involves subpicosecond phenomena. The reciprocal effects can be followed theoretically by implementing, for instance, the generalized SCRF scheme. Such developments have been anticipated in the EVB approach by Warshel and coworkers. However, an ab initio approach is desirable in order to attain a deeper understanding of non-equilibrium phenomena.

The full quantum statistical mechanical approach to solvent effects on dynamic processes has not been analyzed in detail. It is important to note recent developments by Banacky and Zajac [102, 103] on the theory of particle dynamics in solvated molecular complexes. A time-dependent nonlinear equation of motion for the probability density of a proton in a solvated symmetric H-bond system was derived. Earlier work has been overviewed by the present author in a recent paper [10].

Progress is forseen in the study of solvent effects on chemical reactions in liquids, solids, miscelles, and enzymes. Brute force MD simulations of solvent effects on the dynamics properties of protein summarized in this work will serve as benchmark calculations to gauge model representations of solvent effects on biomacro-molecules. The use of realistic dielectric models can be also be seen as a complementary approach to represent solvent effects on biomolecules. In chemical dynamics, important advances have been made with analytical simple model approaches [11, 104, 105]. The conditions are now ripe for including more sophisticated ab initio studies into the description of time-dependent phenomena.

The prospects for future developments in this field are most promising.

Acknowledgments: The author would like to thank the NFR for financial support and R. Cardenas for help in gathering information. Part of this work was made while the author was at the Department of Molecular Biology. The warm and friendly hospitality offered to him by C-I Bränden, H Eklund, B Stransberg, and colleagues is sincerely appreciated.

6 List of Symbols

Ω	capital omega
$\mathbf{r}$	bold low case r
T	capital t in e.g. $T(\mathbf{r} - \mathbf{r}') = 1/|\mathbf{r} - \mathbf{r}'|$

T	bold capital t in e.g. $\mathbf{T}(\mathbf{r}''' - \mathbf{r}')$ indicates a vector quantity.
T	second order tensors are recognized by this style.
C	capital C symbol for a second order tensor
$\boldsymbol{\alpha}$	alpha (tensor of second rank)
$\boldsymbol{\varepsilon}$	epsilon
1	unit tensor
$\mathit{\Pi}$	capital pi
$\mathit{\Sigma}_{\mathrm{i}}$	capital sigma (subindex i)
δ	lowcase delta as in: $\delta(\mathbf{r} - \mathbf{r}_{\mathrm{i}})$
∇_{r}	bold capital delta (subindex r)
$\int \mathbf{dr}$	integral symbol
$\hat{\Psi}$	operator are identified with this symbol on top of the latter.
Ψ	capital psi
T	fat tau
$\gamma(\mathrm{t})$	fat gamma as in page 24
$\Phi(\mathrm{t} - \mathrm{t}')$	fat capital phi
H_{s}	capital H italics
$\mathscr{H}_{\mathrm{s}}$	capital H handwrite (script)

7 References

1. Amis ES, Hinton FH (1973) Solvent effects on chemical phenomena. Academic, New York
2. Tanaka N, Ohtaki H, Tamamushi R (1983) (eds) Ions and molecules in solution. Elsevier, Amsterdam
3. Strauss HL, Babcock GT, Moore CB (1986) (eds) Annual Review of Physical Chemistry, vol 37. Ann. Rev., Palo Alto, CA
4. Dogonadze RR, Kalman E, Kornyshev AA, Ulstrup J (eds) (1986) The chemical physics of solvation; Part B Spectroscopy of solvation. Elsevier, Amsterdam
5. Dogonadze RR, Kalman E, Kornyshev AA, Ulstrup J (eds) (1985) The chemical physics of solvation; Part A theory of solvation. Elsevier, Amsterdam
6. Tapia O (1980) in: Daudel R, Pullman A, Salem L, Veillard A (eds) Quantum theory of chemical reactivity vol 2, Reidel, Dordrecht, p 25
7. Tapia O (1982) in: H. Ratajczak and W. J. Orville-Thomas (eds) Molecular interactions vol 3, Wiley, Chichester, p 47
8. McQuarrie DA (1976) Statistical mechanics. Harper and Row, New York
9. Serra R, Andretta M, Compiani M (1986) Physics of complex systems. Pergamon, Oxford
10. Tapia O (1988) in: Maruani J (ed) Molecules in physics chemistry and biology vol 3, Kluwer Academic, Holland, p 405
11. Hynes JT (1985) in: Baer M (Ed) The theory of reactions in solution vol 4. CRC, Boca Raton, FL, p 171
12. Szabo A, Ostlund NS (1982) Modern quantum chemistry. MacMillan, New York
13. McWeeny R, Pickup B (1980) Rep. Prog. Phys. 43: 1065
14. Craig DP, Thirunamachandran T (1984) Molecular quantum electrodynamics. Academic, London
15. Buckingham AD (1967) Adv. Chem. Phys. 12: 107
16. Jackson JD (1962) Classical electrodynamics, 2nd ed. Wiley, New York
17. Brooks BR, Bruccoleri RE, Olafson BD, States DJ, Swaminathan S, Karplus M (1983) J. Comp. Chem. 4: 187
18. van Gunsteren WF, Berendsen HJC, Hermans J, Hol WGJ, Postma JPM (1983) Proc. Nat. Acad. Sci. USA 80: 4315

19. Berendsen HJC, Postma JPM, van Gunsteren WF (1981) In: Pullman B (ed) Intermolecular forces. Reidel, Dordrecht, p 331
20. Jorgensen WL (1982) J. Chem. Phys. 77: 4156
21. Matsuoka O, Clementi E, Yoshimine M (1976) J & Chem. Phys. 64: 1351
22. Price SL (1988) Mol. Simul. 1: 135
23. Tapia O, Johannin G (1981) J. Chem. Phys. 75: 3624
24. Feynman RP (1972) Statistical mechanics. Benjamin, Reading, MA
25. Sese LM, Botella V, Gomez PC (1986) J. Mol. Liquids 32: 259
26. Angyan J, Silvi B (1987) J. Chem. Phys. 86: 6957
27. Angyan J, Allavena M, Picard M, Potier A, Tapia O (1982) J. Chem. Phys. 77: 4723
28. Karelson MM, Katritzky AR, Zerner M-C (1986) Int. J. Quantum Chem. Symp. 20: 521
29. Sanchez-Marcos E, Terryn B, Rivail JL (1985) J. Phys. Chem. 89: 4695
30. Contreras R, Aizman A (1986) Int. J. Quantum Chem. Symp. 20: 573
31. Rinaldi D, Ruiz-Lopez MF, Martins Costa MTC, Rivail J (1986) Chem. Phys. Lett 128: 177
32. Ruiz-Lopez MF, Rinaldi D, Rivail JL (1986) Chem. Phys. 110: 403
33. Mikkelsen KV, Dalgaard E, Swanstrøm P (1987) Chem. Phys. 91: 3081
34. Mikkelsen KV, Ratner MA (1987) Int. J. Quantum Chem. S21: 341
35. Ågren H, Medina-Llanos C, Mikkelsen KV (1987) Chem. Phys. 115: 43
36. Karlström G (1987) J. Phys. Chem. 92: 1315
37. Mikkelsen KV, Ågren H, Jensen HJA, Helgaker (1988) J. Chem. Phys. 89: 3086
38. Hoshi H, Sakurai M, Inoue Y, Chujo R (1987) J. Chem. Phys. 87: 1107
39. Stamato FML, Tapia O (1988) Int. J. Quantum Chem. 33: 187
40. Constanciel R (1986) Theoret. Chim. Acta 69: 505
41. Gomez-Jeria JS, Contreras R (1986) Int. J. Quantum Chem. 30: 581
42. Langlet J, Claverie P, Caillet J, Pullman A (1988) J. Phys. Chem. 92: 1617
43. Gersten JI, Sapse AM (1985) J. Am. Chem. Soc. 107: 3786
44. Rashin AA, Honig B (1985) J. Phys. Chem. 89: 5588
45. Rullman JAC, van Duijnen PTh (1987) Mol. Phys. 61: 293
46. Morse M, Rice SA (1981) J. Chem. Phys. 74: 6514
47. Morse M, Rice SA (1982) J. Chem. Phys. 76: 6514
48. Townsend M, Morse M, Rice SA (1983) J. Chem. Phys. 79: 2496
49. Nielson G, Rice SA (1984) J. Chem. Phys. 80: 4456
50. Ravishanker G, Mehrotra PK, Mezei M, Beveridge DL (1984) J. Am. Chem. Soc. 106: 4102
51. Clementi E, Corongiu G (1983) Int. J. Quantum Chem. 10: 31
52. Detrich J, Corongiu G, Clementi E (1984) Chem. Phys. Lett. 112: 426
53. Neumann M (1985) J. Chem. Phys. 82: 5663
54. Adelman SA (1980) J. Chem. Phys. 73: 3145
55. Balk MW, Brooks CA, Adelman SA (1983) J. Chem. Phys. 79: 804
56. Berkowitz M, McCammon JA (1982) Chem. Phys. Lett. 90: 215
57. Brooks CL, Karplus M (1983) J. Chem. Phys. 79: 6312
58. Brunger A, Brooks CL, Karplus M (1984) Chem. Phys. Lett. 105: 495
59. Warshel A (1979) J. Phys. Chem. 102: 6218
60. Brooks CL, Brunger A, Karplus M (1985) Biopolymers 24: 843
61. Brunger A, Brooks CL, Karplus M (1985) Proc. Natl. Acad. Sci. USA 82: 8458
62. van Gunsteren WF, Berendsen HJC (1985) In: Hermans J (ed) Molecular dynamics and protein structure. University of North Carolina Printing Dep. Chapel Hill, p 5
63. Jorgensen WL (1983) J. Phys. Chem. 87: 5304
64. Linse P, Karlström G, Jönsson B (1984) J. Am. Chem. Soc. 106: 4096
65. Cournoyer ME, Jorgensen WL (1984) J. Am. Chem. Soc. 106: 5104
66. Clementi E, Cavallone F, Scordamaglia R (1977) J. Am. Chem. Soc. 99: 5531
67. Clementi E, Corongiu G, Ranghino G (1981) J. Chem. Phys. 74: 578
68. Clementi E, Corongiu G (1982) In: Tanaka N, Ohtaki H, Tamamushi R (eds) Studies in theoretical chemistry vol 27, Elsevier, Amsterdam, p 397
69. Clementi E, Corongiu G, Dietrich JH, Khanmohammadbaigi H, Chin S, Domingo L, Laaksonen A, Nguyen HL (1985) in Clementi E, Corongiu G, Sarma MH, Sarma RH (eds) Structure and dynamics: Nucleic acids and proteins. Adenine Press, New York, p 49
70. Beveridge DL, Mezei M, Ravishanker G, Jayaram B (1986) Int. J. Quantum Chem. 29: 1513

71. Jorgensen WL and Buckner JK (1987) J. Phys. Chem. 91: 6083
72. McCammon A, Harvey (1987) Dynamics of proteins and nucleic acids. Cambridge UP, Cambridge
73. Ahlström P, Teleman O, Jönsson B, Forsen S (1987) J. Am. Chem. Soc. 109: 1541
74. van Gunsteren WF, Karplus M (1982) Biochemistry 21: 2259
75. Swaminathan S, Ichiye T, van Gunsteren WF, Karplus M (1982) Biochemistry 21: 5230
76. van Gunsteren WF, Berendsen HJC (1984) J. Mol. Biol. 176: 559
77. Levitt M, Sharon R (1987) In: Moras D, Drenth J, Strandberg B, Dietrich S, Wilson K (eds) Chrystallography in molecular biology. Plenum, New York, p 197
78. Kruger P, Strassburger W, Wollmer A, van Gunsteren WF (1985) Eur. Biophys. J. 13: 77
79. Åqvist J, van Gunsteren WF, Leijonmark M, Tapia O (1985) J. Mol. Biol., 183: 593
80. Åqvist J, Leijonmark M, Tapia O (1989) Eur. Biophys. J.—
81. Chandrasekhar J, Smith S, Jorgensen WL (1984) J. Am. Chem. Soc. 106: 3049
82. Chandrasekhar J, Smith S, Jorgensen WL (1985) J. Am. Chem. Soc. 107: 154
83. Madura JD, Jorgensen WL (1983) J. Am. Chem. Soc. 108: 2517
84. Bergsma JP, Gertner BJ, Wilson KR, Hynes JT (1987) J. Chem. Phys. 86: 1356
85. Tapia O, Lluch JM (1985) J. Chem. Phys. 83: 3970
86. Andres J, Cardenas R, Silla E, Tapia O (1988) J. Am. Chem. Soc. 110: 666
87. Tapia O, Lluch JM, Cardenas R, Andres J (1989) J. Am. Chem. Soc. —
88. Hwang J-K, King G, Creighton S, Warshel A (1988) J. Am. Chem. Soc. 110: 5297
89. Hwang J-K, Creighton S, King G, Whitney D, Warshel A (1988) J. Chem. Phys. 89: 859
90. Warshel A, Sussman F, Hwang J-K (1988) J. Mol. Biol., 201: 139
91. Schowen RL (1988) in: Liebman JF, Greenberg A (eds) Principles of enzyme activity, vol 9, Molecular structure and energerics. VCH Publishers, FL
92. Stamato FMLG, Longo E, Ferreira R, Tapia O (1986) J. Theor. Biol. 118: 45
93. Tapia O (1988) J. Mol. Catal. 43: 199
94. Tapia O, Andres J, Aullo JM, Cardenas R (1988) J. Mol. Struc. 167: 395
95. Tapia O, Cardenas R, Andres J, Colonna-Cesari F (1988) J. Am. Chem. Soc. 110: 4046
96. Naray-Szabo G, Simon K (1987) (eds) Steric aspects of biomolecular interactions. CRC Press, Boca Raton, FL
97. Smit B, van der Put A, Peters CJ, de Swaan Arons J (1988) J. Chem. Phys. 88: 3372
98. Smit B, van der Put A, Peters CJ, de Swaan Arons J (1988) Chem. Phys. Lett. 144: 555
99. Toxvaerd S (1987) J. Chem. Phys. 86: 3667
100. Fleming GR (1986) Ann. Rev. Phys. Chem. 37: 81
101. Demchenko AP (1986) Ess. Biochem. 22: 120
102. Banacky P (1986) Chem. Phys. 109: 307
103. Banacky P, Zajac A (1988) Chem. Phys. 123: 267
104. Ladanyi BM, Hynes JT (1986) J. Am Chem. Soc., 108: 585
105. Gertner BJ, Bergsma JP, Wilson KR, Lee S, Hynes JT (1987) J. Chem. Phys. 86: 1377

Theoretical Models of Chemical Bonding
Ed. Z. B. Maksić

Part 1
Atomic Hypothesis and the Concept of Molecular Structure

1990. XXVIII, 324 pp. 40 figs. 51tabs. Hardcover DM 350,- ISBN 3-540-51578-X

Subscription price (valid only for subscribers to the complete work):
Hardcover DM 280,-

Contents: *B. T. Sutcliffe:* The Concept of Molecular Structure. - *O. E. Polansky:* Topology and Properties of Molecules. - *J. P. Dahl:* Symmetry in Molecules. - *L. D. Barron:* Chirality of Molecular Structures - Basic Principles and their Consequences. - *J. E. Boggs:* Interplay of Experiment and Theory in Determining Molecular Geometries. A. The Experiments. - *J. E. Boggs:* Interplay of Experiment and Theory in Determining Molecular Geometries. B. Theoretical Methods. - *A. Y. Meyer:* Molecular Mechanics alias Mass Points and Elastic Springs Model of Molecules. - *K. B. Wiberg:* Atoms in Molecular Environments. - *Z. B. Maksić:* The Modelling of Molecules as Collections of Modified Atoms.

Part 2
The Concept of the Chemical Bond

1990. X, 643 pp. 181 figs. 88 tabs. Hardcover DM 450,- ISBN 3-540-51553-4

Subscription price (valid only for subscribers to the complete work):
Hardcover DM 280,-

Contents: *W. Kutzelnigg:* The Physical Origin of the Chemical Bond. - *R. G. Pearson:* Absolute Electronegativity and Absolute Hardness. - *K. Jug, M. S. Gopinathan:* Valence in Molecular Orbital Theory. - *J.-P. Malrieu:* The Magnetic Description of Conjugated Hydrocarbons. - *Z. B. Maksić:* Directional Properties of Covalent Bonding in Molecules. - *P. R. Surján:* The Two-Electron Bond as a Molecular Building Block. - *C. Edmiston:* Interpretation of Molecular Behaviour by Localized Molecular Orbitals (LMOs). - *W. L. Luken:* Properties of the Fermi Hole and Electronic Localization. - *P. J. Kuntz:* The Diatomics-in-Molecules Method and the Chemical Bond. - *P. Fulde:* Calculation of Electron Correlations by Using Local Operators. - *M. Grodzicki:* The Concept of the Chemical Bond in Solids. - *E. Kraka, D. Cremer:* Chemical Implication of Local Features of the Electron Density Distribution. - *A. A. Low, M. B. Hall:* Electron Deformation Densities and Chemical Bonding in Transition Metal Complexes. - *W. H. E. Schwarz:* Fundamentals of Relativistic Effects in Chemistry.

Springer-Verlag
Berlin
Heidelberg
New York
London
Paris
Tokyo
Hong Kong
Barcelona
Budapest

Theoretical Models of Chemical Bonding

Ed. Z. B. Maksić

Part 3

Molecular Spectroscopy, Electronic Structure and Intramolecular Interactions

1991. Approx. 660 pp. 172 figs. 126 tabs. Hardcover DM 350.– ISBN 3-540-52252-2

Subscription price (valid only for subscribers to the complete work): Hardcover DM 280,–

Contents: *J. E. Boggs:* Nuclear Vibrations and Force Constants. – *H. P. Figeys, P. Geerlings:* Some Aspects of the Quantum-Chemical Interpretation of Integrated Intensities of Infrared Absorption Bands. – *S. P. McGlynn, K. Wittel, L. Klasinc:* The Orbital Concept as a Foundation for Photoelectron Spectroscopy. – *E. Honegger, E. Heilbronner:* The Equivalent Bond Orbital Model and the Interpretation of PE Spectra. – *M. Eckert-Maksić:* Through-space and Through-bond Interactions as Mirrored in Photoelectron Spectra. – *K. Ohno, Y. Harada:* Penning Ionization – The Outer Shape of Molecules. – *K. Jug, Z. B. Maksić:* The Meaning and Distribution of Atomic Charges in Molecules. – *Z. B. Maksić:* Elec-tron Spectroscopy for Chemical Analysis (ESCA) – Basic Features and Their Model Description. – *K. T. Leung:* Experimental Momentum-Space Chemistry by (e, 2e) Spectroscopy. – *J. Kowalewski, A. Laaksonen:* Theoretical Parameters on NMR Spectroscopy. – *D. Feller, E. R. Davidson:* Theoretical Approaches to ESR Spectroscopy. – *C. J. Jameson:* Rovibrational Averaging of Molecular Electronic Properties. – *M. Klessinger, T. Pötter:* Properties of Molecules in Excited States. – *J. Tomasi, G. Alagona, R. Bonaccorsi, C. Ghio, R. Cammi:* Semiclassical Interpretation of Intramolecular Interactions. – *F. Bernardi, M. Olivucci, M. A. Robb:* The Analysis of Potential Energy Surfaces in Terms of the Diabatic Surface Model.

Part 4

Theoretical Treatment of Large Molecules and Their Interactions

1991. Approx. 470 pp. 104 figs. 52 tabs. Hardcover DM 350,– ISBN 3-540-52253-0

Subscription price (valid only for subscribers to the complete work): Hardcover DM 280,–

Contents: *J. G. Angyán, G. Náray-Szabó:* Chemical Fragmentation Approach to the Quantum Chemical Description of Extended Systems. – *Ch. L. Brooks:* Semiclassical Methods for Large Molecules of Biological Importance. – *G. M. Maggiora, J. D. Petke, R. E. Christoffersen:* Electronic Excited States of Biomolecular Systems: Ab Initio FSGO-based Quantum Mechanical Methods with Applications to Photosynthetic and Related Systems. – *A. J. Stone:* Classical Electrostatics in Molecular Interactions. – *V. Magnasco, R. McWeeny:* Weak Interactions Between Molecules and Their Physical Interpretation. – *S. Scheiner:* An Initio Studies of Hydrogen Bonding. – *J. Tomasi, R. Bonaccorsi, R. Cammi:* The Extramolecular Electrostatic Potential. An Indicator of the Chemical Reactivity. – *S. Shaik, P. C. Hiberty:* Curve Crossing Diagrams as General Models for Chemical Reactivity and Structure. – *R. A. van Santen, E. J. Baerends:* Orbital Interactions and Chemical Reactivity of Metal Particles and Metal Surfaces. – *A. van der Avoird:* Intermolecular Forces and the Properties of Molecular Solids. – *O. Tapia:* Theoretical Evaluation of Solvent Effects.

Springer-Verlag
Berlin
Heidelberg
New York
London
Paris
Tokyo
Hong Kong
Barcelona
Budapest